Springer Series in Wood Science

Editor: T. E. Timell

Springer

Berlin
Heidelberg
New York
Barcelona
Hong Kong
London
Milan
Paris
Singapore
Tokyo

Springer Series in Wood Science

Editor: T. E. Timell

J. F. Siau
Transport Processes in Wood (1984)
R. R. Archer
Growth Stresses and Strains in Trees (1986)
W. E. Hillis
Heartwood and Tree Exudates (1987)
S. Carlquist
Comparative Wood Anatomy (1988)
L. W. Roberts/P. B. Gahan/R. Aloni
Vascular Differentiation and Plant Growth Regulators (1988)
C. Skaar
Wood-Water Relations (1988)
J. M. Harris
Spiral Grain and Wave Phenomena in Wood Formation (1989)
B. J. Zobel/J. P. van Buijtenen
Wood Variation (1989)
P. Hakkila
Utilization of Residual Forest Biomass (1989)
J. W. Rowe (Ed.)
Natural Products of Woody Plants (1989)
K.-E. L. Eriksson/R. A. Blanchette/P. Ander
Microbial and Enzymatic Degradation of Wood and Wood Components (1990)
R. A. Blanchette/A. R. Biggs (Eds.)
Defense Mechanisms of Woody Plants Against Fungi (1992)
S. Y. Lin/C. W. Dence (Eds.)
Methods in Lignin Chemistry (1992)
G. Torgovnikov
Dielectric Properties of Wood and Wood-Based Materials (1993)
F. H. Schweingruber
Trees and Wood in Dendrochronology (1993)
P. R. Larson
The Vascular Cambium: Development and Structure (1994)
M.-S. Ilvessalo-Pfäffli
Fiber Atlas: Identification of Papermaking Fibers (1995)
B. J. Zobel/J. B. Jett
Genetics of Wood Production (1995)
C. Matteck/H. Kubler
Wood – The Internal Optimization of Trees (1995)
T. Higuchi
Biochemistry and Molecular Biology of Wood (1997)
B. J. Zobel/J. R. Sprague
Juvenile Wood in Forest Trees (1998)
E. Sjöström/R. Alén (Eds.)
Analytical Methods in Wood Chemistry, Pulping, and Papermaking (1999)
R. B. Keey/T. A. G. Langrish/J. C. F. Walker
Kiln-Drying of Lumber (2000)
S. Carlquist
Comparative Wood Anatomy. 2nd edition (2001)

Sherwin Carlquist

Comparative Wood Anatomy

Systematic, Ecological, and Evolutionary Aspects of Dicotyledon Wood

Second, completely revised edition

With 103 Figures and 9 Tables

Springer

Prof. Dr. Sherwin Carlquist
Santa Barbara, CA
USA

Series Editor:
T. E. TIMELL
State University of New York
College of Environmental Science
and Forestry
Syracuse, NY 13210, USA

Cover: Transverse section of *Pinus lambertiana* wood. Courtesy of Dr. Carl de Zeeuw, SUNY college of Environmental Science and Forestry, Syracuse, New York

ISSN 1431-8563
ISBN 3-540-41173-9 Springer-Verlag Berlin Heidelberg New York

Library of Congress Cataloging-in-Publication Data
Carlquist, Sherwin John, 1930–
Comparative wood anatomy. systematic, ecological, and evolutionary aspects of dicotyledon wood / S. Carlquist. – 2nd, completely rev. ed.
p. cm. – (Springer series in wood science, ISSN 1431-8563)0
Includes bibliographical references (p.).
ISBN 3540411739 (hardcover)
1. Wood – Anatomy. 2. Dicotyledons, I. Title. II. Series.
OK647.C37 2001
575.4′6 – dc21 00-049216

Springer-Verlag Berlin Heidelberg New York
a member of BertelsmannSpringer Science+Business Media GmbH
http://www.springer.de

Typesetting: Best-set Typesetter Ltd., Hong Kong
Cover design: Design & Production, Heidelberg
SPIN 10676748 31/3130 – 5 4 3 2 1 0 – Printed on acid-free paper

Preface and Acknowledgments for the Second Edition

Earlier decades have brought books dealing with various aspects of wood anatomy, such as identification, dendrology, dendrochronology, and other aspects closely related to comparative wood anatomy. These books are certainly part of the field of comparative wood anatomy and are cited in Chapter 1 and elsewhere in this book. The purpose of this book is not to duplicate those worthy efforts, but to offer a volume with a different focus. This book assumes that the reader will have an elementary knowledge of wood anatomy, such as might be obtained in a course in plant anatomy or even general botany, but will want to become better acquainted with structural diversity in wood of dicotyledons.

Because comparative wood anatomy deals with descriptions of wood features, an attempt has been made to describe and illustrate as many wood features and their diverse expressions as possible. The present compilation attempts to take into account the collective wisdom of various workers. The nature of that wisdom is continually in the process of change, and this book follows significant changes that have occurred in wood anatomy during the past several decades. Terminology and topics selected in this book are influenced by what seems significant to me: chiefly, this is the understanding of the physiology of conduction and through those physiological mechanisms, adaptations of wood features to ecology. The systematic distribution of wood features is also a primary concern of comparative wood anatomy. Such style features as adoption of a particular scheme of terminology or recognition of segregate families (often done here for ease in locating generic groupings) may differ from practices of other botanists, but diversity in a field of science should be welcomed.

Ideally, one wants a synthesis between wood anatomy and wood physiology and thereby one can reveal the adaptation of wood anatomy to ecological factors. Such a synthesis is constantly under construction, and has been aided by such workers in wood physiology as H. J. Braun, Frank Ewers, John Sperry, and Martin Zimmermann. A number of wood anatomists have enthusiastically welcomed this enterprise. I have taken the position that wood should be viewed at least in part in terms of its conductive and mechanical functions. Those interested in systematics and identification of woods need to know the physiological basis (and therefore the likely evolutionary status) of particular wood structures, and therefore terminology should reflect these functions accurately. Science has always benefited from syntheses, not from provincialism.

No two individuals would have selected material for a book similarly, and my selection, although idiosyncratic, hopefully can serve a range of interests. This second edition has been extensively revised and rewritten. Some older papers have, unfortunately, been omitted in this edition in order to make way for interesting new ones (unfortunately, the economic necessities of publication prevented expansion by more than a very modest amount); many older papers retain much of value, however. This edition involves a change of 25 photographs. Fields such as ultrastructure (at the TEM level), fossil woods (except for the relationship between fossil wood and evolutionary concepts) are not covered, nor is this book intended as an introduction to wood identification, although the features are described so as to form a basis for that activity (although not all workers use terms in the same ways).

I take pleasure in acknowledging the many who have aided me. Selecting a few names is difficult, for I have learned directly, and by letters, from many, including Pieter Baas, I. W. Bailey, Frank Ewers, C.R. Metcalfe, Regis Miller, William L. Stern, Elisabeth Wheeler, Martin Zimmermann, and the many knowledgeable persons who have reviewed manuscripts of my papers. Many who have furnished me with material deserve mention, and are cited in my papers on various taxonomic groups and wood topics. Students and former students, as well as colleagues and teachers, have provided valuable assistance, challenging questions, and inspiration.

Santa Barbara, California
Autumn 2000

Contents

1 Methods for Comparative Wood Anatomy Studies

1.1 Techniques for Comparative Wood Anatomy Studies

1.1.1 Sectioning for Light Microscopy and Scanning Electron Microscopy (SEM)

The sliding microtome (sledge) has been the standard instrument for preparation of wood sections, and its use seems likely to continue indefinitely. The sliding microtome is useful because most woods have degrees of hardness suitable for this instrument. Excessively hard woods can be softened by the use of ethylene diamine (Kukachka 1977). The use of hydrofluoric acid for this purpose is disadvantageous because it requires more time than ethylene diamine, and is dangerously corrosive. Ethylene diamine should be used with care because of its strong alkalinity. However, ethylene diamine, used in various dilutions and with the aid of heat (as in a paraffin oven) or at room temperature can soften more effectively than hydrofluoric acid in shorter periods of time. As with hydrofluoric acid, treatment with ethylene diamine can result in swelling of certain cell walls, especially if treatment is excessive.

Truly fine sectioning technique with the sliding microtome is difficult. While a high-quality knife perfectly sharpened is required for the best work, various other precautions and procedures are beneficial. Sectioning defects may not be obvious under the light microscope, but flaws are very apparent under SEM. The discussions of Kucera (1981) are helpful in this regard. The reader familiar with the three-surface preparations illustrated by Butterfield and Meylan (1980) and Meylan and Butterfield (1978a) will appreciate the marvelous results possible.

Although the usefulness of SEM is manifold, and although an enormous amount of detail awaits exploration with SEM, one must remember the value of light microscopy. With the light microscope, one can survey enormous quantities of material in a short time, exploring variability and distribution of features within woods. Moreover, SEM cannot see into cells unless they are cut open. For example, focusing into cells in search of perforation plates (or their absence) permitted me to develop ideas about presence of vasicentric and vascular tracheids. The extensive presence of these cell types in dicotyledons (Carlquist 1985a) could never have been revealed with SEM. A worker who abandons light microscopy loses much valuable information. A slide collection developed for light microscopy (with slides of the collection hopefully prepared in replicates) can always be used as an archive for SEM work: one can soak off the cover slips, dissolve away resins, and use the sections for SEM work quite readily. SEM is an extraordinarily valuable tool in wood anatomy, and many examples could be cited.

1.1.2 Problems in Sectioning and Staining of Wood Sections

Microtechnique is legendary for presentation of precise recipes in print. Yet the variations on these recipes as one finds them used in various laboratories are legion, leading one to believe that microtechnique should be used imaginatively and variations should be tried where results are not at first satisfying.

Safranin is usually recommended as the stain of choice for wood sections. Hematoxylin is applied before safranin staining by many workers to provide contrast between primary walls (purple-gray with hematoxylin) and secondary walls (red with safranin). While safranin is highly effective with secondary walls, several alternative counterstains also achieve differentiation of primary walls. Counterstaining is very desirable because details of pitting, so difficult for beginning students to observe accurately, are maximally rendered by counterstaining.

Although those interested in timbers are not often likely to encounter woods that are too soft to section readily on a sliding microtome, woods of herbaceous families, that is, soft woods or woods that are not hard but that have excessively large vessels, often provide problems in sectioning. Woods in these problem categories benefit, ironically, from being softened further so they can be sectioned not on a sliding microtome, but on a rotary microtome after they have been embedded in paraffin (Carlquist 1982g). Various details of that method other than just softening and paraffin embedding must be followed in order to achieve good results. This method is not effective for woods of medium softness.

The aim of sectioning often seems to be the preparation of sections as thin as possible. While thin sections do reveal some details well, and while they produce photographs of admirable clarity, thicker sections have virtues not evident at first. For example, thicker radial sections often permit one to see entire scalariform perforation plates intact. Walls of all kinds are seen in face view better in thicker sections because larger portions are intact, and one can also see better the context of a particular cell: which cells lie behind or above. In addition, sometimes one obtains less flawed sections when one increases the thickness, and even sections as thick as 30 μm are entirely usable for many purposes (Carlquist 1987d), especially SEM studies.

1.1.3 Macerations

Institutional wood collections often do not include macerations because sections are of basic importance for wood identification purposes. For comparative purposes, macerations are essential for obtaining quantitative data on lengths of vessel elements and imperforate tracheary elements. There are other purposes served by macerations. When viewed by SEM, macerations yield superb three-dimensional images of cell shapes. When attempting to count bars per perforation plate, macerations are essential. The presence of fibriform vessel elements, caudate vessel elements, and vasicentric tracheids can be determined with certainty only by the use of macerations.

1.1.4 Resin-Casting Methods

Although requiring more time than sectioning, the infiltration of woods with resins, followed by removal of the cell walls and examination of the resulting resin casts by SEM, opens fascinating possibilities. The paper by Fujii (1993) is a fine example.

1.1.5 Transmission Electron Microscopy (TEM)

There is no question that any microscopic technique capable of producing high resolution will find many applications in study of woods. While endorsing the use of TEM, I would like to suggest some caution where comparative studies are concerned. TEM procedures are very time-intensive, and thus one cannot view many samples and one can view only a small portion of a sample at a time. Comparative work, by definition, tends to involve reviewing large amounts of material with anatomical methods, and thus is a field where TEM work may be contraindicated in many cases. However, TEM can certainly clarify the nature of small structures in wood as no other method can. Among recent papers using TEM, those by Wheeler (1981, 1983), Nair et al. (1983), Parameswaran and Richter (1984), and Parameswaran et al. (1985) deserve mention. Other examples can be found in a book edited by Côté (1965).

1.2 Sources of Wood Samples and Wood Sample Preservation

Comparative wood anatomists are familiar with the existence of wood sample collections, better known as xylaria (Stern 1978b). Wood samples in xylaria should be cited, where possible, not merely by xylarium accession number, but by a collector's number that applies not merely to a wood sample, but to an herbarium voucher specimen as well. These procedures are discussed by Stern and Chambers (1960) and by Baas (1980). Collection data for illustrations in the present book are listed in cited papers, and are on file in my slide collection (data supplied upon request).

Misidentification of wood samples unfortunately does occur with some frequency despite the best efforts of collectors and curators, and before an investigator reports a mode of wood structure suspiciously unusual for a family or genus, one should question the identification. Verification can often be easily accomplished by sectioning a small amount of twig material from a verified herbarium specimen. Twig material is not recommended for wood studies in general, but it is validly used in this connection. In considering identity of a wood sample, one should be aware of the possibility that the voucher specimen related to it may have been redetermined.

Some investigators use twig materials in order to provide a more complete roster of species in a systematic wood anatomy study. While one empathizes with that effort, for wood is rarely available for as many species in a genus as one would wish, one should be aware of the limitations of twig material. Quantitative features of twigs are certain to differ from those of mature wood (see Stern and Greene 1958), but some qualitative features (e.g., ray type) may also be different. While workers are often very conscientious in citing provenance of materials (and I endorse this), citation of the sizes of the wood samples studied, especially if some are portions of small stems, may be equally important to interpretation of results.

Collection and preservation of wood samples has become more complicated as conservation awareness has increased. A review of collection and preservation techniques has been offered (Carlquist 1996a). Drying has been quite widely used in the past, but preservation in 50% aqueous ethanol is advisable where preservation of cells with nonlignified walls (e.g., phloem) is desired (excess liquid can be drained for shipping). Where alcohol is not available and woods are to be shipped, paraformaldehyde powder can be used.

1.3 Literature on Comparative Wood Anatomy

The literature on comparative wood anatomy prior to 1900 was summarized by Solereder (1908). Metcalfe and Chalk (1950) extended this coverage to the middle of the 20th century. With the appearance of a second edition of *Anatomy of Dicotyledons*, much new information is being made available. The bibliography of Gregory (1994) is indispensable because of its thoroughness and coverage of topics as well as taxonomic groups.

Works organized on a floristic basis can contain enormous amounts of data on comparative wood anatomy. Perhaps the most outstanding example is that of Moll and Janssonius (1906–1936) because of the large number of species involved and because of the detailed and accurate descriptions. Greguss (1959) has offered an important monograph in which native and some introduced woods of Europe and adjacent regions are described and figured both with drawings and with photomicrographs, but a more recent work with greater coverage by Schweingruber (1990) is highly recommended. The monograph by Meylan and Butterfield (1978a) of New Zealand woods is entirely illustrated by SEM photographs of superb quality. A study by Fahn et al. (1986) on the woods of Israel and adjacent region takes note of ecological tendencies.

Very closely allied to the above are works the primary concern of which is wood identification. The review of Wheeler and Baas (1998) is of prime importance; also worthy of interest for wood anatomists are a number of recent works dealing with various aspects of wood identification. Space does not allow citation of the many excellent studies in this field.

The International Association of Wood Anatomists has played an important role in promoting wood anatomy and increasing in size and diversity the membership of the Association. The IAWA Journal is indispensable for comparative wood anatomists.

1.3.1 Use of Computers

The large quantities of information subsumed in studies of wood anatomy, either for purposes of identification or with an interest in systematics and evolution, are potentially ideally served by means of the computer, despite the enormity of work involved in preparation and entry of appropriate databases. The use of computers in wood anatomy is certain to be increasingly significant and helpful. Attempts to create standardized lists of features for computerization have been made (IAWA Committee 1981; Pearson and Wheeler 1981; Wheeler and Pearson 1985). These papers show that there may be more than one kind of ideal format, depending on the interests of the worker. A wood identification manual accompanied by software suitable for various types of computers produced by Wheeler et al. (1986) shows the promise of computerization for those interested in wood identification.

1.4 Terminology

Throughout the present book, concern is paid to careful choice of terminology and to consistent application of terms. The IAWA Committee on Nomenclature (1964) has been followed in all cases except where new information has forced me to adopt alternative terms or new terms. For example, understanding of ontogeny of cambial variants (anomalous secondary thickening) is now much better than it was in 1964, and consequently new categories based on important ontogenetic distinctions (with easily visible end products) must now be employed (Chapter 8). An extensive review of vasicentric and vascular tracheids required new definitions of these terms, but the redefinitions come close to the use of these terms by most workers and these terms have been so vaguely defined by glossaries that they have not served as guidelines in any case. New phenomena require new terms, as in the instances of ray-adjacent parenchyma, interconnected rays, and vessel restriction patterns. In the difficult instance of imperforate tracheary elements, I endorse the IAWA Committee on Nomenclature (1964), with very minor emendations. The 1964 terminology has the advantage of recognizing the conductive nature of tracheids (Braun 1970; Carlquist 1984a), and the nonconductive nature of libriform fibers (which have simple pits) and fiber-tracheids (which have vestigially bordered pits). An alternative scheme has been proposed (Baas 1986a), and others have been used (e.g., Metcalfe and Chalk 1950). In view of the fact that choices may vary among workers (and have always varied), I strongly recommend explicit citation by

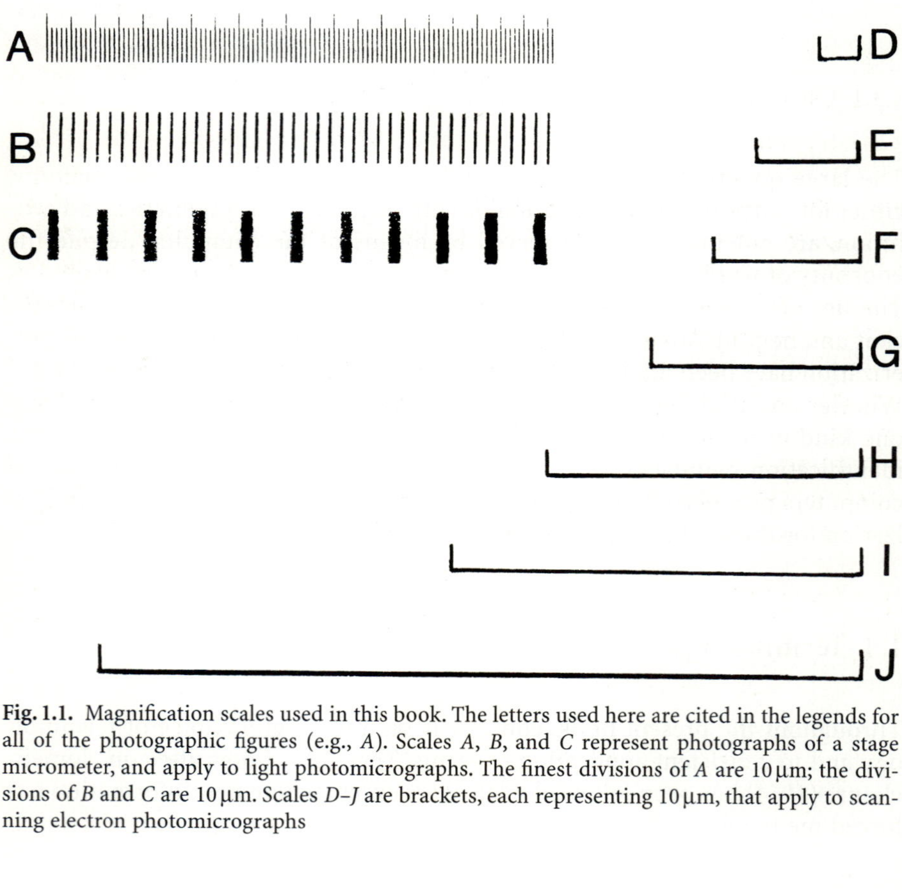

Fig. 1.1. Magnification scales used in this book. The letters used here are cited in the legends for all of the photographic figures (e.g., *A*). Scales *A*, *B*, and *C* represent photographs of a stage micrometer, and apply to light photomicrographs. The finest divisions of *A* are 10 μm; the divisions of *B* and *C* are 10 μm. Scales *D–J* are brackets, each representing 10 μm, that apply to scanning electron photomicrographs

an author of the nomenclature that is being followed. A recent scheme (IAWA Committee 1989) is designed for those who intend to identify woods.

1.5 Sequence of Features and Conventions for Data Presentation

The formats used by wood anatomists vary somewhat, but the sequence of topics used in this book follows the sequence of features typically used in descriptions of family, genus, or species by wood anatomists.

In addition to running text, one can use tables, keys, and cladograms for highlighting data and exploring phyletic significance. The use of these methods and examples of them are presented in Chapter 10 (see Sect. 10.4).

1.6 Magnifications

Work on comparative wood anatomy should give magnifications when photographs or drawings are presented. Magnifications can take the form of a figure in the legend (e.g., ×65). However, calculating the magnification figure can be complicated, particularly when one must project in advance the degree of reduction of one's illustrations when they are published.

Simpler to present and to use are scales of various sorts, such as used in this book and assembled here as Fig. 1.1. One method I have found convenient is to photograph a stage micrometer at the same magnification as the wood photomicrographs, and to enlarge it when the photomicrograph negatives are enlarged. The advantage of this type of scale is that one may compare the length of an object in a photograph (e.g., a vessel diameter) to the scale and immediately find the size of the object. Photographs of a stage micrometer are used for the three magnifications at which light microscope photomicrographs are enlarged in this book (Fig. 1.1A–C).

Another commonly used method is to superimpose a bracket, line, or other linear marking representing a given number of microns on a photomicrograph or to place such a device beside drawn figures. This device is used here in Fig. 1.1 (scales D–J) for SEM photographs in this book. Like the photograph of a stage micrometer, the line or bracket scale has the advantage of not requiring one to project the amount of reduction that will occur when an illustration is published.

2 Growth Rings

2.1 Introduction

Earlier work on growth rings includes the survey of Jost (1891), the often-cited work of Chowdhury (e.g., 1939, 1940a,b) and the key to ring-porous woods by Record (1942a). Chowdhury attempted to compare growth rings in these species to all relevant factors: temperature, light, rainfall, humidity, foliar development, locality (e.g., microclimates), and "internal factors" (e.g., growth-promoting substances). Chowdhury concluded that the factors are interrelated, and one obtains no clear picture of which factors are primary and which are secondary in initiation and differential events within a growth ring.

However, Chowdhury's studies were undertaken in a tropical (but seasonal) area; clearer results can be obtained in a temperate area with more sharply defined seasonal events. Moreover, there are many physiological measures not included by Chowdhury that one would ideally want: transpiration rates, xylem tension, and water availability (soil moisture), for example. Such factors are taken into account in a paper by Schweingruber (1992), as well as in his large monograph on dendrochronology (1996).

Recently, much attention has been devoted to the formation of growth rings or other phenomena that could be used in age determination in tropical trees. Noteworthy in this regard are the symposium papers edited by Baas and Vetter (1989), Eckstein et al. (1995), and Eckstein and Baas (1999).

2.2 Factors Affecting Growth Ring Formation

Onset of drought leads to cessation of cambial activity and is a prime factor in growth-ring formation according to Fahn (1933, 1955, 1958, 1959a,b, 1962), Larson (1960, 1964), Amobi (1973), and Baas (1973). Cold is claimed to induce cambial activity cessation by Waisel and Fahn (1965). Photoperiodic events may play a role in number and size of elements (Waisel and Fahn 1965; Larson 1969). Paliwal and Prasad (1970) stress timing of leaf fall in relation to cessation of growth rings. Ladefoged (1952) and Bosshard et al. (1982) note that cambial activity in ring-porous species begins before bud break. Suzuki et al. (1996) showed that in ring-porous species, earlywood vessels are initiated 2 to

4 weeks before leaf expansion, whereas in diffuse-porous species, vessel initiation follows leaf expansion. De Paolis (1948, 1950) claimed that a period of low light intensity led to formation of wider vessels in *Prunus*. Very low light intensity led to shift to diffuse-porous wood in species that normally have growth rings (Lohr 1969). Priya and Bhat (1999) show that onset of rainfall initiates earlywood in teak. Growth rings are not always present in deciduous species or absent in evergreen ones (Tomlinson and Craighead 1972). However, 35% of rain forest trees in Brazil do have growth rings (Mainieri 1959). In this latter instance, occurrence of growth rings is presumably related to seasonal drought. This example shows how climatic factors may be isolated in the analysis of factors leading to growth ring formation. Other possibilities lie in species that form non-annual growth rings, such as a mangrove species of *Diospyros* reported by Duke et al. (1981). One must not confuse production of successive cambia with formation of growth rings; Bailey (1980) shows that in *Simmondsia* the xylem bands produced by successive cambia are not related to annual or seasonal events, and that growth rings may begin anywhere within these bands. Experimental work is cited later in this chapter (Sect. 2.4), and is mentioned here only by way of noting that correlation between particular factors and cellular events can be found.

2.3 Factors in Classification of Growth Rings

Although much more remains to be learned about origin and function of growth rings, a classification of growth rings was attempted (Carlquist 1980a). That scheme attempted to take into account all known types of cell variations seen within growth rings of angiosperms and gymnosperms. Also, it was based on known physiological functioning of portions of growth rings. For example, narrow latewood vessels tend to embolize much less readily than wide earlywood vessels (Ellmore and Ewers 1985; Hargrave et al. 1994; Jarbeau et al. 1995). The 1980 scheme has been revised here, but most of the types (using Arabic instead of Roman numerals) remain the same. The revised classification is presented here in a semidiagrammatic scheme (Fig. 2.1) and the types, in sequence, are illustrated in Figs. 2.2–2.13.

Prior to presentation of this scheme, woods were termed either ring-porous or diffuse-porous; the term "semi-ring-porous" was used for those woods in which vessels are appreciably but not markedly larger in earlywood than in latewood. Type 1 corresponds, in general, to "semi-ring-porous." Most dicotyledons with ring-porous wood fall into type 3, somewhat fewer into type 5. The other types can be considered as relatively infrequent but distinctive manifestations of seasonal change in growth rings. The traditional terms "earlywood" and "latewood" are used here because they are in general useful, although problems have been mentioned by such authors as Bosshard and Kucera (1973a), who stress that comparative studies of growth ring diversity, as attempted here, are needed. The growth rings discussed below are understood to be annual unless otherwise specified.

A growth ring contains within it contrasting modes of adaptation to different climates, as well as the climatic events of a particular year. Earlywood of a particular species may be likened to wood of a tropical rain forest tree in which wide vessels accommodate large volumes per unit time to suit active transpiration; such wood is low in conductive safety. Latewood may resemble wood of a desert shrub, in which large numbers of narrow vessels offer greater resistance to embolism formation and thus greater conductive safety (Hargrave et al. 1994) and redundancy in case some of them do embolize. Type 5, in which vessels are few or none in latewood, is a logical and interesting exception, but in functional terms it represents ultimate safety: substitution of tracheids for narrow vessels. Interestingly, helical thickenings are present in latewood of woods that often lack them in earlywood vessels; in other cases, helical thickenings are much more pronounced in latewood vessels (see Sect. 3.12.3; Figs. 3.19, 3.20). Preservation of water columns in latewood is more probable than preservation of water columns in earlywood, so location of more prominent helical thickenings in the less expendable latewood vessels is understandable (Carlquist 1983f, 1995a). Latewood vessels, by virtue of their greater number and narrower diameter, have greater friction than earlywood vessels, but this is presumably not a serious detriment because conduction is relatively slow in latewood. Although photographs and drawings of transections illustrate the diminution of vessels during a year's growth, the quantitative graphs (vessel diameter graphed against frequency of those diameters) show that ring-porous woods have some sort of bimodal curve, whereas diffuse-porous or semi-ring-porous woods have curves approaching normal distribution of diameters (Woodcock 1989a,b).

Diminution of vessel size without a corresponding increase in vessel density in latewood would have the effect of increasing the volume devoted to mechanical tissue in latewood. This situation, although a possible one (some examples of type 1E), is in fact not the only source of improvement in mechanical strength. Probably most examples referred to type 1E demonstrate greater thickness of walls in the imperforate tracheary elements of latewood (Fig. 2.3, parts 3 and 4), a more effective way of increasing mechanical strength. Decreasing diameter of imperforate tracheary elements in latewood (Fig. 2.3, part 2) is also theoretically a mechanism for increasing mechanical strength in latewood. These patterns could represent ways for compensating in latewood for the large volume devoted to conduction in earlywood.

The nature of imperforate tracheary elements that form the background of a wood is important with respect to conductive capability and must be taken into account. If, in a given species, the imperforate tracheary element is a tracheid, there is a subsidiary or secondary conductive system that probably remains intact if some or all of the vessels are embolized. In a wood with libriform fibers or with fiber-tracheids, no subsidiary conductive system exists, and the conductive safety is thereby reduced. Evidence has been presented that

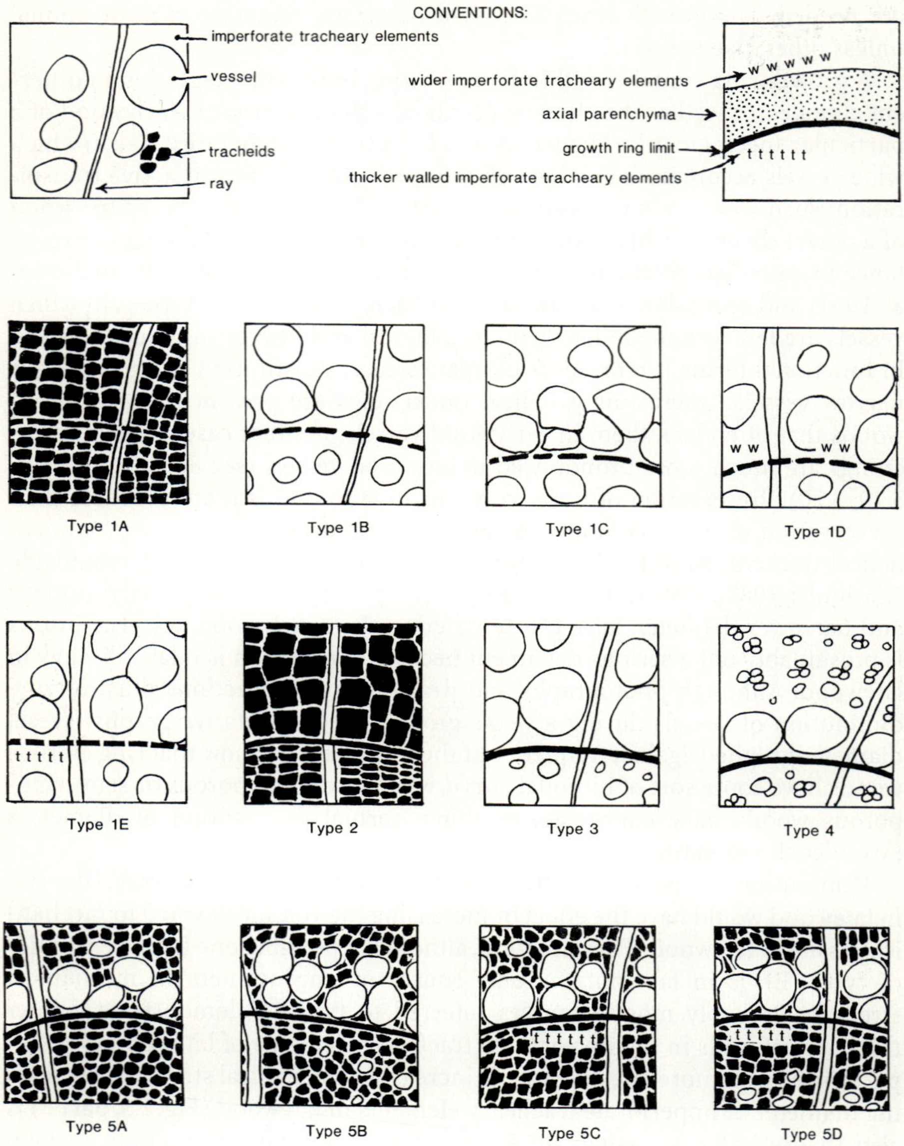

Fig. 2.1. Diagrams of the growth ring types. Conventions are at *top, left*. Subtypes (*A* etc.) are shown for some of the types. For further explanation, see text. All of the growth ring types shown here are also illustrated by one or more photomicrographs in the remaining photographs of this chapter

fiber-tracheids (sensu IAWA Committee on Nomenclature 1964) are essentially nonconductive (Braun 1970; Carlquist 1984a, 1985a). Either true tracheids or appreciable numbers of vasicentric tracheids would form a subsidiary conductive system. In species in which vascular tracheids (here defined as tra-

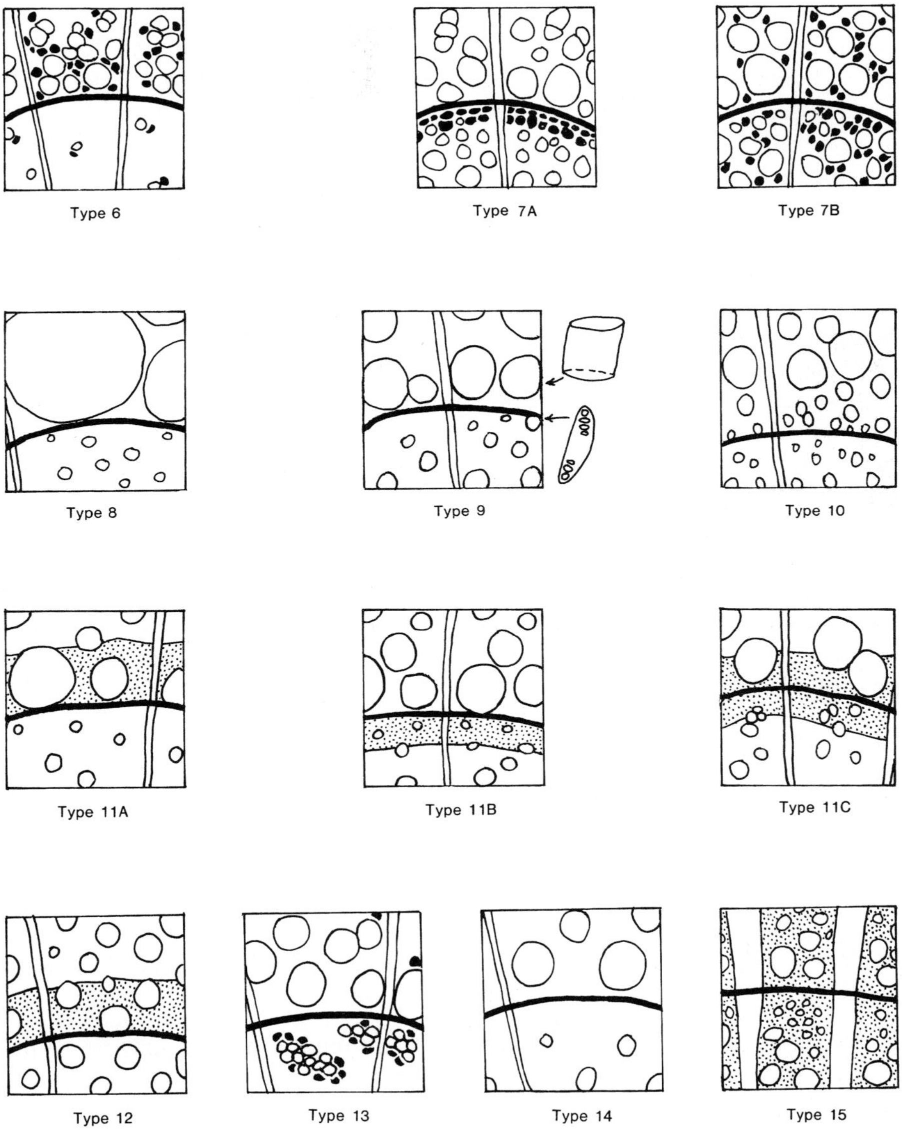

Fig. 2.1. *Continued.*

cheids that terminate a growth ring) are present, the subsidiary conductive system does not protect the three-dimensional nature of the vessel pathways in its capabilities; it could provide the cambium with a continuous water supply and thereby prevent dying of stems, however.

Parenchyma bands at the ends of growth rings ("marginal parenchyma") have been subdivided into terminal parenchyma and initial parenchyma. In my experience, terminal parenchyma is the predominant form (Fig. 2.11, parts 1 and 2). Instances may be found in which both terminal and initial parenchyma are present (Fig. 2.11, parts 3 and 4). Occurrences of initial parenchyma (Fig. 2.10, part4) are comparatively infrequent. Terminal parenchyma may contain starch, which can have either of two physiological functions: a store of photosynthates for rapid production of foliage or flowers; or a reserve from which sugars, by hydrolysis, could be transferred into vessels, altering their osmotic potential. The latter phenomenon has been shown in the sugar maple, *Acer saccharum* (Sauter et al. 1973). Very likely the phenomenon is much more widespread (Carlquist and Eckhart 1984), judging from occurrences of terminal parenchyma in various types of dicotyledons (e.g., woody vines such as *Clematis*; Carlquist 1995a). One must note that starch stored in living fibers scattered throughout the growth ring may also have the same physiological function noted above for terminal parenchyma.

All of the shifts in histology within growth rings mentioned above are incorporated into the classification below. Some of the subtypes (e.g., 5A, 5B) represent minor variants. Also, the reader must recognize that a few growth rings may not fall readily into one of the types, and a choice of either of two types given below is thus possible; obviously evolution does not work in terms of mutually exclusive definitions. Notable in this regard is the transition between diffuse-porous (by definition omitted from the classification below) and semi-ring-porous (chiefly type 1B). Use of growth rings as systematic features is obviously difficult, because anatomy of growth rings is primarily influenced by physiological factors.

2.4 Description of Growth Ring Types

2.4.1 Type 1

Moderate differentiation into growth rings ("semi-ring-porous" if vessels present); changes with seasons gradual.

A Tracheids only present (Fig. 2.2, part 1). Although examples may be found in tropical conifers, such as certain Araucariaceae and Podocarpaceae, examples may be found in vesselless dicotyledons from moderately seasonal climates, including Amborellaceae (Bailey and Swamy 1948) and certain Winteraceae (Bailey 1944a; Carlquist 1981a, 1982a, 1983a,b).

B Vessels wider in earlywood, but not markedly so; other cell types unaffected. Many examples can be cited, because type 1B represents the bulk of woods termed semi-ring-porous. Three examples are shown here (Fig. 2.2, parts 2–4).

C Vessels not appreciably larger in earlywood, but more numerous (Fig. 2.3, part 1).

D Vessels wider in earlywood, but imperforate tracheary elements also are wider (Fig. 2.3, part 2).

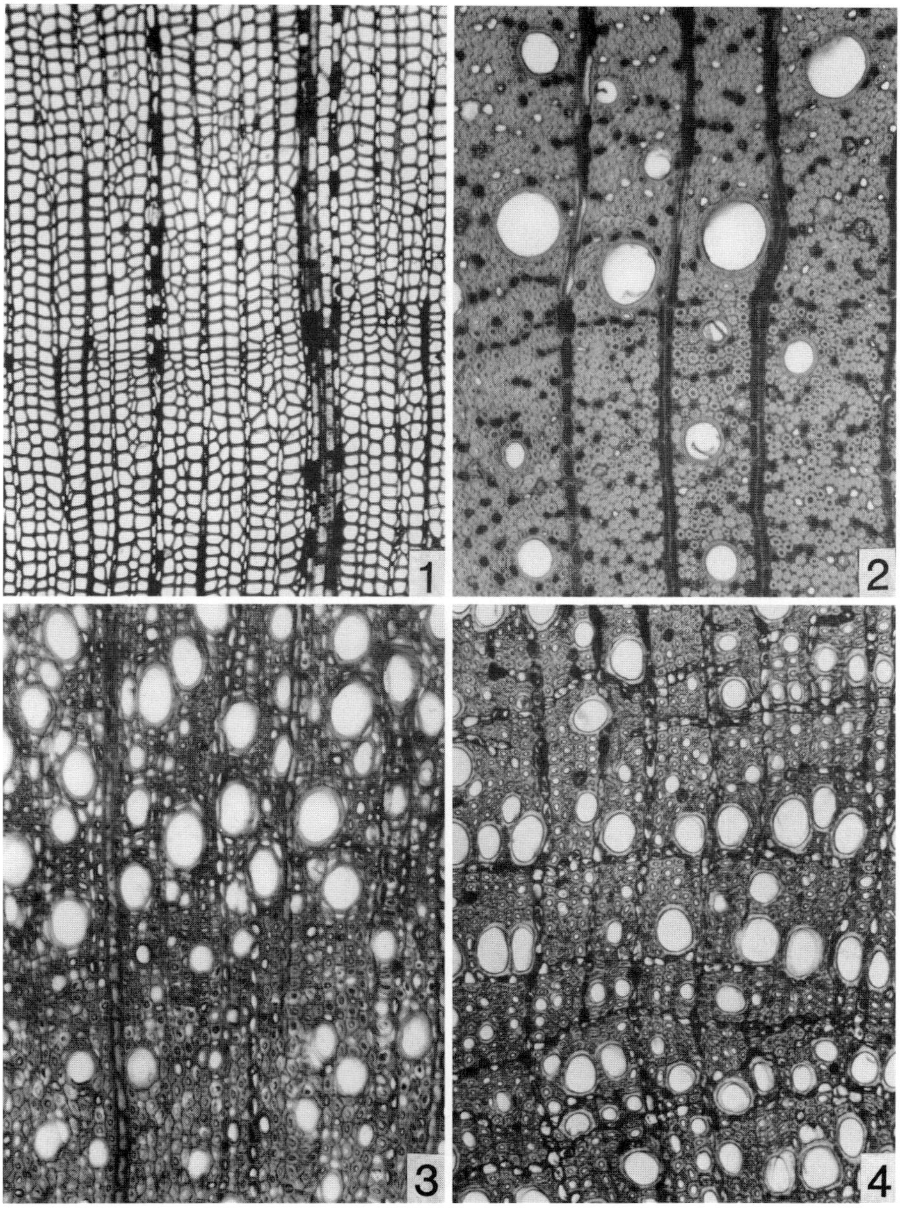

Fig. 2.2. Growth rings, type 1 (growth rings inconspicuous, semi-ring-porous except 1). *1* Type 1A, *Drimys confertifolia* (Winteraceae); *2* type 1B, *Larrea tridentata* Zygophyllaceae; *3* type 1B, *Cistus purpureus* (Cistaceae); *4* type 1B, *Krameria parviflora* var. *imparata* (Krameriaceae). A single growth ring is shown in all except *4*, where five rings are present. In this figure and the remaining figures of this chapter, wood transections are arranged so that outer side (closer to cambium) is *above*, inside (closer to pith) is *below*. (Scales for *1–4:B* in Fig. 1.1)

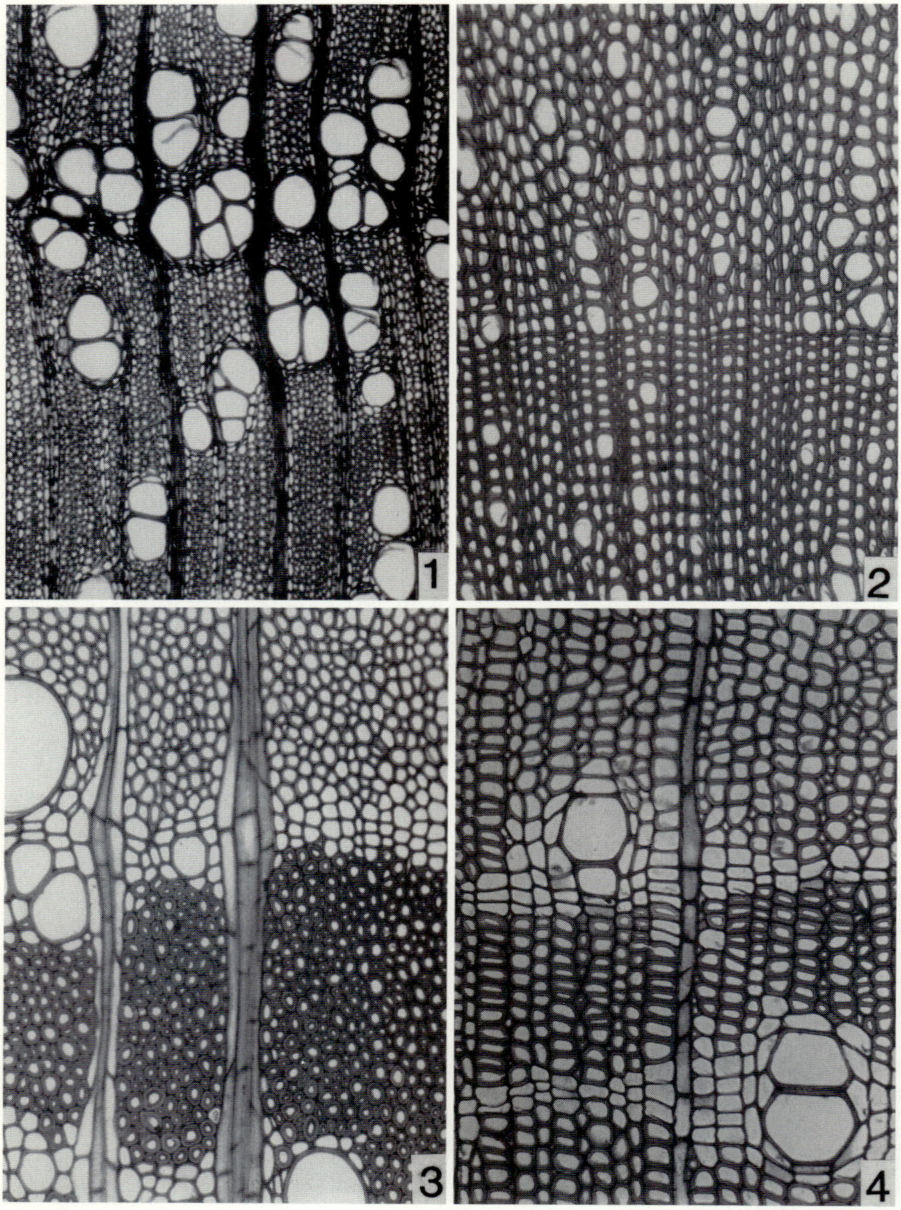

Fig. 2.3. Growth rings, type 1 (growth rings inconspicuous, semi-ring porous). *1* Type 1C, *Vernonia colorata* (Asteraceae); *2* type 1D, *Viola chamissoniana* (Violaceae); *3* type 1E, *Calodendron capense* (Rutaceae); *4* type 1E, *Euodia cucullata* (Rutaceae). (*1–4*:*B*)

E Vessels wider in earlywood, imperforate tracheary elements thinner walled (Fig. 2.3, parts 3 and 4).

Type 1 growth rings exemplify moderate widening of earlywood vessels (tracheids in type 1A); the functional explanation is assumed to be flow of greater volumes of water per unit time in earlywood than in latewood. One can also claim increased safety in latewood if narrower vessels embolize less readily than wider ones, as demonstrated by Ellmore and Ewers (1985) and Hargrave et al. (1994).

2.4.2 Type 2

Vessels absent; tracheids radially wider and thinner walled in earlywood.

Type 2 is familiar as the growth ring type seen in temperate conifers. It is also found in temperate vesselless dicotyledons such as *Trochodendron* (Fig. 2.4, part 1), *Tetracentron*, and in the Winteraceae, *Pseudowintera* (Patel 1974).

Fluctuation of tracheid diameter maximizes conductivity in earlywood, safety in latewood (Denne 1973). Narrower latewood tracheids are less susceptible to embolism than wider earlywood tracheids (Lewis and Tyree 1985). Greater radial width of earlywood tracheids also maximizes area of pitting on overlap areas (Carlquist 1975a). Mechanical strength of the latewood tracheids, conferred by their greater wall thickness, compensates for the thin-walled nature of earlywood tracheids.

2.4.3 Type 3

Vessels wider in earlywood, not markedly less numerous than in latewood; imperforate tracheary elements are other than tracheids (therefore libriform fibers or fiber-tracheids).

The majority of ring-porous dicotyledons fall into this category. In the earlier study (Carlquist 1980a), three subtypes were recognized: A, with ordinary fiber-tracheids or libriform fibers; B, with septate fibers (illustrated here by *Aralia spinosa*; Fig. 2.4, part 2; and *Juglans nigra*; Fig. 2.4, part 3); and C, with thick-walled libriform fibers (illustrated here by *Phoradendron tomentosum*; Fig. 2.4, part 4). These are distinctions of minor importance. Type 8 growth rings have markedly wider earlywood vessels, and therefore the number of vessels in latewood per unit area is inevitably increased markedly compared to type 3 growth rings.

The functional significance of type 3 growth rings is that of markedly enhanced earlywood conductive capacity. Because density of vessels (vessel number per mm^2) is not markedly elevated in latewood, safety is not

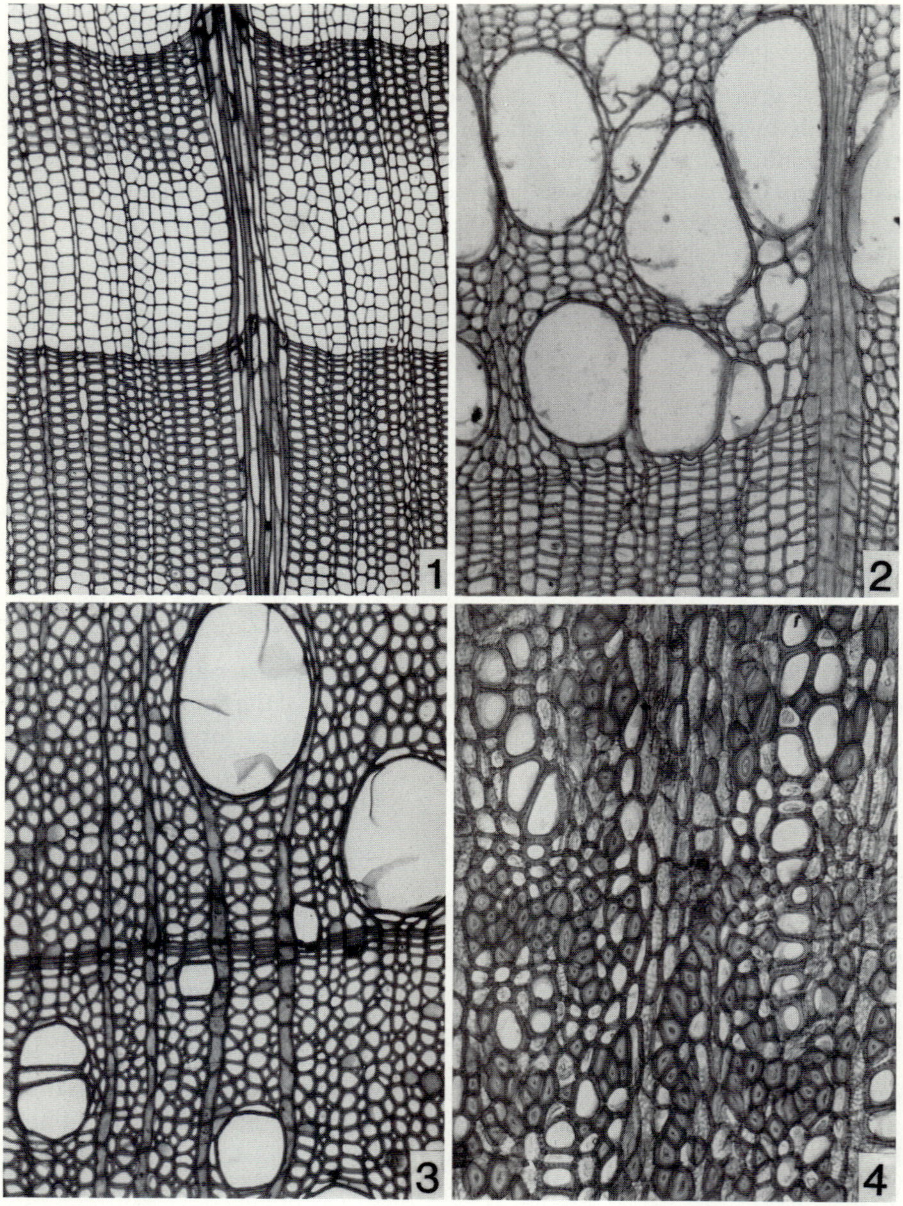

Fig. 2.4. Growth rings, type 2 and type 3. *1* Type 2 (growth rings marked, vessels absent), *Trochodendron aralioides* (Trochodendracecae); *2* type 3B (vessels wider in earlywood, grading to narrower in latewood), *Aralia spinosa* (Araliaceae); *3* type 3B, *Juglans nigra* (Juglandaceae); *4* type 3C, *Phoradendron flavum* var. *microphyllum* (Viscaceae). (*1*:*A*; *2–4*:*B*)

heightened extraordinarily, although as noted earlier, whenever narrower vessels are present in latewood, they probably embolize less easily than the wide earlywood vessels.

2.4.4 Type 4

Vessels wider in earlywood, but instead of grading into narrower latewood vessels, there is a rather marked change in vessel diameter from earlywood to latewood.

Type 4 is illustrated here by *Olearia virgata* (Fig. 2.5, part 1) and *Lepidospartum squamatum* (Fig. 2.5, part 2). The functional significance of type 4 growth rings can be described as essentially the same as with type 3, but with a brief period of peak water conduction each season, followed by a long period during which conduction is slower and latewood conductive safety is of value. Type 4 growth rings may be expected in Mediterranean-type climates, in which a brief rainy season is followed by a prolonged dry season. Type 13 growth rings are similar, but tend not to be found in Mediterranean-type climates; type 13 features vasicentric tracheids or an abundance of extremely narrow vessels (usually both).

2.4.5 Type 5

Vessels are wider in earlywood, narrower in latewood (and thus ring porosity is present); tracheids are the imperforate tracheary element type.

In order to identify this growth ring type, one must examine a longitudinal section (preferably a radial section) to see if pit diameter, density, and border width are sufficient to designate the imperforate tracheary elements as tracheids. Viewing a longitudinal section in order to determine the type of growth rings, ordinarily observed in transections, may seem a procedure that makes designation of growth ring type a laborious task. However, in woods in which vessels are few or none but tracheids are present as the background cell type in latewood, an exceptional degree of potential safety is present. If, as generally assumed, air embolisms expand from one vessel element to another, disabling an indefinite number of vessel elements in a vessel, whereas pit membranes prevent spread of an air embolism from one tracheid to a neighboring tracheid, the safety of tracheids is greater than that of even very narrow vessel elements. Latewood of type 5 growth rings can be, if vessels are few or none, essentially like conifer latewood and comparable to it in the high degree of safety. Therefore, type 5 growth rings are theoretically an ideal way of satisfying peak conductive demands where water availability is seasonal yet meeting requirements for maximal safety during the portion of the year in

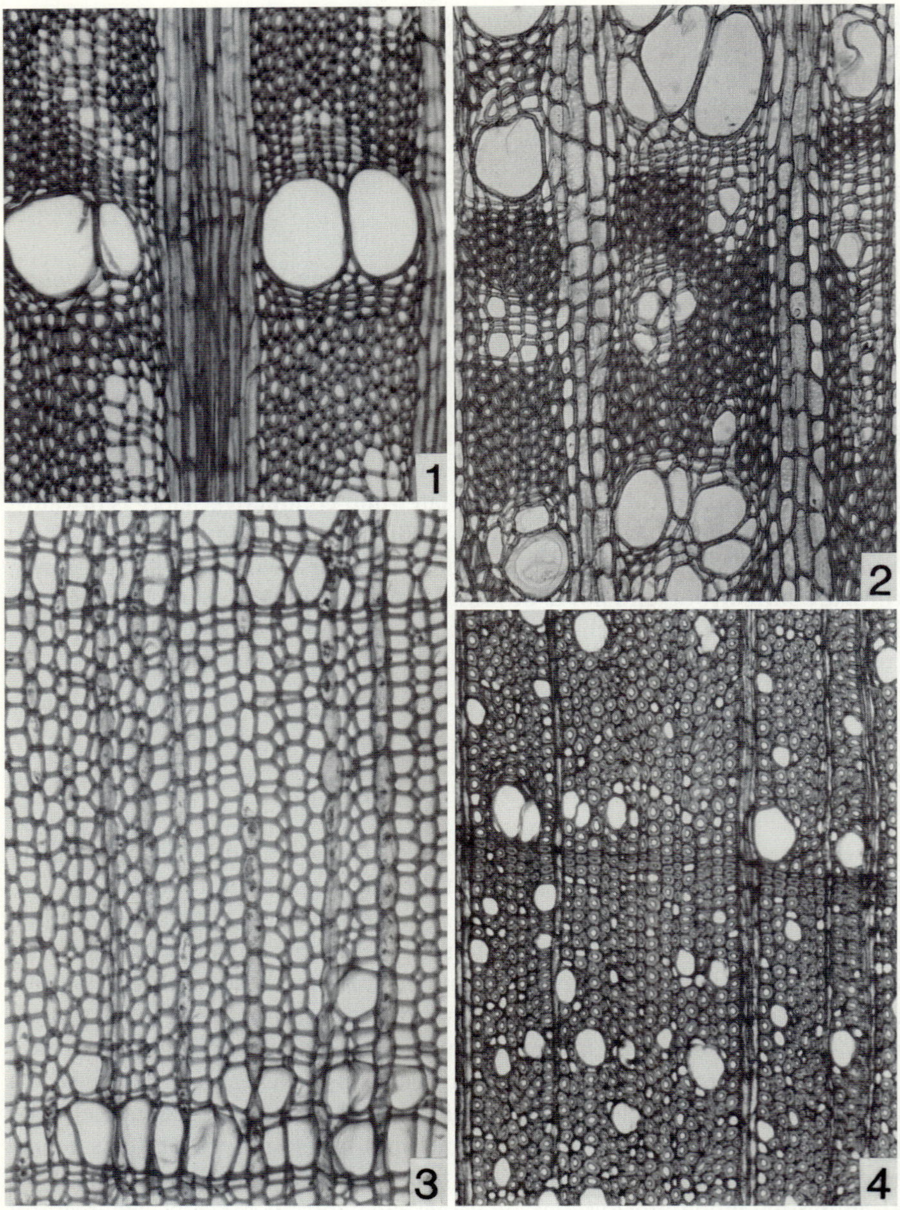

Fig. 2.5. Growth rings, type 4 and type 5. *1* Type 4 (wide earlywood vessels succeeded abruptly by narrow vessels), *Olearia virgata* (Asteraceae); *2* type 4, *Lepidospartum squamatum* (Asteraceae); *3* type 5A, *Myrica hartwegii* (Myricaceae); *4* type 5B, *Cercocarpus intricatus* (Rosaceae). (*1–4:B*)

which water availability is minimal. This formula, theoretically ideal, is limited to phylads that retain tracheids. Presence of vasicentric tracheids scattered throughout a growth ring might confer a similar kind of safety in latewood, although the omnipresence of tracheids in a wood with true tracheids makes the type 5 growth ring unexcelled in degree of potential safety.

Subtypes of type 5 may be recognized on the basis of relative absence of vessels in latewood as well as tracheid wall thickness variations:

A Vessels absent in latewood, tracheids uniform in wall thickness (*Myrica hartwegii*; Fig. 2.5, part 3).

B Vessels fewer in latewood than in earlywood, tracheids uniform in wall thickness (*Cercocarpus intricatus*; Fig. 2.5, part 4; *Cercidiphyllum japonicum*; Fig. 2.6, part 1).

C Vessels absent in latewood, tracheids thicker walled in latewood than in earlywood (*Ephedra californica*; Fig. 2.6, part 2).

D Vessels fewer in latewood, tracheids thicker walled in latewood (*Leptodactylon pungens*; Fig. 2.6, part 3; *L. californicum*; Fig. 2.6, part 4).

Examples of type 5 growth rings are abundant in a number of Bruniaceae, such as *Lonchostoma esterhuyseniae* (Carlquist 1978a).

2.4.6 Type 6

Vessels abundant in earlywood, less abundant in latewood.

This growth ring type was reported by Tupper (1913) in *Eucalyptus delegatensis*. Two examples included here are *Artemisia spinescens* (Fig. 2.7, part 1) and *Ceanothus leucodermis* var. *eglandulosus* (Fig. 2.7, part 2).

If latewood provides a form of safety, type 6 growth rings would seem to be ineffective. Presence of vasicentric tracheids in latewood would provide a form of safety (Carlquist 1985a), and in both of the examples illustrated, vasicentric tracheids are, in fact, present (they also are in *Eucalyptus*). Species with vasicentric tracheids may account for the majority of instances of type 6 growth rings. An additional possibility is present: woods in which all vessels are very narrow, and potentially embolize to a small extent. In such woods, if earlywood vessels remain functional for an entire season, addition of a large number of latewood vessels would increase neither conductive capacity nor safety to a marked extent. In this regard, one should note the type 5 growth rings illustrated, in which latewood vessels are narrow and few in number.

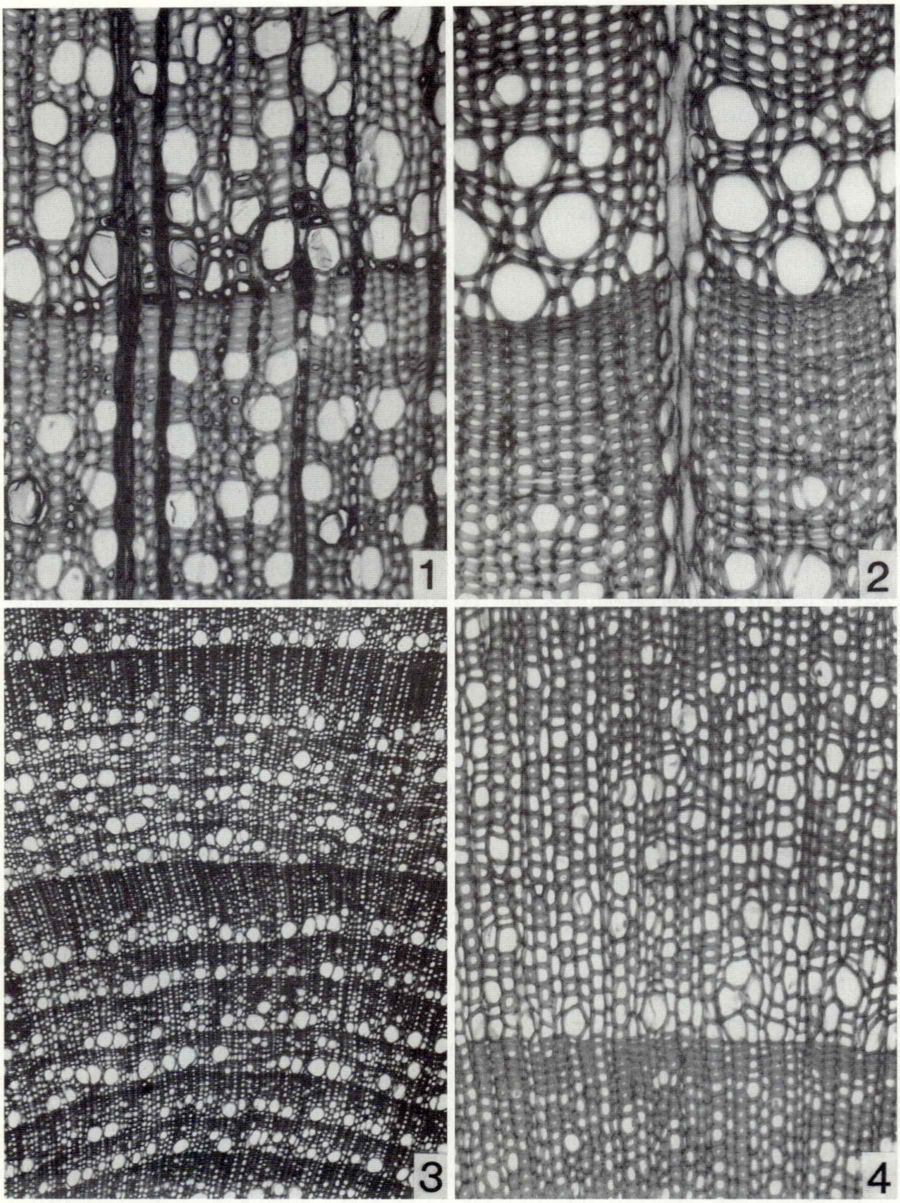

Fig. 2.6. Growth rings, type 5 (tracheids present, vessels few to none in latewood). *1* Type 5B, *Cercidiphyllum japonicum* (Cercidiphyllaceae); *2* type 5C, *Ephedra californica* (Ephedraceae); *3* type 5D, *Leptodactylon pungens* root (Polemoniaceae). *4* Type 5D, *Leptodactylon californicum* stem (Polemoniaceae). (*1,3,4:B*; *2:A*)

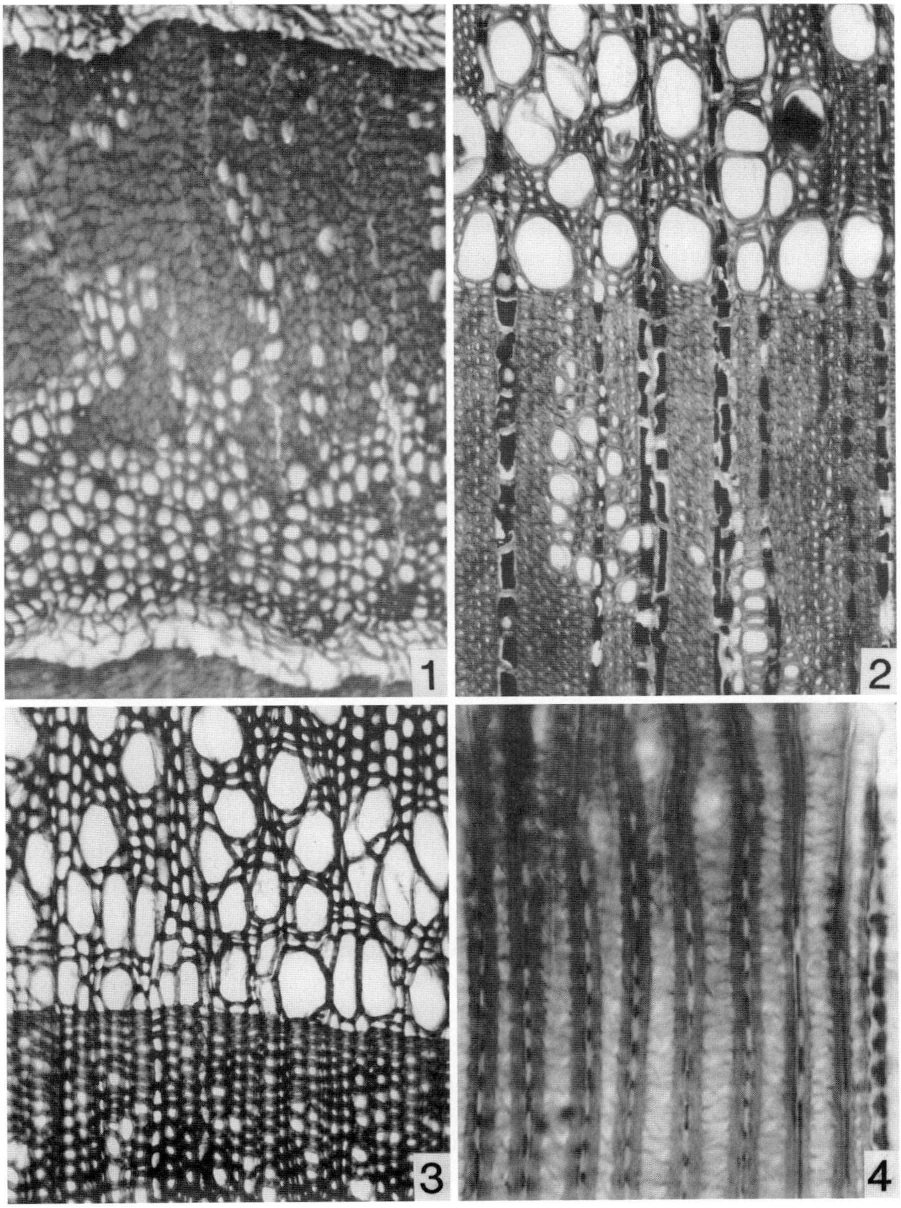

Fig. 2.7. Growth rings, type 6 and type 7. *1* Type 6 (vessels less abundant in latewood), *Artemisia spinescens* (Asteraceae); *2* type 6, *Ceanothus eglandulosus* (Rhamnaceae); *3*, *4* type 7A (vascular tracheids in latewood), *Keckiella ternata* (Scrophulariaceae); *3* transection, *4* radial section (vascular tracheid, left, plus several vessel elements with perforation plates above). (*1–3:B*; *4:C*)

2.4.7 Type 7

Vessels are wider in earlywood than in latewood; latewood vessels are as numerous or more numerous per mm^2 than in earlywood; vasicentric or vascular tracheids are present in addition to libriform fibers and fiber-tracheids.

This type is much like type 6, but with a greater representation of latewood vessels; the examples referred to type 7 here do not have vasicentric tracheids throughout a growth ring, and vasicentric tracheids tend to be absent in earlywood, present near vessels in latewood. Thus the representatives of type 7 show accommodation to increased flow rates in their wider vessels of earlywood, but also enhanced conductive safety but having both numerous narrow vessels and either vascular tracheids (end of growth ring only) or vasicentric tracheids (scattered adjacent to vessels in at least the latter parts of a growth ring).

A Species with vascular tracheids only are recognized here as a subtype; *Keckiella ternat*a (Fig. 2.7, parts 3 and 4) and *Isomeris arborea* Nutt. (Fig. 2.8, part 1).

B Under this heading are recognized species with vasicentric tracheids as defined above: for example, *Arbutus texana* (Fig. 2.8, part 2).

2.4.8 Type 8

Vessels are much wider in earlywood than in latewood. Vessels are variously more numerous in latewood than in earlywood.

Two examples are illustrated: *Quercus borealis* (Fig. 2.8, part 3), with relatively few latewood vessels, and *Q. alba* (Fig. 2.8, part 4) with numerous latewood vessels. Both of these examples have vasicentric tracheids, but presence of vasicentric tracheids is not essential to definition of type 8; *Fraxinus excelsior* also has exceptionally wide earlywood vessels (Yatsenko-Khmelevsky 1954) and is an example. In terms of functional interpretation, type 8 growth rings represent a strongly pronounced adaptation for conducting large volumes of water in earlywood, combined with conductive safety (evidently with much reduced volume handling capability) in latewood. This type is to be expected in deciduous species of wet temperate forest, especially some species with large leaf area.

2.4.9 Type 9

Vessels are wider in earlywood, narrower in latewood; at least some latewood vessels in all species of this type have scalariform perforation plates; earlywood perforation plates are simple.

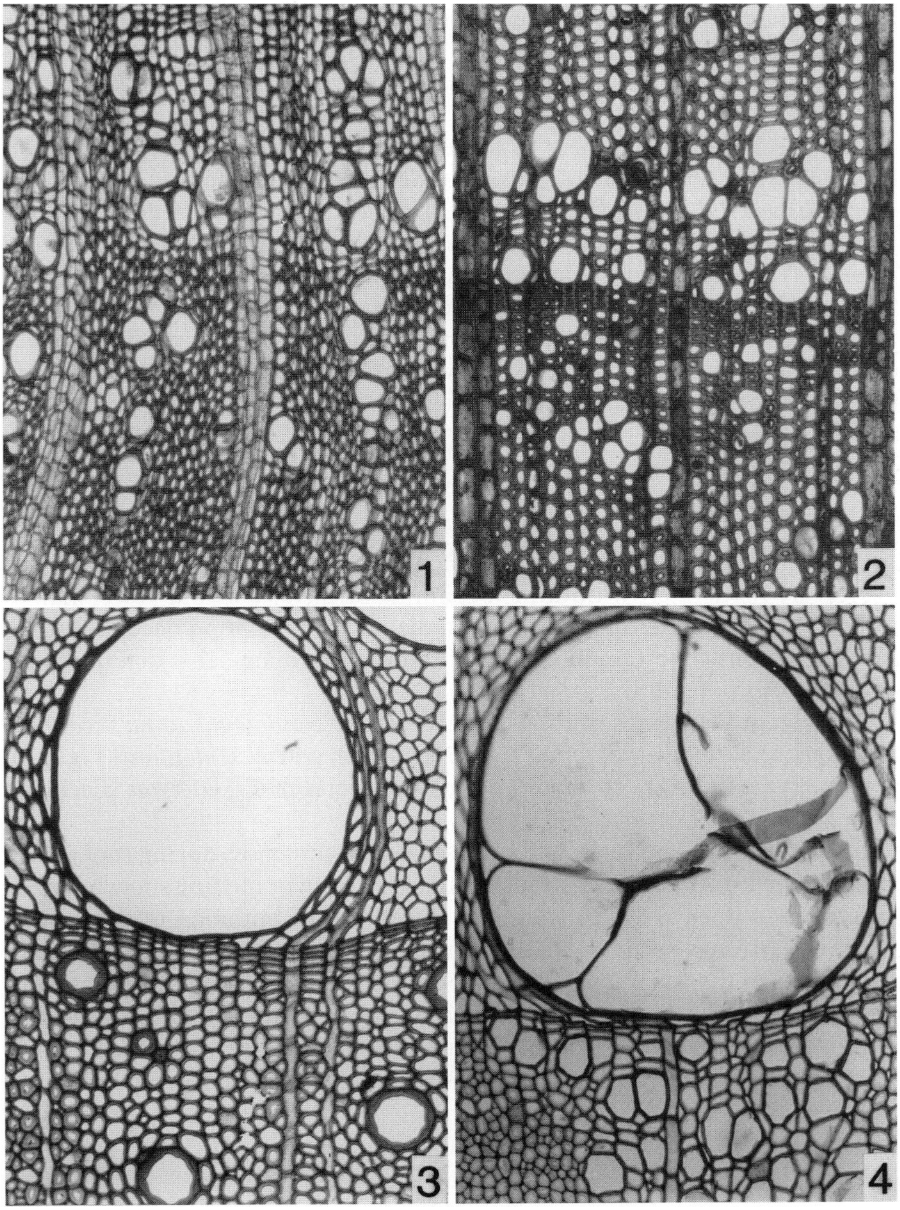

Fig. 2.8. Growth rings, type 7 and type 8. *1* Type 7A (vascular tracheids in latewood), *Isomeris arborea* (Capparaceae); *2* type 7B (vasicentric tracheids throughout wood), *Arbutus texana* (Ericaceae); *3* type 8 (earlywood vessels exceptionally large), *Quercus borealis* (Fagaceae); *4* type 8, *Quercus alba*. (*1–4:B*)

Only the presence of scalariform perforation plates in latewood differentiate this type from type 3. Presence of scalariform perforation plates in latewood must be established from examination of radial sections (Fig. 2.9, parts 2–6).

The species illustrated here as representative of type 9 include *Vaccinium arboreum* (Fig. 2.9, partsl–4), *Sassafras albidus* (Fig. 2.9, part 5), and *Styrax officinalis* var. *fulvescens* (Fig. 2.9, part 6). Other Lauraceae can be cited, such as *Laurus nobilis* (Carlquist 1980a), and various Platanaceae (e.g., *Platanus racemosa*) exemplify type 9. In functional terms, one can hypothesize that type 9 represents primitive phylads in which selection for peak conductive rates in earlywood has led to simplification of the perforation plate, whereas the slower conductive rates characteristic of latewood did not lead to such selection, so that scalariform perforation plates have persisted.

2.4.10 Type 10

Vessels are moderately wide at the beginning of earlywood, but vessel diameter increases thereafter until decrease in diameter in latewood occurs. Type 10 is a ring-porous condition in which optimum vessel diameter is deferred until after the beginning of the growth ring.

Type 10 growth rings are illustrated here for *Nothofagus antarctica* (Fig. 2.10, part 1), *Artemisia filifolia* (Fig. 2.10, part 2), and *A. tridentata* (Fig. 2.10, part 3). Type 10 has been reported for five species of *Eriodictyon* (Carlquist et al. 1983).

In species with type 10 growth rings, growth commences during cool winter months when soil moisture is available, thanks to rain or melting snow, but peak transpiration does not occur until some weeks after initiation of the growth ring. Such regimes occur in Mediterranean-type climates or in climates (such as portions of the Great Basin of the USA) with very cold winters, but hot and dry summers. The climatic regime of *Nothofagus antarctica* localities differs from these in that soil is probably always moist (except when frozen) but the growing season features slow acceleration and deceleration of temperatures.

2.4.11 Type 11

Vessel elements change from wide to narrow in the progression from earlywood to latewood; parenchyma bands present at the interface between one growth ring and the next (marginal parenchyma, "boundary parenchyma" of some authors).

Chowdhury (1964) emphasizes that parenchyma bands formed in a non-annual fashion cannot be cited in connection with growth rings. I agree,

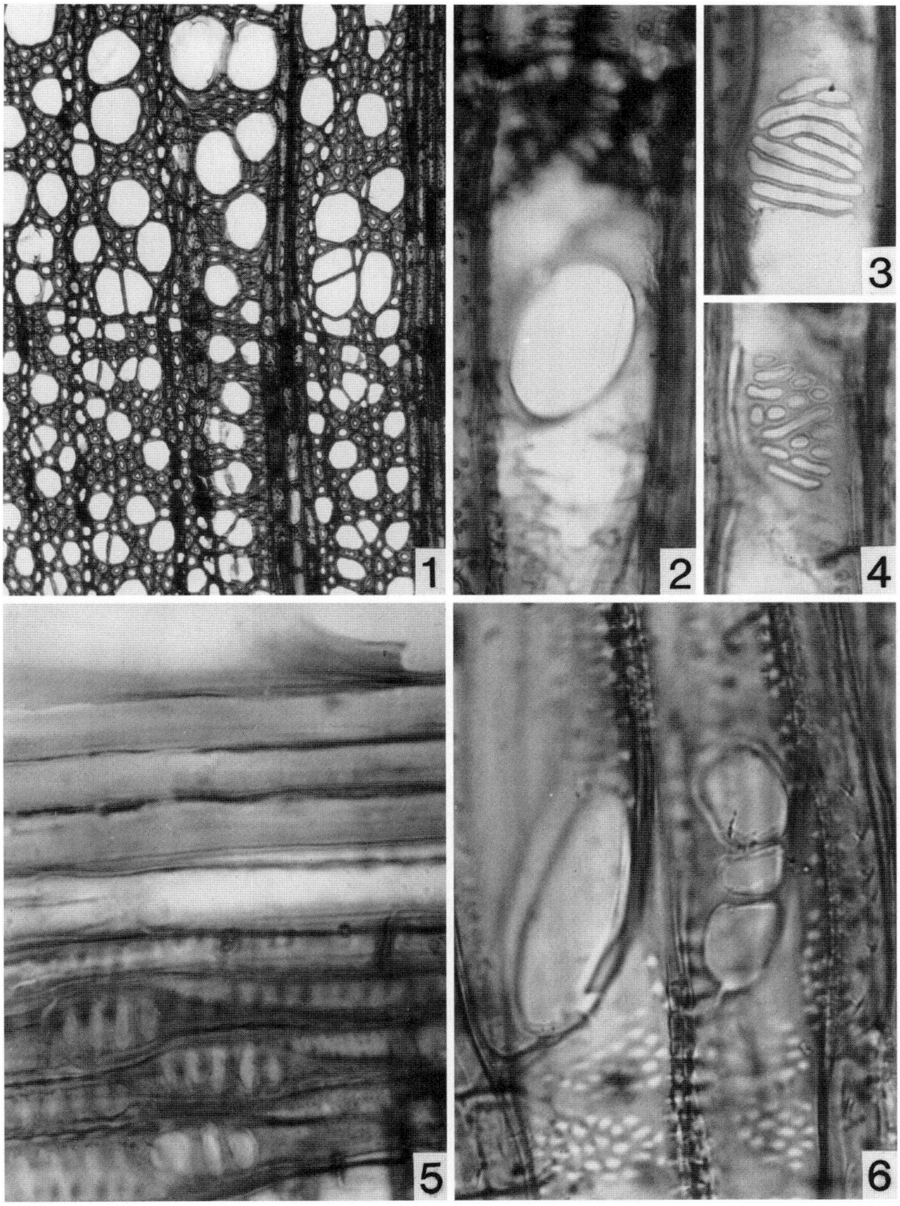

Fig. 2.9. Woods with type 9 growth rings (perforation plates scalariform in latewood, simple in earlywood). *1–4 Vaccinium arboreum* (Ericaceae). *1* Transection; *2–4* radial sections; *2* simple perforation plate from earlywood; *3* portion of scalariform perforation plate from latewood; *4* modified scalariform perforation from latewood; *5 Sassafras albidum* (Lauraceae) radial section – scalariform perforation plates *below*, portion of a simple perforation plate *upper right*. *6 Styrax officinalis* var. *fulvescens* (Styracaceae), radial section – simple perforation plate from earlywood at *left*, scalariform perforation plate from latewood at *right*. (*1:B*; *2–6:C*)

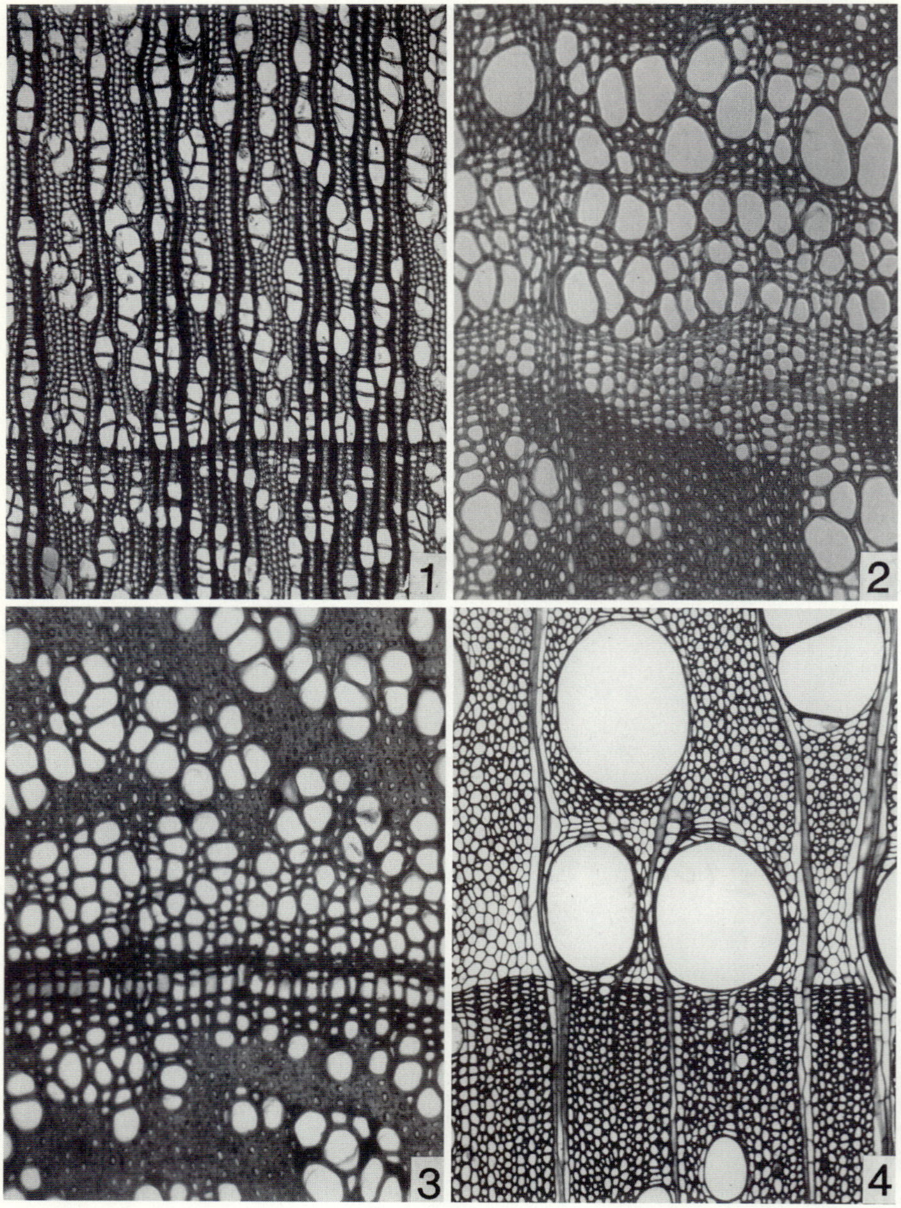

Fig. 2.10. Growth rings, type 10 and type 11. *1* Type 10 (earlywood vessels narrow at first, then wider), *Nothofagus antarctica* (Fagaceae), earlywood vessels few; *2* type 10, *Artemisia filifolia* (Asteraceae), no interxylary cork at end of growth ring; *3* type 10, *Artemisia tridentata*, interxylary cork band at end of growth ring; *4* type 11A (initial parenchyma present), *Tectona grandis* (Verbenaceae). (*1,4:A*; *2,3:B*)

and the examples cited as type 11 all show distinctively annual parenchyma bands.

A Initial parenchyma is present.
Initial parenchyma is less frequent than terminal parenchyma. However, examples can be cited, such as *Cedrela odorata*, *Gmelina* sp., *Tectona grandis* (Fig. 2.10, part 4), and other species from dry or seasonal tropical area (Jane 1934).

B Terminal parenchyma is present.
Terminal parenchyma is the predominant form of marginal parenchyma. It is shown here for *Populus grandidentata* Michaux (Fig. 2.11, parts 1 and 2). This example is representative of one extreme – in other examples one might find as many as ten cell layers in a terminal parenchyma band. Where only a single cell layer is present, as in *Populus grandidentata*, one may have difficulty deciding if the layer represents terminal or initial parenchyma, although observation of a tree in midwinter ought to reveal whether the band is already present or whether it is initiated only at the end of the winter.

C Both terminal and initial parenchyma are present.
This condition is shown here for *Robinia pseudoacacia* (Fig. 2.11, parts 3 and 4). The functional significance of marginal parenchyma appears to lie primarily in its starch storage. Hydrolysis of starch could yield sugars, which, when transferred into vessels, can raise osmotic pressure, promoting sap flow in vessels adjacent to the parenchyma. This idea was pioneered with respect to the sugar maple by Sauter (1966a,b; Sauter et al. 1973), and certainly *Acer saccharum* does have terminal parenchyma in which starch storage and hydrolysis have been observed (Gregory 1978). Braun (1983, 1984) has applied this idea to temperate trees in general, including those without marginal parenchyma. Terminal parenchyma bands often occur conspicuously in certain shrubby or suffrutescent perennial herbs (Carlquist and Eckhart 1984; Carlquist 1995a). Possibly terminal parenchyma bands in shrubby herbs can serve for promoting conduction when growth is renewed during the wet season. An alternative possibility for the function of terminal parenchyma would be storage of starch to support rapid flushes of growth, flowering, and fruiting.

2.4.12 Type 12

Wood diffuse-porous, but with an annually produced parenchyma band.

Chowdhury (1964) presents a table of diffuse-porous species in which initial parenchyma is present. The list includes Combretaceae, Fabaceae, Meliaceae, and Santalaceae. Gottwald and Parameswaran (1966) list terminal parenchyma for the majority of species of *Hopea* (Dipterocarpaceae); wood in this genus is

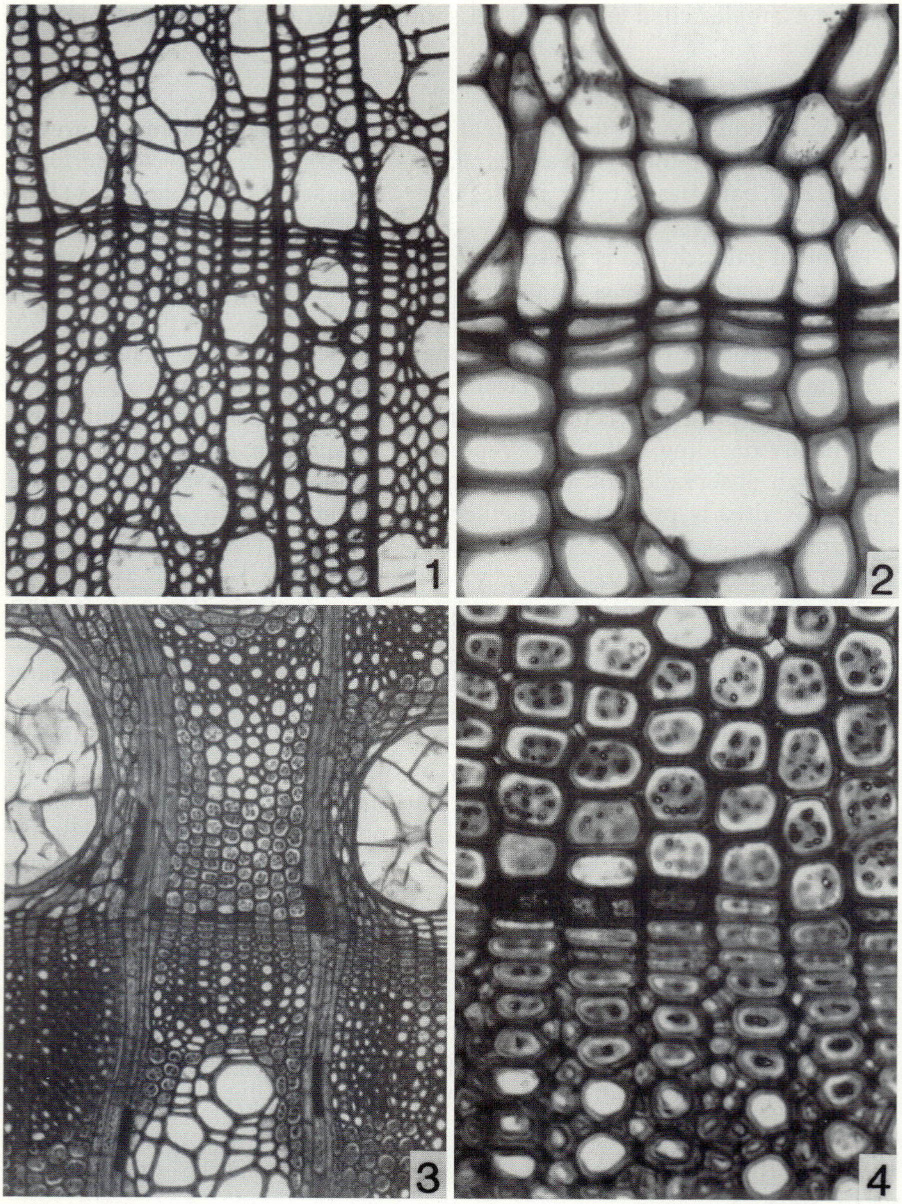

Fig. 2.11. Growth rings, type 11. *1, 2* Type 11B (terminal parenchyma present), *Populus grandidentata* (Salicaceae). *1* Transection showing vessel diameters; *2* transection portion showing that terminal parenchyma is a single cell layer thick; *3, 4* type 11B (terminal plus initial parenchyma present), *Robinia pseudoacacia* (Fabaceae); *3* latewood vessels (above), portions of earlywood vessels (*below*); *4* juncture between terminal and initial parenchyma, showing starch grains in parenchyma. (*1,3:B*; *2,3:C*)

diffuse-porous with respect to vessel diameter. Thus growth rings demarcated only by parenchyma, with little or no variation in vessel diameter or density, are more than a theoretical possibility. Type 12 growth rings are illustrated here for *Thunbergia holstii* (Fig. 2.12, part 1).

Functionally, type 12 growth rings could be explained in terms of seasonal storage of starch related to flushes of growth. Many diffuse porous dicotyledons have annual flushes of growth (e.g., many Araliaceae and Fabaceae).

2.4.13 Type 13

Vessels markedly dimorphic between earlywood and latewood: vessels wide in diameter and not in extensive groups in earlywood; latewood vessels narrow and formed in extensive groupings that traverse rays and may be oriented either diagonally (intermediate between tangential and radial) or tangentially.

Included in this definition are woods with the so-called ulmiform condition (tangential bands of narrow latewood vessels). Some species with this condition are keyed by Record (1942b). For further details on growth rings of Ulmaceae, see Sweitzer (1971). Also included in type 13 are woods with latewood vessels in diagonal bands; keys for genera and some species with this condition are given by Record (1942c). A paper has been devoted to listing and determining the histology and significance of woods with either tangential or diagonal latewood bands (Carlquist 1987c). These two types were combined in a single paper because the anatomical and physiological significance seems the same for both types – in fact, transitions may be found in a single wood, as in *Celtis occidentalis* (Fig. 2.12, part 4). In some woods with type 13 growth rings, vessel grouping appears in the form chiefly of clusters (Fig. 2.12, part 2), whereas in the majority, extensive tangential (Figs. 2.12, part 3 and 2.13, part 1) or diagonal (Fig. 2.12, part 4) bands are present. The latter have been termed "flamelike" or "dendritic" vessel aggregations, terms I am not adopting because their descriptive value is limited and does not tend to describe well some woods that obviously belong to this category.

The functional interpretation of type 13 growth rings (Carlquist 1987c) is obviously wide earlywood vessels that would accommodate peak flow. However, the latewood appears to be an unusually well-developed series of mechanisms for conductive safety: either very narrow vessels or (very commonly) vasicentric tracheids are present in the latewood vessel bands, and thus elements that embolize more tardily (very narrow vessels or vasicentric tracheids) are intermixed with the narrow latewood vessels. Additionally, the diagonal and tangential bands have the effect of forming a network, perhaps interconnecting virtually all vessels within a growth ring, which assures maintenance of the conductive pathways regardless of embolism formation in a large number of vessels.

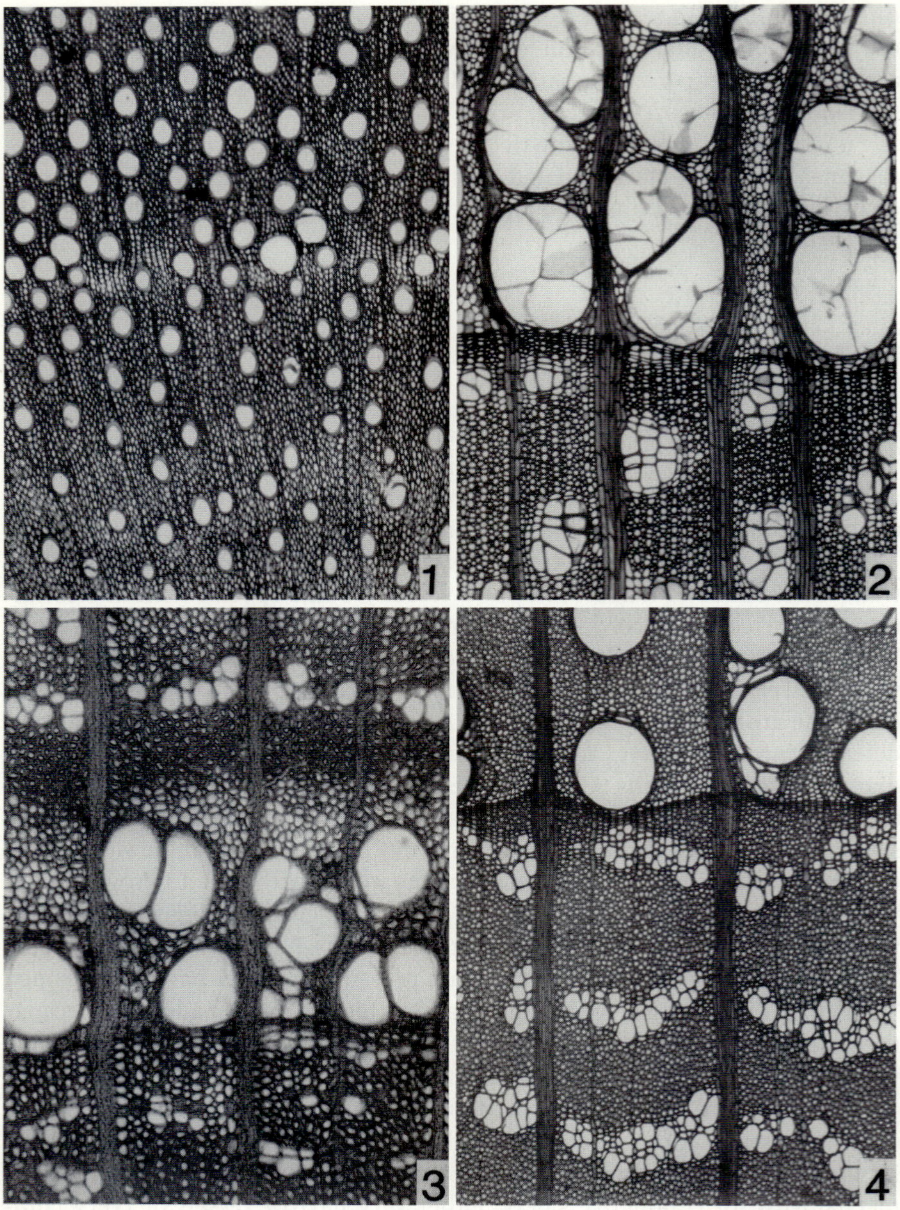

Fig. 2.12. Growth rings, type 12 and type 13. *1* Type 12 (vessels vary little in diameter, marginal parenchyma present), *Thunbergia holstii* (Acanthaceae); *2* type 13 (vessels in diagonal groupings in latewood), *Morus rubra* (Moraceae); *3* type 13, *Ulmus glabra* (Ulmaceae); *4* type 13, *Celtis occidentalis* (Ulmaceae). (*1–4:A*)

2.4.14 Type 14

Vessels are markedly wider in earlywood, and the wood is thus ring-porous. The growth ring is delimited by markedly thicker walled imperforate tracheary elements.

This growth ring type was mentioned by Ingle and Dadswell (1953b) for *Eucalyptus gigantea*. It is illustrated here for another species of *Eucalyptus* (Fig. 2.13, part 2). The thick-walled imperforate tracheary elements in these eucalypts are fiber-tracheids; the latewood fiber-tracheids are distinctive in having pits on tangential rather than radial walls (radial walls decrease in width in latewood), and are, in fact, rather tracheid-like (a phenomenon that can also be seen in some species of *Platanus*). The functional significance of type 14 growth rings does not appear to be one of enhanced safety in the latewood fiber-tracheids; conductive safety in *Eucalyptus* wood is definitively achieved by the presence of vasicentric tracheids, present in all species of the genus according to Ingle and Dadswell (1953b).

2.4.15 Type 15

Vessels grade from wider in earlywood to narrower in latewood, although the difference in diameter is not as marked as in typically ring-porous woods; imperforate tracheary elements are either absent or present, only in latewood, as a pair of fiber strands in each fascicular area.

Type 15 has been devised to recognize the features of growth rings in *Misodendron* subgenus *Angelopogon* (Carlquist 1985c). From that subgenus, *M. brachystachyum* is illustrated here (Fig. 2.13, parts 3 and 4). The absence of imperforate tracheary elements throughout the growth ring or only in localized strands at the end of it constitute the distinctive feature of this type.

2.5 Interxylary Cork

Although presence of interxylary cork was used as a feature for recognizing a growth ring type in an earlier scheme (Carlquist 1980a), it has not been so used here because in a genus where interxylary cork is present, such as *Artemisia*, some species may have interxylary cork, others may lack it, but growth ring types are otherwise alike. Thus, *Artemisia filifolia* (Fig. 2.10, part 2) lacks interxylary cork but *A. tridentata* has interxylary cork; both can be designated type 10. For a discussion of interxylary cork, see Section 5.5.8.

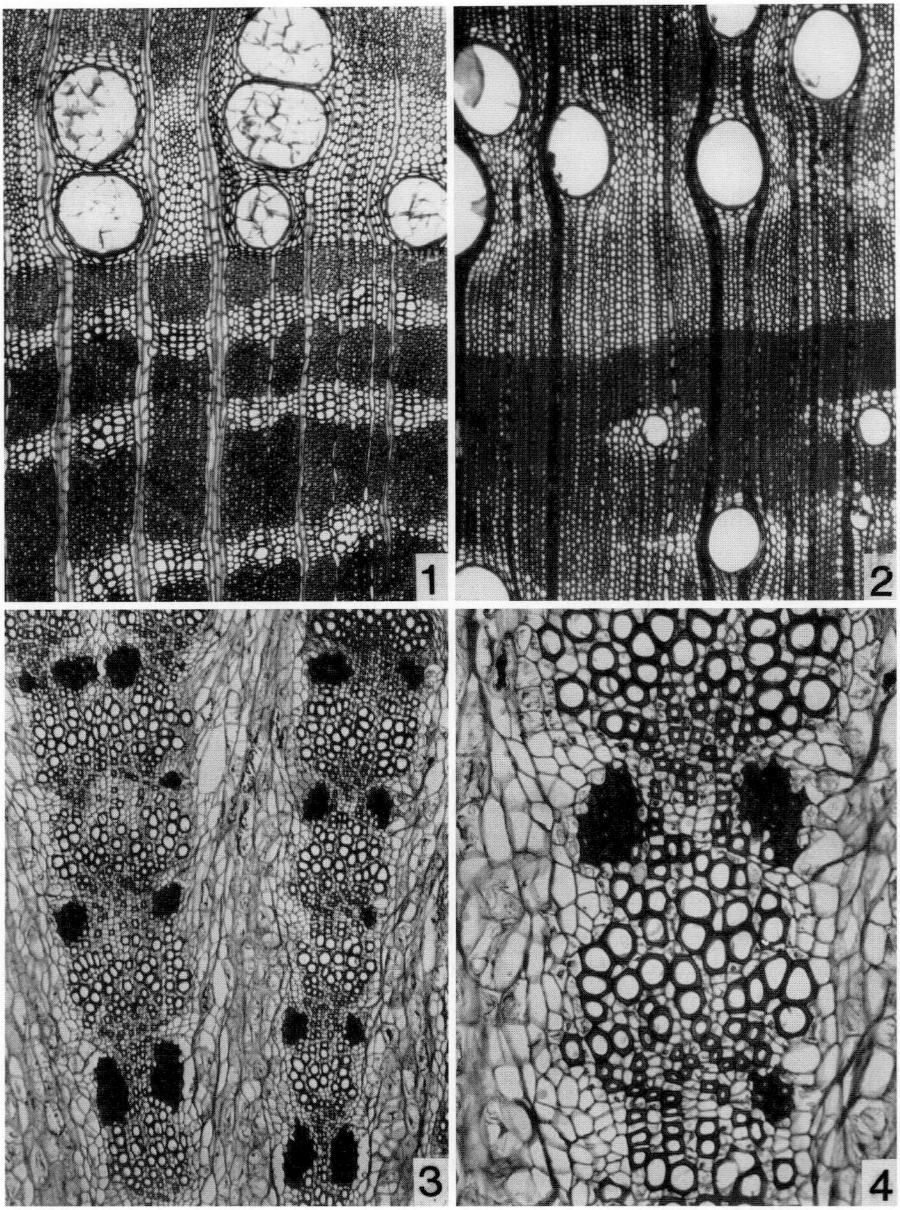

Fig. 2.13. Growth rings, type 13, type 14, and type 15. *1* Type 13 (latewood vessel bands tending to tangential orientation), *Maclura pomifera* (Moraceae); *2* type 14 (imperforate tracheary element wall thickness demarcates growth ring), *Eucalyptus dalrympleana* (Myrtaceae); *3, 4* type 15 (latewood demarcated by thick-walled fibers plus narrow vessels), *Misodendron brachystachyum* (Misodendraceae); *3* five growth rings in two adjacent fascicular areas; *4* growth ring, enlarged, portions of two others. (*1–3:A*; *4:B*)

2.6 Modifiability of Growth Rings

Not all deciduous species have growth rings; not all tropical evergreen species lack growth rings (Chowdhury 1964; Tomlinson and Craighead 1972). Some growth rings are thought to be obligate, some facultatively present, but we have data on modifiability in very few species. Bissing (1976, 1982) compared wood of Californian native species from plants in the Rancho Santa Ana Botanic Garden with wood from the respective natural populations from which seeds of the garden-cultivated plants were collected. Bissing used an aridity index to separate species into two groups: in group I, the garden was more mesic than the wild habitat; in group II, the garden was more xeric. Group I species shifted toward uniform diffuse-porous conditions from some degree of ring porosity; group II species shifted toward narrow ring porosity from broad ring porosity. Bissing's studies show that with increased water availability, earlywood is widened and vessels become wider in earlywood. These findings are in accord with common suppositions about growth ring formation and function. Discontinuous growth rings (those not complete around the circumference of the stem) and false rings (cells like latewood included within earlywood portions) have been recognized by various authors (MacDougal 1939; Brown et al. 1949; Panshin and de Zeeuw 1980). These phenomena are interesting in showing modifiability of growth rings. Width of growth rings has been claimed directly related to water availability (Novruzova 1972), a fact which is basic to the use of growth rings as a way of developing a climatic record suitable for dating (see Fritts 1976).

2.7 Experimental Studies

Growth rings offer a fascinating subject for experimental work because rings may begin with wide vessels like those of tropical trees but end with grouped narrow vessels like those of a desert shrub.

Ewers (in Gould 1984) and Davis et al. (1999) have found that narrower vessels are more resilient (refill after embolisms). Narrower vessels embolize less readily than wide ones in ring-porous species, whether drought or freezing is the cause (Ellmore and Ewers 1985; Hargrave et al. 1994; Jarbeau et al. 1995). As early as 1935, Huber had enunciated the idea that conduction is much more massive and rapid in earlywood, and he presented data to support that concept. Huber (1935) also integrated into his picture of ring porosity the fact that earlywood vessels tend to be longer than latewood vessels. This observation, confirmed by Zimmermann (1983), shows a reason for the contrast between earlywood and latewood with respect to safety: the greater length of earlywood vessels favors efficiency of conduction (fewer vessels end blindly) but represents greater vulnerability. Greater lengths of the conductive system would be disabled if embolisms were to form in earlywood vessels.

2.8 Cell Length with Respect to Growth Rings

Differences within a growth ring have been observed with respect to lengths of component cells. Sometimes these results are expressed in terms of cells in general: Butterfield (1973) records cells shorter in earlywood than in latewood of *Hoheria angustifolia* (Malvaceae). Some studies compare earlywood and latewood lengths for particular cell types separately: Parameswaran (1964) shows vessel elements and fibers (libriform fibers) to be shorter in earlywood than in latewood of *Tectona* (Verbenaceae); Chalk (1970) obtained similar but more strongly marked length fluctuation for vessel elements and libriform fibers of Fraxinus. Swamy et al. (1960) studied fluctuation in length of both vessel elements and imperforate tracheary elements within growth rings of various dicotyledons; their results show, in many species, shortening of elements in latewood just prior to commencement of the succeeding ring. The significance of these patterns has not been explained by these authors. Length of the fusiform cambial initial is not known to change on an annual basis, and therefore we are observing differential elongation of derivatives from the fusiform cambial initials. As a hypothesis, I propose as a general idea that narrower (latewood) cells tend to elongate less than wider (earlywood) cells.

2.9 Systematic Distribution of Growth Rings

Three lists that attempt to cover the world flora have been prepared here. These represent expansions of the lists given by Record (1942a) for American woods. These lists are doubtless quite incomplete. Families are listed if **any** genus has growth rings, and the lists should **not** be regarded as indicative that all or most genera of the families listed have growth rings. This is most frequently true of the species in the third list.

Markedly ring-porous

Adoxaceae (*Sambucus*)
Anacardiaceae
Araliaceae
Asteraceae
Berberidaceae
Betulaceae
Calycanthaceae
Caprifoliaceae
Celastraceae
Crossosomataceae
Eleagnaceae
Fabaceae (in part)
Fagaceae
Juglandaceae
Lamiaceae
Malvaceae
Moraceae
Myricaceae
Oleaceae
Polemoniaceae
Polygonaceae
Ranunculaceae
Rhamnaceae
Rosaceae
Rutaceae
Sapindaceae
Scrophulariaceae
Tamaricaceae
Ulmaceae
Verbenaceae
Vitaceae

Ring-porous at most but commonly semi-ring-porous

Bignoniaceae
Bruniaceae
Buxaceae
Casuarinaceae
Cercidiphyllaceae
Cneoraceae

Cunoniaceae
Daphniphyllaceae
Dipsacaceae
Ebenaceae
Empetraceae
Ericaceae
Eucryphiaceae
Eupteleaceae
Fouquieriaceae
Geraniaceae
Grossulariaceae
Hamamelidaceae
Hydrangeaceae
Hydrophyllaceae
Illiciaceae
Koeberliniaceae
Lardizabalaceae
Leitneriaceae
Misodendraceae
Myrothamnaceae
Pittosporaceae
Platanaceae
Roridulaceae
Rosaceae
Santalaceae
Staphyleaceae
Theaceae
Thymeleaceae
Tremandraceae
Valerianaceae
Zygophyllaceae

Semi-ring-porous or ring-porous in a minority of species

Acanthaceae
Achatocarpaceae
Alangiaceae
Annonaceae
Apocynaceae
Aquifoliaceae
Boraginaceae
Cactaceae
Cistaceae
Combretaceae
Connaraceae
Coriariaceae
Elaeocarpaceae
Euphorbiaceae
Fabaceae (in part)
Krameriaceae
Lauraceae
Loganiaceae
Loranthaceae
Lythraceae
Malpighiaceae
Meliaceae
Myrsinaceae
Olacaceae
Papaveraceae
Passifloraceae
Proteaceae
Rubiaceae
Sapotaceae
Simaroubaceae
Solanaceae
Sterculiaceae
Styracaceae
Tiliaceae
Violaceae
Viscaceae

3 Vessel Elements

3.1 Definition

Vessel elements (vessel members, vessel segments of some authors) may be defined as xylem cells in which one or more pitlike structures lack a pit membrane (which dissolves) at maturity, thus forming perforations. Despite lysis, a perforation may retain membrane remnants. Perforations characteristically occur on end walls ("overlap areas"); the end walls are thus perforation plates. Vessel elements are regarded as specializations of tracheids in which loss of pit membranes on some of the end wall pits has occurred. Exceptions to the definition offered may easily be found, because evolution does not typically provide mutually exclusive categories. Absence of pit membranes may be partial in perforations in primitive vessel elements, as detailed below, and tracheary elements transitional between vessel elements and tracheids will be demonstrated more frequently as work with scanning electron microscopy progresses. Perforations may occur on apparently lateral faces of vessel elements in the case of fibriform vessel elements; this is not so much an exception with regard to location of the perforation plate as it is an example of a less precisely defined end wall.

Perforated ray cells can be cited as a phenomenon in which a cell combines characteristics of a ray cell (located in a ray, shaped like ray cells but usually larger) with characteristics of vessel elements (perforation plates present, lateral walls with bordered pits). Perforated ray cells are considered in this book in Chapter 6 in connection with rays. However, one must note that perforated ray cells can be demonstrated to represent a bridge from a vessel on one side of a ray to a vessel on the other side.

Attention is called to the usage "vessel elements" when referring to a single cell, whereas "vessel" refers to a series of vessel elements – by presumption, the entire series of cells each of which bears perforation plates at both ends except for the uppermost and lowermost vessel elements (each of which lacks a perforation plate at its terminating end). One may use the term "vessel" where one is not referring to a vertical series of vessel elements: for example, transections are said to show vessels. The term "pore" is also used for vessels as seen in transection; this term, although well established in dendrological literature, is avoided in this book except for growth rings because "pore" has a multiplicity of meanings in plant anatomy (e.g., pores in sieve areas) and because it duplicates the term "vessel".

Occasionally, those less experienced in wood anatomy use the term "vessel" where vessel element is intended, and care in using these similar terms is urged.

3.2 Types of Vessel Elements

The term "vessel element" without any modifier may refer to a cell long and fusiform in shape, as long as an end wall, no matter how oblique, is present. In most vessel elements, delimiting an end wall offers no difficulty; the end wall may be transverse, as in the vessel element shown in Fig. 3.1, part 2. However, one may note that the longer and more nearly fusiform the vessel element, the greater the chance that a perforation plate may not be seen in a longitudinal section (which may be sufficiently thin or slightly oblique so that one or both perforation plates of a given vessel element may be not be present). The use of macerations is strongly urged as a way of elucidating the nature of vessel elements.

3.2.1 Fibriform Vessel Elements

Certain groups of dicotyledons characteristically have vessel elements of a fusiform shape, the tips of which extend well beyond perforation plates; perforation plates are sometimes described as lateral, and tend to be vertically oriented rather than oblique or horizontally oriented. They may be very much like fibers or tracheids in shape (Fig. 3.1, part 1, left), or a little wider but still fusiform (Fig. 3.1, part 1, right). Woodworth (1935) was the first to call attention to these vessel elements under the term used here; he reported them in Passifloraceae, in which they have been found also by Stern and Brizicky (1958b) and Ayensu and Stern (1964). Although fibriform vessel elements may occur in some other families in which a scandent habit is common, such as Convolvulaceae (Mennega 1969; Pant and Bhatnagar 1975) and Nepenthaceae (Carlquist 1981b), they can be found in nonscandent families as well. Fibriform vessel elements have been reported in *Corokia* of the Cornaceae (Patel 1973), *Bergia* of the Elatinaceae (Carlquist 1984b), Gentianaceae (Carlquist 1984c), *Eriodictyon* (Fig. 3.1, part 1) of the Hydrophyllaceae (Carlquist et al. 1983) as well as other Hydrophyllaceae (Carlquist and Eckhart 1984), a few Polemoniaceae (Carlquist et al. 1984), Retziaceae (Carlquist 1986c), *Ceratopyxis* of the Rubiaceae (Vales and Babos 1977), the tribe Anthospermeae of the Rubiaceae (Koek-Noorman and Puff 1983), and Stilbaceae (Carlquist 1986c).

Woodworth's (1934) term "perforated fiber-tracheid" should be rejected in favor of his (1935) term "fibriform vessel element." Likewise, such confusing terms as "perforated tracheid" should be rejected.

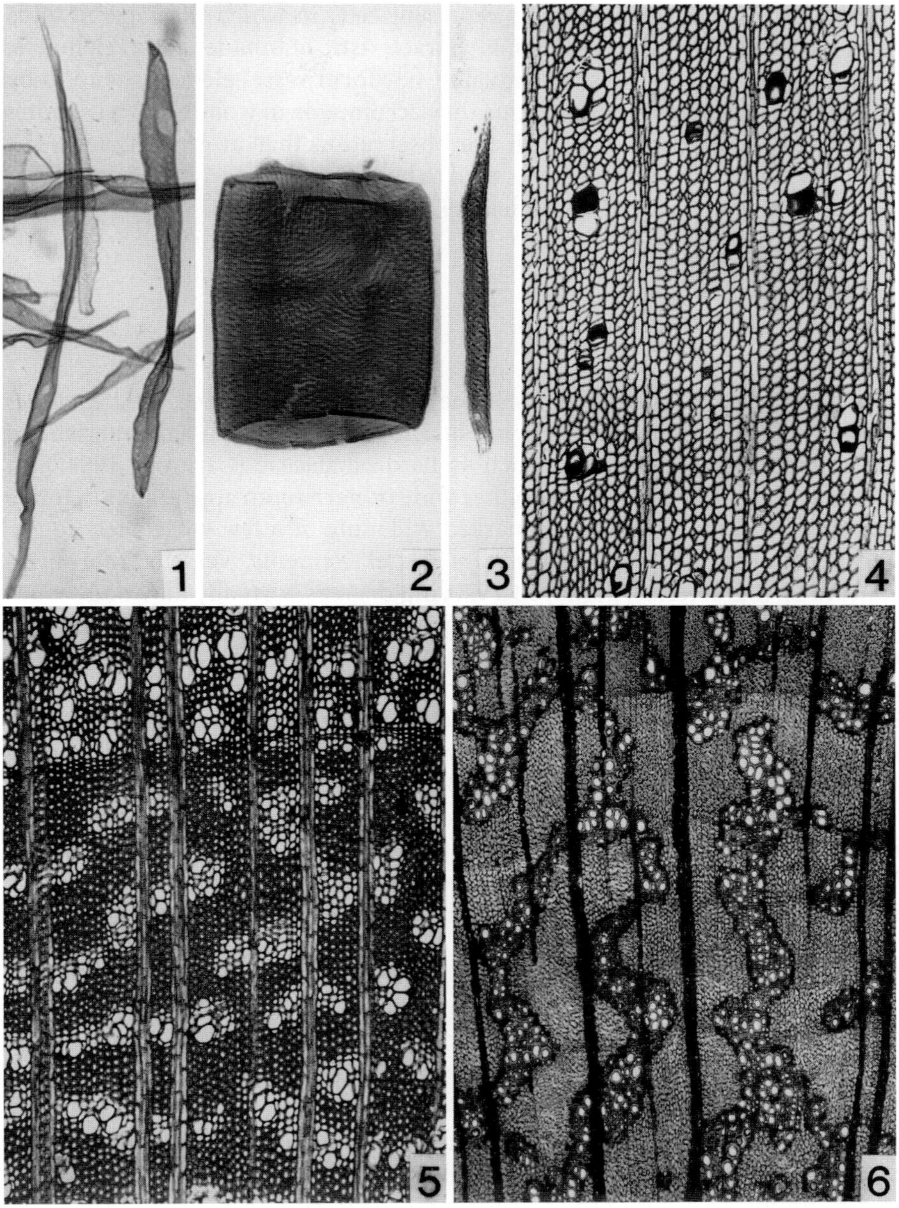

Fig. 3.1. Vessel types shown in macerations (*1–3*); vessel grouping related to ecology shown by means of transections – further explanation in text (*4–6*). *1* Two fibriform vessels (oriented vertically), *Eriodictyon trichocalyx* subsp. *lanatum* (Hydrophyllaceae); *2* wide vessel from *Nepenthes lowii* (Nepenthaceae) wood; *3* narrow (fibriform) vessel from *Nepenthes lowii* wood; *4 Olearia argophylla* (Asteraceae); *5 Olearia avicenniaefolia*. *6 Olearia muelleri*. (*1–3:B*; *4–6:A*)

Fibriform vessel elements may represent cells in which the tapered ends connote greater intrusiveness than is characteristic of broader vessel elements. In fact, where observations are available, fibriform vessel elements tend to be longer than ordinary vessel elements they accompany in woods that have both. This seems to be correlated with the observation that in growth rings, the narrower latewood vessel elements tend to be longer than earlywood vessel elements (Swamy et al. 1960; Butterfield 1973).

3.2.2 Vessel Dimorphism

This term was originated to describe a situation observed in *Nepenthes* (Carlquist 1981b: illustrated here in Fig. 3.1, part 2), but also characteristic of other woody scandent groups, such as Lardizabalaceae (Carlquist 1984f). In woods of these vines, vessels are either wide or narrow, an appearance that one can interpret as a tendency for marked widening of a few vessels tending to forestall widening of the majority of vessel elements. Vessel dimorphism appears to characterize, perhaps to varying degrees, virtually all woody vines observed (Carlquist 1985d). Where vessel dimorphism occurs, the narrow vessel elements can be termed fibriform vessel elements, and are at least a little longer than the wider vessel elements they accompany (Fig. 3.1, part 2). Not all instances of fibriform vessel occurrence involve vessel dimorphism, however: the nonscandent genera mentioned above as having fibriform vessel elements, such as *Eriodictyon* (Carlquist et al. 1983), do not exemplify vessel dimorphism.

Vessel dimorphism confers greater conductive efficiency to wide vessel elements and greater conductive safety to narrow vessel elements (Carlquist 1985d), and thus provides a system of vessels that can conduct if wider vessels embolize. Vessel dimorphism occurs in some growth rings (see graphs of Woodcock 1989a,b).

3.2.3 Imperforate Vessel Elements

This term would be misleading and therefore dubious if applied to vascular tracheids, in which one can see gradual narrowing and finally disappearance of perforation plates toward the end of a growth ring in certain species (see Chap. 4). However, in a very few woods with scalariform perforation plates, one can see that in some perforation plates pit membranes are absent, whereas in others, they are retained; these latter cells could legitimately be called imperforate vessel elements because in every detail of morphology, other than pit membrane presence, the imperforate vessel elements are identical to perforation-bearing vessel elements. A perforation plate in which pit membranes have been retained (original observation) is shown here for *Myrothamnus* (Fig. 3.9, part 4).

3.3 Vessel Dimensions

3.3.1 The Problem of Quantification

Van den Oever et al. (1981) have shown that standard deviation decreases very little if one bases a mean on more than 25 measurements of vessel element length. A mean seems a useful figure to present for vessel dimensions. However, should one give, in addition, a standard deviation, a total range, or a most frequent range? Let us suppose that most cell populations tend to follow a normal distribution curve (Vales and Babos 1977). If this is so, a mean provides the most informative datum, because it closely approximates the majority of the cells. Range appears the least informative, because extremely short or long, narrow or wide cells are almost chance occurrences. If physiological features of the wood are related to cell dimensions (e.g., conductivity related to vessel diameter), the functional characteristics are most accurately reflected by the average cell, not the extreme cell. Likewise, means are more reliable tools than extremes where identification is concerned. One must remember that standard deviations only reflect the material measured, and if derived from vessels at a stem base, for example, do not apply to cell populations of a branch, root, or even those of another stem base in a given species where ecological conditions and ages may vary.

3.3.2 Relation of Vessel Element Dimensions to Location in Plant

As one goes from the pith to the cambium, vessel element length changes markedly during a juvenile period, which may be brief or prolonged. These are discussed with relation to paedomorphosis and other ontogenetic changes in Chapter 9.

Vessel diameter may change markedly as one goes from inside to outside of a woody stem (Davidson 1976; Fukazawa 1984; Khan 1980; Carlquist 1985b; Hayden and Hayden 1994). This may represent accommodation of increased volumes of water as the plant forms a large leafy crown. Vessel element length does not increase proportionately to vessel diameter within a cell population (Davidson 1976).

Although one would expect a decrease in vessel density as vessel diameter increases, such an inversely proportionate increase may not occur (Khan 1980).

Vessel elements are reported to be shorter in lignotubers than in woody stems of a given species (Carlquist 1978a). Vessel elements tend to be shorter in branches than in the main stem (Phelps et al. 1982; Iqbal and Ghouse 1983) unless pedomorphosis occurs (Carlquist 1969a). Vessel elements tend to be wider and longer in roots than in stems of a given species (Patel 1965; Carlquist 1978b), although the reverse trend occurs in lianas (Ewers et al. 1997).

3.3.3 Vessel Element Dimension in Relation to Other Factors

Dwarfing of a plant may result in diminution of vessel element length and diameter (Baas et al. 1984), as was also observed in tracheids (Bailey and Tupper 1918). Within a growth ring, latewood vessels tend to be narrower and longer than those in earlywood (see Sect. 2.8). However, the increase in length is not proportional to the decrease in diameter. In particular cases, vessel element length increases with diameter (e.g., van den Oever et al. 1981; Carlquist and Hoekman 1985a), but growth rings show clearly the independence of the two dimensions. This is to be expected, because vessel element length is dictated by fusiform cambial initial length; intrusive growth can occur during maturation of derivatives, but apparently relatively little in the case of vessel elements. As examination of the various diameters of cells – even vessel elements – in a radial file reveals, increase in diameter fluctuates greatly, probably in relation to a growth factor. Intrusiveness of imperforate elements is thought always to be greater than that of the vessel elements they accompany in a given wood sample, and different degrees of intrusiveness characterize imperforate tracheary elements of particular plants (Bailey and Tupper 1918). A low degree of intrusiveness of imperforate tracheary elements has been said to be, in general, a primitive condition (Carlquist 1975a). There are a few reports of vessel elements longer than the imperforate tracheary elements they accompany (Carlquist 1976a, 1977a); these may be merely statistical anomalies, although one should note that these particular instances represent shrubby plants with very narrow and more intrustive vessel elements than those found in those of most woody dicotyledons.

Polyploids have greater vessel diameters than diploids (Swamy and Govindarajalu 1957). This may, in turn, be related to plant size. Certainly habit is correlated with vessel element length and diameter. Vessel element diameter and length are greater in woodier plants than in their more herbaceous relatives (Cumbie and Mertz 1962); they are greater in species that are taller than in shorter relatives (Gibson 1973). Vessel element length and diameter are greatest in trees, intermediate in shrubs, least in subshrubs (Carlquist 1966a; Wallace 1986). In the southern Californian flora, herbs fall between trees and shrubs in both vessel element length and diameter (Carlquist and Hoekman 1985b). Vessel element diameter and length decrease with aridity (Carlquist 1966a; Carlquist and Hoekman 1985b). The possible significance of trends in vessel element length are examined fully in Chapter 11, but it may be that if air bubbles can be localized within individual vessel elements (Sperry 1985), even those with simple perforation plates (Slatyer 1967; see also Fig. 3.10, part 12), shorter vessel elements localize air embolisms better, and thus shorter vessel elements would be adaptive in more arid situations.

Longer vessel elements have been shown to be correlated with more mesic habitats (Carlquist 1966a, 1975a and the literature cited therein). The correlations between vessel element length and altitude or latitude (Baas 1973) should be traced to factors of water availability and temperature, since altitude and latitude, although readily available from herbarium data, are not ecological factors in themselves. Vessel dimensions are sensitively related to ecology. While there is definitely a heritable component, there is wide latitude for phenotypic modifiability also, as can be demonstrated where a given genetic stock is grown in two or more different localities (Bissing 1976; Akachuku and Burley 1979).

Thus, one must know the source of a specimen in order to deal meaningfully with vessel element dimensions. Unfortunately, the sources of most wood samples in wood collections (xylaria) are not specified at all precisely, nor can one know whether the sample came from the base of a plant or a branch, or often how large or how old a plant was sampled. Where correlations involving vessel element dimensions are concerned, the wood anatomist would be well advised to collect his own material. Qualitative features of vessel elements vary negligibly from one locality to another (Stern and Greene 1958).

3.3.4 Measurement of Vessel Element Dimensions

Some authors measure radial and tangential diameters of vessels separately (Patel 1965). Others have mentioned the vessel at its widest point. In the past, the vessel wall has typically been included in measurement of vessel diameters. If one is concerned with conductive capacity of a vessel, one should measure the lumen diameter and exclude the wall. If one is concerned with conductive capacity, a vessel oval in transection is most accurately measured not by widest or narrowest diameter, but by an average between widest and narrowest diameter, I atttempt to measure mean vessel lumen diameter. If outside vessel diameter is given, one can calculate lumen diameter by subtracting twice the vessel wall thickness. Authors should specify the basis for their vessel diameter measurements.

With respect to length of a vessel element, most workers measure from tip to tip, so that the "tails" are included, even though some vessel elements have them while others lack them in a single sample (Chalk and Chattaway 1934). If this method is to be used, macerations rather than sections (in which the tails are often not evident) should be used. This detail shows that one must use both macerations and sections in comparative wood anatomy.

The length of a vessel (= uninterrupted series of vessel elements) can be measured, although it is a laborious task. Longer vessels are, in general, correlated with wider vessel diameters (Zimmermann and Jeje 1981). Vessels tend to be longest in roots, next longest in stems, and shortest in branches (Zimmermann and Potter 1982). Shorter vessels are claimed to confer greater conductive safety (Zimmermann 1978a).

3.4 Vessel Grouping

3.4.1 Significance of Vessel Groupings

Since the major work of Grew (1682), we have known that in some dicotyledonous woods, vessels are grouped in various ways whereas in others, vessels are solitary or nearly so (Fig. 3.2, part 1). Grew figured solitary vessels in oak (*Quercus*), but grouped vessels in wormwood (*Artemisia*). Elevated grouping of vessels in arid situations was clearly demonstrated more that 20 years ago (Carlquist 1966a), but the reasons why this should occur in some taxonomic groups with a broad ecological range but not in others that also occur in both wet and dry areas remained to be explored. An explanation has now been offered (Carlquist 1984a).

Vessels are more vulnerable than tracheids, since air embolisms are capable of spreading from one vessel element into the next through the perforation plate. When an air bubble forms in a tracheid, it cannot expand into an adjacent tracheid because it is blocked by the intervening pit membrane. Vessels of plants subject to drought or freezing are in danger of becoming embolized. Therefore a subsidiary conducting system becomes imperative if the three-dimensional pathways of water columns are to supply stems and leaves. If tracheids surround a vessel, the tracheids can continue to conduct water if the vessel embolizes. Vessel grouping is not advantageous in this case, because redundancy of vessels is of less functional value (judging from absence of grouped vessels in tracheid-bearing woods) than tracheid presence. If, however, a wood has fiber-tracheids or libriform fibers as the background cell type accompanying vessels, there is no subsidiary cell type that can conduct if vessels do not function: fiber-tracheids and libriform fibers are effectively non-conductive, and in these woods grouped vessels provide safety (if one of a group of adjacent vessels embolizes, the others of the group maintain the water column pathway). This hypothesis and the evidence for it have been detailed elsewhere (Carlquist 1984a). If vessels are very numerous per mm^2 in a wood with tracheids, some grouping will occur merely because of packing, and therefore such instances are not really an exception to the hypothesis.

If one defines tracheids in accordance with Bailey (1936) or the IAWA Committee on Nomenclature (1964), tracheids are cells densely provided with fully bordered pits. Interestingly, this definition turns out to be a functional one (Braun 1970; Carlquist 1986a,b), because presence of tracheids defined in this way does, in fact, markedly depress the degree of vessel grouping expected in a taxon (provided it occurs in a dry or highly seasonal environment), whereas fiber-tracheids and libriform fibers do not depress vessel grouping in taxa of such areas (Carlquist 1984a). Vessel grouping is low in taxa of wet areas regardless of the type of imperforate tracheary element. In a group such as Asteraceae (which have libriform fibers), degree of vessel grouping rises markedly with relation to dryness of the habitat (Carlquist 1966a). This is shown

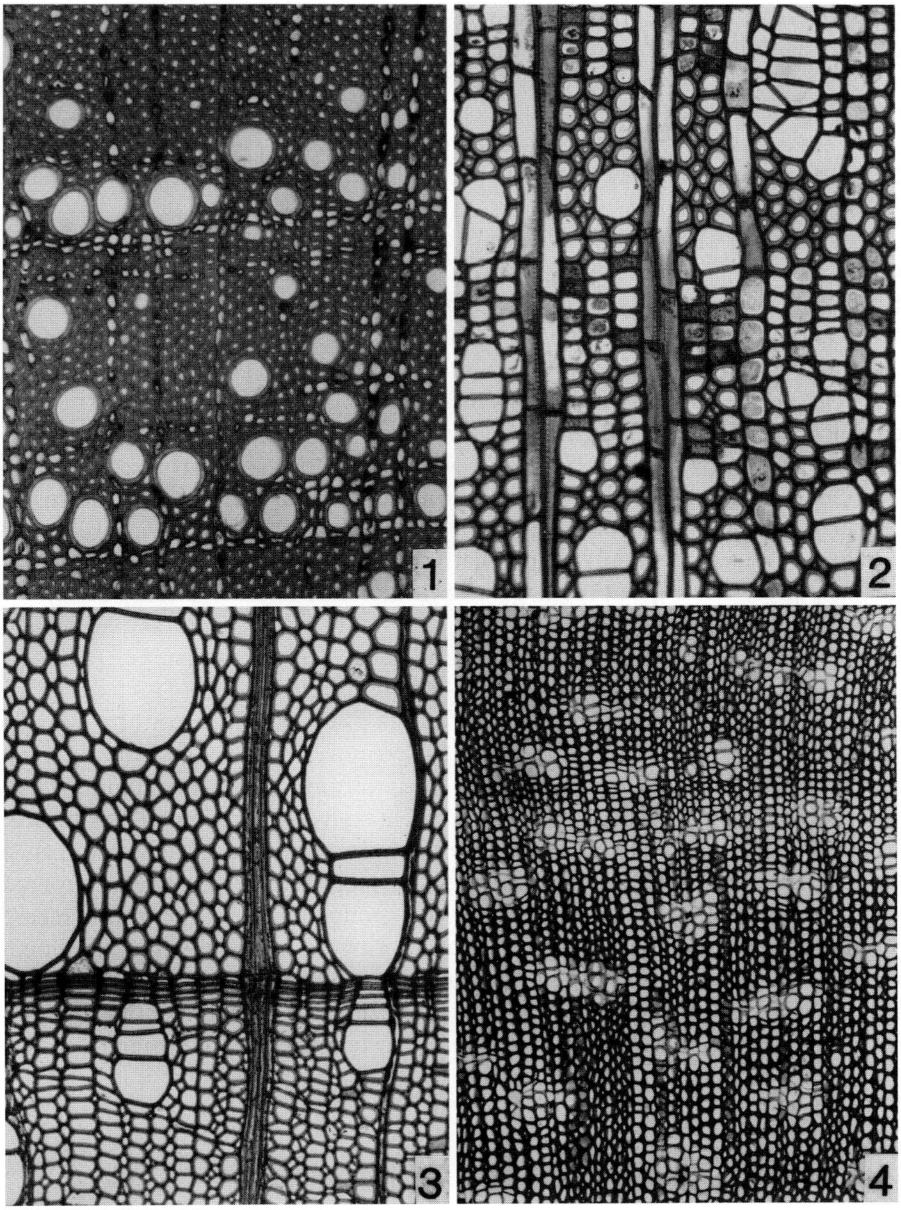

Fig. 3.2. Kinds of vessel (pore) grouping. *1* Vessels solitary, *Krameria grayi* (Krameriaceae); *2* vessels in radial multiples, *Magnolia grandiflora* (Magnoliaceae); *3* vessels in radial multiples, multiple across growth ring at right, *Betula nigra* (Betulaceae); *4* vessels in clusters (slight tendency toward tangential widening of clusters), *Kalanchoe beharensis* (Crassulaceae). (*1–3:B*; *4:A*)

dramatically in Fig. 3.1: *Olearia argophylla* (Fig. 3.1, part 4) occurs in wet forest of coastal southeastern Australia; *O. avicenniaefolia* (Fig. 3.1, part 5) is from sunny but seasonally moist scrub of New Zealand; *O. muelleri* (Fig. 3.1, part 6) grows in dry interior scrub of southeastern Australia. Lists of families with solitary, somewhat grouped, and markedly grouped vessels (subdivided according to whether the family has tracheids or another imperforate tracheary element type) have been offered by Carlquist (1984a).

Species with vasicentric tracheids provide an interesting series of cases with respect to vessel grouping. If vasicentric tracheids are adjacent to vessels, either fiber-tracheids or libriform fibers (depending on the taxon) will occur elsewhere in a wood (as the "ground mass" fibrous tissue). In such a wood, vasicentric tracheids potentially offer a subsidiary conductive tissue that would make vessel grouping superfluous. In fact, in woods in which vasicentric tracheids are quite abundant, as in *Quercus* (Fig. 2.8, parts 3 and 4), vessels are solitary whereas in woods in which vasicentric tracheids are scarce, as in *Calycanthus* (Fig. 3.3, part 3), vessel grouping is proportionate to the xeromorphy of a species.

There is experimental evidence that where a single genetic stock is grown in two places, vessel grouping is greater in the drier habitat (Bissing 1982).

3.4.2 Quantification of Vessel Grouping

The best measure of vessel grouping as seen in transection is mean number of vessels per group. In this latter method a solitary vessel is counted as 1, a pair of vessels in contact as 2, etc., and these figures are averaged. The value of this method lies in its ability to demonstrate accurately degrees of vessel grouping: for example, essentially all vessels may be grouped in a wood with a mean number of vessels per group of 2.7, but another wood with mostly grouped vessels might have 4.5 vessels per group. Care should be exercised in woods with long vessels with overlapping ends because in the area of overlap, in transectional view, one may mistakenly think that two vessels are present.

3.4.3 Types of Vessel Grouping

The appearance, based on orientation in which vessels are grouped as seen in transection, has been used for the recognition of types. The term "aggregation" is used here, as it was in an earlier paper (Carlquist 1987c) for those groupings that are more extensive and often extend across rays as seen in wood transections.

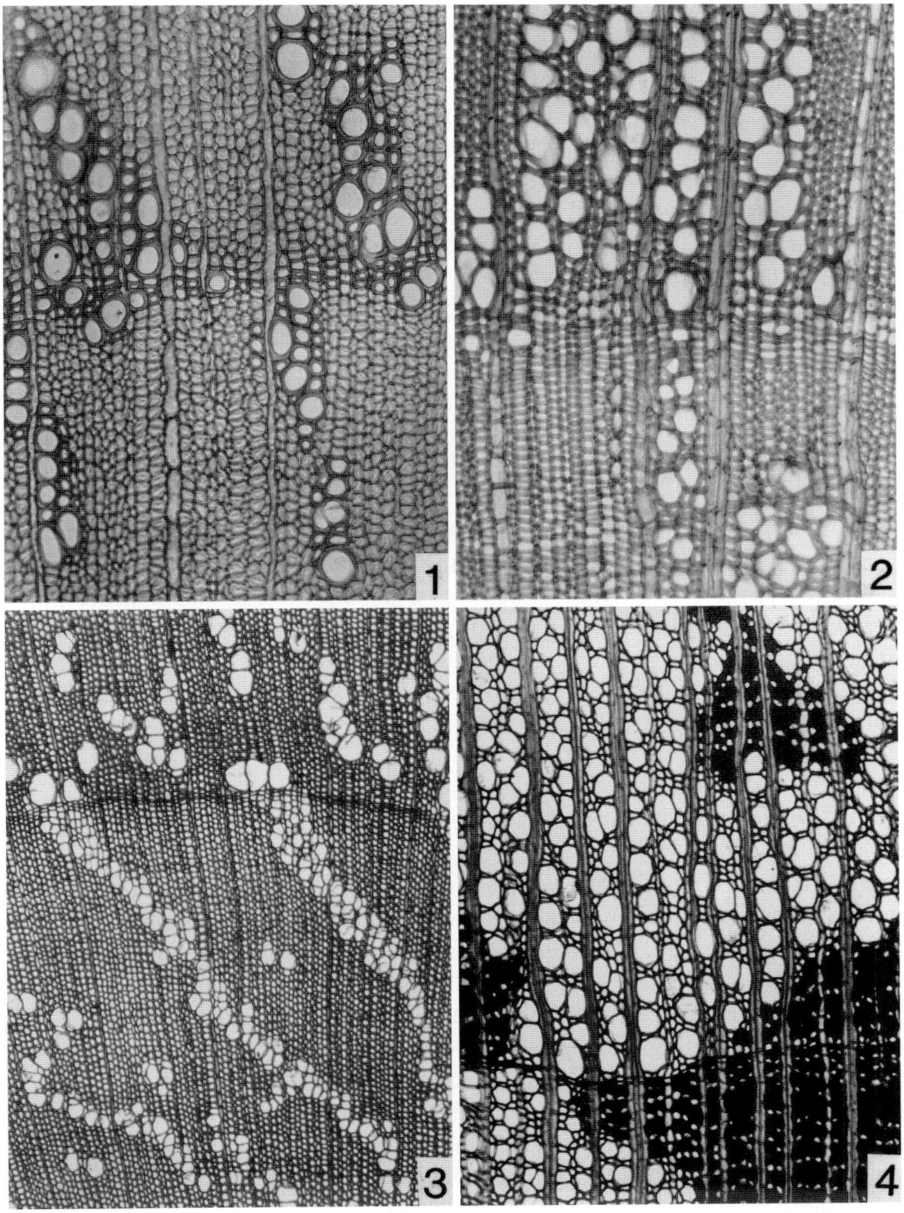

Fig. 3.3. Vessels in diagonal aggregations. *1* Small aggregations, *Cneoridium dumosum* (Rutaceae); *2* large aggregations, *Osmanthus ilicifolius* (Oleaceae); *3* long diagonal aggregations, *Calycanthus floridus* var. *laevigatus* (Calycanthaceae); *4* vessel aggregations more extensive than remaining wood portions, *Bumelia lanuginosa* (Sapotaceae). (*1–2:B*; *3–4:A*)

3.4.3.1 Radial Multiples

Radial multiples (sometimes "vessels in chains") are said to occur when vessels are in contact in radial series (Fig. 3.2, part 2). These series may cross growth rings (Fig. 3.2, part 3). The IAWA Committee on Nomenclature (1964) has noted a distinction between short radial multiples and long radial multiples. Radial multiples theoretically offer a way for the conductive system to form new vessels that can take over the function of earlier-formed vessels without alteration of the conductive pathways (Carlquist 1984a). Radial multiples of particular degrees may characterize species, as in the family Chloanthaceae (Carlquist 1981c).

3.4.3.2 Clusters

The term "clusters" is applied to vessel groupings in which the vessels touching each other form a collection about as wide tangentially as radially. This condition is shown in Fig. 3.2, part 4, although that example shows a slight tendency for tangentially wide groups. Most examples tend to illustrate more radial than tangential width of the group (a fact that may be related to the value of radial groupings hypothesized above). The term "cluster" may, therefore, be used even if some aspect of radial grouping is present and the two types should not be regarded as mutually exclusive.

3.4.3.3 Diagonal Aggregations

Aggregations (groupings often traversing rays) that are oriented in directions midway between radial and tangential are considered under this heading. The terms "arc-porous" (Kukachka 1978), "dendritic or flamelike arrangement of pores" (Record 1942c), or "vessels in echelons" have been employed. The term "diagonal aggregation" is used here as an effort to introduce a topographically more precise wording. Short diagonal bands in which vessels are not mixed with large numbers of vasicentric tracheids are shown for *Cneoridium dumosum* in Fig. 3.3, part 1. The diagonal bands of *Ceanothus* (Fig. 3.3, part 2) appear "looser" in that vessels are intermixed with vasicentric tracheids and very narrow vessels like vasicentric tracheids. Long diagonal bands are illustrated here for *Calycanthus* (Fig. 3.3, part 3). In *Bumelia* (Fig. 3.3, part 4) the bands are massive, bulking larger than the nonvessel-bearing portions of the xylem; vessels are intermixed with vasicentric tracheids in this example. In some examples, diagonal aggregations are transitional to tangential (Fig. 3.4, parts 1 and 2), with different degrees of tangential vs. diagonal orientation in various parts of the wood. The example shown in Fig. 3.4, part 2, is interesting in that the diagonal bands are composites of radial multiples. The significance of diagonal aggregations (Carlquist 1987c, 1987c) is claimed to be a form of

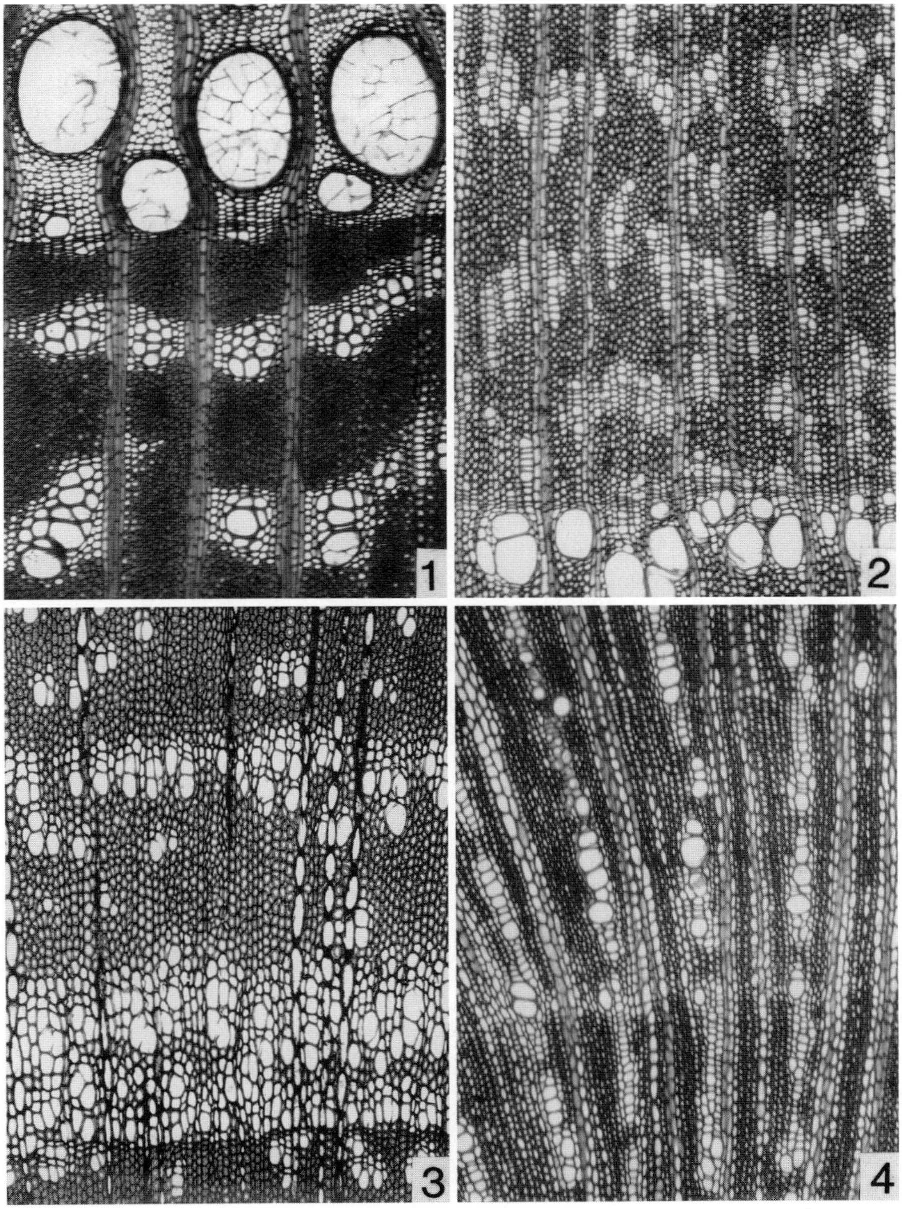

Fig. 3.4. Vessel aggregation types (*1–3*); Vessel restriction (*4*). *1* Latewood vessel aggregations vary between tangential and diagonal, *Robinia pseudoacacia* (Fabaceae). *2* Latewood vessels in diagonal aggregations formed from radial multiples, *Buddleja parviflora* (Buddlejaceae); *3* vessels in tangential bands, *Persoonia longifolia* (Proteaceae); *4* vessels restricted to central portions of fascicular areas (vessels tend not to occur near rays), *Launea spinosa* (Asteraceae). (*1–4*:*A*)

vessel redundancy and therefore conductive safety in taxa that have either vasicentric tracheids (most instances) or very narrow vessels (plus an occasional vasicentric tracheid) mixed with wider vessels. Diagonal aggregations of vessels form bands that often intersect each other; thus, all of the vessels in a stem are potentially linked with each other in a single group (with intermixed narrow vessels and vasicentric tracheids forming a safeguarding subsidiary conductive system for the entirety).

A list of groups with vessels in diagonal bands was offered by Record (1942c). The following list represents an updating of a list offered by Carlquist (1987c). In many of these families, only a few genera and species are known to have diagonal bands.

Araliaceae
Asclepiadaceae
Asteraceae (Fig. 3.1, part 6)
Berberidaceae (Carlquist 1995b)
Boraginaceae
Buddlejaceae (Fig. 3.4, part 2)
Calycanthaceae (Fig. 3.3, part 3; Carlquist 1983c)
Casuarinaceae (Moseley 1948)
Cneoraceae (Carlquist 1987g)
Dipterocarpaceae (several genera: Gottwald and Parameswaran 1966)
Fabaceae (*Genista*, *Laburnum*, *Sarothamnus*, *Spartium*, *Ulex*: Cozzo 1950)
Fagaceae (aggregations present but "loose" because they include vasicentric tracheids)
Goetzeaceae (Carlquist 1988b)
Leitneriaceae
Loganiaceae (*Logania*: Mennega 1980)
Melastomataceae
Moraceae (*Maclura*, *Morus)*
Myrtaceae (several genera, including some species of *Eucalyptus*: Ingle and Dadswell 1953b)
Oleaceae (*Nestegis*: Patel 1978; also other genera)
Pittosporaceae: (some species of *Pittosporum*: Carlquist 1981d)
Rhamnaceae (Fig. 3.3, part 2)
Rosaceae
Rutaceae (Fig. 3.3, part 1)
Sapotaceae (especially *Bumelia*; Fig. 3.3, part 4: Kukachka 1978)
Solanaceae (*Lycium*)
Thymeleaceae (*Passerina*)
Ulmaceae (Fig. 2.12, part 2)
Zygophyllaceae (*Bulnesia*, *Plectrocarpa*, *Porlieria*, *Zuccagnia*: Cozzo 1948)

3.4.3.4 Tangential Aggregations

Vessels predominantly oriented in tangential bands have been termed "festoons" (IAWA Committee on Nomenclature 1964). The term "ulmiform" has been attached to this pattern also.

A listing of families with an "ulmiform" pattern has been offered by Record (1942b), and that list is probably the basis for the list in Boureau (1957). Tangential bands of vessels represent not a single category, but several phenomena. In one of these, there are large, nongrouped earlywood vessels followed by tangential bands of smaller vessels (vasicentric tracheids also occur in such bands in many taxa: Carlquist 1987c). In the writer's opinion, the ulmiform condition is a variant of the diagonal bands described above, and hence the family Ulmaceae appears in the list above. In any given species with the "ulmiform" condition, the latewood bands fluctuate between tangential and diagonal in orientation (Figs. 2.12, parts 3 and 4, 2.13, part 1), and very likely even if tangential bands predominate, the aggregations form an intersecting network in three dimensions. There are also examples of tangential bands of vessels in which wood is diffuse-porous and the tangential bands may occur without respect to position in a growth ring. Examples of this condition include *Persoonia* (Fig. 3.4, part 3) or any of several Gyrostemonaceae (Carlquist 1978b). Another phenomenon that might be cited in this category, although quite different from the above examples, is represented by the tendency for earlywood vessels to be numerous and crowded (e.g., Fig. 2.5, part 3); this phenomenon is best considered under growth rings, not vessel grouping.

3.4.4 Vessel Restriction Patterns

This term has been used to describe woods in which vessels occur in the central portions of fascicular areas and thus tend not to be in contact with rays. This phenomenon can be seen in *Launea* of the Asteraceae (Fig. 3.4, part 4), as well as *Acacia pennata* (Obaton 1960), *Valeriana* (Carlquist 1983e) of the Valerianaceae, several Papaveraceae (Carlquist and Zona 1988b; Carlquist et al. 1994), Berberidaceae (Carlquist 1995b), Plumbaginaceae (Carlquist and Boggs 1996), and *Isomeris* of the Capparaceae (new report).

Another type of vessel restriction is seen in lianoid Convolvulaceae, especially *Ipomoea* (Carlquist and Hanson 1991; McDonald 1992), and lianoid Icacinaceae (Bailey and Howard 1941a; Obaton 1960). In these, patches of vessel-free wood may occur in the earlier-formed secondary xylem, or side by side with vessel-bearing portions of secondary xylem.

3.5 Vessel Density

Vessels as seen in transection may be few to many, and this is commonly recorded in the number of vessels seen per mm^2 of transection. (Earlier workers calculated density of vessels in terms of "vessels or vessel groups," but mixing vessels with vessel groups loses physiological significance, and only the total number of vessels per mm^2 should be calculated.) Low numbers of vessels per mm^2 would be represented by such figures as 3.1 (part 4) and 3.2 (part 1); for example, woods of tropical rain forest trees such as Scytopetalaceae (Carlquist 1987h) fall in this range. Numbers of 100 would be considered numerous. Numbers above 500 are unusual, but have been found in plants of notably dry habitats (Cristiani 1948; Michener 1981, 1983; Carlquist and Hoekman 1985b) or notably cold habitats (H.J. Miller 1975). The highest number thus far recorded (2673) is in a species of *Cassiope*, a boreal alpine shrub (Wallace 1986). Vessel density is an extremely sensitive measure of mesomorphy and xeromorphy, and although one would expect vessel density to be roughly inverse to vessel diameter, the relationship is not close and the two features can vary independently to a large degree (as shown by some lianas in which vessel diameter is large but vessel density is several times that of a tropical forest tree. For this reason, vessel density is one of three quantitative features related to vessels that were incorporated into the Mesomorphy ratio (Carlquist 1977b; Carlquist and DeBuhr 1977). The rationale for this ratio and comparison with other figures relating to water conduction, such as conductivity, are discussed in Chapter 11.

3.6 Origin of Vessels

3.6.1 The Hypothesis

The theories and evidence for vessel origin in angiosperms are examined here as a way of introducing apparently primitive expressions in perforation plates, lateral wall pitting, and other aspects of vessel element morphology. The work of Frost (1930a,b, 1931) on vessel elements, an extension of the ideas of Bailey and Tupper (1918), is basic in this regard. According to these papers, there is a progressive drop in vessel element length with phylogenetic specialization. This concept and the degree to which irreversibility can be hypothesized in wood evolution are discussed further in Chapter 11. For the purposes of the present discussion, we can merely assume that length of vessel elements can be used as a key to vessel element specialization trends. The least specialized vessel element conditions can then be hypothesized to be very similar to the kind of long tracheids ancestral to vessels in dicotyledons. Study of vesselless dicotyledons (e.g., Bailey and Thompson 1918) have played a role in these considerations.

Vessels are claimed to have originated separately in angiosperms and in Gnetales (Bailey and Thompson 1918). The end wall of the gnetalean vessel element represents only small modifications of a series of large, circular bordered pits like those of conifers. By loss of pit membranes on the end wall, foraminate perforations result. On the other hand, primitive vessel elements of dicotyledons have scalariform lateral wall pitting and scalariform end walls (Fig. 3.5, parts 1 and 2). Frost (1930a) finds that a tracheid much like an earlywood tracheid of *Trochodendron* can be hypothesized. To be sure, Muhammad (1984) and Muhammad and Sattler (1982) tried to cite resemblances between certain vessel elements of certain angiosperms and certain vessel elements of *Gnetum* as evidence of a phylogenetic connection between Gnetales and angiosperms. However, the perforation plates that these authors cite for *Comptonia* are unusual in Myricaceae, and allegedly similar plates in *Gnetum* are also exceptionally rare and may be regarded simply as "malformations" (nonrepresentative variants) of scalariform or foraminate perforation plates. Such "scalaroid" plates (like those figured here; Fig. 3.1, part 4) can be found occasionally in phylads in which there are transitions between scalariform and simple perforation plates. The perforation plates they show for *Gnetum* likewise are scarce nonrepresentative formations. The perforation plates of this sort mentioned by Bliss (1921) and MacDuffie (1921) in such plants as *Cydonia*, *Paeonia*, and *Vitis* are not phylogenetically significant. These plates may be regarded as infrequent nonrepresentative formations, as expressions of the fact that transitional forms are not always formed so as precisely to resemble the predominant types of the group concerned. The inability of morphogenesis always to form the typical perforation plate should not be used as evidence of relationship.

Tracheids with scalariform lateral wall pitting and scalariform end wall pitting must be hypothesized as ancestral to angiosperm vessels. Conifers and Gnetales do not qualify in this regard. Scalariform pitting is widespread in metaxylem tracheids of ferns, but these do not have secondary growth. Scalariform pitting may be found in tracheids of secondary xylem in Cycadales, Bennettitales (Cycadeoideales), and Cordaitales (early secondary xylem), and progymnosperms. Thus, scalariformly pitted tracheids did exist widely in seed plants in eras prior to angiosperm origin, and the fossil record is in accord with the ideal that angiosperms had scalariformly pitted tracheids, at least in part (as we will see, where tracheids are narrow, pits may be circular rather than scalariform, and the important feature is that primitive angiosperms must be capable of forming scalariform pits where wider wall faces permit them). Bailey (1925, 1944b) did not overlook such evidence. He believed that in the gymnosperm and gnetophyte lines, large torus-bearing circular bordered pits developed (presumably as a specialization) on secondary xylem tracheids and became so pervasive that circular bordered pits may be found intercalated into

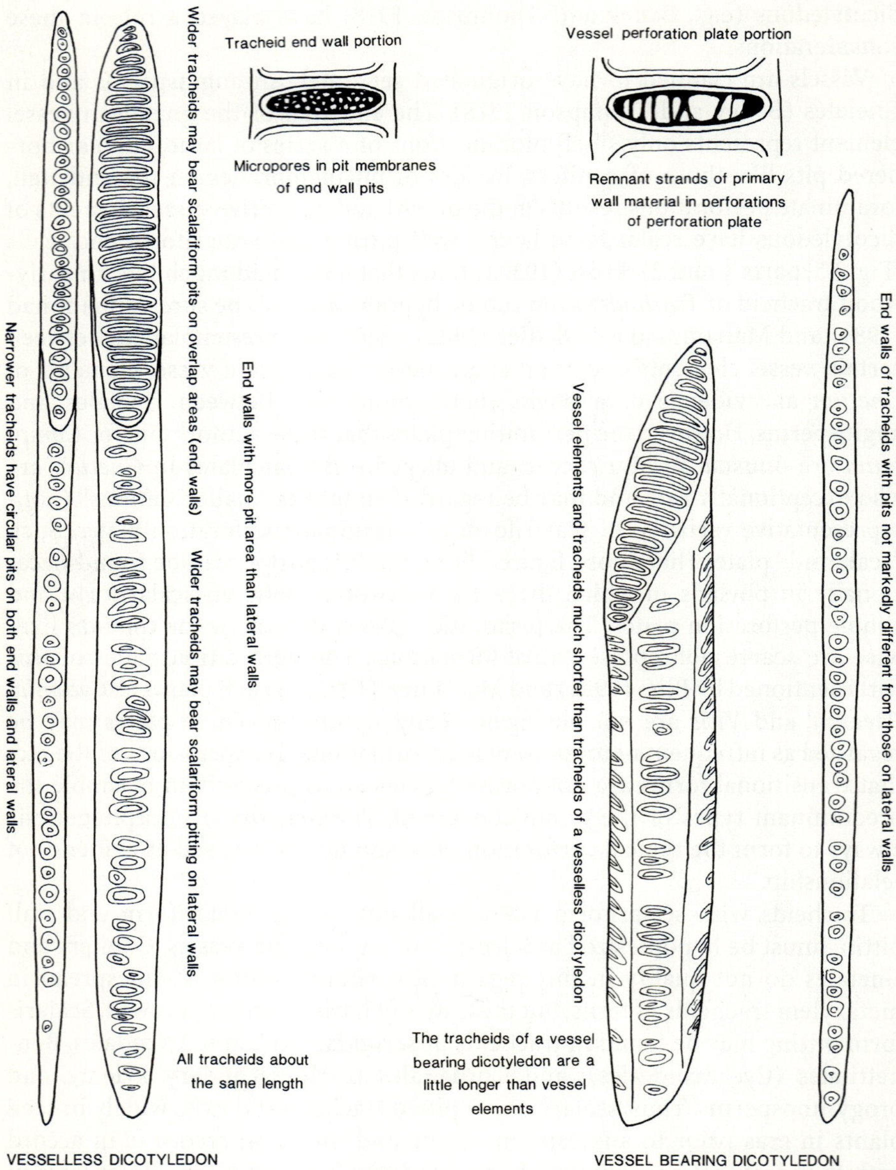

Fig. 3.5. Transition between vesselless and vessel-bearing wood in dicotyledons. This diagram is intended to show the alterations that occur as this transition occurs, primarily in qualitative terms. The quantitative aspects are shown less accurately, but a marked drop in length from tracheid to primitive vessel element occurs; in a primitive vessel-bearing wood, a tracheid is a little longer than a vessel element and vessels are wider than tracheids

the helical bands on protoxylem and metaxylem tracheids in these groups (see Carlquist 1996d). Although in recent years Gnetales have been claimed to be basal to angiosperms, the latest evidence from wood anatomy (Carlquist 1996d) and from analysis of multiple genes (Soltis et al. 1999) shows Gnetales nested among conifers rather than basal to angiosperms.

Let us suppose that the ancestral angiosperm (*Amborella*, according to Soltis et al. 1999) had vesselless wood like that of *Amborella*, *Trochodendron* or *Tetracentron* as ancestral to vessel-bearing dicotyledons. To be sure, *Trochodendron* does have marked growth rings, whereas vessel origin may have occurred in nonseasonal areas. One must concede that the earlywood tracheids of *Trochodendron*, which have scalariform lateral wall and end wall pitting, do not represent the only form of tracheid pitting in the genus; latewood tracheids have circular bordered pits. However, this may be a size constraint (latewood tracheid walls are too narrow to permit lateral elongation of pits), and similar size constraints may be found in some tracheids of the vesselless stems of *Sarcandra* (Bailey and Swamy 1948) and *Amborella* (Swamy and Bailey 1950). In Winteraceae, scalariform lateral wall pitting of tracheids may be found in metaxylem, and although secondary xylem tracheids may have multiseriate circular bordered pits on lateral walls, the ability to form scalariform pits has been retained; they are formed on end walls in *Zygogynum* (Carlquist 1981a), *Belliolum* (Carlquist 1983a), and *Bubbia* (Carlquist 1983b), and ability to form scalariform pitting appears frequently in *Drimys* and *Tasmannia* as well (Carlquist 191988a, 1989c). One must merely hypothesize the ability to form scalariform pitting on tracheids, one need not hypothesize omnipresent scalariform pitting (Fig. 3.5); the latter would be nonadaptive, in fact, for it would provide a weak wall structure, and such xylem does not occur unless some form of sclerenchyma compensates for weakness of the scalariformly pitted tracheids, as in lepidodendrids or ferns (Carlquist 1975a, 1983b).

3.6.2 The Tracheid-Vessel Element Transition

Bailey (1944b) gives evidence that in woody dicotyledons, vessels originated in secondary xylem and progressed backward into primary xylem. This is confirmed abundantly by the data provided by Bierhorst and Zamora (1965). If one imagines a vesselless woody dicotyledon with tracheids that have circular bordered pits when narrower, and scalariform bordered pits when wider, several changes must occur in order for vessel origin to occur (Fig. 3.5). Vessels must become wider; conceivably tracheids could remain about the same in diameter, and one need not necessarily imagine that tracheids must become narrower in a woody plant that has recently acquired vessels. If shorter tracheids are narrower, as appears statistically true in conifers (Bannan 1965), tracheids are, in fact, likely to become narrower after vessel origin because length of tracheary elements (both vessel elements and imperforate tracheary elements) in primitive woody dicotyledons is much shorter than that of equivalent vesselless woody dicotyledons or conifers (Carlquist 1975a; p. 141). The woody root of *Sarcandra* (Carlquist 1987k) shows vessels only a little wider than tracheids, and about the same length; without SEM examination of end walls, one could not, in fact, be sure that the wider tracheary elements of the *Sarcandra* root are vessel elements rather than wide tracheids. The *Sarcandra* root does illustrate that in a primitive wood, vessels may be only a little wider than tracheids.

If vessels are wider than tracheids in a vesselless phylad in which wider tracheids may bear scalariform pitting, vessels (which characteristically are wider than wide tracheids in a given primitive wood) certainly should bear scalariform pitting on both lateral walls and end walls (Fig. 3.5). This idea is stressed because if only by default, one may be tempted to think that if scalariform lateral wall pitting and scalariform perforation plates characterize primitive vessels, the vesselless precursors must have had wood composed wholly of scalariformly pitted tracheids. This supposition is, in fact, unlikely, and one need only hypothesize that *wider* tracheary elements have scalariform pitting, while narrow tracheids may bear circular pits.

The scanning electron microscope provides some fascinating new vistas into the origin of perforations in primitive woody dicotyledons (Figs. 3.6, 3.7). Earlier literature based on light microscopy gives the impression that there is a marked difference between perforations, even in scalariform perforation plates, and pits on tracheid end-walls. There have been reports of species in which some perforation plates have perforations while in other similar perforation plates, pit membranes are still present (*Geissoloma*: Fagerlind and Dunbar 1973). Meylan and Butterfield (1978a) reported "microfibrillar webs traversing the individual openings at each end of the scalariform perforation plate" in *Ascarina* (Chloranthaceae), *Laurelia* (Monimiaceae), *Neomyrtus* (Myrtaceae), and all genera of Cunoniaceae and Escalloniaceae they studied; a similar pattern is shown by Ohtani and Ishida (1978b) for *Cornus kousa*. Other illustrations suggest that such "microfibrillar webs" may not, in fact, be restricted to the ends of perforations, but traverse the entire perforation, thereby approaching intermediacy between pits and perforations (*Pentaphragma*; Fig. 3.6, parts 3 and 4). Abundant membrane remnants can be found in perforations of Aextoxicaceae (Fig. 3.6, part 1), Chloranthaceae (Carlquist 1990a, 1992a,b; Fig. 3.7, parts 2–4), Illiciaceae (Carlquist 1992c; Fig. 3.6, parts 2–4), Paracryphiaceae (Carlquist 1992c; Fig. 3.7, part 1), and Sarraceniaceae (Carlquist 1992c). I have found that in all genera, remnants of pit membranes may occur not merely at the ends of the perforation plates, but throughout them. Attempting to discount artifacts related to preparation, I find that pit membranes may be present in various degrees and locations: intact but containing pores only slightly larger than those in tracheid pit membranes (Fig. 3.7, part 1), intact but markedly porose (Fig. 3.6, part 3), present in some portions of some perforations, absent in other perforations (Fig. 3.7, parts 2 and 4), intact at lateral ends of perforations but not in central portions

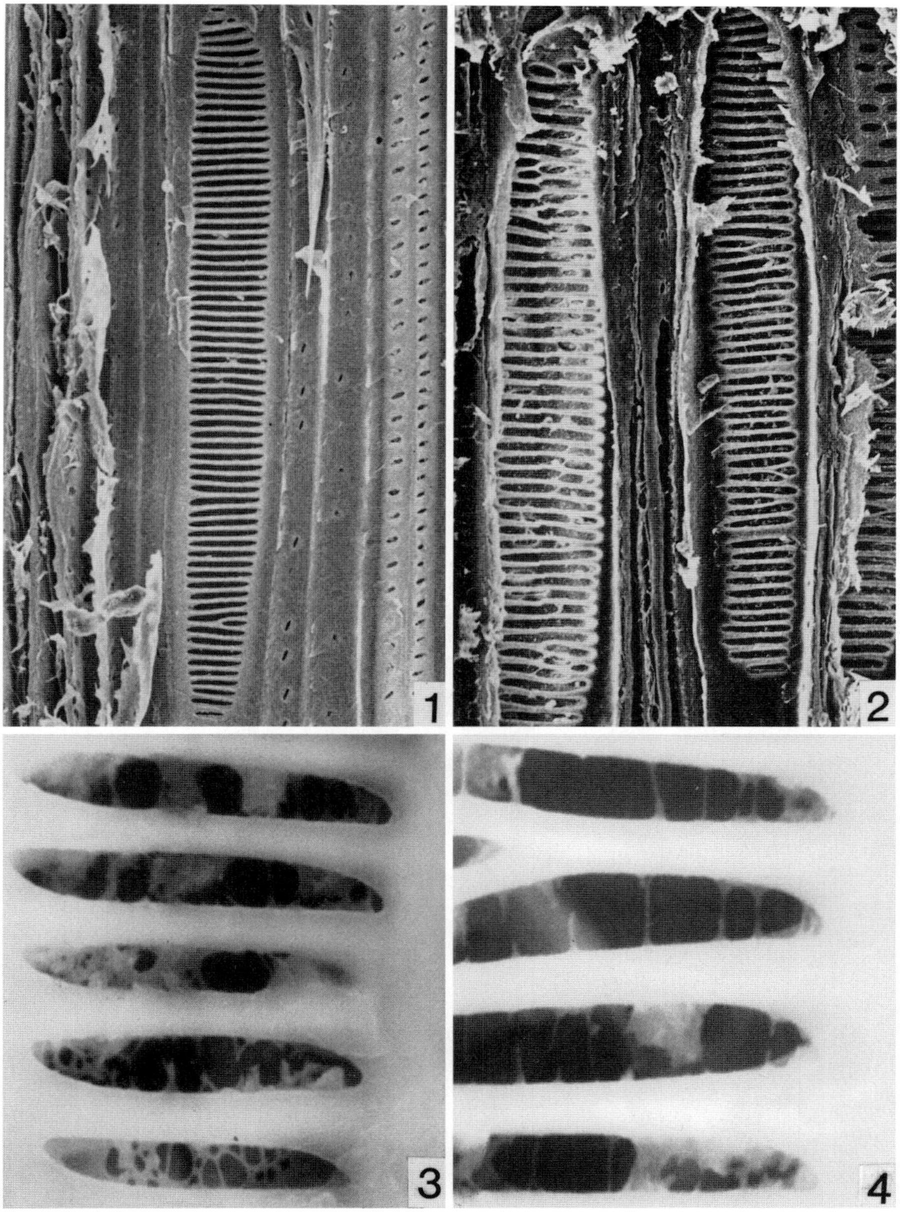

Fig. 3.6. Perforation plates of very primitive vessel elements in dicotyledons; SEM photographs from radial sections. *1* A long perforation plate, tracheids at right, *Aextoxicon punctatum* (Aextoxicaceae); *2* two perforation plates, portion of a third at right; primary wall fragments occur in perforations, *Illicium floridanum* (Illiciaceae). *3*, *4 Pentaphragma* sp. (Pentaphragmataceae). *3* Porose remnants of primary wall in perforations; *4* strandlike remnants of primary wall in perforations. (*1,2:D*; *3,4:G*)

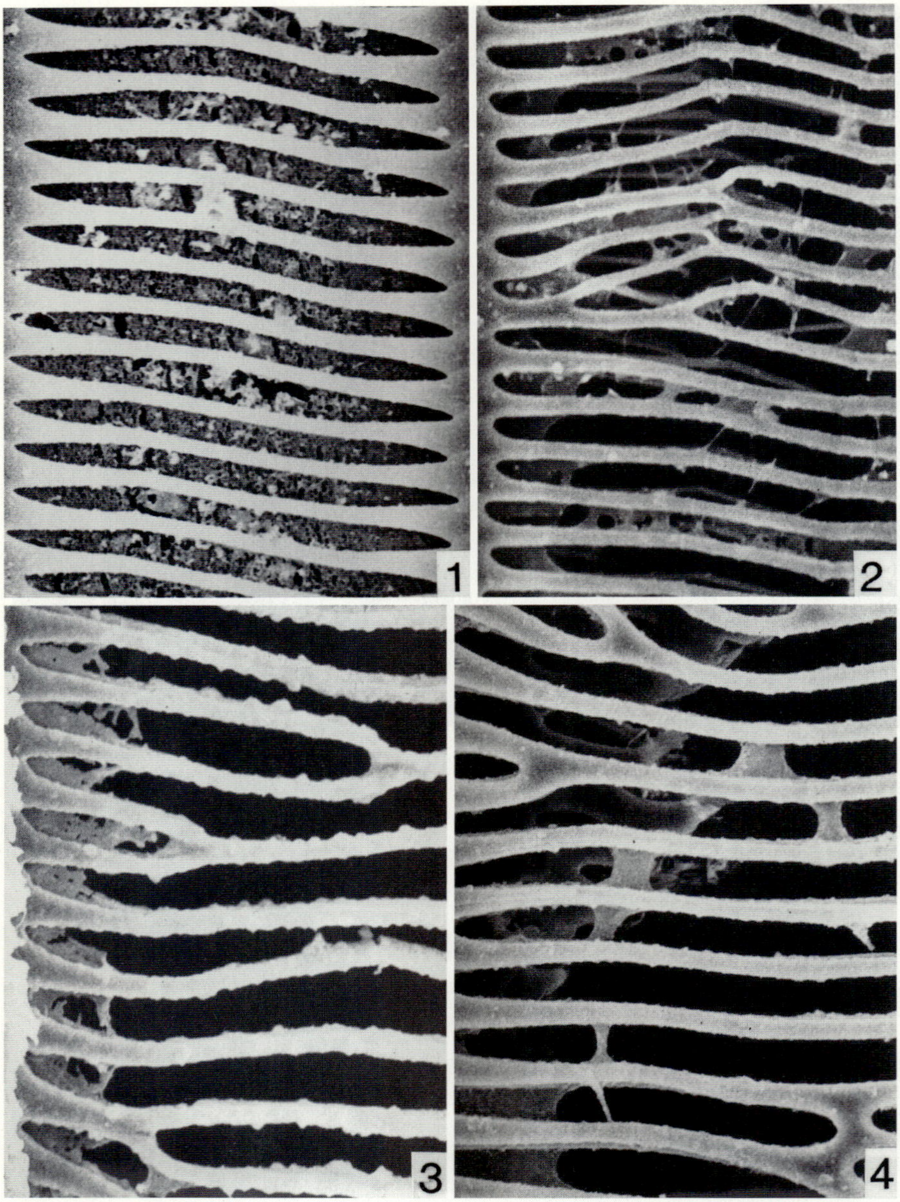

Fig. 3.7. Portions of perforation plates showing remnants of primary walls, from radial sections. *1* Perforations bear porous pit membranes that are nearly intact, *Paracryphia alticola* (Paracryphiaceae); *2* strands and sheets of primary wall remnant in perforations, *Ascarina rubricaulis* (Chloranthaceae). *3*, *4 Hedyosmum nutans* (Chloranthaceae). *3* Primary wall remnants at ends of perforations; *4* primary wall sheets and strands in perforations. (*1*,*2*:*E*; *3*,*4*:*F*)

(Fig. 3.7, part 3), present as distinctive strands between which rather well-defined large pores occur (Fig. 3.6, part 4), or present only as threadlike strands (*Sarcandra*: Carlquist 1987k). In these examples, one must not assume that all perforation plates are alike throughout a wood. In at least some of the examples just cited, one can find some perforation plates in which pit membrane remnants are absent or nearly so and, in the same section, perforation plates containing various degrees of pit membrane presence in the perforations. In end-wall pits of a vesselless dicotyledon, *Tetracentron*, end-wall pit membranes are porose (Fig. 4.3, part 1), much like the perforation pit membranes shown here for *Paracryphia* (Fig. 3.7, part 1). A review of pit membrane remnants in scalariform perforation plates, showing some of the more striking examples, has been offered (Carlquist 1992c).

One would expect that in primitive woods, perforation plates should be composed of bordered rather than nonbordered perforations, and in fact, all of the examples shown in Figs. 3.6 and 3.7 have bordered perforations.

The information reviewed here concerning origin of vessels demonstrates that the vessel-tracheid transition is not a sharp one, and that all degrees of intermediacy may be found, although these intermediate expressions are admittedly not abundant in extant dicotyledons (see, however, *Trochodendron*; Fig. 4.3, part 1). This conclusion is detailed to a greater degree in a recent paper (Carlquist 1996c). Cladograms suggest a presence-absence contrast between tracheids and vessel elements, but this should be regarded skeptically in basal angiosperm groups.

3.7 The Perforation Plate

3.7.1 Scalariform Perforation Plates

Frost (1930b) finds that scalariform perforation plates in which perforations are completely bordered occur in woods with an average vessel element length of 1,340 μm. If shortening of vessel elements is a reliable indicator, the concepts that longer vessel elements are primitive and that bordered bars on perforation plates are primitive are reinforced. Frost (1930b) does, however, note that in some species, such as *Hamamelis virginiana*, one can find in a given section a continuum from completely bordered perforations to apparently nonbordered perforations. Simple perforation plates are commonly bordered.

In my experience, many species with scalariform perforation plates show various degrees of presence of borders on bars of perforation plates, as in the genus *Roridula* (Carlquist 1976b). Families in which borders are characteristically present include Bruniaceae (Carlquist 1978a), Chloranthaceae, Geissolomataceae (Carlquist 1975b), and Sarraceniaceae (DeBuhr 1977). In

Austrobaileya, a liana, perforation plates on wider vessels have bars with narrower borders, whereas perforations on narrower vessels bear wider borders (Bailey and Swamy 1949). This is true in another lianoid genus with primitive wood, *Piptocalyx* (Carlquist 1984e). Some Rhizophoraceae have few but wide and conspicuously bordered bars (Metcalfe and Chalk 1950; Vliet 1976a). A hypothesis that such bars have selective value to prevent deformation and prevent vessel deformation was offered earlier (Carlquist 1975a), although Vliet (1976a) demurred, claiming that such Rhizophoraceae grow in wet situations. Mangrove Rhizophoraceae, however, are known to endure high tensions in vessels (Scholander et al. 1962), and the abundance of water in which mangroves grow is irrelevant. Vliet (1976a) does not take the high tensions in rhizophoraceous vessels into account or offer any alternative hypothesis for presence of these massive bars.

Transition between lateral wall pitting and scalariform perforation plates is examined by Frost (1930a). Frost recognizes two types of lateral wall pitting in species with scalariform perforation plates: (1) those in which scalariform lateral wall pitting grades imperceptibly into the scalariform perforation plate; and (2) those in which lateral wall pitting is clearly different (e.g., opposite or alternate) from the scalariform perforation plate. In Frost's sample, 29 species with type 1 had an average vessel element length of 1,270 μm, whereas 22 species exemplifying type 2 had an average vessel element length of 870 μm. This is yet another way of validating the primitiveness of scalariform lateral wall pitting, and thus the tracheidlike nature of primitive vessels.

3.7.2 Systematic and Ecological Distribution of Scalariform Perforation Plates

A listing of dicotyledonous families with scalariform perforation plates has been offered by Metcalfe and Chalk (1950, 1983) and Carlquist (1975a). In the latter reference, the families are listed according to habitat, and that list, in an updated fashion, is offered here. Citations are given for recent additions to this listing. Families in which only scalariform plates have few bars and are infrequent (e.g., Corylaceae, Eucommiaceae, Himantandraceae, Juglandaceae) are omitted. Although the list seems large, the reader will note that those families that characteristically have long scalariform plates are perhaps only about half of the list or less. Moreover, the families in the list are mostly small ones, with only a few medium-sized families represented (e.g., Theaceae). Abbreviations: F, few bars (typically fewer than ten, sometimes with simple plates also present); LS, long scalariform (typically more than ten bars); R, (with a range in bar number from more than ten to few).

Trees of moist forest or riparian habitats, some with tendencies toward understory:

Actinidiaceae: *Saurauia* (LS)
Aextoxicaceae (LS)
Akaniaceae (R; Carlquist 1996b)
Alangiaceae (F)
Aquifoliaceae (in part; LS mostly)
Araliaceae (R)
Betulaceae (LS)
Bretschneideraceae (R; Carlquist 1996b)
Canellaceae (R)
Cercidiphyllaceae (LS)
Chloranthaceae (LS) in part (Carlquist 1992a)
Clethraceae (LS)
Cornaceae (LS) in part (Noshiro and Baas 1998)
Cunoniaceae (R)
Cyrillaceae (LS)
Daphniphyllaceae (LS)
Davidsoniaceae (F)
Degeneriaceae (F)
Dilleniaceae (in part; LS)
Eucryphiaceae (R)
Euphorbiaceae (in part; R)
Eupteleaceae (LS)
Fagaceae (R; Ohtani and Ishida 1978b)
Flacourtiaceae (in part; R)
Gomortegaceae (LS)
Goupiaceae (LS)
Hamamelidaceae (R)
Humiriaceae (LS)
Icacinaceae (in part; R)
Lacistemaceae (LS)
Lauraceae (F, R)
Lecythidaceae (in part; R)
Magnoliaceae (F)
Monimiaceae (R)
Myristicaceae (R)
Nothofagaceae (F)
Nyssaceae (R; Noshiro and Baas 1998)
Octoknemataceae (F)
Olacaceae (in part; R)
Pentaphylacaceae (LS)
Platanaceae (in part; F)
Sabiaceae *Meliosma* (R; Carlquist et al. 1993)
Scytopetalaceae (in part; F; Carlquist 1987h)
Sphenostemonaceae (LS)
Staphyleaceae (in part; LS)
Strasburgeriaceae (LS)
Styracaceae (in part; F)
Symplocaceae (LS)
Theaceae (LS)
Ticodendraceae (LS; Carlquist 1991)
Trimeniaceae: *Trimenia* (LS; Carlquist 1984e)
Violaceae (in part; R)

Mangroves:

Rhizophoraceae (some genera; F)

Shrubs of moist forest:

Alseuosmiaceae (LS; Dickison 1986)
Aquifoliaceae (in part; LS)
Buxaceae (R)
Caprifoliaceae (LS; Ogata 1988)
Celastraceae (in part; R)
Chloranthaceae (LS; in part; Carlquist 1990a, 1992b)
Columelliaceae (R; Stern et al. 1969)
Cornaceae (in part; R; Noshiro and Baas 1998)
Desfontaineaceae (LS; Mennega 1980)
Dichapetalaceae: *Tapura* (F)
Epacridaceae (in part; R)
Ericaceae (in part; R)
Escalloniaceae (R)
Eupomatiaceae (LS; Carlquist 1992d)
Flacourtiaceae (in part; R)
Grossulariaceae (F)
Hydrangeaceae (R)
Icacinaceae (in part; R)
Illiciaceae (LS)
Marcgraviaceae: *Norantea* (LS)
Myricaceae (F)
Paeoniaceae (F)
Passifloraceae: *Soyauxia* (LS)
Stachyuraceae (R)
Staphyleaceae (in part; R)

Shrubs of nonforest habitats but with underground water or other mitigating circumstance:

Bruniaceae (LS)
Dilleniaceae (in part; F)
Empetraceae (F; Carlquist 1989a)
Epacridaceae (in part; F)
Garryaceae (F; Moseley and Beeks 1955)
Geissolomataceae (LS)
Grubbiaceae (LS)
Myrothamnaceae (LS)
Pterostemonaceae (F)
Retziaceae (F; Carlquist 1986c)
Roridulaceae (R)
Saxifragaceae (R)
Stilbaceae (F; Carlquist 1986c)

Woody herbs of understory or moist habitats:

Campanulaceae subfam. Campanuloideae (in part; F; Shulkina and Zikov 1980)
Pentaphragmataceae (F; Carlquist 1975a, 1997a)
Penthoraceae (LS; Haskins and Hayden 1987)
Saururaceae (R; Carlquist et al. 1995)
Valerianaceae: *Patrinia* (F; Fig. 3.11.1–3.11.5; Carlquist 1983e)

Woody small vines or small lianas:

Austrobaileyaceae (F)
Dilleniaceae: *Tetracera* (F)
Marcgraviaceae (in part; F)
Schisandraceae (F, R; Carlquist 1999c)
Trimeniaceae: *Piptocalyx* (F; Carlquist 1984e)

The above listing of systematic and ecological distribution of scalariform perforation plates was used to support the concept that phylads with scalariform perforation plates tend to be restricted to relatively nonseasonal mesic habitats, like tropical cloud forest, summer-wet temperate forest, or boreal habitats in which the soil never dries. In Mediterranean-type habitats, phylads with scalariform perforation plates appear to persist by modification of foliage and other mechanisms, but one notes that inevitably the seasonality of these habitats makes loss of bars on the perforation plate advantageous as a mechanism for accommodating brief periods of peak flow; shrubby Epacridaceae and Dilleniaceae from southwestern Australia exemplify this. A series of families from South Africa appear in this list: Bruniaceae (Carlquist 1978a), Geissolomataceae (Carlquist 1975b), Grubbiaceae (Carlquist 1977a), Myrothamnaceae (Carlquist 1976b), Retziaceae (Carlquist 1986c), and Stilbaceae (Carlquist 1986c). All of these occur on cool north-facing slopes and can often be observed to colonize areas where seeps provide moisture throughout the dry season; moist species in these families also exemplify microphylly.

Decrease in number of bars has been noted in a number of tropical lowland phylads in which warm temperatures accompanied by abundant soil moisture doubtlessly lead to high transpiration rates (Versteegh 1968; Carlquist 1975a, 1976c, 1981d; Baas 1976; Dickison et al. 1978; Dickison 1979; Schmid and Baas 1984). Despite the massive evidence on the basis of comparative studies, one finds the following statement by Zimmermann (1983): "It makes no sense to argue that certain habitats require higher flow rates than others and thereby exert a selection preessure that eliminates scalariform perforation plates. Flow rates depend on too many other facts . . . to permit such an assumption." One is surprised that Zimmermann would expect dicotyledons to evolve all other anatomical means of accommodating flow rates but yet leave the perforation plate completely unaltered. The examples of Mediterranean shrubs or lianas in phylads in which primitive wood is characteristic (see above list) abundantly show selective pressure for simplification of the perforation plate. Zimmermann is very likely trying to discredit simplification of the perforation plate so that he can cite presence of scalariform perforation plates as a means for sieving out bubbles that result from thawing of water frozen within vessels (Zimmermann 1983). However, the distribution of scalariform perforation plates (most abundant in cloudy tropical uplands such as montane New Caledonia, Malayan highlands, etc., where frost never occurs), contradicts the Zimmermann hypothesis. Scalariform perforation plates may be found in boreal shrubs, but many lack them (e.g., *Salix*, Arctic Lamiaceae and Rosaceae: Miller 1975; Zimmermann 1978 cited *Salix* erroneously as having scalariform perforation plates) and boreal genera with scalariform perforation plates probably reflect the tendency for woods with such primitive vessels to occur on sites that never dry rather than in sites where freezing occurs.

Scalariform perforation plates have also been cited as a potential mechanism for localizing bubbles in vessels (Sperry 1985). The results of Sperry are undoubtedly valid, but Slatyer (1967) claimed that bubbles tended to be localized within vessel elements even if they have simple perforation plates. In some simple experiments with *Ceratostigma* (Carlquist, unpubl.; one photograph is shown here as Fig. 3.11, part 12), I was able to confirm Slatyer's claim.

A few groups of aniosperms characteristically have scalariform perforation plates with few bars. One can cite Buxaceae except for *Styloceras* (Carlquist 1982d), several genera of Magnoliaceae such as *Michelia* (Fig. 3.9, part 3), *Paeonia* of the Paeoniaceae (Keefe and Moseley 1978), *Cinnamomum* of the Lauraceae (Ohtani and Ishida 1978b), and certain Rhizophoraceae such as *Ceriops* (van Vliet 1976a). As noted earlier, I have entertained the possibility (Carlquist 1975a) that in such groups, the bars may have some selective advantage. Otherwise, one would be hard pressed to explain not only why these groups have stabilized with small numbers of bars, but, more significantly, why bars in these perforation plates are unusually thick and well bordered, not at all like a vestigial manifestation. These perforations could aid in localizing embolisms, in accordance with the idea of Sperry (1985), or they could serve some mechanical function, such as resistance to deformation under tension (slender bars would serve just as well as thick ones if bubble localization were their only function). The fact that very few groups typically have few bars per perforation plate suggests that whatever the function of these bars, that function probably can also be accomplished by other mechanisms.

3.7.3 Mixed Scalariform and Simple Plates

One might expect that with increasing selection for simplification of the scalariform perforation plate, there would be taxa in which simple and scalariform perforation plates would coexist within a given wood sample. Lauraceae (Stern 1954; Richter 1981) exemplify this. Other interesting examples may be found in taxa with type 9 growth rings (see Fig. 2.9; and Sect. 2.4.9); such growth rings have simple perforation plates except in some or all latewood vessels (Carlquist 1980a; Dickison and Phend 1985), sometimes in much of the growth ring except the first-formed vessels (*Platanus*: original data).

In certain diffuse-porous woods, scalariform and simple perforation plates co-occur. In a few examples, as in *Hieronyma andina* of the Euphorbiaceae (Giraud 1981a,b), about equal numbers of simple and scalariform plates may be found. In a few examples, a few simple perforation plates may be found mixed with a large number of scalariform perforation plates, as in *Balanops australiana* (Carlquist 1980b). However, in the majority of cases, one finds a few scalariform plates mixed with a much larger number of simple perforation plates: most species of *Hieronyma* (Giraud 1981b), *Pterocephalus dumetorum* (Carlquist 1982b), and *Casearia obliqua* (Teixeira 1983), for example. Examples are illustrated by Meylan and Butterfield (1978a) for *Mida salicifolia*, *Nothofagus fusca*, and *N. solandri*. Mostly simple plates mixed with a few plates with wide bars – appearing almost like more than one simple perforation plate – can be found in *Byblis gigantea* (Fig. 3.11, part 10; Carlquist 1976b) and *Nepenthes lowii* (Fig. 3.11, part 11; Carlquist 1981b). The probable explanation for the above examples in some phylads with primitive woods is modification and simplification of scalariform perforation plates. In such phylads, the plates like those of *Byblis* are much less frequent than scalariform perforation plates or simple perforation plates, respectively. In the ring-porous examples, one can hypothesize that selection has acted only on earlywood, where conductive rates and volumes are great, whereas in latewood, either no such selection exists, or else some other force accounts for retention of the bars on plates – localization of air embolisms is such a possibility. However, the examples are so few that the most probable explanation would be that of relictual and virtually functionless persistence of the bars on plates in these taxa. One can cite, however, examples in which a family characteristically has simple plates, but in an exceptional species, occasional scalariform plates with numerous bars (*Brachyglottis repanda* of the Asteraceae) or an altered scalariform conformation (*Vitex lucens* of the Verbenaceae) are present (Butterfield and Meylan 1975). Examples such as these two may involve paedomorphosis (see Chap. 9). One example that deserves investigation is Patel's (1965) report of simple plates exclusively in stems of *Aesculus* but presence of a few scalariform ones in roots.

Table 3.1. Vessel element length of dicotyledon vessel categories. (Frost 1930b)

End wall type	No. of species	Length of vessel elements (μm)
Entirely scalariform	52	1,090
Scalariform + simple	19	810
Simple but oblique	34	690
Simple and transverse	169	410

3.7.4 Statistical Correlations of Perforation Plate Type

In his pioneering paper on vessel evolution, Frost (1930b) developed correlations between perforation plate type and vessel element length (Table 3.1).

If indeed shortening of vessel elements is a pervasive, polyphyletic, and essentially irreversible trend, then disappearance of bars on a perforation plate is a clear phyletic tendency also.

3.7.5 Scalariform Perforation Plate Variants

In the majority of species with scalariform perforation plates, relatively little aberration occurs. However, one feature that is sufficiently common so that it cannot be regarded as an aberration is forking of bars (Fig. 3.8): this is seen once in Fig. 3.6, part 1, and several times in Fig. 3.6, part 2, several times in Fig. 3.7, parts 2–4. Forking of bars is reported to be fairly common in *Eucryphia mulligani* (Dickison 1978), *Ixerba brexioides* (Meylan and Butterfield 1978a), and *Illicium parviflorum* (Carlquist 1982c).

Perhaps a heightened form of bar forking, but perhaps a different kind of formation is a tendency for perforations to be subdivided (Fig. 3.8), much like transitional lateral wall pitting. This is illustrated here for *Nothobuxus natalensis* (Fig. 3.9, part 4; Carlquist 1982d). Other examples include *Clethra ovalifolia* (Giebel and Dickison 1976), *Archeria racemosa* (Meylan and Butterfield 1978a), *Canella alba* (Wilson 1960), *Balanops sparsifolia* (Carlquist 1980b), *Illicium anisatum* (Carlquist 1982c), and *Akania bidwillii* (Carlquist 1996b). A heightened form of this is seen in *Myodocarpus fraxinifolius* (Fig. 3.10, part 2), in which two or three rows of oval bordered pits comprise the perforation plate, and a scalariform nature is not readily evident.

Forked (or, viewed inversely, fused) bars may be seen in the perforation plates of *Roridula dentata* (Fig. 3.9, part 1; Carlquist 1976c), but occasionally meshworklike perforation plates occur in this species (Fig. 3.9, part 2). Meshworklike plates of that sort have also been illustrated for *Quintinia acutifolia*

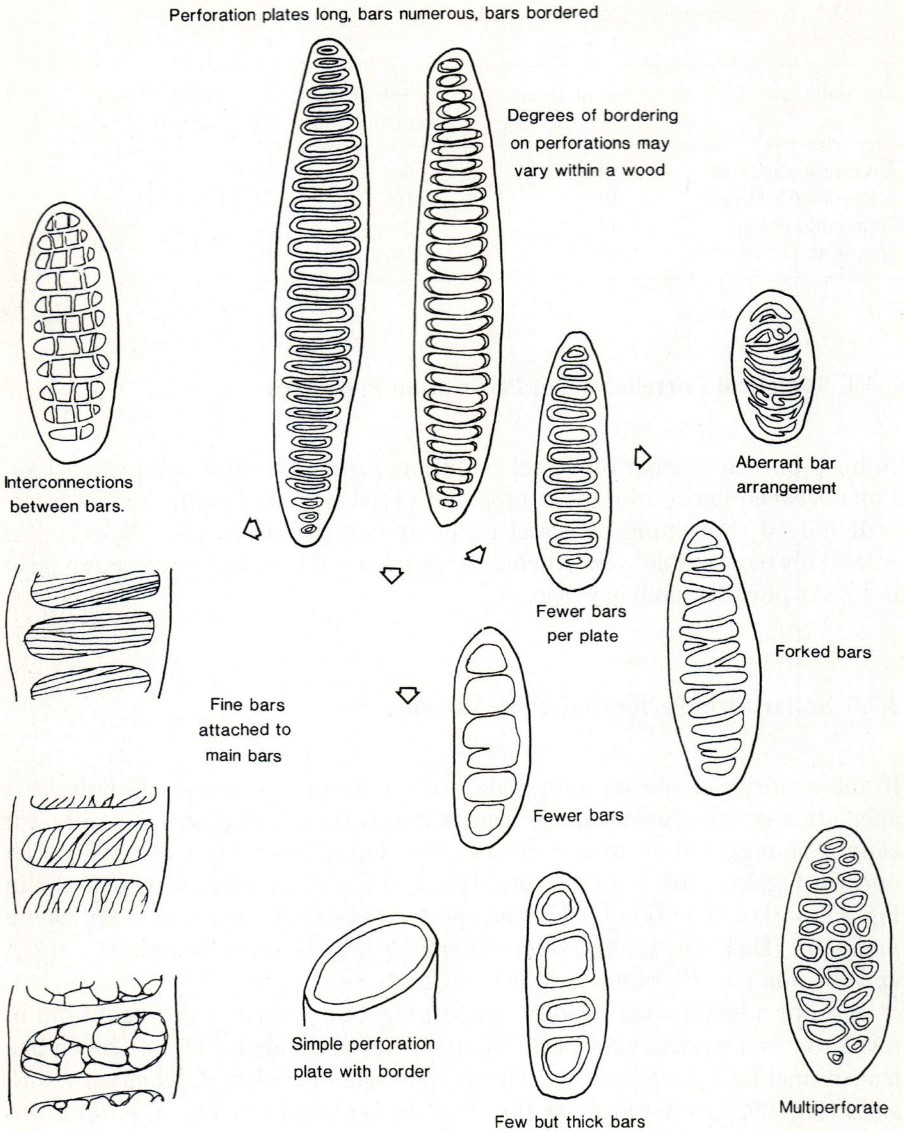

Fig. 3.8. Types of simplification and alteration of scalariform perforation plates. Variants of scalariform perforation plates are numerous, and only some of the more distinctive modes are shown here

(Meylan and Butterfield 1975). Possibly such a plate results from a combination of opposing diagonal patterns which could, in turn, result from similarly-oriented cyclosis in adjacent vessel element tips.

Perforation plates with bars that are curved, parallel to a degree but variously oriented, with some plates resembling fingerprints, have been figured for

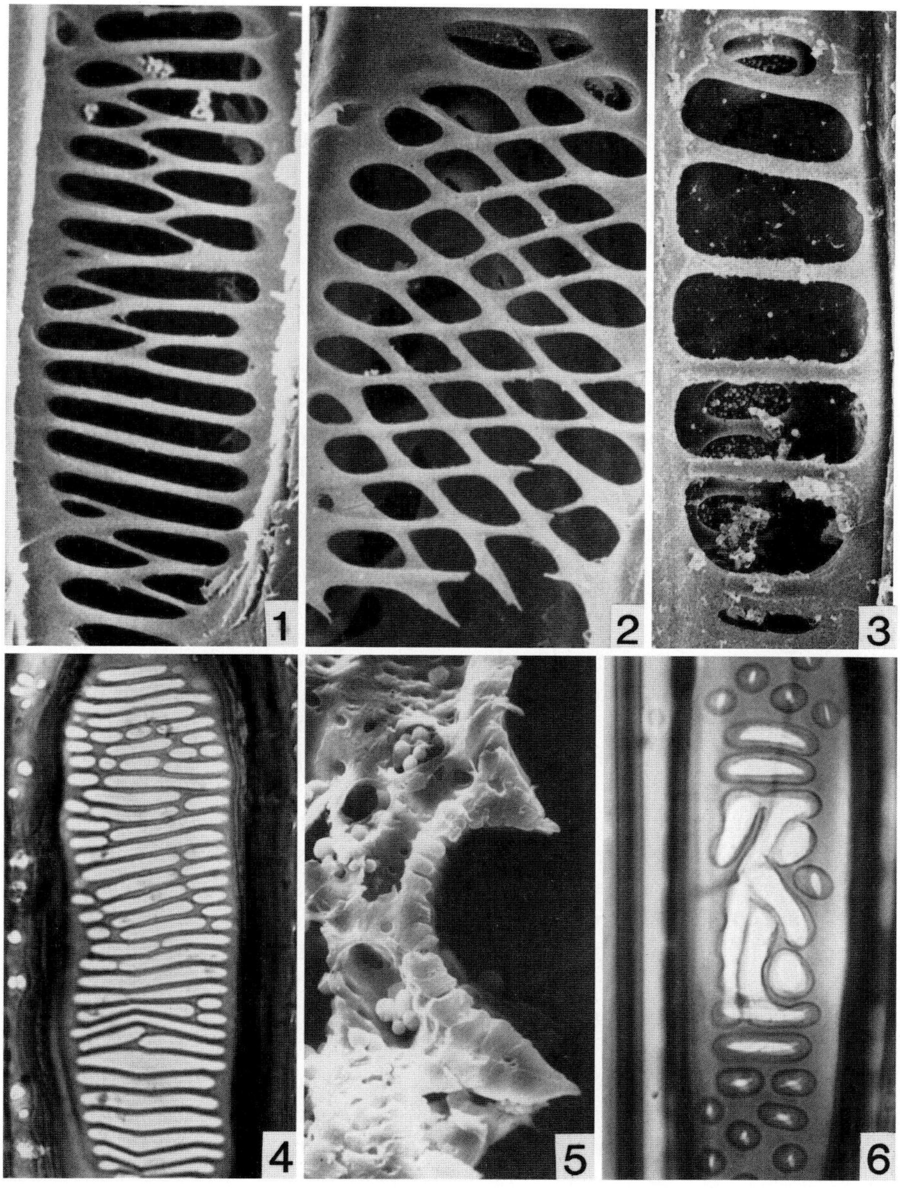

Fig. 3.9. Variations in perforation plates, from radial sections. *1*, *2* *Roridula dentata* (Roridulaceae). *1* Scalariform plates with some bars normal, some bars forked; *2* bars form a meshworklike pattern; *3* scalariform plate with bars few and wide, *Michelia fuscata* (Magnoliaceae); *4* perforation plate with some elongate, some ellipsoidal perforations, *Notobuxus natalensis* (Buxaceae); *5* SEM of tangential section of vessel in wood, showing two vessel junctures that lack borders; the constrictions between vessels form points in sectional view, *Barbeuia madagascariensis* (Barbeuiaceae). *6* Intergradation between (irregular) pit shape and scalariform perforation plate, from latewood of *Castanopsis chrysophylla* (Fagaceae). (*1–3:E*; *4,6:C*; *5:F*)

Brickellia multiflora (Carlquist 1965a), *Euodia lunu-ankeula* (Sharma et al. 1985), *Phoenicoseris berteriana* (Carlquist 1960b), *Scalesia crockeri* (Carlquist 1982e), *Photinia* and *Sorbus* (Zhang and Baas 1992), *Tropaeolum* (Carlquist 1996), *Sonneratia* (Rao et al. 1989), and *Citharexylum* (Gomes et al. 1989). Although not as elaborate as those examples, the latewood perforation plate of *Castanopsis chrysophylla* figured here (Fig. 3.9, part 6) falls into this category with respect to irregular orientation of bars. Although small, the perforation plate figured here for *Loasa picta* (Fig. 3.11, part 9) can be referred to this catgegory also.

Perforation plates in which all apertures are nearly circular, and a so-called foraminate condition (but differing from the gnetalean condition by much smaller size of the perforations and their borders, as well a greater number of perforations per plate) have been figured for *Canthium barbatum* (Rudall 1982), *Coprosma arborea* (Meylan and Butterfield 1978a), *Ugni candollei* (Schmid and Baas 1984), *Gmelina* (Ohtani et al. 1989), some Sabiaceae (Carlquist et al. 1993), various Bignoniaceae (the paper by Chalk 1933 is devoted to this phenomenon), and various Dipterocarpaceae (Gottwald and Parameswaran 1964). This type of perforation plate (Fig. 3.8, lower right) is shown here for a perforated ray cell rather than an ordinary vessel element) in Fig. 3.10, part 1.

Scalariform perforation plates that appear to have strands of interconnecting secondary wall material between the bars are illustrated here for *Myrothamnus flabellifolia* (Fig. 3.10, part 3). This condition with should not be confused with presence of primary wall strands in perforations (e.g., Fig. 3.6, part 4).

In some herbaceous plants of wet habitats, scalariform as well as simple perforation plates may be seen. An example of this is shown for *Patrinia villosa* (Fig. 3.11, parts 1–5). Some of the perforation plates shown are markedly aberrant versions of scalariform perforation plates. Valerianaceae, to which *Patrinia* belongs, is a family that typically has simple perforation plates in the secondary xylem but scalariform perforation plates in the primary xylem (Bierhorst and Zamora 1965). In such a group, where selection for simple perforation plates is not maximal because of constantly moist habitat, scalariform and modified versions of scalariform perforation plates may occur in secondary xylem by means of paedomorphosis (Carlquist 1983e). This was the explanation given for the scalariform perforation plates of *Pentaphragma* (Carlquist 1975a, 1997a; Fig. 3.6, parts 3 and 4), as well as those of *Crepidiastrum* (Carlquist 1983d) and the Campanulaceae with perforation plates other than simple ones figured by Shulkina and Zhikov (1980). This phenomenon, however, is not a common one and is applicable only in a few special cases.

If a vessel forks, at one end a single perforation plate is to be expected, but a pair of perforation plates at the other end where the forking occurs. In this case the pair of perforation plates where the vessel forks will be spacially separated and not likely to be confused with a perforation plate crossed by one bar. However, a pair of (or three) well-spaced perforations not related to a

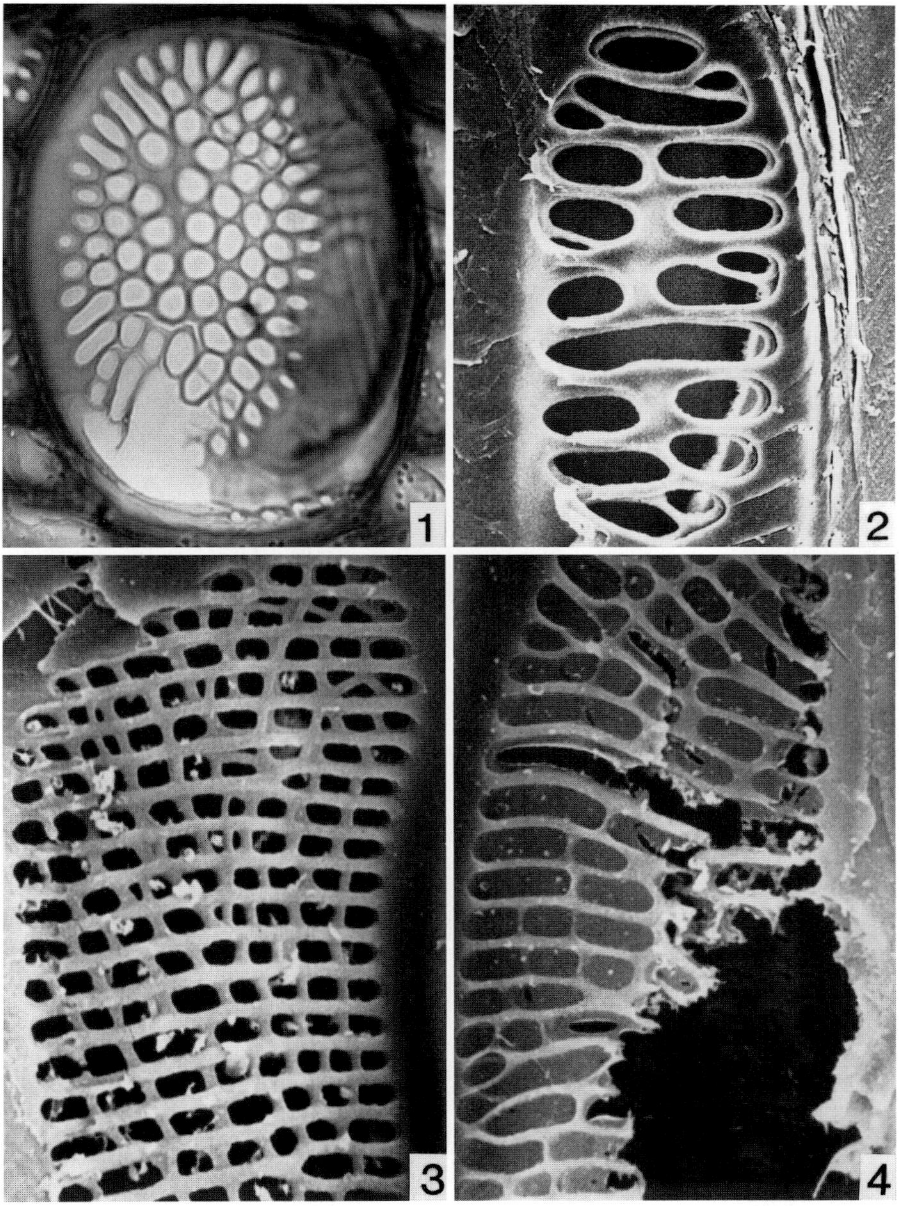

Fig. 3.10. Variant types of scalariform perforation plates, from radial sections. *1* Multiperforate type with mostly circular perforations, perforated ray cells, *Staphylea bumalda* (Staphyleaceae); *2* perforations elliptical to oval but large, *Myodocarpus fraxinifolius* (Apiaceae). *3, 4 Myrothamnus flabellifolia* (Myrothamnaceae). *3* Interconnections between bars, apparently formed from secondary wall material; *4* perforation plate in which pit membranes have not been lost from perforations. (*1:C*; *2:E*; *3,4:G*)

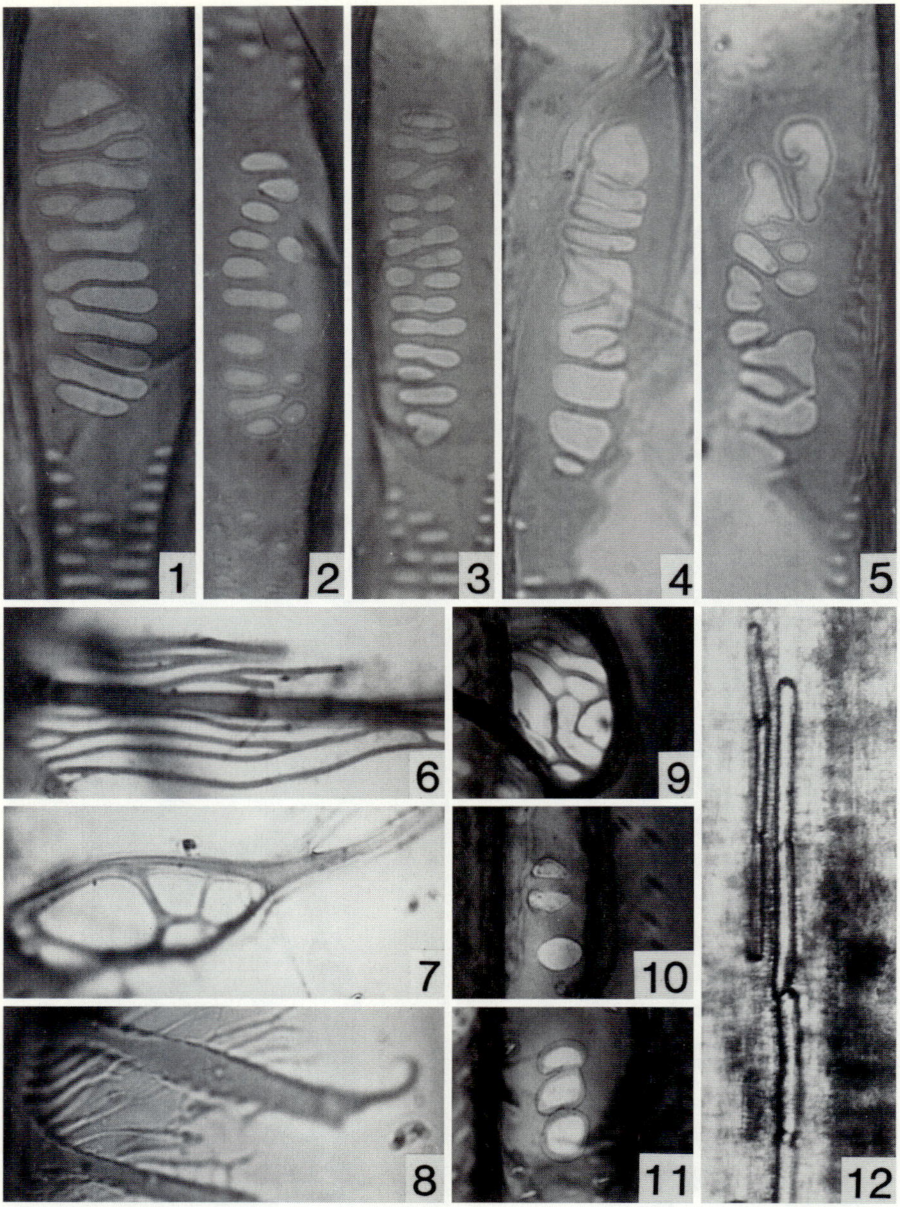

Fig. 3.11. Variant types of perforation plates (*1–11*); simple perforation plates (*12*); from radial sections. *1–6 Patrinia villosa* (Valerianaceae), perforation plates from a single section, showing a range of types from near-scalariform (*1*) to much altered with pitlike perforations (*2, 3*) and large perforations (*4, 5*). *3–5* Small portions of perforation plates of Myristicaceae, showing portions of the large bars with lesser bars connected to them. *6* Finer bars run parallel to major bars, *Iryanthera junius*; *7* finer bars form a network, *Iryanthera laevis*; *8* finer bars run diagonally to major bars, *Knema heterophylla*; *9* networklike perforation plate, *Loasa picta* (Loasaceae); *10* well-separated small perforations on fibriform vessel element, *Nepenthes lowii* (Nepenthaceae); *11* three pitlike perforations comprising a perforation plate, *Byblis gigantea* (Byblidaceae); *12* air bubbles confined within vessel elements of a wood with simple perforation plates; bubbles end at perforation plates: *Ceratostigma wilmottianum* (Plumbaginaceae). (*1–11*:*C*; *12*:*B*)

vessel forking and on the same side of a vessel element rather than on opposite sides can occasionally be seen in woods of some dicotyledons. I am terming these double or multiple perforation plates. Such double perforation plates were reported for Polemoniaceae with fibriform vessel elements (Carlquist et al. 1984) and for *Beilschmeidia tarairi* of the Lauraceae (Meylan and Butterfield 1975). The perforation plate of *Byblis gigantea* illustrated here (Fig. 3.11, part 10) is referable to this phenomenon. Perforations of *Pterostemon* are separated by rather wide bars, but may still be termed scalariform; vessel elements are fibriform (Wilkinson 1994).

A curious series of variations of the scalariform plate characterize certain Myristicaceae (Garratt 1933; Metcalfe and Chalk 1950; Ohtani et al. 1992). Three of these are illustrated here (Figs. 3.8, lower left, 3.11, parts 6–8). In these, there are large bordered bars, typical of a large scalariform perforation plate, but attached to these bars are strands (like bars of a second order of magnitude in some cases) that run parallel to (Fig. 3.11, part 6), diagonal to (Fig. 3.11, part 8), or in a series of networklike formations adjacent to the major bars (Fig. 3.11, part 7).

3.7.6 Simple Perforation Plates

Simple perforation plates require little mention because we are so familiar with them. The simple perforation plate almost always represents a marked constriction in the vessel and rarely approximates the width of the vessel at its widest place in diameter. This fact has been little stressed, but the fact that such constrictions occur may account for the habit of air bubbles to be confined to individual vessel elements (Fig. 3.11, part 12: original data, but like the statement in Slatyer 1967 for which no data are offered).

Simple perforation plates much narrower than the diameter of the vessel in which they occur – often less than half the diameter of the vessel – have been reported in Acanthaceae, in which they characterize the genera *Bravaisia*, *Pseuderanthemum*, and *Sanchezia* (Carlquist and Zona 1988a).

Simple perforation plates that bear vesturing around their rims have been reported in *Coprosma* (Rubiaceae) and *Leptospermum* (Myrtaceae) by Kucera et al. (1977).

3.7.7 Nonbordered perforation plates

The existence of perforation plates in which the adjacent simple plates of adjacent vessel elements meet either in a rounded or pointed (Fig. 3.9, part 5) pairing, as seen in sectional view, may be more widespread than reports indicate. The SEM work of Ohtani and Ishida (1978b) calls attention to the former of these types in *Populus* and *Prunus*; the latter type (e.g., Fig. 3.9, part 5) is illustrated by Ohtani and Ishida (1978b) for *Clerodendrum*, *Sapindus*, and *Zelkova*. This latter type seems to characterize most of the order Caryophyllales (although probably not Cactaceae), and has been reported in Achatocarpaceae (Carlquist 2000a), Barbeuiaceae (Carlquist 1999a), Phytolaccaceae (Carlquist 1998a, 2000b), and Stegnospermataceae (Carlquist 1999b); original observations in other caryophyllalean families indicate the pervasiveness of this feature in the order. Caryophyllales in a revised sense has been claimed to include such families as Plumbaginaceae, Polygonaceae, and Santalaceae, and these three families also have nonbordered perforation plates (unpublished data). Frost (1930b) early reported nonbordered perforation plates for *Ehretia* and *Hamamelis* (the latter has scalariform rather than simple plates).

3.7.8 Angle of Perforation Plates (End Walls)

In earlier decades, the angle of the end wall of the vessel element was often cited, and Frost (1930b, 1931) uses this in his statistical comparisons of vessel element features (e.g., vessels with simple but oblique perforation plates are placed in a separate category from those with transverse simple perforation plates). Although end-wall angle has been used by a few authors, the wide ranges of fluctuation in angle and the difficulty of measuring the angle have mitigated against its use, and this feature is now little mentioned. Certainly long scalariform perforation plates are highly oblique, but so are the perforation plates of fibriform vessel elements, which are not phylogenetically comparable: the latter may merely represent narrow vessels in an instance of vessel dimorphism. There is a tendency for wider vessels to have more nearly transverse perforation plates than narrow vessels within a single sample, although this is by no means always true. For these various reasons, one cannot recommend using angle of perforation plate by itself as a comparative feature, although when other factors (e.g., fibriform vessel elements) are taken into account, it may be worth considering. Despite the lack of utility of this feature, one may still agree with Frost that phylogenetically, there has been a tendency for the perforation plate to shift from highly oblique to oblique to nearly transverse.

3.8 Lateral Wall Pitting of Vessels

3.8.1 Definitions

A lateral wall of a vessel may contact other vessels (in which case the pitting between them is called intervascular, or vessel-vessel pitting) or a vessel wall may be in contact with ray cells (interconnected by vessel-ray pitting) or axial parenchyma (interconnected by vessel-axial parenchyma pitting). Pitting between a vessel and imperforate tracheary elements (which may also be termed "intervascular" according to the IAWA Committee on Nomenclature 1964) may be dense, much like vessel-vessel pitting, or pitting may be sparse. Dense pitting tends to occur between vessels and tracheids, but dense pitting can also be found on interfaces between vessels and fiber-tracheids or between vessels and libriform fibers.

If vessels are solitary or nearly so, little vessel-vessel pitting occurs (e.g., *Quercus*; Wheeler and Thomas 1981), and obviously vessel-vessel pitting becomes more common with increased degrees of vessel grouping. If vessels are solitary, vessel-vessel pitting may be found on the overlapping ends between a pair of vessel elements.

3.8.2 Types of Lateral Wall Pitting

Lateral wall pitting will be discussed here first in terms of vessel-vessel pitting, but these types may be found, often with some modification, on vessel-parenchyma interfaces as well. The types and their phylogenetic interrelationships are illustrated in Fig. 3.12.

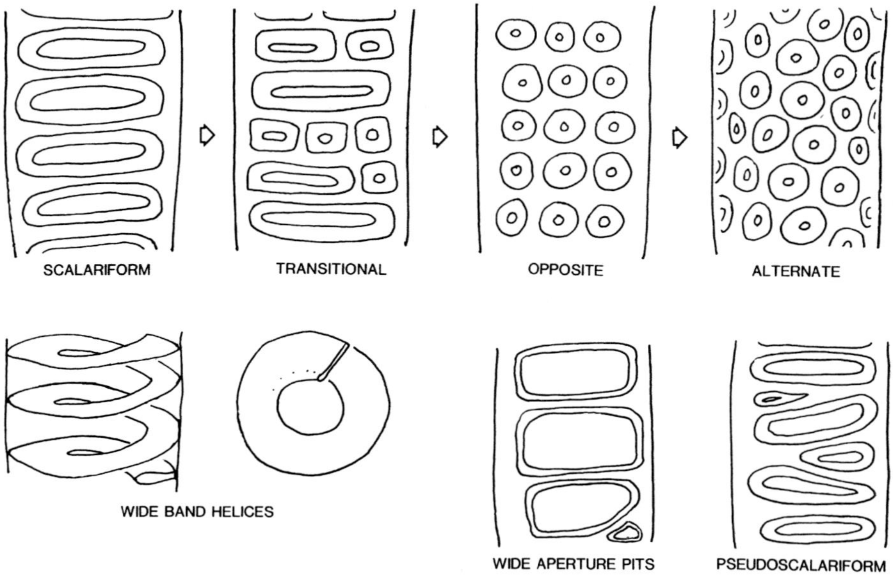

Fig. 3.12. Evolutionary trends in lateral wall pitting of vessels, based in part upon Frost (1931). In addition to the main phylesis from scalariform to alternate, paedomorphic alterations of lateral wall pitting are shown

Scalariform pitting (Fig. 3.13, part 1) consists of pits the lateral length of which equals a wall face (facet). The term "wall face" is, by coincidence, applicable here because scalariform pitting tends to occur on vessels angular (polygonal) in section (and thus with distinctive faces as opposed to the comparatively "faceless" walls of vessels round in transection). Scalariform pitting also occurs on vessel-ray interfaces (Fig. 3.14, part 5); in this instance, the contact with a ray can provide a facet on a vessel round in transection, accounting for the requirement of a distinct wall face for scalariform pitting to occur. Scalariform vessel-vessel pitting is not common in dicotyledons, and one can easily cite species in which vessels have long scalariform perforation plates combined with alternate lateral wall pitting.

Transitional pitting (Fig. 3.13, part 2) consists of a scalariformlike pattern in which some of the pits do extend the full width of a vessel face whereas at other points two or three pits are present instead of a single long one. It can be envisioned in phylogenetic terms (Sect. 3.12) as the breakup of some pits in the scalariform pattern, and this visualization appears to represent a true evolutionary picture. Transitional pitting is not common in dicotyledons at large.

Opposite pitting (Fig. 3.12, part 3) is said to occur when pits form lateral series on vessel walls. Characteristically opposite pits are round in outline, but some may be oval or even elliptical. Frequently students mistake instances of alternate pitting, in which they tend to scan lateral lines of pits instead of seeing them in helices, for opposite pitting. For opposite pitting to be present, one must see clearly defined horizontal lines of pits, and often in a species with opposite vessel-vessel pitting, some portions of intervascular pitting may be transitional as well. Some species do very characteristically have opposite pitting. For example, in Magnoliaceae, *Magnolia* characteristically has scalariform intervascular pitting, whereas *Liriodendron* has opposite intervascular pitting.

Alternate pitting (Fig. 3.13, parts 4 and 5) is by far the most common type of pitting in dicotyledons, so one should expect to find this type in a given species unless one of the other types is clearly present. Even if some pits can be paired as lateral to each other in viewing such a pattern, the overall pattern is usually helical, and clear horizontal lines of pits do not predominate.

Pseudoscalariform pitting (Fig. 3.14, parts 1 and 2) has not been commonly cited, but must be recognized because in a number of dicotyledons where paedomorphosis or some other kind of evolutionary phenomenon occurs, it is present. To designate pseudoscalariform pitting in such groups as true scalariform pitting would result in confusion, and, in fact, criteria do exist for separating scalariform from pseudoscalariform pits. Pseudoscalariform pitting looks like (and in phylogenetic terms probably also represents) the product of lateral elongation of pits in an alternate pattern. Thus, as shown in Fig. 3.14, parts 1 and 2, a pseudoscalariform pattern can appear intermediate between alternate and scalariform, with some pits less than the full width of the wall. Pits shorter than the wall face are not in lateral series (as in transitional pitting)

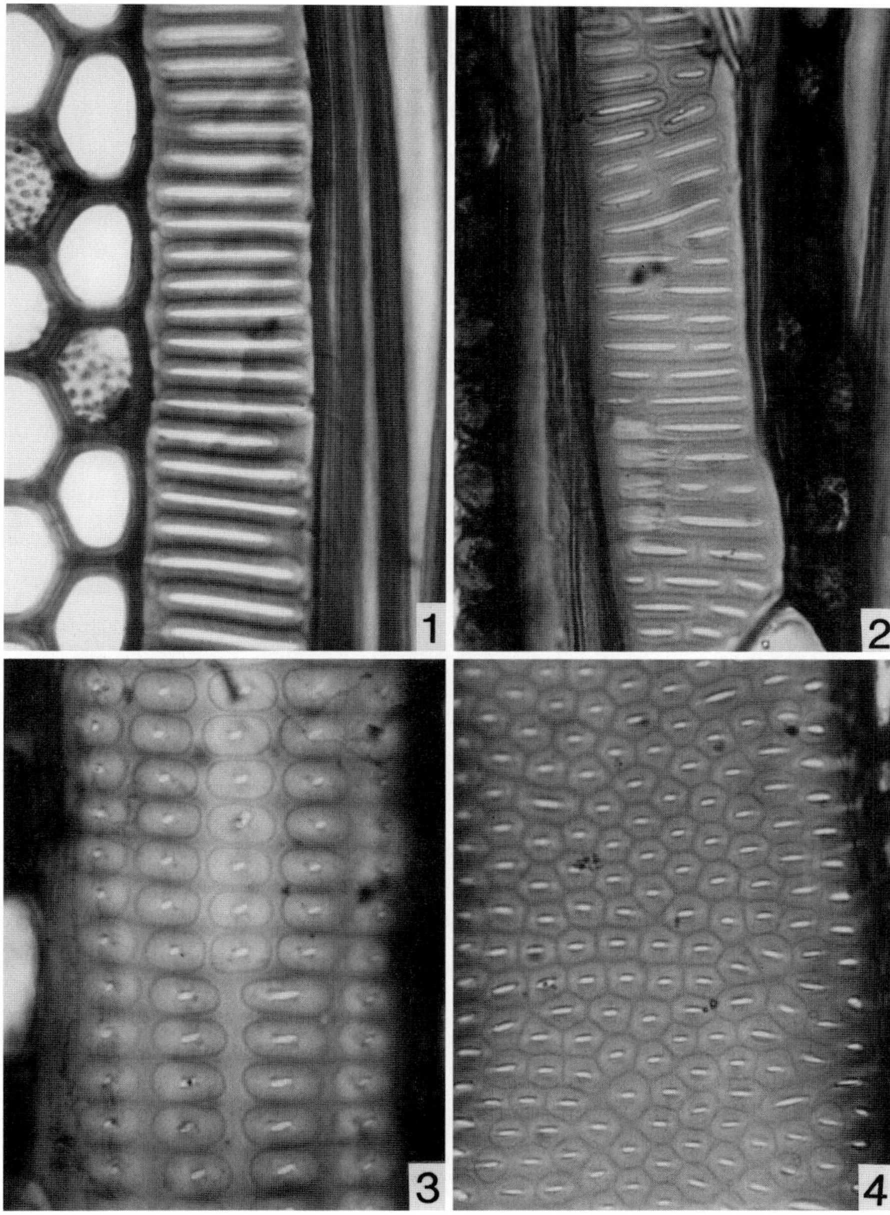

Fig. 3.13. The main types of lateral wall pitting of vessels, from tangential sections. *1* Scalariform, *Magnolia grandiflora* (Magnoliaceae); *2* transitional, *Nothofagus antarctica* (Fagaceae); *3* opposite, *Peridiscus lucidus* (Peridiscaceae); *4* alternate (pits polygonal in outline), *Brosimum* sp. (Moraceae). (*1–4:C*)

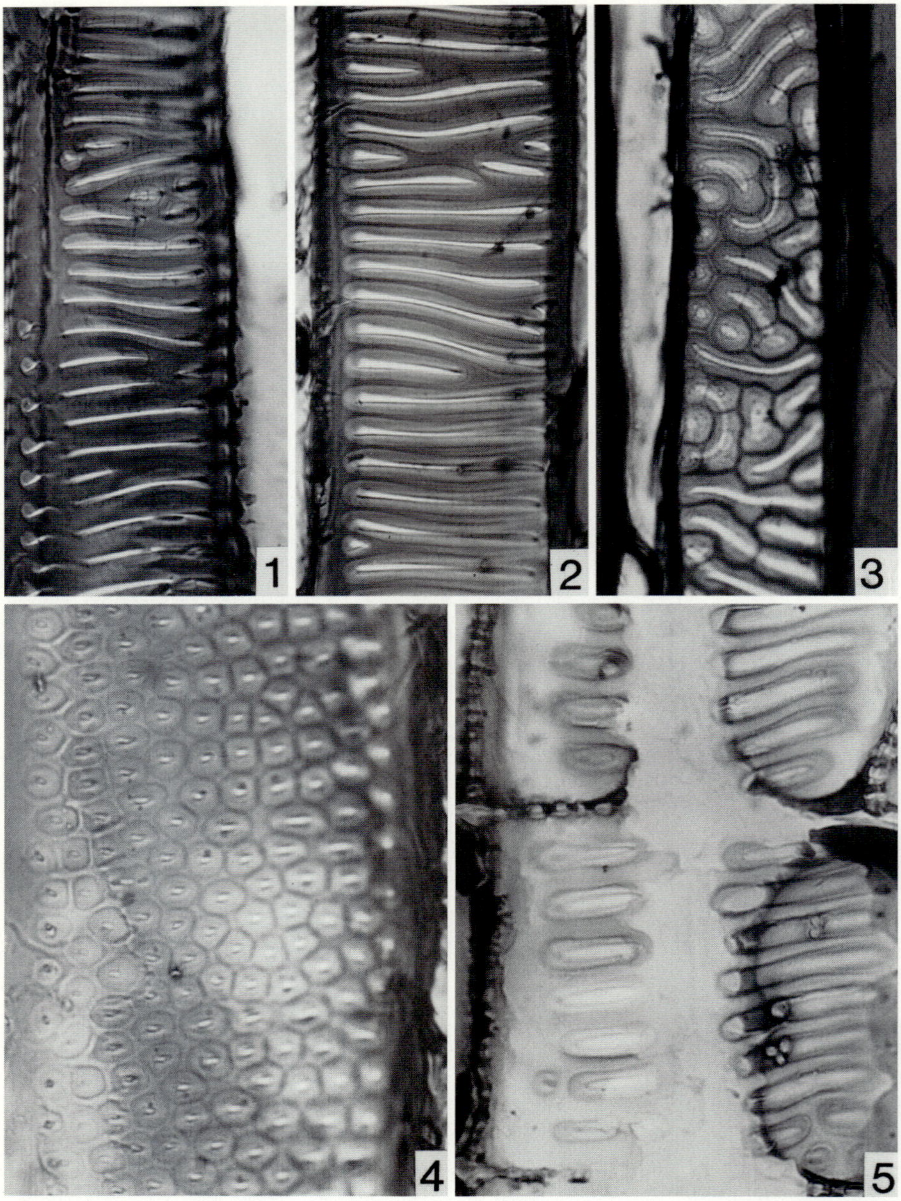

Fig. 3.14. Types of lateral wall pitting of vessels, from tangential (*1–4*) and radial (*5*) sections. *1* Pseudoscalariform pitting, *Euphorbia lactea* (Euphorbiaceae); *2* pseudoscalariform pitting, *Euphorbia candelabrum*; *3* anomalous pit shape and pit-aperture interconnection patterns, *Ludwigia anastomosans* (Onagraceae). *4*, *5*, pitting alternate on vessel-vessel contacts (*4*) but scalariform on vessel-to-ray contacts (*5*), *Rhaptopetalum roseum* (Scytopetalaceae). (*1–5:C*)

but occur like "wedges", spreading apart the laterally longer pits. One can expect pseudoscalariform pitting in groups that show other evidences of paedomorphosis, and these other phenomena may thus condition one to expect pseudoscalariform pitting.

Pseudoscalariform pitting may be found in woods with abundant axial parenchyma (Fig. 3.15, part 1), and some of these instances may closely resemble true scalariform pitting. One of the types of pseudoscalariform pitting that is pertinent in this regard involves the wide-aperture pits that can be found in seasonal bands of parenchyma in herbacous or herblike species (Carlquist and Eckhart 1984). Because the bands of secondary wall material are about as thick as those of primary xylem elements, these wide-aperture pits have sometimes been considered like those of primary xylem tracheary elements intercalated into secondary xylem, as in *Alyssum spinosum* (Metcalfe and Chalk 1950; p. 85), *Phoradendron* (Ashworth and Dos Santos 1997), and *Gypsophila* (Carlquist 1995c; Fig. 3.15, part 4) but they are merely markedly pseudoscalariform in pitting. Such extreme pseudoscalariform pitting with some degree of transition to true helical thickenings occurs in *Lewisia* (Carlquist 1995c), *Hectorella* (Carlquist 1998b), and *Nastanthus* of the Calyceraceae (Carlquist and DeVore 1998). In the instances in which a close approach to helical bands (or actual helical bands) occur, the wood has little or no fibrous tissue and the parenchyma expands and shrinks with seasonal change in water availability.

In addition to examples of pseudoscalariform lateral wall vessel pitting cited in a study of paedomorphosis (Carlquist 1962a), instances include *Chimantaea mirabilis* (Carlquist 1957a), any of the Juan Fernandez Cichorieae (Carlquist 1960b), several Senecioneae (*Gynoxys*, *Liabum*, *Senecio* sect. *Dendrosenecio*; Carlquist 1962b), several lobelioid Campanulaceae (notably *Cyanea tritomantha* and *Delissea undulata*; Carlquist 1969a), *Scaevola glabra* (Carlquist 1969b), any of the succulent species of *Euphorbia* (Carlquist 1970a), *Lecocarpus* (Carlquist and Eckhart 1982), and *Mentzelia humilis* (Carlquist 1984d). These could all be called succulent rosette trees or rosette shrubs. However, similar pseudoscalariform pitting has been reported in vessels in the seasonal parenchyma bands of woody herbs such as *Iva axillaris* (Carlquist 1966b), *Ipomopsis aggregata* (Carlquist et al. 1984), and various species of *Phacelia* (Carlquist and Eckhart 1984).

Extreme examples of pseudoscalariform pits with wide apertures are shown here for Crassulaceae (Fig. 3.15, parts 2 and 3). The pseudoscalariform nature of these pits is evident in occasional deviations from a true scalariform pattern: some alternate pits (some laterally widened) appear intercalated into the pattern at various points. In the examples from Crassulaceae, (a) the pit apertures are very wide, so that one has the impression of widely-separated bands of secondary wall material rather than pits; (b) the pits may extend laterally beyond a single face, even completely around the cell (slender vertical strands of wall material may or may not interconnect the bands at angles of the vessel elements). The wide separation between the bands permits vessel elements with what may be called wide-aperture pseudoscalariform pitting to expand

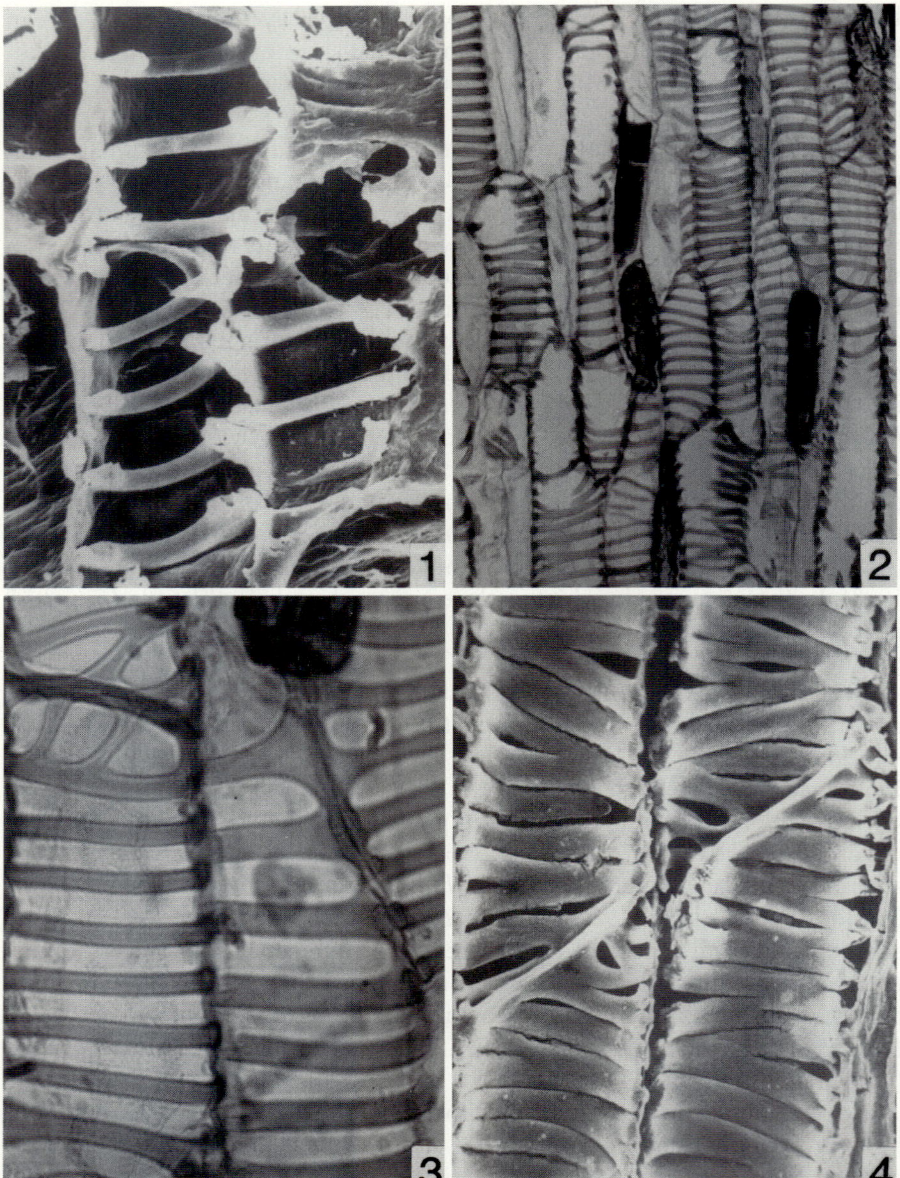

Fig. 3.15. Helices and pseudoscalariform lateral wall pitting of vessels, from tangential sections. *1* Secondary wall material of vessels in the form of wide helices, *Anacampseros marlothii* (Portulacaceae); *2* pitting on the short vessel elements of *Crassula arborea* (Crassulaceae); *3* view showing large apertures of pits, *Crassula arborea*; *4* pseudoscalariform pitting transitional to a helical pattern, *Gypsophila patrinii* (Caryophyllaceae). (*1:E*; *2:B*; *3:C*; *4:D*)

and contract as the parenchyma in which they are embedded expands and shrinks, and this provides a key to the curious morphology of these vessels, which have not been explained clearly in wood literature.

An extreme form of vessel lateral wall modification related to succulence can be called wide-helix bands, as in Portulacaceae (Carlquist 1998b; Fig. 3.15, part 1). Preston (1901) first figured these, and they have also been figured by Gibson (1973, 1978a) and Mauseth et al. (1995). Wide-helix bands are common in vasicentric tracheids of globular cacti, and consequently have been illustrated in the chapter on imperforate tracheary elements (Fig. 4.9, parts 1 and 2). However, wide-helix bands may also occur on vessel elements; the wood of globular cacti may have greater numbers of vasicentric tracheids than vessel elements. Gibson (1978a) notes that in a species in which some xylem is more fibrous, some more parenchymatous, the wide-helix bands can be found in vessels and vasicentric tracheids of the more parenchymatous wood, whereas pitted vessels (albeit with laterally widened pits verging on a pseudoscalariform pattern) occur in the more fibrous wood. This example, as well as the tendency for wide-helix tracheary elements to occur in globular cacti subject to marked seasonal shrinkage and expansion, shows that the wide-helix tracheary elements represent an accommodation to changes in volume of the secondary xylem. The fact that the helical bands are wide laterally, not vertically, permits the wall strength to be maximized along with the space between the gyres of the helix (allowing for expansion and contraction).

Occasionally aberrant types of vessel lateral wall pitting not referable to any of the above types may be found. Such an aberrant pattern is shown here for *Ludwigia* (Fig. 3.14, part 3). This pattern may be considered a modification of an alternate pattern in which pit apertures are elongate and interconnect pits in various directions.

In a particular wood sample, vessel-ray (or vessel-axial parenchyma) pitting may be roughly the same as vessel-vessel pitting. In other species, one finds a marked discrepancy between vessel-vessel and vessel-parenchyma pitting. An example of the latter situation is shown in Fig. 3.14, parts 4 and 5, in which vessel-vessel pitting is alternate but vessel-ray pitting is scalariform. This example is from Scytopetalaceae (Carlquist 1987h), but a number of other families characteristically show this, such as Cephalotaceae (Carlquist 1981e) and Melastomataceae (van Vliet et al. 1981).

A characteristic difference between vessel-vessel and vessel-parenchyma pitting is not one of type but of size and of aperture wideness. Vessel-parenchyma pits often are somewhat larger than vessel-vessel pits and have wider apertures. These features have been noted by Braun (1970), who has invented the term "contact pits" for comparatively large vessel-parenchyma pits. Braun (1970, 1983, 1984) has shown that phosphatase activity, indicating conversion of starch into sugar in parenchyma cells and transmission of the sugar into vessels (thereby accelerating conduction in the vessel), is associated with this kind of pit.

3.8.3 Evolutionary Status of Lateral Wall Pitting

Frost (1930b) compared vessel element lengths of woody dicotyledons to the types of lateral wall pitting in the species he sampled (Tables 3.2, 3.3).

One can see that these two tables are very similar. If one compares these tables with the one for perforation plate types above, one can see that evolution of lateral wall pitting types has progressed at a rate similar to that involved in perforation plate morphology. The central theme of progression in lateral wall pitting from scalariform to transitional to opposite to alternate is evident.

Frost's data also inferentially show that scalariform pitting (or, as noted above, ability to form scalariform pitting where tracheary elements are wider) occurred at the tracheid-vessel element transition in dicotyledon phylesis.

3.8.4 Systematic Distribution of Lateral Wall Pitting Types

As with scalariform perforation plates, particular families often tend characteristically to have particular lateral wall pitting types, although there is a range in some families. Opposite pitting, although not often characteristic of a family, tends often to be found in the families Buxaceae (Carlquist 1982d),

Table 3.2. Vessel element length compared to vessel-vessel pitting. (Frost 1930b)

Vessel-vessel pitting type	No. of species	Vessel element length (μm)
Scalariform	15	1,130
Transitional	28	1,070
Opposite	33	790
Alternate	183	460

Table 3.3. Vessel element length compared to vessel-ray pitting type. (Frost 1930b)

Vessel-ray pitting type	No. of species	Vessel element length (μm)
Scalariform	13	1,110
Transitional	42	960
Opposite	49	740
Alternate	156	430

Lardizabalaceae (Carlquist 1984f), and Platanaceae, and tends to occur in some species of *Ilex* (Baas 1973).

Alternate pits are probably more common than the figures above for Frost's survey indicate. Frost's sample was based largely upon woody dicotyledons; herbaceous and woody-herbaceous dicotyledons belong mostly to families rich in specialized features, and therefore tend to have alternate vessel-vessel pits.

Listing families for all four nonpaedomorphic pitting types seems impractical for reasons inferred above, but a listing of families with scalariform pitting can be offered. This listing attempts to omit instances of pseudoscalariform pitting. Also, families in which scalariform pitting occurs only on vessel-ray contacts are omitted. Scalariform pitting occurs in some, not necessarily all, of the species in families listed below.

Actinidiaceae (incl. *Saurauia*)
Aextoxicaceae
Anacardiaceae
Annonaceae
Aquifoliaceae
Araliaceae
Aristolochiaceae
Balanopaceae
Begoniaceae
Betulaceae
Brunelliaceae
Bruniaceae
Buxaceae: *Styloceras*
Canellaceae
Caprifoliaceae
Chloranthaceae
Clethraceae
Clusiaceae
Cornaceae
Cunoniaceae
Cyrillaceae
Daphniphyllaceae
Dilleniaceae
Elaeocarapaceae
Ericaceae
Escalloniaceae
Eucryphiaceae
Fagaceae
Flacourtiaceae
Fouquieriaceae
Geissolomataceae
Goodeniaceae
Grossulariaceae
Grubbiaceae
Hamamelidaceae
Hydrangeaceae
Icacinaceae
Illiciaceae
Lacistemaceae
Lardizabalaceae
Lissocarpaceae
Loasaceae (*Eucnide*; Carlquist 1984d)
Magnoliaceae
Malpighiaceae
Monimiaceae
Myricaceae
Nyssaceae
Pentaphragmataceae
Piperaceae
Platanaceae
Rhizophoraceae
Roridulaceae
Rubiaceae
Santalaceae
Staphyleaceae
Styracaceae
Symplocaceae
Theaceae
Trimeniaceae
Violaceae
Vitaceae

3.8.5 Pit Outline Shapes

Scalariform pits ordinarily form smooth ellipses. This is true of transitional pits, although where several form a lateral series, their ends may be blunted, as illustrated in Fig. 3.13, part 2. Opposite and alternate pits are most often circular to slightly oval (usually laterally widened) in outline (Fig. 3.13, part 4). However, pits that are markedly angular are common on vessel-vessel interfaces, as shown here for *Brosimum* (Fig. 3.13, part 5) and *Rhaptopetalum roseum* (Fig. 3.14, part 3). Such angular pits have been illustrated for *Hymenoclea salsola* (Carlquist 1958a) and *Fuchsia excorticata* (Carlquist 1975a).

One may ask why there has been a phyletic shift from scalariform, ending with alternate. An explanation that has been offered (Carlquist 1975a) relates to the conflicting requirements for pits as contact areas between cells and for mechanical strength. Pits represent a source of loss of strength because they interrupt the secondary wall. Scalariform pits represent maximal contact areas, but they also represent maximal loss of mechanical strength because of the weakness of the long axis of these pits – like bridge girders with no diagonal cross-members between them. Alternate pits, whether circular or polygonal in outline, represent a maximal strength configuration: they are often hexagonal in outline, and the strips of secondary wall between the pits are thus arranged, like the struts of geodesic domes, in a way that represents a combination of maximal wall strength with maximal contact area between vessels. Conceding this evolutionary idea, one may ask why vessel wall strength is so essential, since one usually thinks of imperforate tracheary elements as contributing mechanical strength to a woody stem. Such plants as succulents show vessel wall patterns with suboptimal strenth characteristics. These plants attain mechanical strength with parenchyma cell turgor rather than imperforate tracheary element wall strength, so that could account for limited mechanical strength of vessel walls. However, negative pressures are also not exceptionally strong in succulents, so that one cannot rule out the role of tension in vessels as a selective factor promoting vessel wall strength. Vessel walls tend to be thicker in plants of arid areas (Carlquist 1980a; Baas et al. 1983); this can be seen here in *Larrea* (Fig. 2.2, part 2) and *Krameria* (Fig. 2.2, part 4).

Pit apertures of pits roughly isodiametric in shape, such as most alternate pits, most commonly tend to be oval to somewhat elliptical in shape (Figs. 3.13, parts 4 and 5; 3.14, part 3). However, in some families, pit apertures tend to be circular, as in Calycanthaceae (Carlquist 1983c), Convolvulaceae (Mennega 1969), or Elaeocarpaceae (Meylan and Butterfield 1978a). Pit apertures may be very narrow and slitlike, in contrast; pit apertures of this type characterize most Acanthaceae (Carlquist and Zona 1988a). If pit apertures of pits adjacent in a helix tend to be elongate in such a fashion that they coalesce into grooves, a helical aperture pattern, described below in the section devoted to wall sculpture, is formed.

3.8.6 Pit Size

Size of vessel wall pits (diameter of pit cavity) has been used as a diagnostic feature in dicotyledon woods (Record and Chattaway 1939). The categories offered are: coarse, pits more than 10 µm in diameter; medium, 7–10 µm; fine, less than 7 µm. Record (1943d) provided a key to genera with pits of small diameter. These dimensions are based primarily on circular alternate pits. The fact that taxa differ from each other in pit size is shown by taxa of Inuleae (Carlquist 1961b). The large circular pits borne in a single series on the walls of fibriform vessel elements of *Dicranostyles* (Mennega 1969) are distinctive. Pit size on vessel-vessel contacts may approximate the same size as vessel-ray pits in some species, while in others, such as *Vernonia baccharoides* (Carlquist 1964a), the vessel-ray pits may be much larger than vessel-vessel pits.

An unusual condition is seen in vessels of Menispermaceae, in which pits close to junctures between vessel elements may be much larger than pits not close to the ends of vessel elements (Carlquist 1996e).

Vessel-vessel pits have notably small apertures in a few families, such as Calycanthaceae (Carlquist 1983c) or Elaeocarpaceae (Meylan and Butterfield 1978a). In the majority of dicotyledons, pit aperture diameter averages mostly 25–35% of pit cavity diameter. "Contact" pits (e.g., vessel-ray pits) may have, on the vessel side, much larger apertures than those of vessel-vessel pits, as in Scytopetalaceae (Fig. 3.14, parts 4 and 5; Carlquist 1987h). Pits with large apertures also characterize vessels of dicotyledons characterized by paedomorphosis, such as many cacti (Carlquist 1962a) as well as vessels in seasonal parenchyma bands of woody herbs (Carlquist and Eckhart 1984).

3.8.7 Tori

The torus is regarded as a feature characteristic of pits of conifer tracheids. However, tori occur in vessels of a few dicotyledons. Reports include those of Ohtani and Ishida (1981), Parameswaran and Gomes (1981), Wheeler (1983), Dute and Rushing (1988, 1990), and Dute et al. (1996).

3.8.8 Crassulae

Crassulae, also known as Bars of Sanio, are thickenings in the primary wall, located midway between pairs of pits. These are characteristic of conifer tracheids. However, as Bailey's (1919) review of this structure shows, they also occur in vessels of *Cercidiphyllum japonicum*, *Magnolia macrophylla*, and *Asimina triloba*. Crassulae also occur in libriform fibers of *Centaurodendron* (Carlquist 1965b).

3.9 Crateriform Pits

Cozzo (1953) discovered that in two species of *Cercidium*, a ring of secondary wall material surrounding the pit aperture projects into the vessel lumen. The most pronounced example of crateriform pit formation is shown here in Fig. 3.16, part 1. Crateriform pits are reviewed with the aid of SEM in a recent study (Carlquist 1989b). Crateriform pits have not yet been reported in genera other than *Cercidium*. No function has yet been hypothesized for crateriform pits, but they may have the same function as other forms of wall relief, as in the verrucae of *Cercidium* discussed below.

3.10 Vesturing

3.10.1 Definition and Types

The reviews of Ohtani et al. (1984a) and Jansen et al. (1998) are detailed and should be consulted by anyone interested in this topic; the survey of Nair and Mohan Ram (1989) covers vesturing in Indian woods. Ohtani et al. (1984a), after reviewing literature on warts and warty layers in tracheids and vessels, concluded that "warts" and "warty layers," terms that have been commonly applied to minute protuberances on wall surface facing the lumen in conifer tracheids, should be rejected. They suggest, instead, adoption of the terms "vestures," "vesturing," and "vestured walls." The reasons of Ohtani et al. (1984a) are that what have been called warts are the same kind of phenomenon as what have been called vestures in pits of dicotyledons with vestured pits. Moreover, findings on systematic occurrence of vesturing have revealed that such minute protuberances are not confined to pit cavities, but in some taxa extend out onto the vessel wall surface also. That these structures are not confined to pit cavities in angiosperms had been realized by Bailey (1933). Bailey figured vesturing that occurs on the vessel wall in addition to the pit cavity for *Vochysia hondurensis*. Bailey's drawings and light microscope photographs are as instructive as today's SEM photographs (e.g., Fig. 3.16, parts 2–4), and his drawings continue to be reproduced (e.g., van Vliet 1978; Zimmermann 1983). Vestures and warts are not structures superimposed on walls but integral portions of the secondary wall (Castro 1991).

Many of the instances of vestured pits correspond to coralloid outgrowths that extend from the inner surface of the pit cavity near the aperture and terminate just short of the pit membrane. In this type of vesturing, one can see with SEM the vesturing as knobs within the pit aperture if one views vesturing from the vessel lumen side (Fig. 3.16, parts 2 and 4). If one views a vestured pit with SEM from the outside of the vessel, one can see the coralloid tips that face a pit membrane if the pit membrane is sectioned away (Fig. 3.16, part 3).

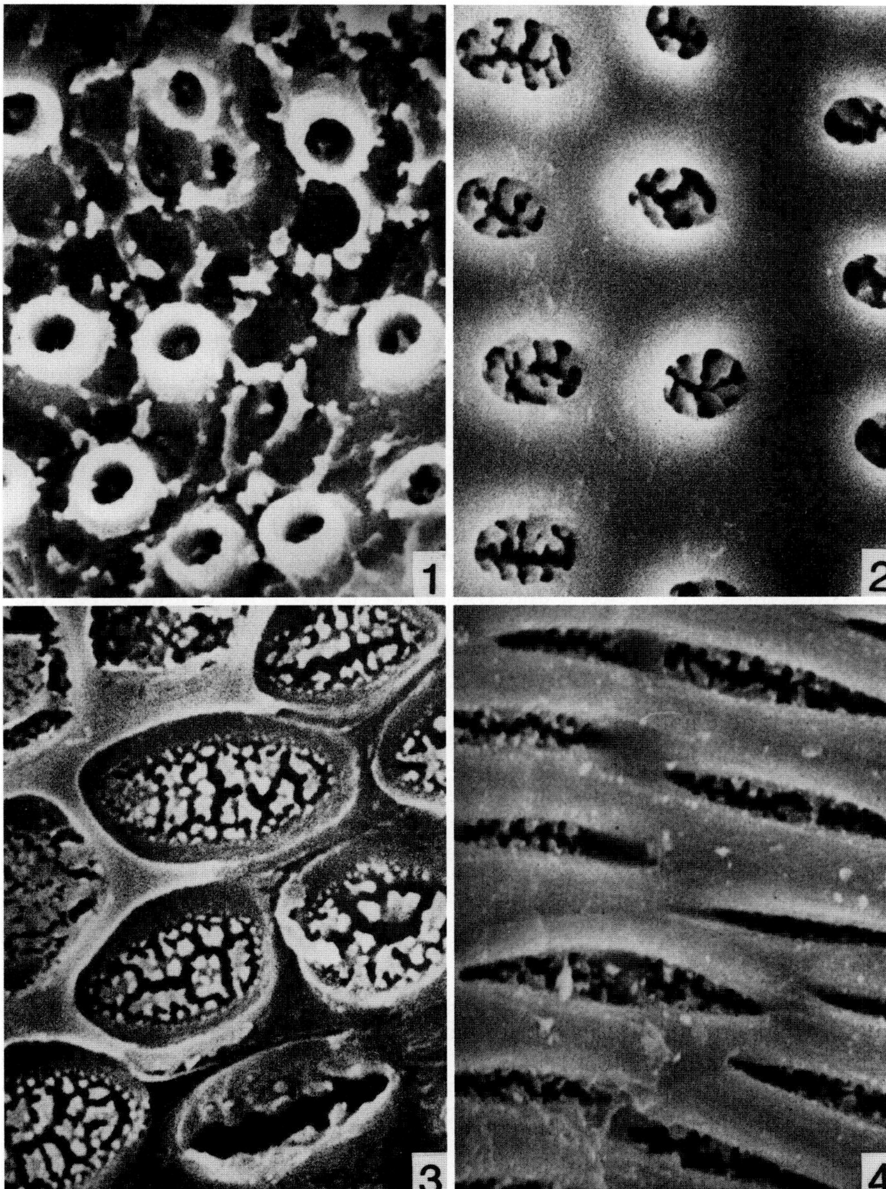

Fig. 3.16. Internal sculpture on vessel walls, from tangential sections, SEM photomicrographs. *1* Crateriform pits, *Cercidium australe* (Fabaceae); *2* pits vestured, wall smooth, *Cercidium floridum* var. *peninsulare* (Fabaceae); *3* vestured pits, seen from outside of a vessel from which the pit membanes have been stripped, *Parkinsonia aculeata* (Fabaceae); *4* vestured pits on elongate pit apertures, *Epilobium caucasicum* (Onagraceae). (*1–3:I*; *4:H*)

This type of vesturing occurs in Combretaceae, in which van Vliet (1978) recognizes some subtypes, or in Fabaceae (e.g., Quirk and Miller 1983, 1985). The type of vesturing figured by Bailey (1933) for *Vochysia hondurensis* is much finer, and the vesturing both projects from the pit border toward the pit membrane and also is present on the vessel wall, where it projects into the lumen. In *Parashorea plicata*, Bailey (1933) figures a mass of filiform secondary wall material that fills the pit cavity and extends into the lumen as a mound rising into the lumen from the pit cavity. Vesturing may be present merely as a series of minute warts around the aperture of a pit (Wheeler 1981). Obviously, there is a wide range of vesturing within vessel pits of dicotyledons, and we are likely with the aid of SEM to discover a greater diversity than is now evident. Under these circumstances, designation of types or even tendencies in vesturing of pits would be premature.

Moreover, as demonstrated by recent workers, vesturing may be present in any given taxon both within pits and on the inner surface of the vessel wall, or only on the inner surface of the vessel wall (Côté and Day 1962; Meylan and Butterfield 1974, 1978a; Parham and Baird 1974; van Vliet 1981; Butterfield et al. 1984; Ohtani et al. 1984a). Côté and Day suggested the term "vestured wall" to denote these occurrences, and Ohtani et al. (1984a) recommended "vestured" and "vestured layer" on account of the complete intercontinuity of vestured pits with vestured wall manifestations. These terms have been accepted.

In addition, vesturing may occur on groovelike pit apertures (Gottwald 1983), on helical thickenings of vessel walls (Ohtani et al. 1984b), on bars of scalariform perforation plates (Parham and Baird 1974), on remnant pit mebranes of scalariform perforation plates (Meylan and Butterfield 1975), on simple perforation plates (Kucera et al. 1977; Vales 1983), on the inner surface of walls of imperforate tracheary elements of dicotyledons (tracheids of Winteraceae; Meylan and Butterfield 1974, 1978a), and even in "tracheoid cells" of seeds (Lersten 1982).

3.10.2 Systematic Distribution of Vesturing

A list is given below of families in which vesturing has been reported. Unless otherwise indicated, vesturing is in pits of vessels. References are cited for taxa in which vesturing has been reported subsequent to the listing of Metcalfe and Chalk (1983; p. 204), and the reader is referred to that source for citations documenting reports prior to 1976.

Apocynaceae
Aquifoliaceae (Baas 1973: possible warts in *Ilex chiapensis*)
Araliaceae (Meylan and Butterfield 1978a; Ohtani et al. 1983; Butterfield et al. 1984: vestured vessel walls)
Asclepiadaceae

Asteraceae
Balanitaceae (Parameswaran and Conrad 1982)
Boraginaceae (Carlquist 1970c; Miller 1977; Gottwald 1980, 1983)
Brassicaceae (Carlquist and Miller 1999)
Capparaceae
Chloranthaceae (*Ascarina* vessel walls; Ohtani et al. 1983)
Cistaceae (Baas and Werker 1981)
Clusiaceae (Guttiferae)
Combretaceae (van Vliet 1978)
Cornaceae
Crypteroniaceae
Dipterocarpaceae
Escalloniaceae (vestured vessel walls of *Ixerba brexioides*; Meylan and Butterfield 1978a)
Euphorbiaceae: *Bridelia* (Nair and Mohan Ram 1989)
Fabaceae (Cassens 1980; Quirk and Miller 1983, 1985; Ohtani et al. 1983, 1984a; vestures on helical thickenings: Ohtani et al. 1984b; vestued simple perforation plates, *Carmichaelia* and *Sophora*: Kucera et al. 1977)
Fagaceae (Parham and Baird 1974)
Gentianaceae (Jansen and Smets 1998)
Gonystylaceae
Hamamelidaceae
Hippocastanaceae
Lauraceae: *Sassafras* (Parham and Baird 1974)
Loganiaceae: vestured simple perforation plates of *Geniostoma* (Kucera et al. 1977)
Lythraceae: *Alzatea* (Baas 1979b); other genera, Baas and Zweypfenning (1979); Baas (1986b)
Malpighiaceae
Melastomataceae (Koek-Noorman et al. 1979; ter Welle and Koek-Norman 1981)
Montiniaceae (Carlquist 1989d)
Myrtaceae: Meylan and Butterfield 1974, 1978a; vestured walls in *Leptospermum* and *Metrosideros* (Ohtani et al. 1983); vestured simple perforation plates (Kucera et al. 1977).
Ochnaceae
Oleaceae (Parameswaran and Gomes 1981; Wheeler 1981)
Oliniaceae
Onagraceae: Carlquist 1975b, 1977b, 1983f, 1987i); vestured simple perforation plates: Kucera et al. (1977)
Penaeaceae (Carlquist and DeBuhr 1977)
Platanaceae: vesturing on walls and perforation plates of latewood vessels: Parham and Baird (1974)
Polygonaceae: vestured simple perforation plate in *Muhlenbeckia* (Kucera et al. 1977)

Proteaceae: vestured walls in *Persoonia* (Meylan and Butterfield 1974, 1978a); vestured simple perforation plates in *Persoonia* (Kucera et al. 1977)
Psiloxylaceae (van Vliet and Baas 1984)
Punicaceae (Bridgewater and Baas 1978)
Rhamnaceae
Rosaceae
Rubiaceae: Meylan and Butterfield 1974, 1978a; vestures on simple perforation plates (Kucera et al. 1977; Vales 1983)
Scrophulariaceae
Sonneratiaceae (Rao et al. 1989)
Thymeleaceae
Verbenaceae: vestured pits and walls (Mathew and Shah 1983)
Vochysiaceae
Winteraceae: vestured (warted) tracheid walls in *Pseudowintera* (Patel 1974; Meyland and Butterfield 1978a), *Drimys* (Carlquist 1988a), *Tasmannia* (Carlquist 1989c): see also Chapter 4.

One must be cautious, reviewing the above list, because in some cases what appears to be vesturing proves to be incrustations of secondary plant products. Bailey (1933) warned against this, and Wheeler (1981) and Gale (1982) have demonstrated instances of such deposits.

In *Bocconia* (Papaveraceae), lateral wall pitting of vessels does not appear to be vestured, but pits have irregular outlines, a phenomenon that might be like an incipient version of vesturing (Fig. 3.17, part 1: original data).

3.10.3 Functional Significance of Vesturing

Zweypfenning (1978) offered a hypothesis for function of vestured pits. He claimed that in the case of a pressure drop between adjacent vessel elements caused by an air embolism, pit membrane aspiration is safer (the membrane does not rupture) when vestures are present because they prevent excessive deflection of the pit membrane.

If vestures occurred only within pit cavities, Zweypfenning's appealing hypothesis might have been tenable. However, the numerous types of vesturing occurrence cited above, plus other vessel wall sculpture manifestations very likely allied to vesturing cited below, indicate that if vesturing is functional, it must be related to something other than pit aspiration. I noted this earlier (Carlquist 1983f) and offered several alternative possibilities; the kinds of vessel wall sculpturing and their distribution within woods were considered in this connection. For example, in *Platanus*, latewood vessels bear vesturing but earlywood vessels do not (Parham and Baird 1974). Alternative possibilities for function of vessel wall sculpturing cited (Carlquist 1983f) include: (1) a mechanism for increasing rate of flow in vessels and tracheids; (2) a mechanism for

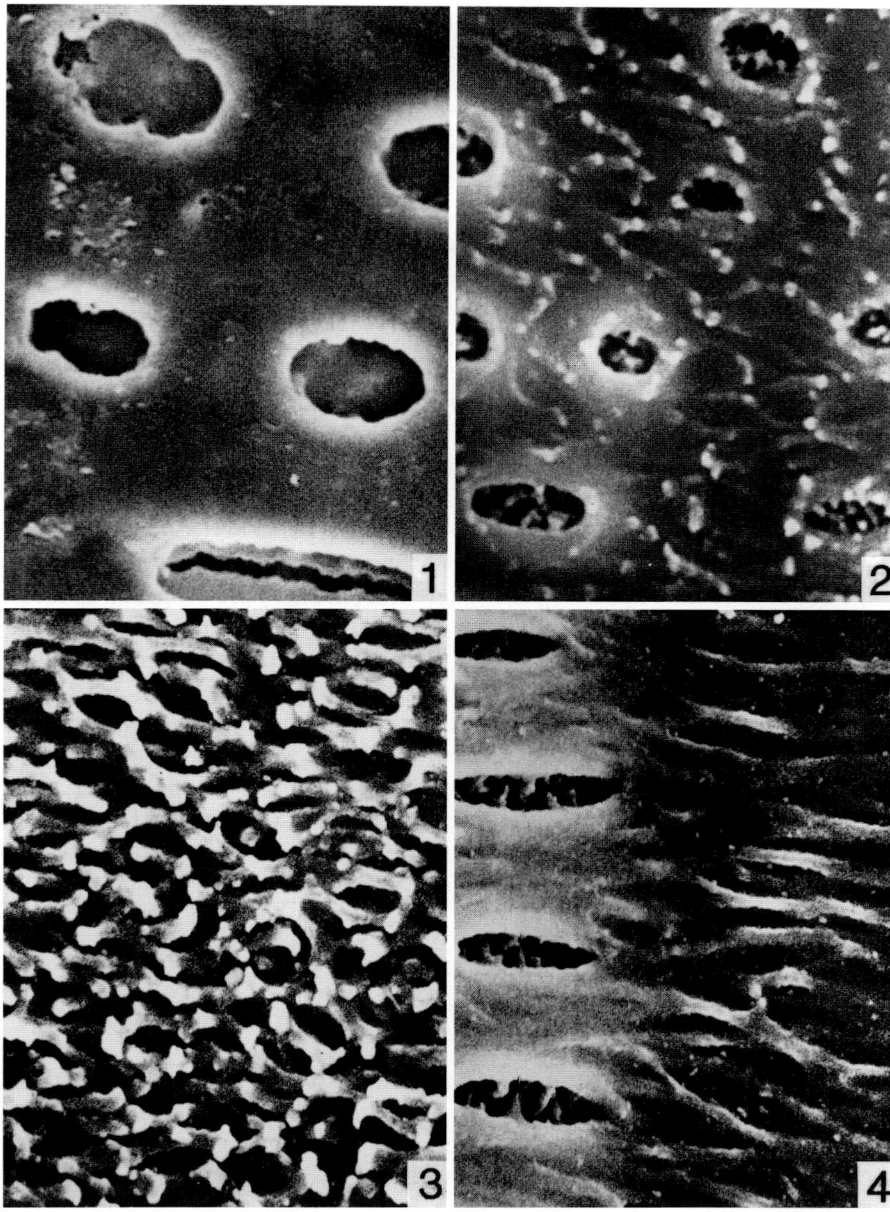

Fig. 3.17. Sculpturing on vessel walls, from tangential sections. *1* Irregular but not truly vestured margins of pit apertures, *Bocconia vulcanica* (Papaveraceae). *2–4* Sculpturing from various places on vessel walls of *Cercidium praecox* (Fabaceae). *2* Small verrucae, with minimal ridges interconnecting them; *3* prominent verrucae, with maximal interconnection into ridgelike formations; *4* ridgelike sculpture (*right*), plus vestured pits (*left*). (*1:H*; *2–4:I*)

removal of air embolisms; (3) a means for preventing formation of air embolisms. The third of these hypotheses was selected as the most likely, although a combination of (2) and (3) is conceivable: both of these two would result in maintenance of the integrity of water columns. It explains why conifers in both dry and cold climates would benefit from a vestured layer facing the lumen in tracheids (Jansen et al. 1998). Tracheids in the vesselless family Winteraceae are vestured only in those species from areas that experience frost: Carlquist 1983b, 1988a, 1989c). Helical sculpture exhibits ecological distribution and distribution within the wood (e.g., more prominent in latewood) that also suggests such a function (Webber 1936; Carlquist 1966a, 1982c). Presence of vesturing in the "tracheoid" cells of legume seeds may be related to development of high water tensions. Zweypfenning (1978) seems to doubt his own hypothesis because he notes that some plants of wet habitats (e.g., *Fuchsia*) have vestured pits, but a phylad can retain a structure even though it is no longer of vital survival value provided it does not require a large expenditure of energy. Other principles to keep in mind are that if vesturing has a function, one need not expect it in all phylads of dicotyledons where that function could be served: development of genetic information for vesturing may not occur readily in all phylads. There is a tendency for evolution to favor mechanisms for preventing permanent damage, rather than repairing damage once it has happened: the latter causes greater loss of biomass and reproductive ability. Ohtani (1987) has found vestured pits in septate fibers, which are nonconductive cells, but this does not really counter evidence that vestures are related to conductive processes; vesturing may be expected occasionally to extend from conductive cells into nonconductive cells in woods that characteristically have vesturing. The very rarity of vesturing in nonconductive cells tends to prove that vesturing is, in fact, related to maintenance of the integrity of water columns in vessel elements and tracheids.

3.11 Verrucae on Vessel Walls

The term "verrucae" is being used to denote coarse types of wall sculpture, such as are present in most species of *Cercidium* (Figs. 3.17, parts 2–4; 3.18, parts 1–4). These verrucae are, where smallest in size, still larger than vesturing (Fig. 3.17, part 2). Although they may be seen under a light microscope, they are much more easily illustrated by SEM. The verrucae may be aggregated in various ways: aligned into knobby wall outgrowths (Fig. 3.17, part 3); occasionally (where they fade out on a wall) present as laterally oriented thickening bands (Fig. 3.17, part 4), grouped into polygons (Figs. 3.16, part 1, 3.18, part 1); present both as irregular knobs and bands uniting the bases of the knobs (Fig. 3.18, part 2); and present in deep, wide depressions uniting pit apertures (Fig. 3.18, part 3). Verrucae were observed in vessels of one collection of *Cercidium floridum* (Fig. 3.18, part 3) but were absent in another collection

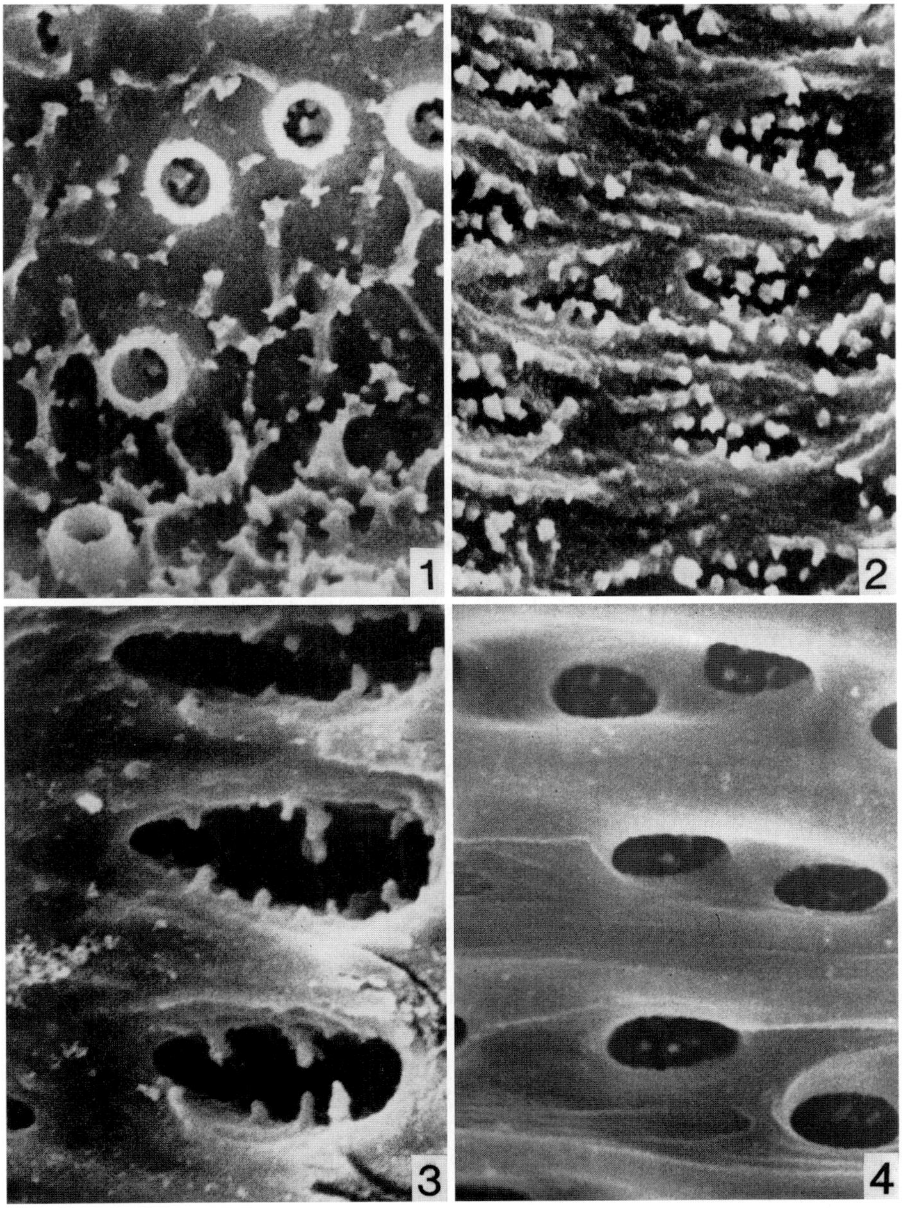

Fig. 3.18. Internal vessel wall sculpturing in *Cercidium* (Fabaceae), SEM photomicrographs. *1* Verrucae tending to be interconnected into polygonal patterns by ridges, *C. australe*; *2* verrucae irregular in outline together with ridges related to the verrucae, *C. andicola*: *3* deep grooves interconnecting pit apertures, lined with verrucae, *C. floridum* (Bissing 180); *4* shallow grooves interconnecting pit apertures, verrucae absent, *C. floridum*. (*1–4:1*)

(Fig. 3.18, part 4). The "bud-like wall outgrowths" reported and figured in Oleaceae by Parameswaran and Gomes (1981) should doubtless be termed "verrucae" in the sense of the above discussion.

3.12 Helical Sculpture on Vessel Walls

3.12.1 Terminology, Types

Under this heading one may include various phenomena that represent uneven depositions on a wall surface or other relief that follows a roughly helical pattern (some with variations). In earlier literature, the term "tertiary helical thickening" was used, but ultrastructural studies showed that the thickenings did not comprise a wall layer additional to the secondary wall, but were part of the secondary wall. The term "spiral" is sometimes seen in connection with vessel wall sculpture; although both spiral and helical are used by the IAWA Committee on Nomenclature (1964), I prefer helical because, in geometry a spiral is two-dimensional, a helix three-dimensional (note the correct usage where DNA structure is concerned). The term "helical bands" can be reserved for the secondary wall pattern of primary xylem.

Although "helical thickening" is used in a number of books and papers as the collective term for all forms of helical relief on vessel walls, I am forced to reject that term in the collective sense in favor of "helical sculpture" (or "helical sculpturing") because it does not include an important and widespread phenomenon that has been confused with true thickenings: grooves interconnecting pit apertures, also termed coalescent pit apertures. In many instances where these grooves occur, no thickenings occur and the wall is smooth except for the depression, and therefore the term "thickening" is inappropriate.

Grooves interconnecting pit apertures may be seen in various legumes: for example, *Parkinsonia* (Fig. 3.19, parts 1 and 2). The pair of figures just cited demonstrates that the elongate grooves do not relate to elongate pit cavities; they overlie pits polygonal in outline, as the second of these figures shows where the wall is shaved away, revealing the pit cavities. SEM views of helical sculpture are less likely to reveal instances of grooves than are light microscope studies because with the light microscope, one can see the outlines of pit cavities underlying grooves (e.g., Fig. 3.19, part 4). Where portions of the wall are shaved away, the grooves are more evident (Fig. 3.19, part 2). Grooves may be short, interconnecting only two or three pit apertures, as in *Wilkesia* (Fig. 3.19, part 5), or the grooves may be nearly continuous from one end of a vessel to another (Fig. 3.19, part 4).

There are a number of instances in which both grooves and thickenings are present. One such example is illustrated here: in *Clematis*, earlywood vessels have only grooves interconnecting pit apertures; latewood vessels have

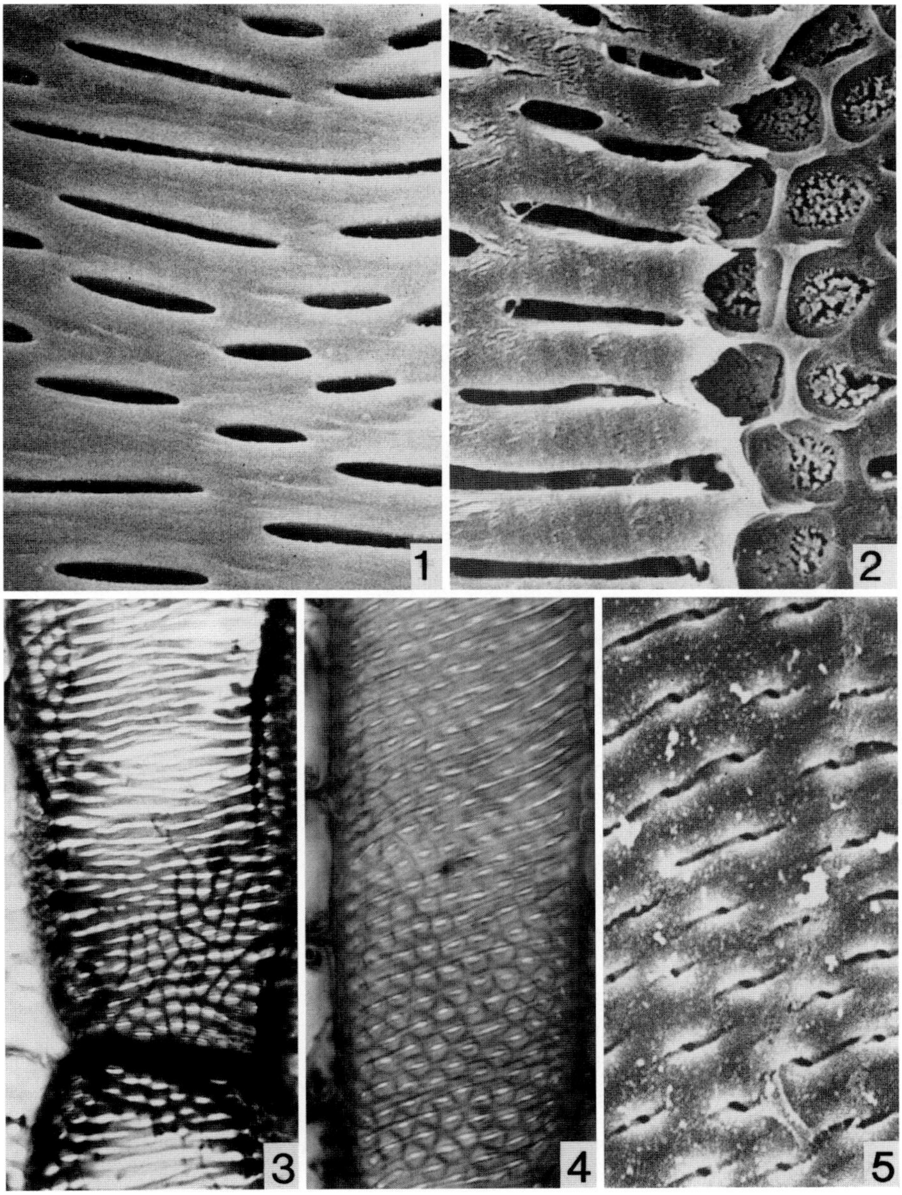

Fig. 3.19. Grooves interconnecting pit apertures, but with no helical thickenings, on vessels walls, from tangential sections; SEM photomicrographs (*1*, *2*, *5*) and light microscope photomicrographs (*3*, *4*). *1*, *2 Parkinsonia aculeata* (Fabaceae). *1* Intact surface of vessel wall, showing grooves only; *2* sectioned vessel wall, wall sliced away so that pit cavities (polygonal in outline, bearing vestures) underlying the grooves are evident; *3* vessel wall partly intact (below), partly shaved away (*above*) so that grooves interconnecting pit apertures are clearly evident, *Coriaria arborea* (Coriariaceae); *4* both grooves and pit cavities visible by virtue of depth of focus, *Hibiscus* sp. (Carlquist 6088; Malvaceae); *5* short grooves interconnecting two or three pit apertures each, *Wilkesia gymnoxiphium* (Asteraceae). (*1*,*2*,*5*:*G*; *3*,*4*:*C*)

grooves, but on either side of the grooves, a ridge (or thickening) is present also (Fig. 3.20, part 2).

3.12.2 Systematic Distribution of Helical Sculpture

Families in which vessels have grooves interconnecting pit apertures, i.e., grooves not accompanied by ridges in at least some taxa from each family listed, are compiled below. More families are likely to be added.

Acanthaceae: *Aphelandra*, *Beloperone* (Carlquist and Zona 1988a)
Asteraceae: numerous genera (see tables in Carlquist 1958a,b, 1959, 1960a,b, 1961a,b, 1962b, 1964a, 1965a,b, 1966b, 1982e)
Brassicaceae: woody genera (Carlquist 1970a)
Coriariaceae: Carlquist 1985e (Fig. 3.19, part 3)
Cucurbitaceae: Carlquist 1992g
Fabaceae: *Parkinsonia* (Fig. 3.19, parts 1 and 2) and very likely many other genera (note wide shallow grooves in *Cercidium* (Carlquist 1989b; Fig. 3.18, parts 2–4)
Geraniaceae: *Viviania* (Carlquist 1985f)
Gesneriaceae: Carlquist and Hoekman 1986a
Goodeniaceae: Carlquist 1969b
Lamiaceae: Carlquist 1992e
Lauraceae: *Umbellularia* (new report)
Leitneriaceae: *Leitneria* (new report)
Malvaceae: *Hibiscus* (new report; Fig. 3.19, part 4)
Myoporaceae: Carlquist and Hoekman 1986b
Nolanaceae: Carlquist 1987a
Polemoniaceae: Carlquist et al. 1984
Papaveraceae: Carlquist et al. 1994
Ranunculaceae: *Clematis* (new report; Fig. 3.20, part 2)
Rosaceae: *Amygdalus* (Zhang and Baas 1992)
Sapindaceae: *Serjania* etc. (Klaassen 1999)
Setchellanthaceae: Carlquist and Miller 1999

Families in which one or more genera with true helical thickenings may be found in vessels are listed below. This list was been altered from Record (1943b) and Metcalfe and Chalk (1983) so as to be a reliable listing for this type of helical sculpture in vessels. In some families, helical thickenings may occur only in tips of vessel elements: Pentaphylacaceae and Theaceae, for example (Fig. 3.22, parts 2 and 3). In some species with helical thickenings, the thickenings are very faint and have been termed striae or striations (Carlquist 1958a). Selected references on helical thickenings in vessels, chiefly since 1972, are cited.

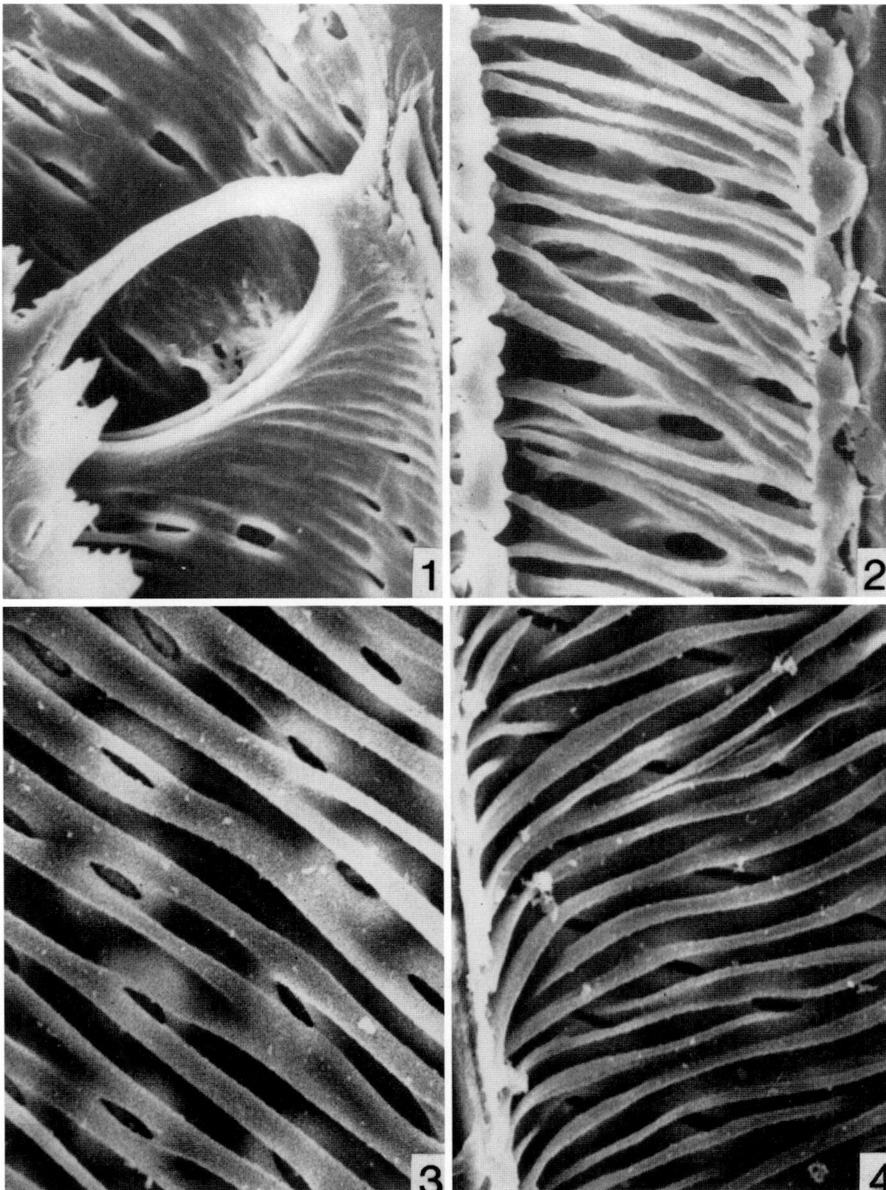

Fig. 3.20. Types of helical sculpture on vessel walls, SEM photomicrographs. *1* Grooves extended the pit apertures; pit apertures are accompanied by pairs of thickenings (*pale*); *Poliomintha longiflora* (Lamiaceae). *2* Vessel wall of latewood vessel, grooves accompanied by thickening bands, *Clematis lasiantha* (Ranunculaceae); *3* helical thickenings running parallel to pit apertures, *Dendromecon rigida* (Papaveraceae); *4* helical thickenings running counter to pit aperture direction, *Melia azedarach* (Meliaceae). (*1:D*; *2:E*; *3,4:F*)

Aceraceae
Actinidiaceae (including *Saurauia*)
Anacardiaceae
Annonaceae
Apiaceae: *Bupleurum* (new report)
Apocynaceae
Aquifoliaceae: Baas 1973; Parham and Kaustinen 1973
Araliaceae: Meylan and Butterfield 1978a
Aristolochiaceae
Asteraceae: Carlquist 1957a, 1958a,b, 1959, 1960a, 1961b, 1962b, 1964a, 1965a, 1966b
Berberidaceae: Carlquist and Hoekman 1985b
Bignoniaceae
Boraginaceae
Bretschneideraceae: Carlquist 1996b
Buddlejaceae: Carlquist 1997b; Fig. 3.22, part 1
Buxaceae: Carlquist 1982d
Calycanthaceae: Carlquist 1983c
Capparaceae
Casuarinaceae: Moseley 1948
Celastraceae
Cercidiphyllaceae
Chenopodiaceae: Carlquist and Hoekman 1985b
Clethraceae
Clusiaceae (*Hypericum*)
Cneoraceae: Carlquist 1987g
Combretaceae: *Maruma* (van Vliet 1978)
Connaraceae: Dickison 1972
Cornaceae: *Corokia* (Patel 1973)
Corynocarpaceae
Dipsacaceae: *Pterocephalus* (Carlquist 1982b)
Elaeagnaceae
Elaeocarpaceae
Epacridaceae
Ericaceae
Escalloniaceae: Stern 1974
Eucommiaceae
Euphorbiaceae
Fabaceae: Parham and Kaustinen 1973; Meylan and Butterfield 1978a
Fagaceae
Garryaceae: Moseley and Beeks 1955
Geraniaceae
Hamamelidaceae
Himantandraceae
Hippocastanaceae

Hydrangeaceae: Stern 1978a
Icacinaceae
Illiciaceae: Carlquist 1982c
Juglandaceae: Miller 1976b
Krameriaceae: Carlquist and Hoekman 1985b
Lamiaceae: Carlquist 1992e; Fig. 3.20, part 1
Lauraceae: *Sassafras* (Stern 1954)
Linaceae
Loganiaceae: Mennega 1980
Loranthaceae
Magnoliaceae: Canright 1955; Parham and Kaustinen 1973
Malpighiaceae
Malvaceae
Meliaceae (Fig. 3.20, part 4)
Monimiaceae
Moraceae
Myrsinaceae
Myrtaceae
Nyssaceae: Parham and Kaustinen 1973
Ochnaceae
Olacaceae
Oleaceae: Meylan and Butterfield 1978a; *Olea* (Fig. 3.21, part 4)
Oxalidaceae
Papaveraceae (Carlquist and Zona 1988b; Fig. 3.20, part 3)
Passifloraceae
Pentaphylacaceae: Carlquist 1984g (Fig. 3.22, parts 2 and 3)
Phytolaccaceae
Pittosporaceae: Meylan and Butterfield 1978a; Carlquist 1981d
Polygalaceae: *Polygala* (Carlquist and Hoekman 1985b)
Polygonaceae: *Eriogonum* (Carlquist and Hoekman 1985b)
Proteaceae
Ranunculaceae *Clematis* (Carlquist 1995a; Fig. 3.20, part 2)
Resedaceae: *Reseda* (Carlquist 1998c)
Rhamnaceae
Rosaceae: Zhang and Baas 1992
Rubiaceae: Ohtani et al. 1984a
Rutaceae
Sabiaceae *Sabia* (Carlquist et al. 1993)
Santalaceae
Sapindaceae: Klaassen 1999
Sapotaceae
Scrophulariaceae: Michener 1981, 1983
Selaginaceae: Carlquist 1992f
Simaroubaceae
Solanaceae: Carlquist 1992h

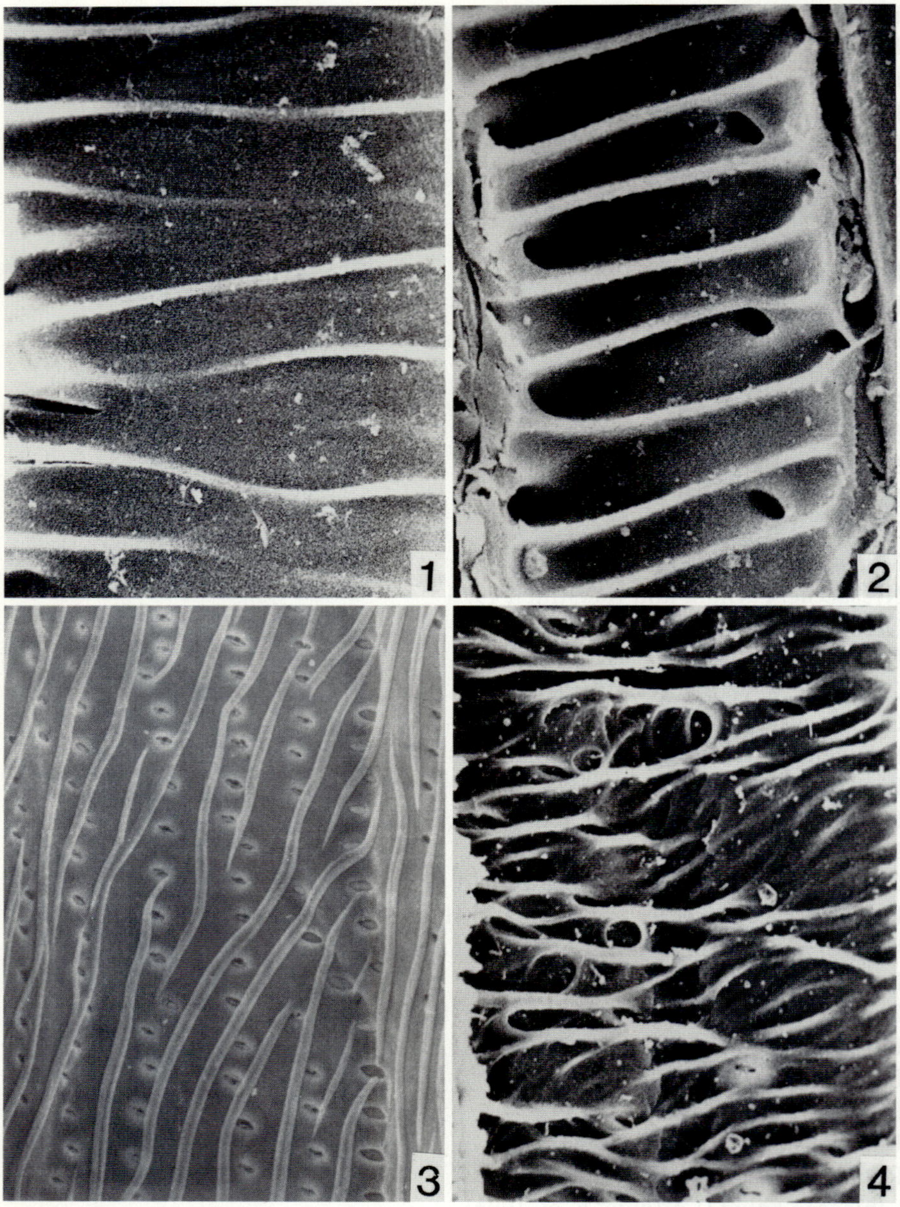

Fig. 3.21. Types of helical thickenings in vessel walls, SEM photomicrographs. *1, 2 Passerina vulgaris* (Thymeleaceae). *1* Less prominent thickenings on earlywood vessel; *2* more prominent thickenings on latewood vessel; *3* helical thickenings some of which taper, oriented in an axial direction, *Sabia japonica* (Sabiaceae); *4* helical thickenings with some tendency to reticulate appearance, *Olea cunninghamii* (Oleaceae). (*1,2:H*; *3:E*; *4:F*)

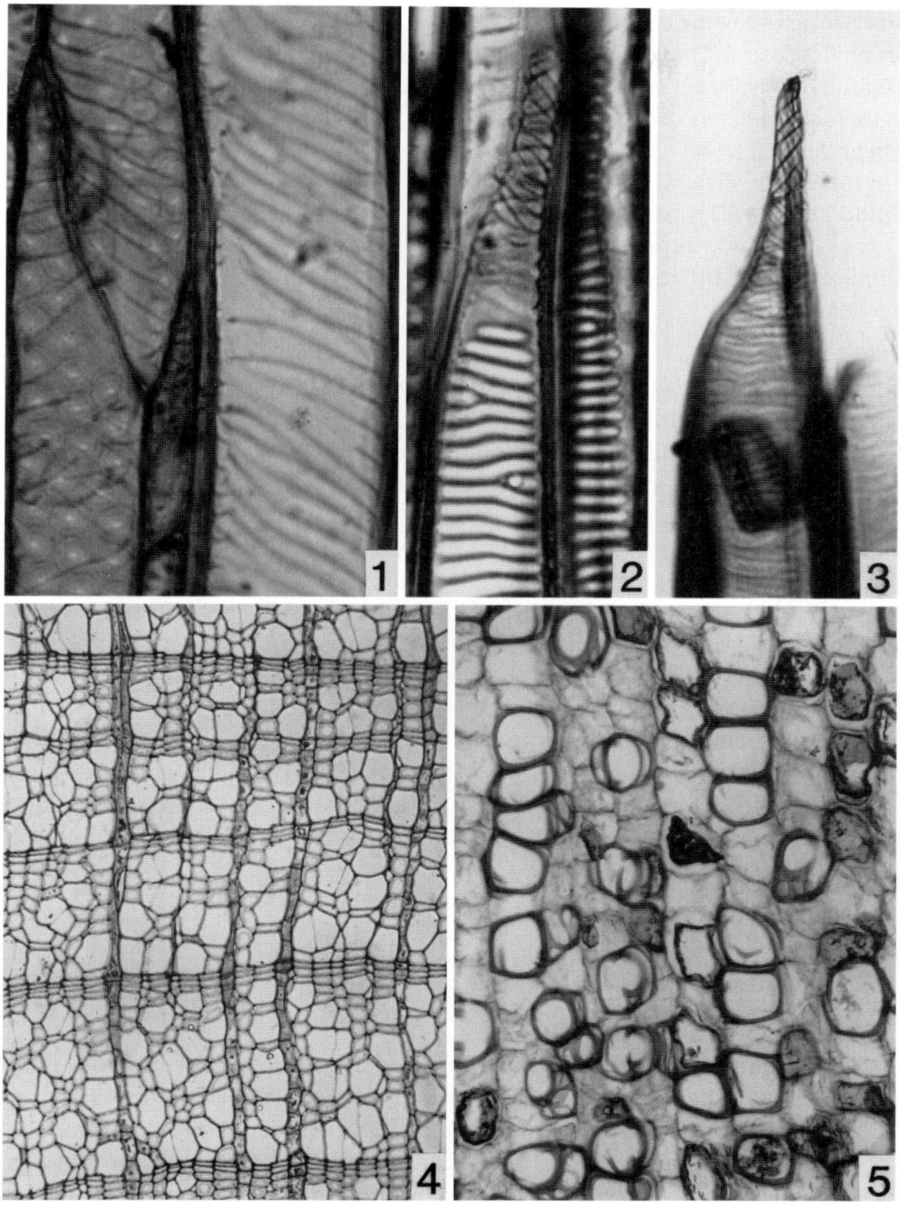

Fig. 3.22. Helical sculpture in vessels, light photomicrographs (*1–3*), angular nature of vessels as seen in transection. *1* Narrow thickenings overlying pits with circular pit apertures (pits on vessel at right shaved away), *Buddleja globosa* (Buddlejaceae). *2*, *3* Helical thickenings confined to tips of vessel elements, *Pentaphylax arborea* (Pentaphylacaceae). *2* Thickenings in vessel element from radial section; *3* thickenings in vessel in maceration; *4* angular vessels in a primitive wood, *Cercidiphyllum japonicum* (Cercidiphyllaceae); *5* angular vessels in a specialized wood, *Crassula arborea* (Crassulaceae). (*1–3:C*; *4,5:B*)

Stachyuraceae (new report): *Stachyurus*
Staphyleaceae: *Staphylea* (Carlquist and Hoekman 1985a)
Sterculiaceae
Symplocaceae: van den Oever et al. 1981
Theaceae
Thymeleaceae (Fig. 3.21, parts 1 and 2)
Tiliaceae (Fig. 3.21, part 3)
Turneraceae (new report): *Turnera*
Ulmaceae: Parham and Kaustinen 1973
Verbenaceae
Violaceae

Other types of wall sculpture, essentially nonhelical, may be found. Miller (1976a) has called attention to such a form, which he calls reticulate thickenings, in *Juglans*. In addition, one should note, however, that helical thickenings are not a unitary phenomenon at all. For example, illustrated here are paired ridges beside grooves (Fig. 3.20, parts 1 and 2), prominent thickenings running parallel to pit apertures (Fig. 3.20, part 3), prominent thickenings running contrary to pit apertures (Fig. 3.20, part 4), widely-spaced thickenings (Fig. 3.21, parts 1 and 2), thickenings with tips fading out (Fig. 3.21, part 3), and numerous fine thickenings with anastomosing tendencies (Fig. 3.21, part 4). The pair of figures, Fig. 3.21, parts 1 and 2, has been selected to show that helical thickenings, if present throughout a wood, are more prominent in latewood than in earlywood. Helical thickenings in latewood have been compiled separately from those in earlywood for each of the species in Carlquist and Hoekman (1985b), and this compilation shows that characteristically some species lack helical thickenings in earlywood but have them in latewood. Such distributions should be recorded.

3.12.3 Functional Significance of Helical Sculpture

If one groups all forms of helical sculpture together, one can see that they tend to be more abundant in areas that are drier (Webber 1936; Carlquist 1966a) or colder, subject to freezing (Carlquist 1982c, 1984f). Obviously ecological factors are the significant features with relation to helical sculpture; one obtains incomplete correlations when one compares helical sculpture with altitude or latitude. Latitude and altitude have been cited because they are conveniently located from herbarium labels. For example, Baas (1973) finds that helices (helical sculpture) characterize a higher proportion of species of *Ilex* in latitudes above 38°, but a greater total number of species of *Ilex* with helices occurs in latitudes between 13° and 37°30′ – relatively frost-free latitudes in general. Likewise, van den Oever et al. (1981) find helices (spirals) increasing with latitude in *Symplocos*, but the latitude category highest with respect to helices begins with 25° – a zone that includes both frost-free and some very cold localities. Baas et al. (1983) are surprised that helical thickenings are not as common in the arid flora of Israel and adjacent regions, but this is at least partly explained by the abundance in this flora of species in families in which helical thickenings are not characteristic (e.g., Moraceae, Tamaricaceae). These families play a small part in the floras of other Mediterranean-climate regions.

The geographical occurrence of species with well-developed helical sculpture seems to emphasize water stress created by either drought or cold; the latter produces physiological drought while soil water is frozen. Either condition can result in high tensions in vessels, and ultimately air embolisms can be induced. Helical thickenings or grooves might have the effect of forestalling air embolism formation and spread, or else they might aid in refilling of embolized vessels (Carlquist 1983f): both have the result of maintaining integrity of water columns, and both functions might result from presence of these structures. The observation that helical thickenings accelerate water flow in vessels under experimental conditions (Jeje and Zimmermann 1979) may, in fact, be demonstrating that helical sculpture could accelerate refilling of embolized vessels. The fact that helical sculpture in some groups, such as Asteraceae, is related to pit apertures (grooves interconnecting pit apertures in many taxa of this family in arid areas) suggests an analogue to vestured pits. Perhaps in some groups, genetic information for vesturing can originate, in others, genetic information for one type or another of helical sculpturing (in a few, both: Ohtani et al. 1984b). In this connection one should take into account vesturing and helical sculpture in tracheid walls of gymnosperms and vesselless angiosperms. Baas et al. (1983) think that compartmentalization of bordered pits or the tips of vessels is already so great that vestures in the former and helical thickenings in the latter (see Fig. 3.22, parts 2 and 3) do not speak for the function of increased bonding of water by wall relief in those sites. However, if vesturing and helical sculpture served to aid refilling of vessels (as well as perhaps to diminish danger of cavitation), any increase in wall relief would be of value. Increase in vessel wall strength is another theoretically possible explanation for helical sculpture (Carlquist 1975a; Zimmermann 1983).

3.13 Vessel Wall in Transection

When one views vessels in transection, one can observe in some species that vessels are angular or polygonal in outline, and with uniformly thin walls, not thickened in the angles (Fig. 3.22, part 4). In other species, vessels are round in outline with the wall irregular to rather uniform in thickness (e.g., Fig. 3.2, part 1). This distinction was utilized by Frost (1930b) in his consideration of vessel phylesis (Table 3.4).

The results of this table have never received extensive comment. Obviously the angular vessel has yielded to the round vessel, although there are also

Table 3.4. Vessel outline compared to perforation plate categories. (Frost 1930b)

Perforation	No. of species	Diameter (μm)	Angular (%)	Thin walled (%)	Evenly thickened (%)
Scalariform	40	67	100	100	97
Simple and transverse	40	120	15	15	22

instances of angular vessels in specialized phylad. Although primitiveness is involved, another factor, narrowness, seems clearly correlated with angularity of vessels. Where vessels are narrower, they tend to be in contact with fewer cells, and thereby by compression a more angular form is achieved (even in specialized taxa; e.g., Fig. 3.23, part 4).

The near doubling in diameter shown in Frost's table with shift to round form of vessels would correlate with roughly a 16-fold increase in conductivity (2^4) according to flow theoretics (Zimmermann 1983; p. 14). Increasing of conductive capacity and efficiency is certainly a persistent theme in dicotyledons; increase in vessel diameter is the simplest means available to achieve this, as growth ring phenomen show. Because vessel diameter is so easily reversible, stress was not laid upon this feature when Bailey and his students, Frost and Kribs, were considering the major trends (some of which have an essentially irreversible nature). We may say that *capability to produce* vessels round in transection rather than unexceptionable production of such vessels marks specialization in dicotyledons.

Mechanical strength of vessels must tend to increase along with increase in vessel diameter. Vines and lianas tend to have markedly thick-walled vessels (Carlquist 1985d) especially where vessels are wider. Thicker vessel walls may relate to increased mechanical strength of the stem (or root) in which they are located, or they may relate to conduction or safety characteristics. Thicker-walled vessels often characterize dryland shrubs (Baas et al. 1983).

Vessel wall thickness is characteristically greater in some taxa than in others, and is now often specified in monographs on wood anatomy.

3.14 Tyloses

Zimmermann (1983) reviews tylosis formation and concludes that Klein's (1923) conclusion is the correct one, namely, that tyloses are formed in relation to wounding. However, the immediate cause for their formation is not trauma but loss of water pressure in vessels. This would explain why tyloses would be characteristically formed in unwounded vessels that fill with air, such as the large earlywood vessels in many species (Figs. 2.4, part 3; 2.8, part 4; 2.11, part 3; 2.12, part 2; 2.13, part 1). Bonsen (1991) finds that tyloses tend to occur in species in which vessel-parenchyma pits are relatively large (larger pits

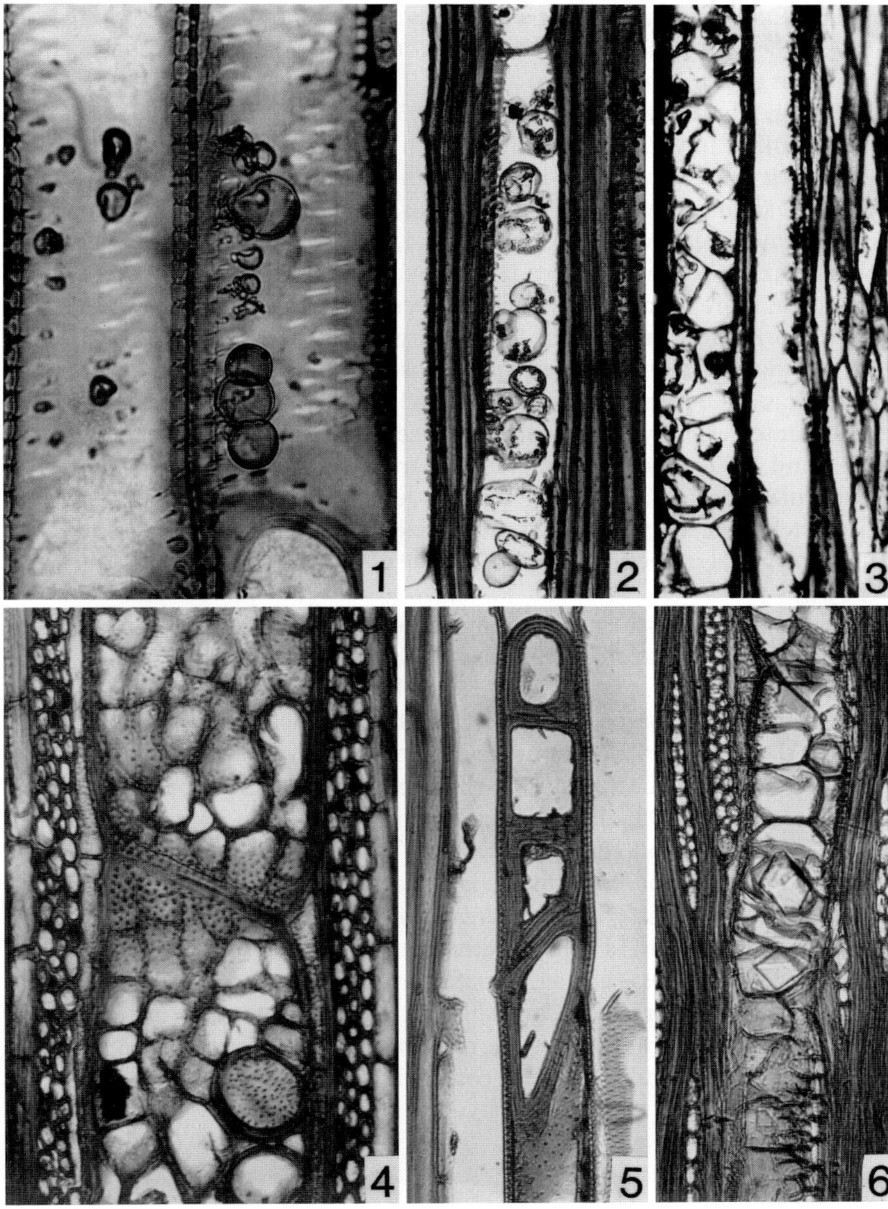

Fig. 3.23. Tyloses in vessels, from radial (*1, 2, 5*) and tangential (*3, 4, 6*) sections. *1* Small tyloses with dark-staining contents, *Stylobasium lineare* (Stylobasiaceae); *2, 3 Begonia parviflora* (Begoniaceae); *2* small tyloses, not yet touching. *3*, larger tyloses, showing forms resulting from mutual compression; *4* tyloses lightly sclerosed, *Scytopetalum klaineanum* (Scytopetalaceae); *5* prominently sclerosed tyloses, *Fitchia speciosa* (Asteraceae); *6* crystal-containing tyloses, *Astronium balansae* (Melastomataceae). (*1:C*; *2–6:B*)

would facilitate breakthrough of a parenchyma cell into the vessel), whereas in species with small vessel-parenchyma pits, deposition of gums may be an alternative form of achieving vessel occlusion.

Although Metcalfe and Chalk (1983; p. 203) and Saitoh et al. (1993) offer listings of families in which tyloses have been reported, there appears to be no phylogenetic pattern of distribution, and the reasons appear to be essentially physiological. In some species, e.g., *Quercus*, one can see tyloses in earlywood vessels but not in latewood vessels (Figs. 2.4, part 3; 2.8, part 4). This phenomenon appears to relate to the tendency of earlywood vessels, which are wide, to embolize more readily than latewood vessels, a phenomenon validated by the work of Ellmore and Ewers (1985), Hargrave et al. (1994), and Davis et al. (1999) Of those taxa with tyloses, most tyloses have thin primary walls. Origin of tyloses as ballooning of adjacent parenchyma cells into adjacent vessels is best illustrated when tyloses are small and separate from each other (Figs. 3.23, parts 1 and 2). The tannin contents of parenchyma cells, present in tyloses of particular species, also reveal this mode of origin well, as in *Cephalotus* (Carlquist 1981e). As tyloses increase in size, they become crowded and form shapes polygonal in sectional view by virtue of mutual compression (Fig. 3.23, part 3).

Even though presence or absence of tyloses is not a precise systematic character, ability to form sclerosed tyloses (walls thick and lignified) characteristically occurs in a few species (Fig. 3.23, parts 4 and 5). Sclerosed tyloses have been reported in Asteraceae (Carlquist 1957b; Carlquist and Grant 1963), Connaraceae (Dickison 1972), Lauraceae (Stern 1954), Myrtaceae (Foster 1967), and Scytopetalaceae (Carlquist 1987h). Numerous other cases could doubtless be cited (see Record 1925c). No special function has been claimed for sclerosed tyloses. Rather, the wall characteristics and contents of tyloses may simulate what happens in the parenchyma cells of particular taxa. This would explain the presence of starch in tyloses of *Pereskia aculeata* (Bailey 1962). Tyloses may contain crystals (Gottwald 1983), as illustrated here (Fig. 3.23, part 5).

All aspects of tyloses – pitting, sclerification, systematic distribution, causes, and whether tylosis presence correlates with other anatomical features (e.g., axial parenchyma or ray type) – are reviewed by Zürcher et al. (1985). There appears to be a moderate degree of correlation with ray types.

3.15 Trabeculae

The term "trabecula" (also "trabecula of Sanio") refers to a rod of secondary wall material crossing the lumen of a vessel or imperforate tracheary element. Trabeculae often occur in radial series of cells, suggesting a temporary cambial anomaly. Trabeculae have been discussed and illustrated in dicotyledon vessels by Butterfield and Meylan (1972b) and Meylan and Butterfield (1973, 1978a). For an illustration of trabeculae here, see Fig. 4.9, part 4.

4 Imperforate Tracheary Elements

4.1 Definitions and Terminology

4.1.1 Definition of Imperforate Tracheary Element

Imperforate tracheary elements can be defined clearly for the most part, but they do intergrade with other cell types in some cases. Narrow vessel elements can be said to intergrade with vascular tracheids, vasicentric tracheids, or true tracheids, but the presence of a perforation plate can be used to draw a distinction (a few vessel elements definable on the basis of morphology retain pit membranes in perforations, however). Nucleated libriform fibers can intergrade with axial parenchyma where fiber dimorphism occurs, although little difficulty in definition is encountered in these cases. Axial parenchyma cells that are not subdivided have sometimes been termed "substitute fibers" (IAWA Committee on Nomenclature 1964).

The fact that wood cells intergrade in some cases may be an annoyance to those in search of mutually exclusive definitions. However, the inherent interest of the fact that intergrades between cell types do occur is that wood cells have evolutionary flexibility. With intermediate cell types, or cell types with a great range of potential manifestations, a phylad can very easily make shifts in structural composition, shifts that can permit entry to new functions, and therefore possibly new habitats.

For convenience, an imperforate tracheary element can be defined as a cell with a secondary wall, derived from a fusiform cambial initial (in secondary xylem; derived from procambium in primary xylem) that neither has perforations (or a single perforation) nor is subdivided into a strand of cells each surrounded by a secondary wall. The last item in this definition permits strand parenchyma to be distinguished from septate fibers.

4.1.2 Types, Terminology

Bailey (1936) recognizes the concept of imperforate tracheary elements. Under this category, he distinguishes tracheids, fiber-tracheids, and libriform fibers. The IAWA Committee on Nomenclature (1964) also recognizes these categories and follows Bailey's definitions. I endorse these categories (Carlquist 1986a,b) although other definitions have been proposed (Baas 1986a; IAWA Committee 1989).

Some might think that libriform fibers should not be termed "tracheary elements" at all, because they feel "tracheary elements" connotes cells that conduct water, whereas libriform fibers are essentially nonconductive. The term "libriform fiber" is a convenient way of distinguishing fibers in the xylem from those outside the xylem (extraxylary fibers), and the term is also a convenient way of denoting that the tracheid has, phyletically, given rise to a wide range of cells, ranging from vessel elements to fibers with simple pits (libriform fibers). Fiber-tracheids are probably not active in conduction, judging from inferential evidence (Carlquist 1984a) and evidence from uptake of fluorescent dyes (Braun 1970).

Instead of the term "imperforate tracheary element," Metcalfe and Chalk (1950) used the term "fiber" and also modified it (e.g., "fiber with simple pits" to denote a libriform fiber). The Metcalfe and Chalk usage has not been widely followed, although their important work does give prominence to this usage. Their usage is one way of handling the fact that imperforate tracheary elements represent a continuum from tracheids, densely covered with fully bordered pits, through fiber-tracheids, to libriform fibers with simple pits. The lack of sharp breaks in this continuum is what has provided difficulty in defining types of imperforate tracheary elements.

Baas (1986a) suggests recognition of only two categories – fiber-tracheids and libriform fibers – and drawing the line between these at a pit cavity about 2 μm in diameter. One could defend such an arbitrary limit if other distinctions are not available or less good. In some Myoporaceae, pits on fiber-tracheids can have pit cavities (borders) a little less that 2 μm in diameter where they abut other fiber-tracheids, but about 3 μm in diameter on fiber-tracheid to ray contacts (Carlquist and Hoekman 1986b). Baas claims that because he has found vestigial pit borders in a few families of dicotyledons where they have not been reported before, truly simple pits are probably not common in imperforate tracheary elements of dicotyledons, and that therefore limiting the term "libriform fiber" to elements with simple pits is not realistic. However, my lists (Carlquist 1984a) show that libriform fibers with simple pits are reported in 95 families of dicotyledons, and parts of many others, based on the observations of Metcalfe and Chalk (1950), whose work is quite accurate in my experience. Even if there are some instances in which taxa thought to have simple pits on imperforate tracheary elements actually have vestigially bordered pits, the number of taxa with libriform fibers bearing simple pits is still larger. Therefore following the 1964 IAWA definitions seems entirely feasible.

At the other end of the spectrum, no distinction between what is termed "tracheid" here and fiber-tracheids was recommended by Reinders (1935) and Baas (1986a). These workers are forced to rectrict the term "tracheid" to secondary xylem of conifers, Gnetales, cycads, *Ginkgo*, various fossil plants, and vesselless angiosperms; Metcalfe and Chalk (1950) similarly restrict the definition of tracheid thus, although Cutler and Gregory (1998) recognize the term "tracheid" in numerous families of Saxifragaceae.

The irony of this disallowing the term "tracheid" in vessel-bearing dicotyledons is that all workers allow the terms "vascular tracheid" and "vasicentric tracheid". However, if cells *identical to vasicentric tracheids* occur in woods of *Rosa*, *Krameria*, or *Empetrum* (to mention only a few of many examples) as the ground mass of the wood, one must call them fiber-tracheids according to the Reinders-Baas concept. In that system *Ephedra* or *Quercus* can be said to have tracheids, but cells identical to those vasicentric tracheids that occur in the ground mass of *Rosa* or *Cornus* wood are termed fiber-tracheids in the Reinders–Baas scheme. *Quercus* is probably a phyletic derivative of Rosales in which libriform fibers have merely been added to the tracheids of *Rosa*-like ancestors – furthering the inconsistency in the use of the word "tracheid" by Baas.

Bailey's (1936) distinction between "tracheid" and "fiber-tracheid" was based on phylogeny (tracheids more primitive than fiber-tracheids), and tracheids as defined by Bailey for angiosperms are, in fact, identical to the cells termed tracheids in gymnosperms and vesselless angiosperms, so Bailey's definition does have the advantage of consistency. However, does it permit a distinction to be drawn between tracheids and fiber-tracheids? Curiously, a distinction unknown to Bailey (1936) is available: tracheids in the Baileyan definition are evidently conductive cells, fiber-tracheids (and libriform fibers) are not.

The nonconductive nature of fiber-tracheids and libriform fibers, in contrast to the conductive capability of tracheids (including vasicentric tracheids), is evident from the experimental work of Braun (1970) with fluorescent dyes. The libriform fibers of *Fraxinus*, when studied in sectioned material frozen and studied under SEM with frozen water intact, prove to be nonconductive (Utsumi et al. 1996); water is released into these fibers only when adjacent vessels cavitate, and water is present in the fibers when they are newly formed, but no conduction is occurring in the fibers themselves, because they are empty most of the year, as shown by the sensitive cryo-SEM techniques (Utsumi et al. 1996). The nonconductive nature of fiber-tracheids and libriform fibers is also clearly evident on the basis of a comparative study (Carlquist 1984a). If one defines tracheids as cells relatively densely covered with fully-bordered pits, one finds circumstantial evidence that tracheids are conductive, fiber-tracheids nonconductive or essentially so:

1. Presence of tracheids as the sole imperforate tracheary element type that depresses vessel grouping (Carlquist 1984a). Tracheids form a subsidiary conductive system that overrides vessel grouping in degree of effectiveness in maintaining integrity of conductive pathways.
2. Presence of vasicentric tracheids, where they are abundant, depresses vessel grouping (*Quercus*; Fig. 4.7, part 1).

3. In the type 5 growth ring (Figs. 2.5, part 3, 2.6, part 4), vessels are scarce or absent in latewood, indicating that tracheids are an effective conductive system; no such growth rings occur in woods with fiber-tracheids or libriform fibers (unless vasicentric tracheids are present).
4. In vesselless gymnosperms and vesselless angiosperms, there is no tendency toward production of fiber-tracheids in addition to or instead of tracheids; if fiber-tracheids were effective in conduction, one might expect some representation of fiber-tracheids in vesselless groups. Fiber-tracheids have been evolved in Gnetales – *Ephedra* and *Gnetum*, which have vessels and tracheids Carlquist (1996d).
5. An unexpectedly high proportion of woody dicotyledons in desert, Mediterranean-type, and alpine climates have tracheids (Carlquist and Hoekman 1985b); evidently these cells have been retained preferentially in plants of these regions because of their conductive safety (embolisms do not spread from one tracheid to another), and indirectly this shows they do conduct. If fiber-tracheids were similarly conductive, they should bulk large in such floras, but they are underrepresented.
6. In woods with tracheids, vessel elements do not shorten phyletically so markedly in woods that have tracheids as in woods that have fiber-tracheids or libriform fibers (Carlquist and Hoekman 1985b). If shortening of vessel elements is a way of achieving greater conductive safety, it is overridden if a more effective way of achieving conductive safety – tracheid presence – is available.
7. Woods with fiber-tracheids as their basic imperforate tracheary element type may have vasicentric tracheids in addition (e.g., Asclepiadaceae, Dipterocarpaceae, Grossulariaceae, Sapotaceae). If fiber-tracheids were effective in conduction, vasicentric tracheids would be superfluous in these woods.

Fiber-tracheids, therefore, are defined here as imperforate tracheary elements that are reduced below the pit density and/or pit border diameter characteristic of tracheids. While precise quantitative guidelines cannot be offered to the definition of tracheid followed here, workers in wood anatomy can quickly see, with some practice, the difference between the two categories. This is particularly evident because the pit diameter, density, and border presence of fiber-tracheids and that of tracheids do not form a real continuum, but show a marked drop (in terms of number of species with intermediate cells). Evolution has not removed intermediate types in this case, as in others, and some genera with cells seemingly intermediate between tracheids and fiber-tracheids may be found (e.g., *Ilex*; Baas 1973). I see value in use of such wordings as "tracheidlike fiber-tracheids" (Carlquist and Wilson 1995) or "fiber-tracheids with minutely bordered pits" or in specifying pit border diameter. There is no terminological solution that is entirely satisfying to all workers, and I have attempted to choose distinctions that demonstrate wood phylesis, diversity and function most accurately and therefore I do not follow the definitions of these cells by the IAWA Committee (1989).

Vasicentric tracheids are defined here as tracheids that occur adjacent to vessels in woods in which either fiber-tracheids or libriform fibers also are present. Vascular tracheids are considered to be tracheids formed as a few cell layers at the end of latewood, with narrow vessels transitional to them, in woods in which fiber-tracheids or libriform fibers are present. These definitions of vasicentric and vascular tracheids do not correspond exactly to the rather scanty definitions offered by the IAWA Committee on Nomenclature (1964), although they do correspond with the usages of Metcalfe and Chalk (1950, p. 1351). In my opinion, the nature of these cells was not understood well in the past, and a recent survey (Carlquist 1985a) has necessitated a revision, as mentioned below. If one wishes to contrast vasicentric tracheids or vascular tracheids with tracheids in those primitive woods that lack either fiber-tracheids or libriform fibers, one may use the term "true tracheid" for the latter. If no contrast needs to be demonstrated, the term tracheid by itself suffices.

4.2 Nature of the Bordered Pit

The functional effectiveness of the pit border is certified by the organographic distribution of the bordered pit in the plant body: the pattern of occurrence is surely not accidental, but represents a definite optimal construction. Understanding of the functional nature of the bordered pit is essential to the understanding of why tracheids (sensu Bailey 1936 and the above discussion) occur and have persisted in primitive phylads (or have been retained in groups with vascular and vasicentric tracheids). The bordered pit is an intricate counterbalancing of structural and physiological requirements (Fig. 4.1).

The bordered pit has a relatively great area devoted to pit membrane. This is indicative of adaptation for passage of relatively large volumes of water as befits a conductive cell. However, the solid portion of the pit membrane represents an obstacle to water conduction. The micropores in the pit membrane do conduct water, but their small size dictates high friction. A perforation – a pit in which a pit membrane is lost in maturation of the cell (vessel element) – offers much greater efficiency in conduction than the pit membrane with its micropores.

The value of retention of the pit membrane in the bordered pit lies in its ability to restrict passage of air bubbles. The minimal size of an air bubble in water is larger than the diameter of the micropores. Thus, if air bubbles form in cell A in Fig. 4.1, they cannot traverse the pit membrane into cell B and thus the water column of cell B remains intact.

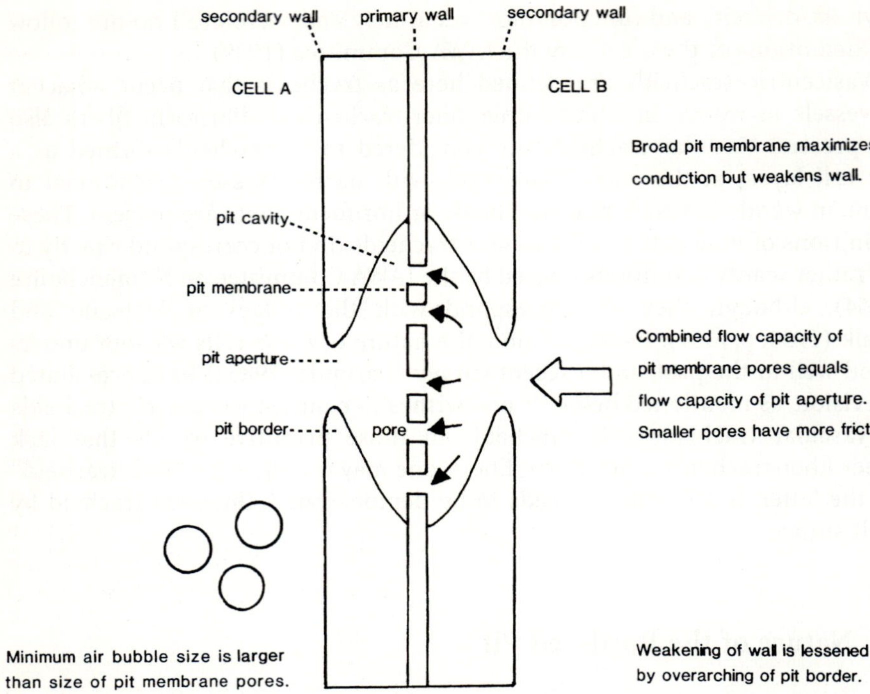

Fig. 4.1. Diagram showing the functional nature of the bordered pit. The bordered pit is shown to be a structure in which a compromise among maximal conductivity, maximal wall strength, and maximal conductive safety is reached. Further explanation in text

The broad pit membrane area of the bordered pit provides adequate passage of water. However, such a large pit membrane area runs counter to requirements for mechanical strength – the larger the area devoted to the pit membrane, the weaker the wall. The only effective way of countering this potential loss of wall strength is formation of a border, which combines a wide pit cavity with a narrow pit aperture.

Obviously the pit aperture exists to permit passage of water: how narrow should it be? The collective water-conducting capacities of the micropores of the pit membrane (small arrows in Fig. 4.1) equal the water-conducting capacity of the pit aperture.

4.3 Evolution from Tracheids to Libriform Fibers

Vesselless woods have densely placed, fully bordered pits on tracheids (Fig. 4.3, parts 3 and 4). With origin of vessel, tracheids are released from a function as the primary conductive system of the plant. Tracheids have been retained for reasons of conductive safety (Carlquist 1980a, 1985a,d; Carlquist and Hoekman 1985b; see also Chap. 11). If, however, a woody dicotyledon grows in an area where moisture availability is not a problem, the conductive safety of the tracheid is not of selective value and evolution of imperforate tracheary elements may be expected to lead to libriform fibers via fiber-tracheids (Fig. 4.2).

Mechanical strength is improved during this phylesis because the pit membrane is reduced in area (one can also state this same change in terms of reduction of a pit border). There is, in addition, a shift to a narrow slitlike pit aperture. Slitlike pit apertures represent openings in the wall parallel to the cellulose microfibrils that compose the wall. Because the microfibrillar structure is merely pushed apart slightly by the aperture and the microfibrils are not broken at this point, little wall strength is lost, giving fiber-tracheids and libriform fibers greater mechanical strength than tracheids (elliptical pit apertures may be seen even in tracheids; Figs. 4.3, part 3, 4.4, part 2, 4.5, part 3). Within a given family with basically primitive wood, the more arboreal genera might be expected to have shifted toward fiber-tracheids, whereas the less woody ones would be expected to retain tracheids. This appears to be true in Aquifoliaceae (Baas 1973), Icacinaceae (Bailey and Howard 1941b), Lardizabalaceae (Carlquist 1984f), and Trimeniaceae (Carlquist 1984e). Other examples could be cited.

Tracheids may be retained for reasons of conductive safety. Mechanical strength of tracheids can be enhanced by development of thick walls (Fig. 4.5, part 1), and even seemingly thick-walled tracheids are apparently still quite effective in conduction (the lumen can be rather narrow, because its conductive capacity need be only at most the sum of conductive capacities of the pit apertures on a cell).

Consequently, tracheids can be retained phyletically if conditions demand conductive safety. Such conditions include aridity (e.g., Casuarinaceae, Hydrophyllaceae, Krameriaceae, Rosaceae) or physiological drought due to freezing (e.g., Cornaceae, Empetraceae). Woody vines and lianas also appear to have retained tracheids preferentially – presumably tracheids can conduct should the large and vulnerable vessels of scandent plants be disabled temporarily or even permanently (Carlquist 1985d).

The selective value of phyletic retention of tracheids is also shown by type 5 growth rings (Chap. 2), as discussed above. The retention of tracheids, as well as their "reinvention" as illustrated by vascular tracheids, also illustrates the selective value of tracheid presence.

If tracheids have been retained in some phylads, they have been lost in many others, as the presence of fiber-tracheids and libriform fibers (see lists below) demonstrates.

A peculiar situation for which an explanation has recently been offered is the coexistence of tracheids and fiber-tracheids in certain families – notably Austrobaileyaceae (Fig. 4.5, parts 2–4), Ericaceae (subfamilies Arbutoideae and Vaccinoideae) and Celastraceae (Metcalfe and Chalk 1950; Carlquist 1988c). Development of septation in tracheary elements of these groups has

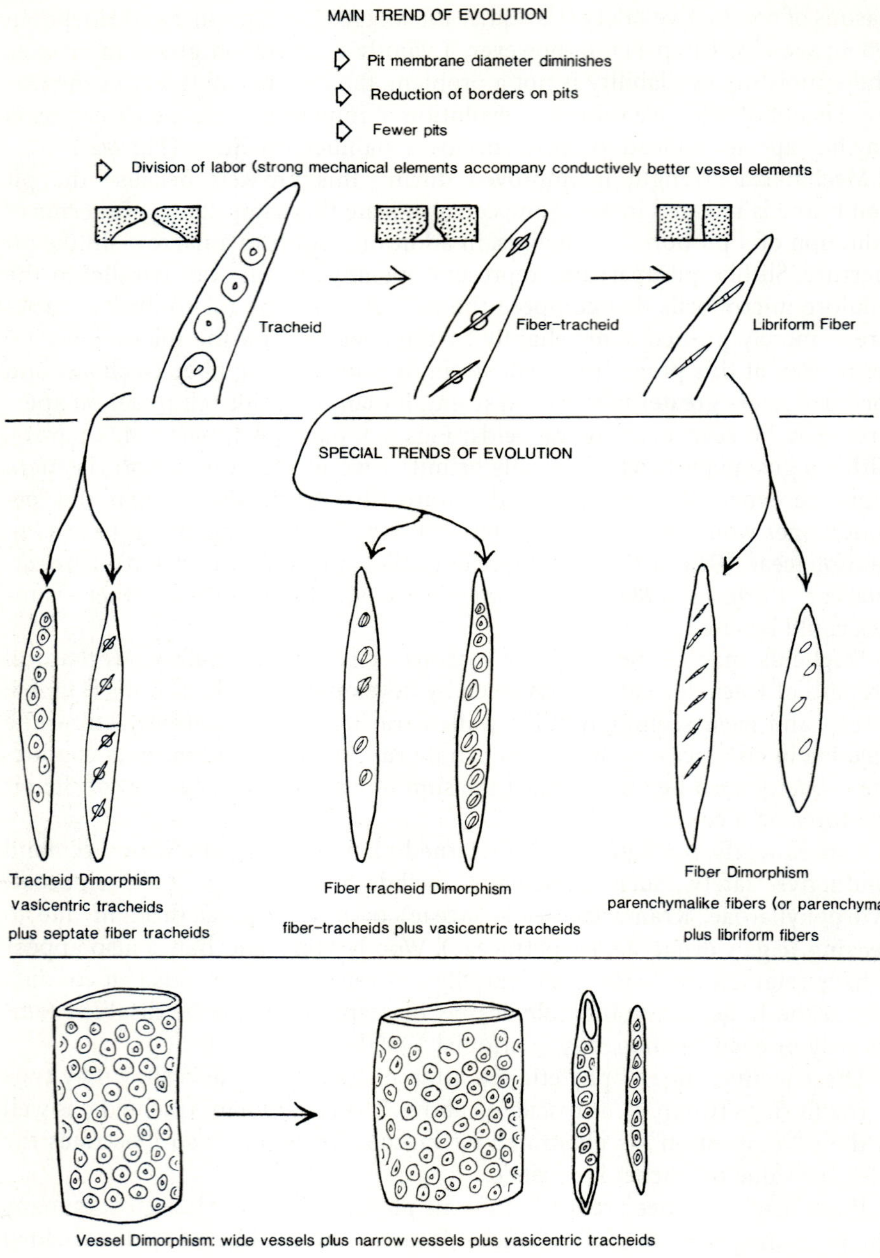

Fig. 4.2. Trends in evolution of imperforate tracheary elements. The trend that leads from tracheids to libriform fibers and represents a division of labor between conductively more efficient vessel elements and mechanically better elements (libriform fibers) is basic to this scheme. However, a number of other trends are hypothesized. Fiber dimorphism leads to formation of parenchyma bands in wood. Vessel dimorphism leads to fibriform vessel elements; an eventual outcome of this process is vasicentric tracheids. Tracheid dimorphism and fiber-tracheid dimorphism represent other ways in which vasicentric tracheids may be achieved

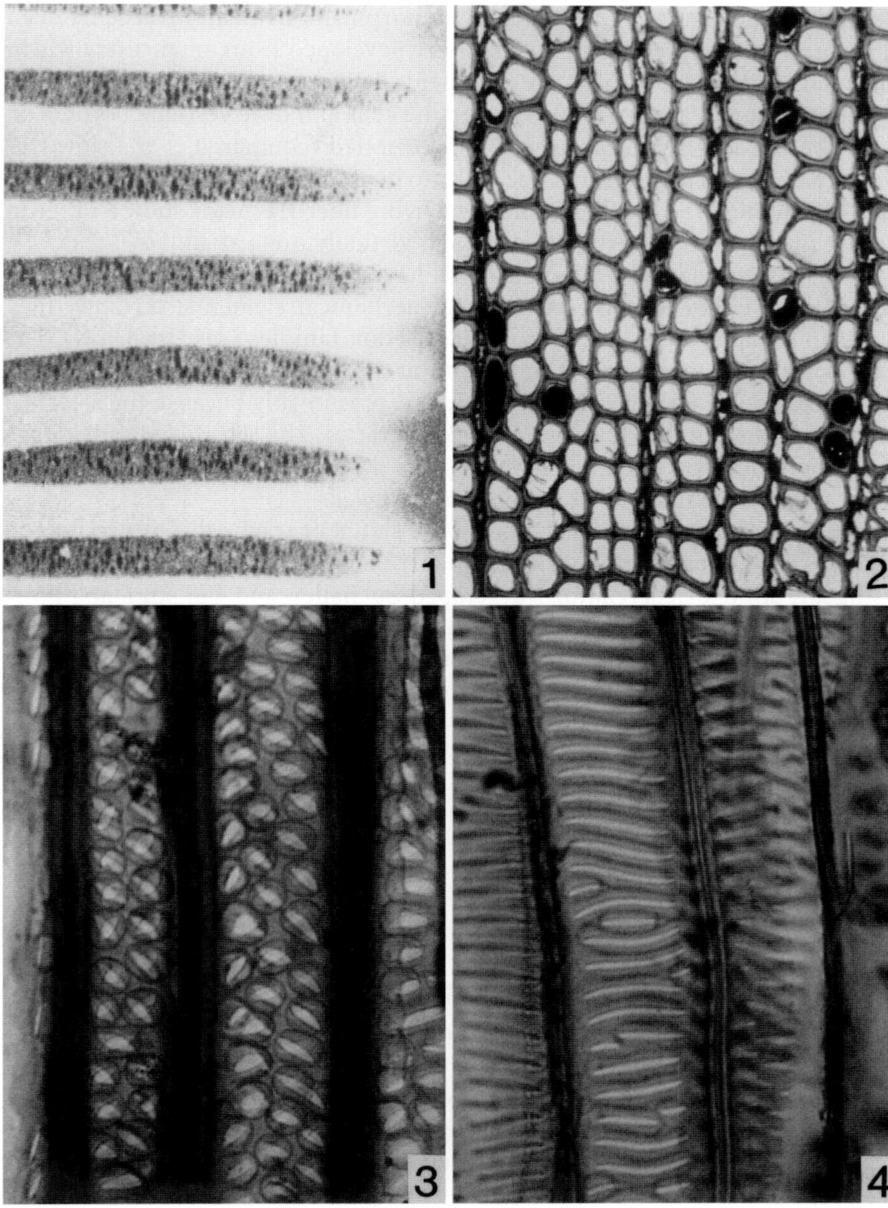

Fig. 4.3. Tracheids in vesselless dicotyledons, from radial sections (*1*, *3*, *4*) and transections (*2*). *1* Porose pit membranes in pits of end wall (overlap area) of tracheid, *Tetracentron sinense* (Tetracentraceae); *2* uniform appearance of tracheids in a vesselless dicotyledon from a relatively seasonless climate, *Tasmannia piperita* (Winteraceae); *3* multiseriate circular pits on an end wall of a tracheid, *Tasmannia piperita*; *4* scalariform pitting from the end walls of tracheids, *Tasmannia membranea*. (*1*:*G*; *2*:*B*; *3*,*4*:*C*)

permitted what one may call tracheid dimorphism. Evidently tracheids dead at maturity and septate fiber-tracheids are developed from an ancestry which included only tracheids. The fiber-tracheids, by virtue of having living protoplasts at maturity, are not water-conducting cells and have vestigial borders on pits – border loss has obviously been accelerated (compared to the slow tracheid to libriform phylesis that one tends to imagine as the predominant mode; Fig. 4.2). In the three families named, the living fiber-tracheids show a range in border presence from vestigial borders to virtually none at all.

One may wonder why fiber-tracheids have been retained at all if they are not optimal either as conductive or as mechanical elements; certainly they occur in an appreciable number of dicotyledon families (see list below). The explanation that appears most compelling at the moment is that fiber-tracheids are, in fact, more fiberlike than tracheidlike, presumably because the selective advantage of a bordered pit is no longer operative. The selective disadvantage of the presence of a small border on a pit of a fiber-tracheid is presumably so small that final extinction of the border is slowed.

There can be no doubt of the primitive status of tracheids with respect to the other cell types, and to the directionality of the trends suggested in Fig. 4.2. Using the kind of correlations developed by Bailey and his students, Frost (1930a,b, 1931), and Kribs (1935, 1937). Metcalfe and Chalk (1950, p. xlv) have correlated tracheid presence with perforation plate morphology: progressively fewer tracheids are present in woody dicotyledons with progressively more specialized perforation plates.

4.4 Tracheids of Vesselless Dicotyledons

4.4.1 Tracheid Length

Mean tracheid length of vesselless dicotyledons ranges from a minimum of 1,750 μm in *Tasmannia lanceolata* (Carlquist 1975a) to a maximum of 5,758 μm in *Exospermum stipitatum* (Carlquist 1982a) and 6,290 μm in *Belliolum gracile* (Carlquist 1983a). These dimensions confirm the earlier conclusion (Carlquist 1975a) that tracheid length is related to plant size in vesselless angiosperms and gymnosperms. My sample of *Exospermum stipitatum* was from a trunk 24 cm in diameter, whereas the trunk studied of *Tasmannia lanceolata*, a Tasmanian alpine shrub, was 3 cm in diameter.

Tracheid length increases with age (Bailey and Tupper 1918), and this may connote greater mechanical strength. In conifers there is an increase in tensile strength of tracheids from inside to outside of a stem, related to tracheid length increase and the concomitant more nearly vertical angle of cellulose microfibrils (Wardrop 1951; Wellwood 1962). There is a similar increase from inside to outside in proportion of cellulose in tracheids in conifers (Wardrop 1951; Hale and Clermont 1963), as well as an increase in crystallinity of cellulose (Lee 1961; Mark 1965). All of these factors seem to demonstrate that longer tracheids are stronger (more resistance to shear, greater tensile strength) in conifers (Mark 1965). There is no reason to believe that vesselless angiosperms are different.

4.4.2 Tracheid Diameter

Tracheid diameter parallels tracheid length in vesselless dicotyledons (Carlquist 1983a). This suggests that the greater potential strength of longer tracheids in vesselless dicotyledons is paralleled by greater conductive ability which increase in diameter would connote.

Narrower tracheid diameter may permit withstanding greater tension in water columns, judging from the data of Rundel and Stecker (1977). Winteraceae with notably narrow tracheids do grow in areas subject to frost (Carlquist 1988a, 1989c).

4.4.3 Tracheid Pitting

Pitting of tracheids in vesselless angiosperms can be considered under four categories:

1. *Tetracentron* (Fig. 4.3, part 1) and *Trochodendron* have scalariform pitting on earlywood tracheids, both on overlap areas and on lateral walls (Thompson and Bailey 1916; Bailey and Thompson 1918). The latewood tracheids tend to have only circular bordered pitting. There are sharply demarcated growth rings in both *Trochodendron* (Fig. 2.4, part 1) and *Tetracentron*.
2. In *Sarcandra* (Chloranthaceae) roots have vessels (Carlquist 1987k) although stems are vesselless. There is scalariform pitting on both overlap and lateral wall areas of *Sarcandra* stem tracheids (Bailey and Swamy 1950). Narrower tracheids, however, predominate and these tend to have pits that are circular or oval. Growth rings may occur, but these are not as sharply demarcated as are those of *Tetracentron* and *Trochodendron*.
3. In *Amborella* (Amborellaceae), tracheid-tracheid pitting was found to consist of circular bordered pits; Bailey (1957) found scalariform pits on some overlap areas. My sections (Carlquist 1975a) came from the periphery of a larger log, and proved to have more abundant scalariform pitting on overlap areas. Growth rings are not evident.
4. In Winteraceae, tracheids of some species have multiseriate (wider tracheids) or uniseriate (narrower tracheids) circular bordered pits on both overlap areas and lateral wall areas (*Drimys* exemplifies this condition

commonly). Scalariform pitting may be present to various extents on overlap areas (Bailey 1944a). I have found scalariform pitting most abundant on end walls of tracheids of *Zygogynum* (Carlquist 1981a), *Belliolum* (Carlquist 1983a), and *Bubbia* (Carlquist 1983b). A few radially elongate pits occur on overlap areas of *Exospermum* (Carlquist 1982a). Although pits have been thought to be circular and multiseriate where tracheids are wider in *Drimys, Pseudowintera*, and *Tasmannia* (Fig. 4.3, part 3), exceptions may be found. The three genera have scalariform pitting on metaxylem tracheids. Bailey and Thompson (1918) figured scalariform pitting on overlap ares of *Drimys winteri* tracheids, and scalariform pitting has also been illustrated for tracheids of *Drimys brasiliensis* (Carlquist 1988a). In *Tasmannia*, instances of scalariform pitting on end walls are rather more frequent (Fig. 4.3, part 4; Carlquist 1989c).

Scalariform pitting in overlap areas of tracheids may be demonstrated to increase with age in the genera *Bubbia* (Carlquist 1983b) and *Zygogynum* (Carlquist 1981a). This may relate to increased transpiration as a tree reaches canopy status (Carlquist 1983b).

The above detailed summaries are offered to clarify certain points. The contention of Bailey and Thompson (1918) that tracheids of vesselless angiosperms seem basically of a scalariform type rather than of a coniferous type seems eminently justified. Paedomorphosis does not seem basic to the pervasive scalariform patterns in vesselless angiosperms. The idea that woody angiosperms derive from an herbaceous stock in which metaxylem pitting has been perpetuated by juvenilism into secondary xylem (Young 1981) has been claimed to be unlikely (Carlquist 1983b). The cladistic results of the multiple gene analysis of Soltis et al. (1999) place a clearly woody dicotyledon (*Amborella*) as basic to all other angiosperms. The idea that the angiosperms are secondarily vesselless (Young 1981) has been rejected, at least in part (Soltis et al. 1999). The continuum between tracheids and vessel elements in primitive phylads (Fig. 3.5) is not widely appreciated.

The pervasive scalariform patterns of the vesselless groups seems evident. Attention is called to the discussion of how this figures in origin of vessels (Section 3.5). The occurrence of circular bordered pits in a multiseriate fashion, as in *Drimys*, may be a secondary phenomenon in Winteraceae, an adaptation to increasing wall strength without decreasing conductive capacity. In *Tetracentron* and *Trochodendron* mechanical strength is achieved by formation of radially narrow latewood tracheids, which bear circular pits only by virtue of their limited width; thus, in these two genera there has been no selection for development of the multiseriate bordered-pit pattern characteristic of *Drimys*.

The pit membranes of tracheids of Winteraceae exhibit micropores that suggest transition to vessels would require only a moderate degree of lysis or diminution of fibril presence to achieve perforations (Fig. 4.3, part 1; Meylan and Butterfield 1982; Carlquist 1983b), particularly when one takes into account the presence of pit membranes in perforations of vessels of primitive dicotyledons (Figs. 3.6, 3.7).

4.4.4 Internal Sculpture in Tracheids of Vesselless Dicotyledons

Helical sculpture occurs occasionally in tracheids of both species of *Pseudowintera* (Patel 1974; Meylan and Butterfield 1978a) and one cool temperate species of *Tasmannia* (Carlquist 1989c). A vestured layer can be seen on some tracheids of *Pseudowintera* (Meylan and Butterfield 1978a), but no such sculpture is present in tracheids of *Bubbia* (Carlquist 1983b). The significance of the vesturing inside winteraceous tracheids is evident from the distribution of this feature within *Drimys* and *Tasmannia* (Carlquist 1988a, 1989c). Vesturing is absent in tracheids in these genera in plants from localities not subject to frost (Fig. 4.4, parts 1, 3), but vesturing becomes progressively more evident in plants from cold localities subject to frost (Fig. 4.4, parts 2, 4, and 5).

4.4.5 Crassulae in Tracheids of Vesselless Dicotyledons

Bailey (1919) reports crassulae ("bars of Sanio," "Querleisten") on tracheids of *Trochodendron aralioides*.

4.5 True Tracheids in Vessel-Bearing Dicotyledons

4.5.1 Length

Mean tracheid length is greater than that of the vessel elements they accompany (e.g., Bailey and Tupper 1918; Swamy et al. 1960). If one computes a ratio between imperforate tracheary elements and the vessel elements that accompany them, one finds that woods with tracheids have a lower ratio than do woods with libriform fibers (Carlquist 1975a, p. 141, 1975c, 1976a,c, 1977a, 1978a; Mennega 1980). Although occasionally a ratio of lower than 1.00 is found (Carlquist 1978a), perhaps more numerous measurements would reveal a ratio of 1.00 or higher. Most often, the ratio for woods with tracheids lies between 1.00 and 1.30 (Carlquist 1982f).

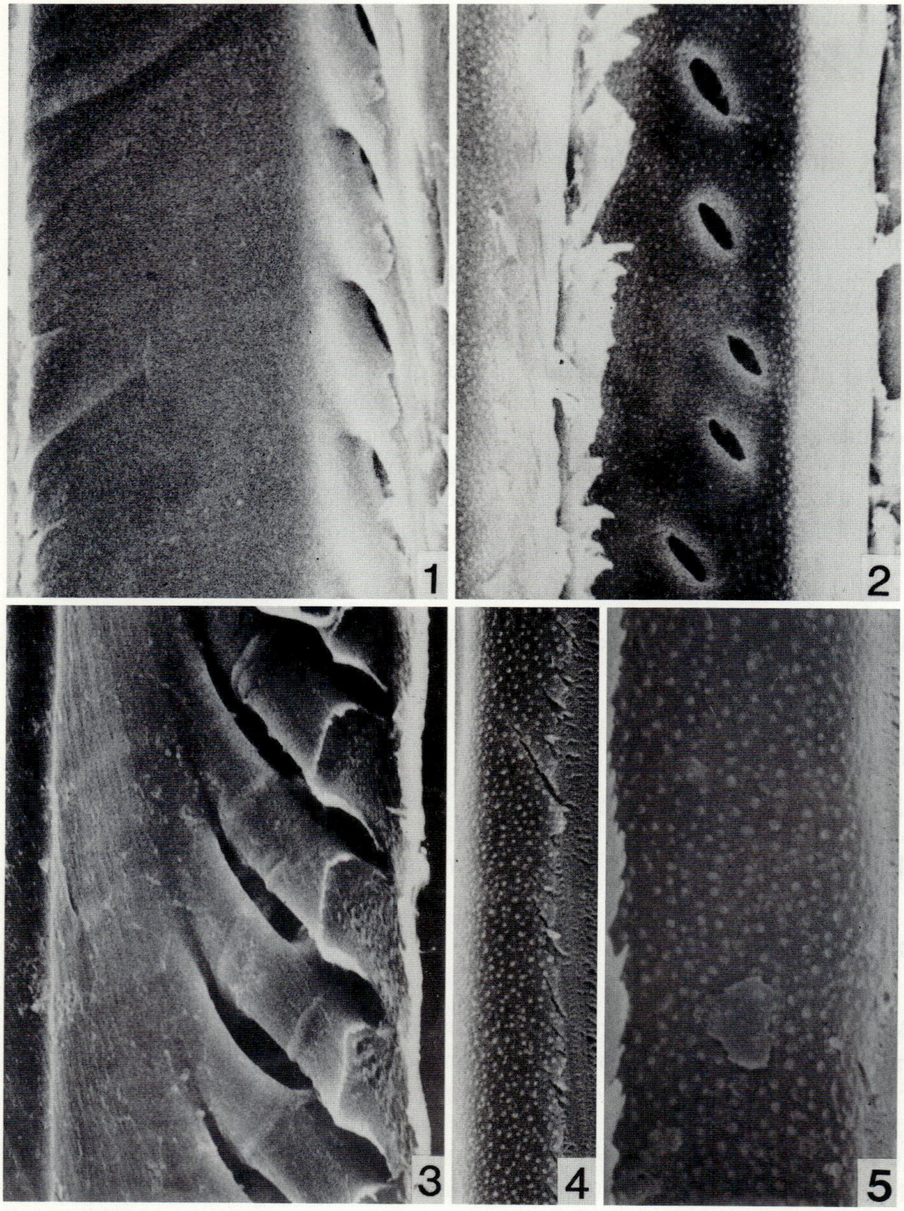

Fig. 4.4. Comparison, using SEM photomicrographs of wood sections, of internal surfaces of tracheids of *Drimys* (*1, 2*) and *Tasmannia* (*3–4*) of the Winteraceae. Species pairs from frost-free (*1, 3*) and colder (*2, 4, 5*) climates are represented. *1* Wall smooth or nearly so, *D. brasiliensis*; *2* wall warted (vestured), *D. granadensis*; *3* wall smooth, *T. piperita*; *4* wall warted, *T. lanceolata*; *5* wall warted, *T. xerophila*. (*1–4:G*; *5:I*)

4.5.2 End Walls

Tracheids in vessel-bearing angiosperms tend to show little or no difference between end walls and lateral walls, whereas tracheids in vesselless angiosperms do show some differentiation. Several factors may account for this. First, if elongation is appreciable, new pit fields would have to be created on the tips of elongating immature tracheids. However, this is likely not the major factor. Tracheids are relatively slender and fusiform compared to the tracheids in vesselless angiosperms. The narrowness of conifer tracheids is correlated with enhanced strength, despite the cause of failure under stress being apparently intercellular (Boatwright and Garrett 1983). Narrowness and the fusiform tip, which touches a number of neighboring cells (rather than overlapping essentially only the tracheid above or below as in vesselless angiosperms), are sufficient to account for lack of differentiation of an end wall. However, one should stress that there is no shortage of bordered pit area on the tips of tracheids in vessel-bearing angiosperms.

4.5.3 Pitting

Pitting of tracheids in vessel-bearing angiosperms typically consists of circular bordered pits in which the aperture is elliptical and diagonal (Fig. 4.5, part 1); the angle of the ellipse parallels the angle of the cellulose microfibrils and creates a maximal strength configuration. Although not an item used primarily in definition of tracheids, the pit aperture tends to be no longer (on its long axis) than the diameter of the pit border, and in many instances is shorter. Where the pit aperture appears to be appreciably longer than the outline of the pit cavity, there may be splitting related to gelatinous walls (e.g., Fig. 4.10, part 4). One should keep in mind that in some groups, gelatinous fibers are much more common than in others (e.g., Bruniaceae; Carlquist 1978a).

Tracheids may vary somewhat within a species. In *Gale palustris* (*Myrica gale*) of the Myricaceae, fiber-tracheids and tracheids coexist (Greguss 1959). In *Platanus racemosa*, latewood imperforate tracheary elements can be called tracheids because they are densely pitted; earlier in the growth ring, tracheary elements have about the same number of pits or somewhat fewer spread over cells much wider in diameter, and thus the earlywood elements can be called fiber-tracheids. The example of a wide range of imperforate tracheary elements cited by Ragonese (1976) in *Myrceugenella apiculata* seems likely merely a case of vasicentric tracheid presence. However, the discussion and review of imperforate tracheary element terminology by Ragonese are excellent.

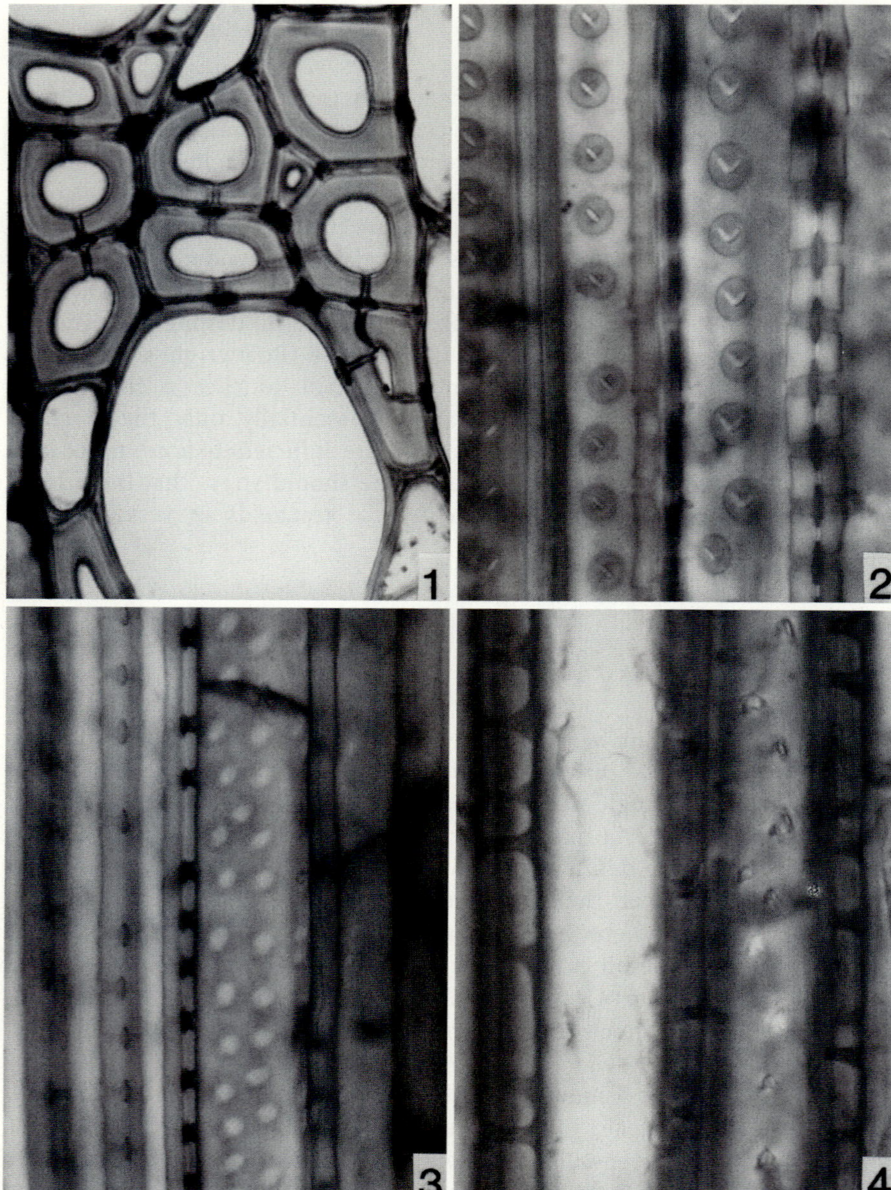

Fig. 4.5. Imperforate tracheary elements (and extraxylary fibers, *4*), illustrated so as to emphasize pitting. *1* Tracheids (vessel, *below center*), with bordered pits partially outlined by the dark-staining deposits, *Pentaphylax arborea* (Pentaphylacaceae). *2–4* Radial sections, showing pits in face and in sectional view, *Austrobaileya scandens* (Austrobaileyaceae); *2* tracheids from secondary xylem; *3* septate fiber-tracheids from secondary xylem; *4* fibers (extraxylary fibers) from inner margins of pith. (*1–4:C*)

4.5.4 Recording Tracheid Data

Obviously, tracheid length should be recorded, preferably as a mean. Diameter of tracheids in vessel-bearing dicotyledons can be given as a mean, and either the tangential or radial, or widest diameter can be chosen (one must specify how the measurement is made). Bordered pits on tracheids are sometimes more abundant on radial walls than on tangential walls, and whether or not this is true should be noted. The diameters of the pit cavity (pit border) and pit aperture should be given, together with information about the shape of the pit cavity as seen in face view and the shape of the pit aperture in face view.

Tracheid wall thickness is often recorded, but one must estimate this rather than present it as a series of precise measurements because the wall thickness varies so much within a single cell. Tracheid wall thickness may vary considerably between earlywood and latewood. Tracheid wall thickness may correlate with taxonomic groupings. In Icacinaceae (Bailey and Howard 1941b), tracheid wall thickness ranges from thin, as in *Platea*, to very thick, as in *Apodytes*, *Citronella*, *Emmotum*, and *Ottoschulzea*. Such a range may occur within a single genus, such as *Illicium*, in which *I. arborescens* has tracheids appreciably thicker walled than those of the other species. Some may wonder if such thick-walled tracheids can in fact be functional when the lumen is so narrow. If conductivity is equal to the fourth power of the diameter (Zimmermann 1983), one need only have a lumen twice the diameter of a pit aperture for it to conduct the water purveyed into that tracheid by 16 pits.

4.5.5 Helical Sculpture

Tracheids of vessel-bearing angiosperms may bear helical sculpture like that seen in vessel elements. In fact, in species in which helical sculpture occurs in vessels, it tends to occur in tracheids often. Tracheids bearing helical sculpture have been figured by Greguss (1959) in *Cistus salvifolius* (Cistaceae), *Helianthemum nummularium* (Cistaceae), *Myrtus communis* (Myrtaceae), *Philadelphus coronarius* (Saxifragaceae), and *Rosa canina* (Rosaceae). Helical sculpture has been shown photographically for *Philadelphus* (Styer and Stern 1979a) and *Sarcococca ruscifolia* (Carlquist 1982d). The listing by Metcalfe and Chalk (1983) for "spiral thickening in fibres" includes not merely true tracheids, but vasicentric tracheids and vascular tracheids. This is also true of Hess's (1946b) key to American woods with helices in "fibers." In all likelihood, because helices are associated with conductively active cells, the majority of the taxa cited by Hess represent tracheid occurrence.

4.5.6 Crassulae

Crassulae were reported for tracheids of *Berzelia abrotanoides* and *Staavia dodii* of the Bruniaceae (Carlquist 1978a).

4.5.7 Vesturing

Vestured pits are found in cells that may be presumed to be tracheids (they may possibly be vasicentric tracheids) of *Lophomyrtus* and *Neomyrtus* of the Myrtaceae and *Persoonia* of the Proteaceae (Meylan and Butterfield 1974). Vestured tracheid walls are found in a number of species of the New Zealand flora (Ohtani et al. 1983).

4.5.8 Systematic Distribution of Tracheids

The following list is a corrected version of the list given in Carlquist (1984a). Unlike some other lists in this book, the families listed below can be presumed to have tracheids throughout (or nearly so) unless an exception ("part," meaning tracheids in only some genera) is indicated. This list includes what is termed true tracheids, in distinction to vascular tracheids (Sect. 4.11) and vasicentric tracheids (Sect. 4.12).

Actinidiaceae (including *Saurauia*)
Aextoxicaceae
Alseuosmiaceae
Apocynaceae
Barbeuiaceae
Betulaceae (part)
Bruniaceae
Buxaceae
Byblidaceae
Canellaceae
Caprifoliaceae (excluding Adoxaceae)
Caryophyllaceae (part)
Celastraceae (part)
Cephalotaceae
Chrysobalanaceae
Clethraceae
Clusiaceae (part)
Columelliaceae
Convolvulaceae
Cornaceae
Crossosomataceae
Daphniphyllaceae
Desfontaineaceae
Diclidantheraceae
Didymeleaceae
Dilleniaceae
Dioncophyllaceae
Dipsacaceae
Droseraceae (*Drosophyllum*)
Elatinaceae
Elaeagnaceae
Empetraceae
Epacridaceae
Ericaceae (some)
Escalloniaceae
Eucommiaceae
Eucryphiaceae
Eupteleaceae
Garryaceae
Globulariaceae
Gomortegaceae
Gonostylaceae
Goodeniaceae
Goupiaceae
Grubbiaceae
Hamamelidaceae (part)
Humiriaceae
Hydrangeaceae
Hydrophyllaceae (*Eriodictyon*, *Wigandia*)
Icacinaceae (genera with scalariform

perforation plates plus *Cantleya*)
Illiciaceae
Juglandacecae (part)
Koeberliniaceae
Lardizabalaceae
Lecythidaceae (*Allantoma*)
Linaceae (except *Lepidobotrys*)
Loganiaceae (*Gelsemium*)
Magnoliaceae (part)
Medusagynaceae
Menispermaceae
Monimiaceae
Myricaceae (except *Gale*)
Myrothamnaceae
Myrtaceae (part)
Nepenthaceae
Olacaceae (part)
Oleaceae (part)
Oncothecaceae
Opiliaceae (except *Lepionurus*)
Paracryphiaceae
Penaeaceae
Pentaphylacaceae (Fig. 4.5, part 1)
Phellinaceae
Polemoniaceae (*Eriastrum*, *Leptodactylon*, *Phlox*)
Polygalaceae (part)
Proteaceae (part)
Quiinaceae
Rhizophoraceae
Rosaceae (except Prunoideae)
Rubiaceae
Sabiaceae (*Sabia*)
Sarcolaenaceae
Sarraceniaceae
Schisandraceae
Simmondsiaceae
Solanaceae (a few genera)
Sphaerosepalaceae
Sphenostemonaceae
Stachyuraceae
Staphyleaceae (part)
Stegnospermataceae
Strasburgeriaceae
Symplocaceae
Theaceae (part)
Trigoniaceae
Tropaeolaceae

Record (1944e) offered a key to American woods in which one finds "fibers with conspicuous bordered pits." Probably the majority of these represent tracheid presence.

4.6 Fiber-Tracheids

4.6.1 Nature of Fiber-Tracheids; Problems in Observation

Little needs to be said about fiber-tracheids as a separate category, because they are transitional between tracheids and libriform fibers and thereby have no characters restricted to them. One may find various degrees of border presence on pits of fiber-tracheids within a single genus, such as *Ilex* (Baas 1973) or *Stackhousia* (Carlquist 1987f). One expects species with fiber-tracheids to have a higher imperforate tracheary element/vessel element length ratio than do species with tracheids, and this appears to be true. Baas (1973) gives the ratio for *Ilex* as typically between 1.4 and 1.8, whereas the ratio for tracheid-bearing species falls mostly between 1.0 and 1.3 (Carlquist 1978a).

As mentioned earlier, fiber-tracheids may bear bordered pits like those of tracheids, merely smaller and/or less densely placed, as in *Duckeodendron* (Fig. 4.6, part 1) or *Stackhousia*. Pits are more common on radial walls (Fig. 4.6, part 1) than on tangential walls (Fig. 4.6, part 2) of fiber-tracheids rather often, judging from a mention of this phenomenon for numerous families in

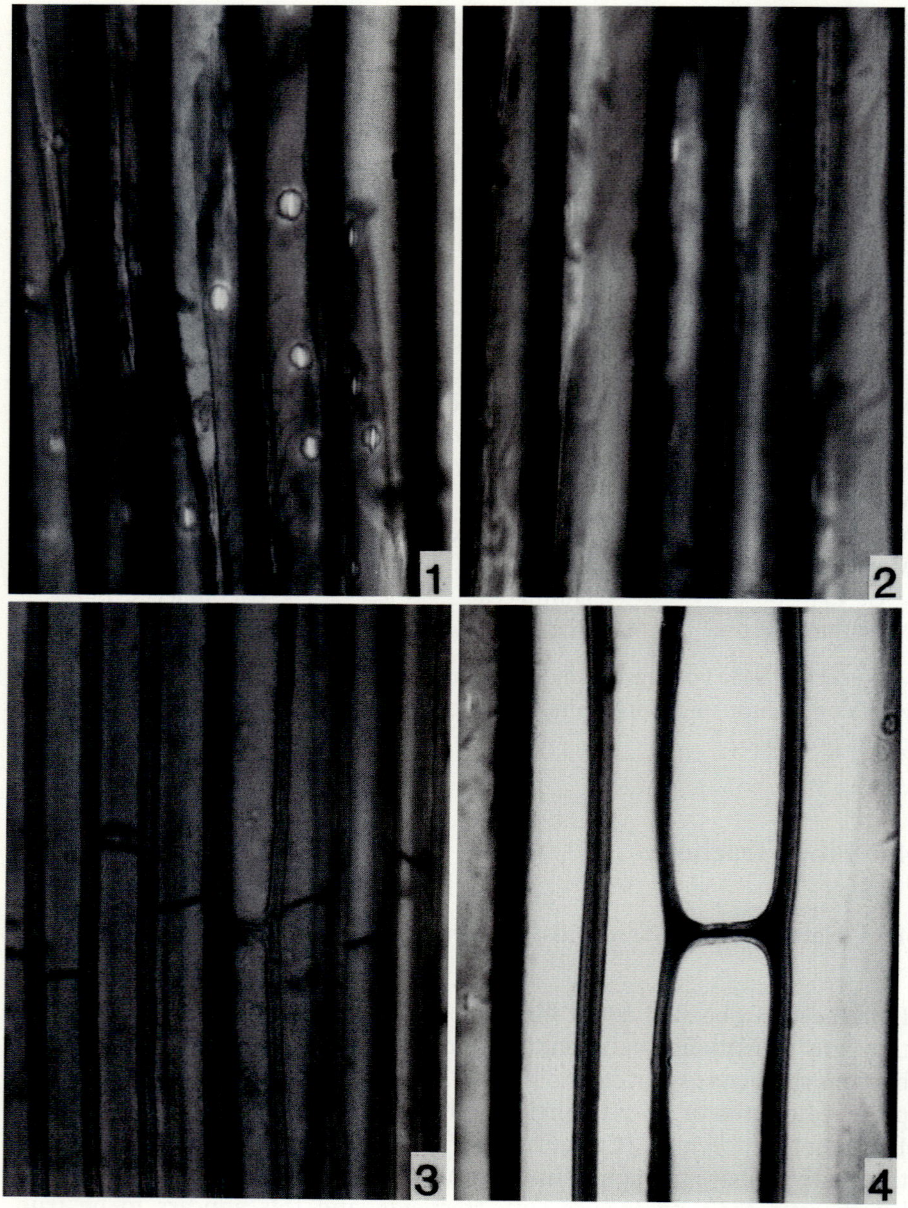

Fig. 4.6. Fiber-tracheids (*1, 2*), septate fibers (*3*), and a pair of strand parenchyma cells (*4*). *1, 2* *Duckeodendron cestroides* (Duckeodendraceae). The radial section (*1*) shows that pits are more abundant on radial walls than on tangential walls (*2*, a tangential section); *3* septate fibers from radial section, showing a thin primary wall, characteristic of septate fibers, in each libriform fiber, *Beloperone ornata* (Acanthaceae); *4* strand parenchyma, showing that each of the pair of cells has a secondary wall, and thus may be differentiated from a septate fiber, *Castanopsis chrysophylla* (Fagaceae). (*1–4:C*)

Metcalfe and Chalk (1950). The largest number of instances of fiber-tracheid presence appears to be represented by species with vestigial and therefore not easily observed borders on pits. This circumstance is troubling to beginning students of wood anatomy, who find borders difficult to locate on pits of tracheary elements, even in some cases where borders are wide.

The problem of pit border detection deserves special comment, for this feature is signifcant in comparative work and cannot be avoided. Scanning electron microscopy aids this task very little, for prolonged observation of sectioned material is necessary to determine vestigial border presence, and this may not be a reasonable expenditure of time and facilities if a number of species must be surveyed. In light microscopy, pit borders can be confused with refractive pit cavities, which produce ringlike appearances that can simulate borders. Interference optics or phase contrast may only exaggerate this artifact.

The most reliable solutions to determining border presence often lie in staining technique. A stain that darkens the pit membrane or otherwise provides a contrast with the secondary wall is helpful. Effective counterstaining is therefore recommended. Hematoxylin as a counterstain to safranin has most commonly been used because it produces a purple-black on primary walls that contrasts with the rose color of safranin-stained secondary walls. Other stain combinations can be used, such as safranin and fast green. Stains that darken pit membranes can sometimes stain a small ring of primary wall adjacent to the pit membrane, suggesting the existence of a border where one may not be present.

Another approach that one can take involves looking at pits in sectional view instead of face view. Thus, if most pits occur on radial walls of a fiber-tracheid, one would look at a tangential section to view pits in sectional view. Sectional views of fiber-tracheid pits in Fig. 4.5 (part 3) as well as the sectional views of tracheid pits in Fig. 4.5 (part 2) illustrate the effectiveness of this approach. Presence of amorphous deposits in imperforate tracheary elements, such as the "gums," "resinlike compounds," etc., that commonly occur in many woods can serve to advantage when they fill pit cavities and thereby outline them in sectional view. Sometimes in wood section preparations air is trapped within a few imperforate tracheary elements. This troubling inclusion can serve usefully, to outline pit cavities at times (see fiber-tracheid of *Batis maritima*; Carlquist 1978b).

4.6.2 Recording Fiber-Tracheid Data

Fiber-tracheid length should be recorded, preferably as a mean. Pit border diameter, pit aperture dimensions and shape, and pit density (specifying radial and tangential walls where appropriate) may be pertinent items. In most instances the pit aperture will be very narrow and slitlike, so conformations other than this are especially worthy of note. Data on wall thickness and diameter of fiber-tracheids should be noted, and comments above on tracheids (Sect. 4.4.4) are applicable to fiber-tracheids.

4.6.3 Septate Fiber-Tracheids

Septate fiber-tracheids are shown here for *Austrobaileya* (Fig. 4.5, part 3). Septate fiber-tracheids have been recorded for *Hypericum* (Vestal and Vestal 1940) *Ribes* (Carlquist 1988c), and, in Lardizabalaceae, *Boquila* and *Lardizabala* (Carlquist 1984f). Fahn and Leshem (1963) report nucleated fiber-tracheids for *Gymnocarpus fruticosus*. Panshin and de Zeeuw (1980, p. 183) claim that fiber-tracheids can grade into axial parenchyma (strand parenchyma), but I have not observed this, and such cases must be very infrequent. The contrast between a septum and cell walls separating a strand of cells, shown in Fig. 4.6 (parts 2 and 3) is a reliable indicator in my opinion. This is discussed further below in connection with septate fibers (septate libriform fibers).

4.6.4 Systematic Occurrence of Fiber-Tracheids

The list of families below represents a corrected version of the list offered in Carlquist (1984a). Unlike some other lists in the present book, the list below represents families in which all or most taxa can be said to have fiber-tracheids unless otherwise indicated ("part" meaning some genera have fiber-tracheids, others have another type of imperforate tracheary element).

Aceraceae (part)
Aitoniaceae
Akaniaceae
Ancistrocladaceae
Annonaceae
Apocynaceae (part)
Aquifoliaceae
Aristolochiaceae (part)
Aclepiadaceae (part)
Austrobaileyaceae (tracheids present in wood also in addition to fiber-tracheids)
Balanitaceae
Balanopaceae
Bataceae
Betulaceae (part)
Bixaceae
Boraginaceae (part)
Brassicaceae
Brunelliaceae
Calycanthaceae
Capparaceae (part)
Caryocaraceae (part)
Caryophyllaceae (part)
Casuarinaceae
Celastraceae (part)
Cercidiphyllaceae
Chloanthaceae
Chloranthaceae (part)
Clusiaceae (part)
Cochlospermaceae
Combretaceae
Convolvulaceae (part)
Corylaceae (part)
Cunoniaceae
Cyrillaceae
Degeneriaceae
Dichapetalaceae
Dioncophyllaceae
Dipterocarpaceae
Duckeodendraceae
Ebenaceae

Elaeocarpaceae (part)
Ericaceae (part)
Erythroxylaceae
Eupomatiaceae
Flacourtiaceae (part)
Fouquieriaceae
Gesneriaceae (*Coronanthera*)
Grossulariaceae
Gyrocarpaceae
Hamamelidaceae (part)
Himantandraceae
Icacinaceae (part)
Juglandaceae (part)
Lardizabalaceae (part)
Lauraceae (part)
Lecythidaceae (part)
Linaceae (part)
Loasaceae
Loganiaceae (part)
Loranthaceae (part)
Lythraceae
Magnoliaceae (part)
Malvaceae (part)
Melastomataceae (part)
Meliaceae (part)
Monimiaceae (part)
Montiniaceae
Moraceae (part)
Myoporaceae
Myricaceae (*Gale* only)
Myrsinaceae (part)
Myrtaceae (part)
Nolanaceae (part)
Nyssaceae
Ochnaceae (part)
Olacaceae (part)
Oleaceae (part)
Opiliaceae (part)
Passifloraceae
Pentaphragmataceae
Penthoraceae
Phytolaccaceae
Pittosporaceae (part)
Platanaceae
Polemoniaceae (part)
Polygalaceae (*Securidaca*)
Proteaceae (part)
Ranunculaceae
Retziaceae
Rubiaceae
Sabiaceae (except *Sabia*)
Santalaceae
Sapotaceae
Scrophulariaceae (part)
Scytopetalaceae (part)
Simaroubaceae (part)
Solanaceae (part)
Stackhousiaceae
Staphyleaceae
Stilbaceae (part)
Stylidiaceae
Styracaceae
Theaceae (part)
Thymeleaceae
Trimeniaceae
Zygophyllaceae

4.7 Libriform Fibers

4.7.1 Length of Libriform Fibers and Its Function

Libriform fiber length parallels vessel element length in any given species (Bailey and Tupper 1918; Chattaway 1936; Swamy et al. 1960; Carlquist 1962a; van den Oever et al. 1981). The figures of Bailey and Tupper (1918) show that some families have notably greater degree of fiber intrusiveness: that is to say, in some species libriform fibers are much longer than the vessel elements they accompany – three or four times as long. Families with such exceptional elongation of libriform fibers include Bignoniaceae, Bombacaceae, Boraginaceae, Moraceae, Proteaceae, Santalaceae, Scrophulariaceae, Solanaceae, Sterculiaceae, Ulmaceae, Urticaceae, and Verbenaceae. The reader familiar with systematic groupings may have noted that most of these belong to Malvales and Scrophulariales. From the data of Bailey and Tupper (1918), one can extract some principles, such as that length of libriform fibers is not proportional to plant height and size as is tracheid length in vesselless angiosperms (Carlquist 1975a). To be sure, libriform fibers of tall trees do not tend to fall in short size classes, but they do not tend to fall in the exceptionally long size classes either.

Obviously greater degree of intrusiveness of libriform fibers (during their elongation phase) has been evolved in particular groups, but the reasons are not entirely apparent. Some of the plants with notably long libriform fibers are not notably large plants. For example, *Boehmeria rugulosa*, which has libriform fibers 1,400 μm in length, could be regarded as as a woody herb. There is a correlation between length and strength in tracheids of conifers and vessel elements, but in those tracheids, greater length also runs parallel to greater conductive ability (greater end wall areas). Longer libriform fibers may have greater strength, but if so, it is only one of several mechanisms that can lead to enhanced mechanical strength in vessel-bearing dicotyledons, and no simple correlation can be expected. If one looks at particular groups, one can note some correlation between libriform fiber length and plant size. Van den Oever et al. (1981) found the longest imperforate element length not in lowland trees but in scrubby trees of uplands. However, correlation between libriform fiber length and habit does occur in *Sophora* (Cumbie and Mertz 1962). If length of tracheary elements increases over time in a plant, as appears typical of woody plants (Bailey and Tupper 1918), plants with larger stems ought in general to have longer libriform fibers than related species with smaller trunks.

Gibson (1973) claimed that libriform fiber length is positively correlated with plant height in cereoid cacti. Shorter cacti may lack fibers altogether (Fig. 4.9, parts 1 and 2). Greater length of libriform fibers characterizes higher levels in rosette trees (Carlquist 1970b,c), and this may also have mechanical significance. In some groups of dicotyledonous herbs, as in globular cacti (Mauseth et al. 1995), libriform fibers may be partly or wholly replaced by parenchyma: *Brighamia* and *Carica* (Carlquist 1962a, 1998d) can be cited in this connection, but Crassulaceae offer many other examples. Turgor pressure can substitute for libriform fiber presence as a means of mechanical support in stem succulents (Caricaceae have large phloem fiber strands). Obviously succulence and turgor pressure will not support a large plant body, but it can suffice for shrubby growth forms. Water storage is, in such plants, a major factor in reduction of quantity of libriform fibers, and libriform fibers become progressively more important with arborescence in a phylad of succulent plants such as cacti or succulent *Euphorbia* species.

Raylessness can also be regarded as a means of achieving a rapid substitution of mechanical tissue for ray tissue. The substitution is not of long duration in some plants (e.g., *Artemisia*), and rays eventually develop in some groups (in others, raylessness continues for the duration of the stem). The fact that raylessness is lost ontogenetically in such plants as *Artemisia* shows that the value of radially-oriented parenchyma tissue increases as a stem increases in diameter, and that ray parenchyma is evidently not of great selective value compared to libriform fibers in some small stems. *Stylidium* is perhaps not truly an example of raylessness, because the vascular cambium is not fascicular in origin; mechanical tissue is produced internally by a special cambium. However, this action, and the topographically rayless woody tissues that result have been interpreted as an adaptation for enhanced mechanical strength (Carlquist 1981f).

4.7.2 Pitting of Libriform Fibers

Libriform fibers are notable for sparsity and small size of pits as compared to other tracheary elements. Typically, pit apertures in libriform fibers are narrow or slitlike ("lenticular" of Meylan and Butterfield 1978a). However, pit apertures may be circular, especially in thin-walled fibers such as those of *Centaurodendron* (Carlquist 1965b). In this instance, one can correlate wall thickness with pit shape. The elliptical shape of pits in libriform fibers represents an opening along the predominantly parallel microfibril orientation. The fact that this opening is parallel to the microfibril orientation reduces loss of wall strength maximally, and the small size of the pit also prevents loss of wall strength. The pit size required for entry of materials necessary for wall formation is evidently much smaller than the pit size required for water conduction, as in tracheids.

4.7.3 Libriform Fibers in Transection: Wall Thickness, Diameter, Abundance

Wall thickness of libriform fibers is very characteristic of some taxa. In *Ochroma* (balsa), libriform fibers have very thin walls, whereas in *Diospyros* (ebony, persimmon), the fibers are so thick walled that the lumen is almost replaced by wall material.

Fiber diameter is reported to be narrower in stems than in roots (Patel 1965); this is of considerable interest, because narrower fibers tend to be stronger than wide ones (Boatwright and Garrett 1983). A correlation between both fiber length and fiber diameter with vessel element length within a species may be expected (Giraud 1980).

The proportion of a wood that is composed of fibers has been recorded by some authors (Akachuku and Burley 1979; G.E. French in Panshin and de Zeeuw 1980). From the data of French, we find that fibers may compose a minimum of 26.6%, a maximum of 68.0% of a wood, depending on the species. We have as yet no idea what the significance of relative fiber abundance is, since the relative proportion of vessels, rays, and axial parenchyma must simultaneously be affected.

4.7.4 Recording of Data on Libriform Fibers

Data on libriform fiber length must be obtained from macerations; only very short libriform fibers can be measured in longitudinal sections, and even in these, small portions of the tips may be cut away. Use of means is recommended.

Libriform fiber diameter is difficult to measure because it changes along the length of a fiber. One can measure the diameter of a fiber most easily when one sees it in transection, but one cannot be sure that one is viewing the widest point of any given fiber, and fibers do vary in diameter, so measuring a fiber of wider diameter in a transection may be measuring of an exceptionally wide fiber, not a representative one. One must therefore choose fibers subjectively rather than randomly when one is attempting to measure fiber diameter. Nevertheless, modally species do tend to have somewhat different fiber diameters, although this feature is not used much systematically.

Wall thickness must likewise be measured subjectively, because it varies along the length of a libriform fiber and the wall thickness of a fiber as seen in a given transection may show greater thickness in angles. Selecting a typical fiber in a section and finding its characteristic wall thickness may be better than making a large number of measurements, some of which may be unrepresentative. If fiber wall thickness differs with respect to position in a growth ring, that fact should be recorded.

Pit size and shape can be recorded; because density of pits on libriform fibers varies widely, pit density may be worth noting (e.g., if pits are exceptionally sparse and hard to find on libriform fibers or relatively abundant).

4.7.5 Helical Sculpture in Libriform Fibers

Helical sculpture is very scarce in libriform fibers. One should expect this, because if helical sculpture is related to the conductive process in some way (Sect. 3.12), it would be of no functional value in libriform fibers, which are essentially nonconductive. Weakly defined helical manifestations in libriform fibers might be related to gelatinous walls, which in turn may indicate the formation of reaction wood. The listing below includes septate as well as nonseptate fibers.

4.7.6 Systematic Distribution of Libriform Fibers

The listing below represents a corrected version of several listings for libriform fiber-bearing families (Carlquist 1984a). The listing below is unlike some other lists in the present book in that the families (or parts of families listed) all characteristically have libriform fibers, not single genera. Both septate and nonseptate fibers are included. For minor exceptions to the below list, see Baas (1986a).

Acanthaceae
Aceraceae (part)
Adoxaceae (*Sambucus*)
Agdestidaceae
Aizoaceae
Akaniaceae
Alangiaceae
Amaranthaceae
Anacardiaceae (including Julianiaceae)
Apiaceae
Araliaceae
Asteraceae
Avicenniaceae
Begoniaceae
Berberidaceae
Bignoniaceae
Bombacaceae
Boraginaceae (part)
Bretschneideraceae
Burseraceae
Cactaceae
Calyceraceae
Campanulaceae (sensu lato)
Cannabaceae
Capparaceae
Caryocaraceae
Caryophyllaceae (part)
Celastraceae (part)
Chenopodiaceae
Chloranthaceae (*Hedyosmum*)
Cneoraceae
Combretaceae (part)
Connaraceae
Coriariaceae
Corynocarpaceae
Crassulaceae
Datiscaceae
Didiereaceae
Elaeocarpaceae
Ericaceae (part)
Euphorbiaceae
Fabaceae
Fagaceae (*Nothofagus*)
Flacourtiaceae (part)
Frankeniaceae
Gentianaceae
Geraniaceae
Gesneriaceae (except *Coronanthera*)
Globulariaceae
Greyiaceae
Halophytaceae
Hernandiaceae
Hippocastanaceae
Lacistemaceae
Lactoridacceae
Lamiaceae
Lauraceae (part)
Lecythidaceae (part)
Leitneriaceae
Loganiaceae (part)
Loranthaceae (part)
Malesherbiaceae
Malpighiaceae
Malvaceae (part)
Marcgraviaceae
Meliaceae (part)
Melianthaceae
Misodendraceae
Moraceae (part)
Moringaceae
Myristicaceae
Myrsinaceae
Nolanaceae (part)
Nyctaginaceae
Ochnaceae (part)
Octoknemataceae
Olacaceae (part)
Oliniaceae
Onagraceae
Oxalidaceae
Papaveraceae
Passifloraceae (part)
Pedaliaceae
Phytolaccaceae (part)
Piperaceae
Pittosporaceae (part)
Plantaginaceae
Plumbaginaceae
Portulacaceae
Primulaceae
Punicaceae
Ranunculaceae (part)
Resedaceae
Rhamnaceae
Rutaceae
Sabiaceae (except *Sabia*)
Salicaceae
Salvadoraceae
Sapindaceae
Scrophulariaceae
Scyphostegiaceae
Scytopetalaceae (part)
Simaroubaceae
Siphonodontaceae
Solanaceae (part)
Sonneratiaceae
Sterculiaceae
Stilbaceae (part)

Stylobasiaceae	Tiliaceae	Verbenaceae
Surianiaceae	Tovariaceae	Violaceae (part)
Tamaricaceae	Tremandraceae	Viscaceae
Theophrastaceae	Turneraceae	Vitaceae
Thymeleaceae (part)	Valerianaceae	Vochysiaceae

4.8 Septate Fibers and Living Fibers

4.8.1 Definition and Description

Septate fibers are libriform fibers in which one or more transverse primary walls occur (Fig. 4.6, part 3). The term is occasionally applied to fiber-tracheids bearing septa (Fig. 4.5, part 3). The secondary wall of the libriform fiber is not involved in the septum because the septum is formed only after the formation of the secondary wall is complete (Parameswaran and Liese 1969). Libriform fibers are with few exceptions easily differentiated from strands of axial parenchyma, in which a fusiform cell is subdivided into a strand of cells before secondary wall formation has ceased, so that each of the strand cells is surrounded by its own secondary wall (Fig. 4.6, part 4).

The presence of one or more septa in a libriform fiber signifies prolonged longevity, and one may reasonably assume that these cells are alive in functioning wood. One can demonstrate presence of nuclei if wood samples are preserved in a pickling fluid rather than dried. There can be other kinds of evidence of prolonged longevity, such as presence of starch. Libriform fibers in which septa may or may not be present but in which nuclei or starch are present can be termed "nucleated fibers" or "living fibers."

All or virtually all of the libriform fibers in a species known to have libriform fibers may be septate. However, only a small portion of fibers in a wood may be septate in some groups, as in Stilbaceae and Retziaceae (Carlquist 1986b). In *Cyrtandra* of the Gesneriaceae, fibers are septate adjacent to vessels but nonseptate farther away (Carlquist and Hoekman 1986a). In this genus, the septate fibers are longer than the nonseptate fibers. Septate fibers that are shorter than accompanying nonseptate fibers have been found in *Tectona* (Parameswaran 1964) and *Elaeodendron* (Ghosh and Purkayastha 1960). Instances of this sort may in part be referable to the phenomenon of fiber dimorphism (see below).

4.8.2 Functional Significance of Septate and Living Fibers

Septate fibers have repeatedly been shown to contain starch, at least in some species and at some seasons (Frison 1948; Harrar 1946; Wolkinger 1969).

Therefore the function of septate fibers may be regarded as like that of axial parenchyma. Species with septate fibers might be expected, therefore, not to have abundant axial parenchyma, and that expectation is often realized (Harrar 1946), although Frison (1948) objects to that generalization. The fact that septate fibers may surround vessels is cited by Spackman and Swamy (1949) as evidence of parenchymalike function.

Wolkinger (1969, 1970a,b, 1971) has studied various aspects of living (nucleated) fibers. He found that they are typically greater in length if nonseptate than if septate, as reported by Govidarajalu and Swamy (1955) in *Pithecellobium dulce* (although Chowdhury et al. 1956 did not find this in other species of *Pithecellobium*). Wolkinger (1970a) thought of living fibers as a storage system, not related to the hydrosystem. However, the living fibers of *Acer saccharum* are chiefly paratracheal (Gregory 1978), whereas dead fibers are more distal from vessels. A connection with conductive proceses has been found (Sauter et al. 1973). Starch in living fibers is hydrolyzed in *Acer saccharum* just prior to initiation of growth in the spring; the resultant sugars are translocated into vessels and, by increasing the osmotic pressure of water in the vessels, renew the flow of water in the conductive system. Fahn and Leshem (1963) found that shrubs and subshrubs of the eastern Mediterranean region have nucleated fibers to a markedly high degree, and these cells are usually viable throughout the sapwood.

In Onagraceae, many of the taxa lack living fibers but possess interxylary phloem, which is a photosynthate-conducting channel. However, the two woodiest genera, *Fuchsia* and *Hauya*, lack interxylary phloem but have starch-rich living fibers (Carlquist 1975b). This pattern suggests that living fibers represent an alternative photosynthate storage and conduction system in taxa where growth and flowering events are constant, rather than in sudden flushes.

In some dicotyledon phylads, living fibers have evolved so as to be relatively parenchymalike. In *Pachycormus* (Gibson 1981), living fibers have starch and are relatively short (mean = 435 μm); the living fibers have thin, nonlignified walls (maximum thickness = 2.5 μm) with sparse pits. In such taxa, living fibers might represent a means of water storage as well as starch storage.

In *Aeschynomene* (Fabaceae), living fibers are only a little longer than vessel elements (Butterfield 1972). The fibers and axial parenchyma cells are very similar in length, and the latter are mostly nonseptate; the living fibers are slightly thicker walled. The living fibers make a very close approach to parenchyma cell morphology (Nair and Mohan Ram 1992).

One must not think that development of parenchymalike characteristics in libriform fibers represents the only way in which a wood ground tissue functioning like that of parenchyma can be achieved. A process in which libriform fibers are apparently replaced by axial parenchyma cells is especially common in succulents (Fig. 4.9, parts 1 and 2) such as *Brighamia* and *Carica* (Carlquist 1962a, 1998d); some cereoid cacti (Gibson 1973), and some Crassulaceae can be cited.

4.8.3 Systematic Distribution of Living and Septate Fibers

The following list represents a revision of the lists offered by Wolkinger (1970b, 1971); Wolkinger reported living fibers in 71 families, 224 genera, and 413 species of dicotyledons, although he did not report the genera and species in print. The list of families offered by Wolkinger is augmented from Harrar's (1946) listing of families and genera with septate fibers and from Metcalfe and Chalk's (1950) listing of septate fibers, and these are designated "(s)" in the listing below (families common to both lists are not specially indicated). Not all of the taxa in the families listed have septate or living fibers. Families listed in italics in Metcalfe and Chalk (1950), implying septate families are a rare occurrence in those families, have not been included here. References are cited for recent reports representing first instances of living or septate fibers in a family.

Acanthaceae (s)
Aceraceae
Adoxaceae (*Sambucus*)
Akaniaceae (s)
Alseuosmiaceae (Dickison 1986)
Anacardiaceae (s)
Apiaceae
Apocynaceae (s)
Araliaceae (s)
Asteraceae
Begoniaceae
Berberidaceae
Bignoniaceae (s)
Bixaceae (s)
Bombacacecae (four genera: s)
Boraginaceae
Brassicaceae
Bretschneideraceae (s)
Brunelliaceae (s)
Buddlejaceae (Mennega 1980)
Burseraceae (s)
Cactaceae (s)
Capparaceae (*Morisonia*; s)
Celastraceae (s)
Chenopodiaceae
Clusiaceae (s)
Combretaceae (s)
Connaraceae (s)
Coriariaceae
Cornaceae (*Marlea*; s)
Cunoniaceae (*Berlangera*; s)
Elaeocarpaceae (s)
Ericaceae (including *Vaccinium*; s)
Euphorbiaceae (s)
Eupomatiaceae (s)
Fabaceae (s)
Flacourtiaceae (s)
Geraniaceae
Gesneriaceae (s)
Grossulariaceae (s)
Hippocrateaceae (s)
Hydrangeacceae (s)
Lacistemaceae (s)
Loasaceae (*Plakothira*; Carlquist 1987d)
Loganiaceae (s)
Lythraceae (s)
Malvaceae
Melastomataceae (s)
Meliaceae (s)
Melianthaceae
Monimiaceae (s)
Moraceae (s)
Moringaceae
Myristicaceae (s)
Myrsinaceae (s)
Myrtaceae (two genera)
Oleaceae (s)
Oliniaceae (s)
Onagraceae (s)
Oxalidaceae (s)
Papaveraceae (*Dendromecon*; s)
Passifloraceae (three genera; s)
Phytolaccaceae (*Rivina*; s)
Piperaceae (s)
Pittosporaceae
Plumbaginaceae
Polemoniaceae
Polygonaceae (s)
Punicaceae (s)
Ranunculaceae
Retziaceae (Carlquist 1986c; s)
Rhizophoraceae (s)

Rosaceae (*Holodiscus*, *Rubus*, *Spiraea*; s)
Rubiaceae
Rutaceae (three genera; s)
Sabiaceae (*Meliosma*; s)
Sapindaceae (s)
Scrophulariaceae (s)
Simaroubaceae (s)
Solanaceae (s)
Sonneratiaceae (s)
Stilbaceae (Carlquist 1986c; s)
Tamaricaceae
Thymeleaceae
Turneraceae (s)
Ulmaceae
Urticaceae (s)
Verbenaceae (s)
Violaceae (s)
Vitaceae (s)
Vochysiaceae (s)
Zygophyllaceae (s)

Record (1944c) has provided a key to American genera of woody dicotyledons with septate fibers.

4.9 Fiber Dimorphism

The concept of fiber dimorphism was originated in a study of wood anatomy of the tribe Heliantheae of Asteraceae (Carlquist 1958a). In the genus *Dubautia*, species such as *D. plantaginea* show virtually no differences among libriform fibers within a given wood section (Carlquist 1998e). In other species, such as *D. laxa*, there are bands of libriform fibers that have greater diameter but shorter length than libriform fibers elsewhere in the wood. In a few species, such as *D. menziesii*, the bands of what appear to be fibers are sharply demarcated, and the fibers are short, wide, thinner-walled, ocasionally subdivided horizontally, and often nucleated, whereas the fibers elsewhere in the wood conform to typical libriform fiber characteristics. These bands must now be described as parenchyma rather than fibers. Because the species of *Dubautia* show degrees of intermediacy between monomorphic fibers and the condition just mentioned for *D. menziesii* (Carlquist 1998e), the interpretation seems clear, and was the basis for a phylogenetic diagram (Carlquist 1961a, p. 50) like a portion of the more comprehensive scheme shown here as Fig. 4.2. The family Asteraceae is rich in examples of fiber dimorphism, and other species in which photographic illustrations of this phenomenon are offered include *Hesperomannia arborescens* (Carlquist 1957a), *Fitchia speciosa* (Carlquist 1957b), *Ambrosia hispida* (Carlquist 1958b), *Baccharis angustifolia* and *B. rosmarinifolia* (Carlquist 1960a), *Stephanomeria guadalupensis* (Carlquist 1960b), *Robinsonia evenia* (Carlquist 1962b), *Vernonia bockiana* (Carlquist 1964a), *Artemisia kauaiensis* (Carlquist 1966b), *Argyroxiphium* (Carlquist 1997c), and *Wilkesia* (Carlquist 1998f); the two last-named genera are very close to *Dubautia*. Other cases in which fiber dimorphism has been demonstrated include Lythraceae (Baas and Zweypfenning 1979), Melastomataceae (van Vliet 1981), Moringaceae (Mark Olson, unpublished data), Pedaliaceae (Carlquist 1987b), and Urticaceae (Bonsen and ter Welle 1984).

Hess (1946c) lists genera with "special fibers in parenchyma-like arrangement." Hess states that these fibers "are usually septate but wider than ordinary fibers and thinner walled, and that they range from very distinct to indistinct and poorly demarcated from the ordinary fibers." Inclusion of families such as Lythraceae and Melastomataceae in his list suggests that Record had at least some instances of fiber dimorphism in mind. Other families in his list include Capparaceae (*Capparis*), Clusiaceae (*Marila*), Connaraceae (*Connarus, Rourea*), Fabaceae (*Poeppigia*), Hippocrateaceae (see Mennega 1997), Polygonaceae (*Triplaris*), Sapindaceae (*Allophyllus*), and Simaroubaceae (*Alvaradoa*). Some of these exemplify fiber dimorphism, but reinvestigation of others is desirable.

There are instances in which septate fibers of two size classes have been found. This has been reported for *Capparis* and *Physocalymma* (Record 1944c), *Allophyllus, Paranephelium*, and *Triplaris* (Purkayastha et al. 1959) and *Averrhoa* (Ghosh and Purkayastha 1960).

Instances in which there has been phyletic replacement of libriform fibers by parenchyma – "parenchymatization" – are problematic with respect to fiber dimorphism. For example, different parts of the same plant of *Maihuenia* have different proportions of fibers to parenchyma in the wood.

Likewise, the secondary xylem of *Aeschynomene* (Butterfield 1972) could be considered a case of parenchymatization because the proportion of the wood devoted to wide, short living fibers (like the parenchyma cells in the wide axial parenchyma bands of *Erythrina*) is very great, with cells resembling ordinary libriform fibers restricted to areas near vessels (Nair and Mohan Ram 1992).

4.10 Fiber Distribution Anomalies

In *Misodendron* subgenus *Misodendron* (Carlquist 1985c), libriform fibers are short and sclereidlike, and appear in both rays and fascicular areas. This could be considered a kind of raylessness. One species of this subgenus, *M. gayanum*, has fibers in ray areas but fascicular areas consist only of parenchyma plus vessels, thereby reversing the expected pattern of fiber occurrence. Fiberlike ray cells convey mechanical strength, and occurrence in ray areas is understandable.

In *Misodendron* subgenus *Angelopogon* (Fig. 2.13, parts 3 and 4), fibers occur in fascicular areas, but in each fascicular area these can occur as pairs of distinctive strands in latewood; the fascicular area is otherwise composed of vessels and parenchyma. Are the strands of fibers equivalent to libriform fibers? Apparently so. *Misodendron linearifolium* has the distinctive thick-walled fibers in latewood areas, but in less discrete strands that thereby look like more normal zones of fascicular xylem. Also, in *Loranthus europaeus* (Greguss 1959), the pattern of secondary xylem (large grouped earlywood vessels mixed with parenchyma; latewood vessels mixed with fibers and parenchyma) is close to that of *Misodendron* subgenus *Angelopogon*, but not markedly different from a normal wood pattern. Misodendraceae are close to Loranthaceae, and this kind of distribution of fibers may be one line of evidence in that regard.

4.11 Vascular Tracheids

4.11.1 Definition, Comparison to Vasicentric Tracheids

Because the topics of vascular and vasicentric tracheids had not been investigated in detail, definitions of these cell types have necessarily tended to be rather vague. Panshin and de Zeeuw (1980) contrast vascular tracheids with vasicentric tracheids in that the former form vertical series. The IAWA Committee on Nomenclature (1964) defines a vascular tracheid as "an imperforate cell resembling in form and position a small vessel member," whereas a vasicentric tracheid is "a short, irregularly formed tracheid in the immediate proximity of a vessel and not forming part of a definite axial row." The "irregular form" probably derives from *Quercus*, in which enlargement of vessels deforms adjacent vasicentric tracheids, but that should not be used to define vasicentric tracheids as a whole. Although these are not clear guides to the phenomena one encounters, one can, after viewing a number of examples, see what may be the intent behind these definitions. In the usage of most authors, and in my definition, a vascular tracheid is an imperforate cell like a reduced vessel element, formed at the end of a growth ring and lacking a perforation plate. Tracheids are absent earlier in the growth ring of a species that has vascular tracheids, and earlier formed portions of the growth ring contain, as tracheary elements, only vessel elements and either fiber-tracheids (in the sense used in this book) or libriform fibers and vessel elements. Vascular tracheids are thus like the last-formed vessels in a growth ring, and could be regarded as so narrow that they lack perforation plates. Vasicentric tracheids, by contrast, may occur in earlier portions of a growth ring, not just in the terminal cell layers of a growth ring, and are formed in contact with vessel elements. This distinction proves to have strong systematic and functional significance (Carlquist 1985a), and I believe that this is really what experienced workers in wood anatomy have intended. As a confirmation of this, my listing of vasicentric tracheids (Carlquist 1985a) includes all of those families included by Hess (1946a) and by Metcalfe and Chalk (1950, 1983), although my list adds more families. The additional taxa typify the conditions shown by the families in the Metcalfe and Chalk listing for vasicentric tracheids, whereas families with vascular tracheids in my definition do not occur in that listing.

4.11.2 Functional Aspects of Vascular Tracheids

The function of vascular tracheids is to provide the last-formed and safest conductive tissue within a growth ring. Because vascular tracheids are imperforate, like true tracheids, they have maximal protection against spread of embolisms from one cell to another and thus are the conductive cells least likely to cavitate. If dicotyledons with vascular tracheids are in regions with water stress conditions at the end of a growing season (Carlquist and Hoekman 1985b, Carlquist 1987j), most of the vessels in a growth ring may be disabled. Thus, the viable portion at the end of a growing season may be little more than the vascular tracheids. The vascular tracheids would represent a way in which the vascular cambium can be preserved by retention of intact water columns adjacent to the cambium at the end of a growing season. Thus, a stem could survive the dry season, even though the leaves, supplied by the three-dimensional network of vessels earlier in the growth ring, would not survive and foliage would be lost as the vessels cavitated. Vascular tracheids would be expected in drought-deciduous shrubs for this reason.

4.11.3 Systematic and Ecological Distribution of Vascular Tracheids

A listing of the families in which vascular tracheids have been observed is probably not an informative device, because ecology and habit of plant are of greater significance than systematic position of a genus in this connection. Moreover, because the term vascular tracheid has been applied to quite different situations by various authors, wood anatomists should clearly specify exactly what they mean by the term "vascular tracheid".

Shrubby Asteraceae from the southwestern USA and other dry regions offer a number of examples. One may find vascular tracheids in Mutisieae (*Flotovia*, *Hecastocleis*, *Proustia*, *Trixis*; Carlquist 1957a), Heliantheae (*Argyroxiphium*, *Borrichia*, *Dubautia*, *Encelia*; Calquist 1958a,b), Helenieae (*Chrysactinia*; Carlquist 1959), Astereae (*Baccharis* spp., *Chrysothamnus* spp., *Gutierrezia microcephala*, *Haplopappus* spp., *Olearia* spp.; Carlquist 1960a), Cichorieae (*Dendroseris regia*, *Stephanomeria blairii*; Carlquist 1960b), Senecioneae (*Lepidospartum* spp., *Tetradymia* spp.; Carlquist 1962b), Cynareae (*Centaurea ragusina*; Carlquist 1965b), Anthemideae (*Artemisia arbuscula*; Carlquist 1966b), and Ambrosieae (*Ambrosia dumosa*; Carlquist 1966b). Vascular tracheids have not been reported in Inuleae (Carlquist 1961b; *Inula viscosa* has vasicentric tracheids), Vernonieae (Carlquist 1964a), or Eupatorieae (Carlquist 1965a). These three tribes are more nearly herbaceous (Inuleae) or typically grow in more humid tropical localities (Vernonieae), circumstances under which vascular tracheids tend not to occur.

Some species in the Californian flora that have vascular tracheids (Carlquist and Hoekman 1985b; Carlquist 1987j) have soft drought-deciduous leaves, such as *Artemisia californica*, *Keckiella ternata* (Fig. 2.7, parts 3 and 4; Michener 1981) or *Lotus scoparius*. Other taxa have leaves that tend to absciss prematurely as the dry season advances, so that a few, often smaller, leaves remain during the dry season: *Fremontodendron*, *Isomeris* (Fig. 2.8, part 1), *Sambucus*, and *Toxicodendron*. Shrubs with vascular tracheids tend to have frequent innovations from the base in comparison with the evergreen shrubs that have vasicentric tracheids.

Some woody vines with deciduous leaves have vascular tracheids. Notable among these are *Parthenocissus* and *Vitis* of the Vitaceae (Carlquist 1985d). One should not forget that the narrow vessels formed in latewood formed just prior to vascular tracheids are very nearly as resistant to embolism formation as the vascular tracheids: narrower vessels embolize less easily than wider vessels (Ellmore and Ewers 1985). Thus a graded series of embolism resistance is available in the cells within each growth ring. In discussions of functions of vascular tracheids and vasicentric tracheids, attention should be paid to narrow vessels because they are very nearly as effective as tracheids in promoting conductive safety.

4.12 Vasicentric Tracheids

4.12.1 Definitions, Types, Origins

As mentioned above (Sect. 4.10.1), vasicentric tracheids may be defined simply as tracheids intermixed with and adjacent to vessels in a wood that also has libriform fibers or fiber-tracheids as an imperforate tracheary element type. Woods in which vasicentric tracheids occur occasionally have what appear to be vascular tracheids as well. One of the modes of origin of vasicentric tracheids appears to be formation of vascular tracheids not just in the last few cell layers of a growth ring, but earlier – perhaps in the latter half of the growth ring, where vessels are inevitably intermixed with some of these tracheids (Fig. 4.2). In this case, I recommend that the term vasicentric tracheid be used, because functionally the important criterion of the vasicentric tracheid – adjacence to a vessel – is being satisfied. Only a few cases fall into this category. Asteraceae, Cactaceae, and Lamiaceae are large families in which this type of phyletic pathway has resulted in development of vasicentric tracheids.

Another series of families involves woods in which vasicentric tracheids have likely arisen by means of what is termed tracheid dimorphism (Fig. 4.2). Woods in this category contain tracheids plus living (sometimes septate) fiber-tracheids or libriform fibers; these instances are further distinguished by the fact that the families in which they occur include, in addition to such woods, only woods with tracheids, so that it is difficult to envision an evolutionary stage, now completely vanished, in which fiber-tracheids were present (Carlquist 1988c). Instead, the most logical hypothesis is that in a primitive phylad with tracheids, some tracheids stayed nucleated at maturity while others lost protoplasts as in normal tracheids; in the nucleated tracheids, borders were lost in pits rapidly because the cells were nonconductive (bordered pits are, in fact, present in the septate "fibers" of *Arctostaphylos* and *Rubus*). If, in such woods in this category, tracheids are adjacent to vessels, I recommend use of the term "vasicentric tracheids." The fact that not all tracheids are in contact with vessels in these woods that have undergone tracheid dimorphism should not disqualify use of the term vasicentric tracheid. In the best-known instance of vasicentric tracheid occurrence (which probably was phyletically a case of tracheid dimorphism), *Quercus*, each vessel is sheathed by many layers of vasicentric tracheids (Fig. 4.7, part 1). Another instance of vasicentric tracheid occurrence probably resulting from tracheid dimorphism is represented here by *Arctostaphylos* (Fig. 4.7, parts 3 and 4). In the species illustrated, each earlywood vessel is surrounded by tracheids, whereas fiber-tracheids occur more distally to the vessels (resinlike deposits in pit cavities of this specimen permit one to determine whether cells have fully bordered or vestigially bordered pits). The example of *Arctostaphylos* satisfies the definition of vasicentric tracheid as used by various authors perfectly, so instances of tracheid dimorphism cannot be excluded from the concept of vasicentric tracheids because a different pathway is involved. Some species of *Prunus* have only tracheids in the wood ground mass, some have vasicentric tracheids plus libriform fibers; wood of the latter species likely exemplifies tracheid dimorphism.

Another presumptive pathway for vasicentric tracheid origin is that of vessel dimorphism (Fig. 4.2). Vessel dimorphism is easy to see in woody vines (see Sect. 3.2.2) such as *Clematis* (Fig. 4.7, part 2). If in vessel dimorphism vessels of every conceivable width are produced, the narrowest will be very tracheidlike, and a few of these may, in fact, lack perforation plates. In certain vines one can see such situations, in which only a very small number of vasicentric tracheids, but large numbers of narrow vessels, are present (e.g., *Polygonum baldshuanicum*; Carlquist 1985d). This can also be observed in nonscandent species (e.g., *Rhus ovata*; Carlquist 1985a). Conceivably a portion of instances of vasicentric tracheid origin may have resulted from this pathway, because they do not fit any of the other patterns described here and because in many of the species with vasicentric tracheids, a wide range of vessel diameters, grading imperceptibly into the vasicentric tracheids as one sees them in a transection, are present. *Hedera* (Fig. 4.8, parts 1 and 2), *Ceanothus* (Fig. 4.8, part 3), and *Buddleja* (Fig. 4.8, part 4) may be in this category. Woods in this category might be expected to have libriform fibers rather than fiber-tracheids as the background imperforate tracheary element type (libriform fibers should be present in related species without vasicentric tracheids also). If fiber-tracheids were present, the process of fiber-tracheid dimorphism mentioned below might be

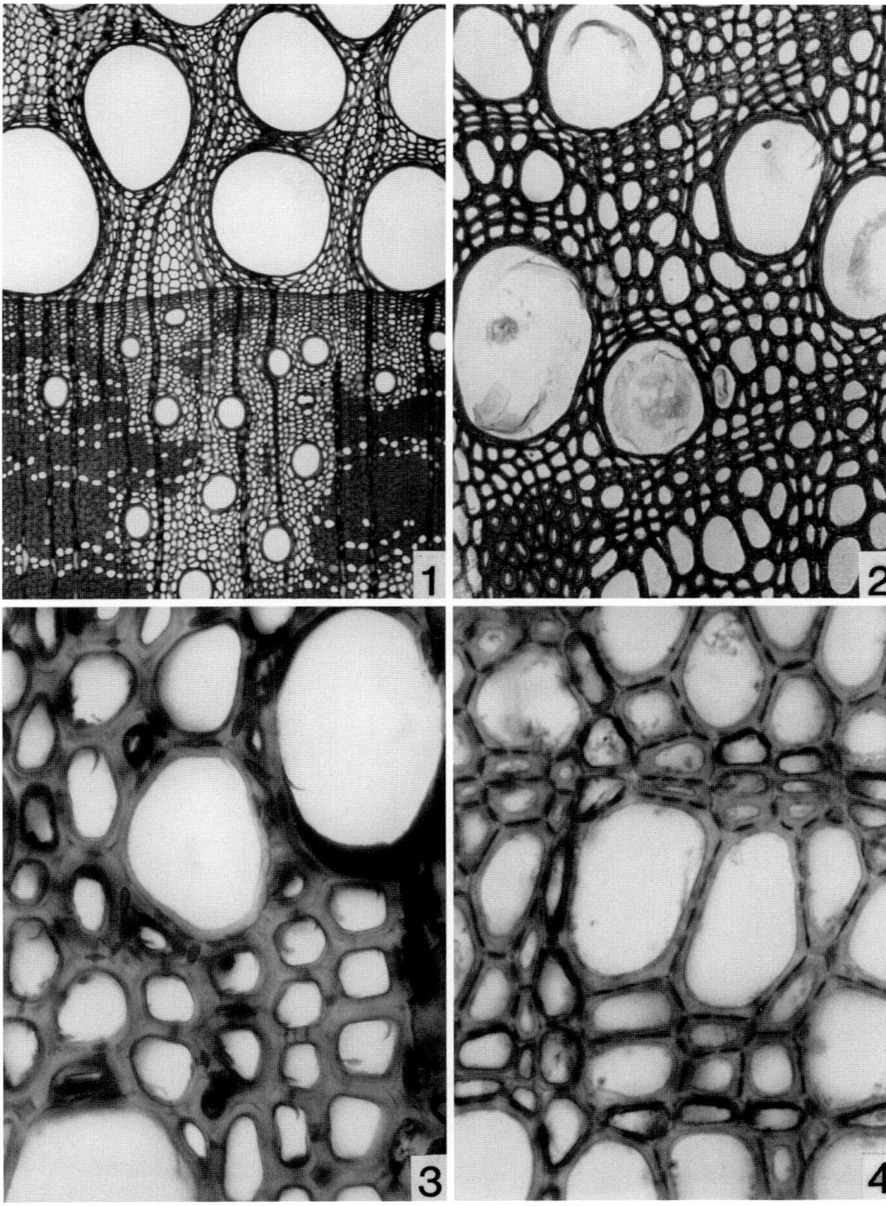

Fig. 4.7. Vasicentric tracheids, compared with other imperforate tracheary elements, as seen in wood transections. *1* Vasicentric tracheids (around vessels) and libriform fibers (*dark areas, lower left and right*, containing parenchyma cells), *Quercus rubra* (Fagaceae); *2* transection in which only a small area of libriform fibers (*lower left*) is present, but in which the remainder of the tissue consists of wide vessels, narrow vesels, vasicentric tracheids (like narrow vessels in diameter), and axial parenchyma, *Clematis lasiantha* (Ranunculaceae); *3*, *4 Arctostaphylos auriculata* (Ericaceae), dark-staining deposits outline pit cavities of some pits, permitting one to see whether cells are tracheids, or fiber-tracheids; *3* earlywood, showing vessels, a few tracheids adjacent to vessels, and (most cells) fiber-tracheids; *4* latewood; aside from a few ray cells and axial parenchyma cells, the cells shown are vessels, narrow vessels, and tracheids (which are vasicentric tracheids). (*1*:*A*; *2*:*B*; *3*,*4*:*C*)

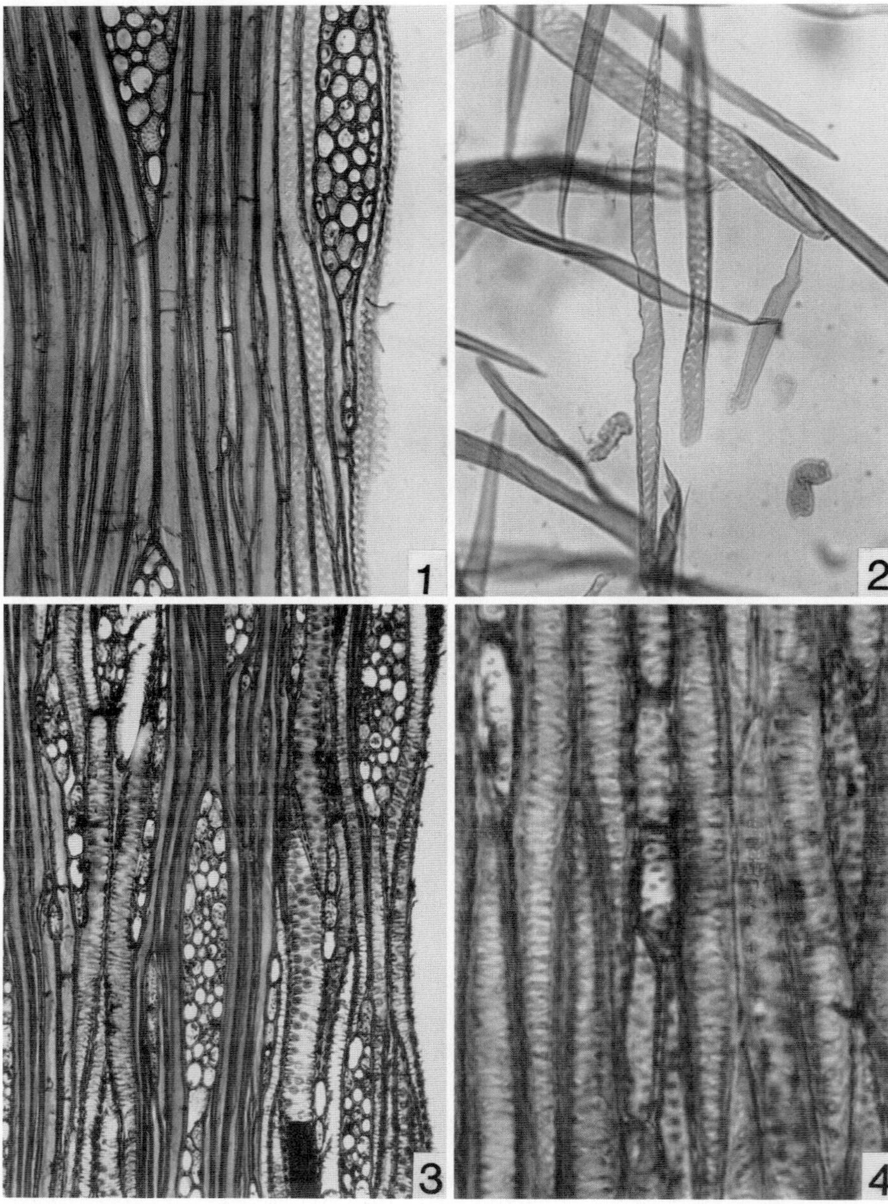

Fig. 4.8. Vasicentric tracheids compared with other imperforate tracheary elements. *1*, *2 Hedera helix* (Araliaceae). *1* Tangential section, vasicentric tracheids near vessel at *right*, and libriform fibers in remainder of wood; *2* maceration, two vasicentric tracheids vertically-oriented, *center*; other cells are vessel elements (*diagonal*, *above*) and libriform fibers; *3* tangential section, with vasicentric tracheids mixed with vessels, *right*; libriform fibers, *center*; *Ceanothus thyrsiflorus* (Rhamnaceae); *4* tangential section showing vasicentric tracheids (with helical thickenings) in latewood, *Buddleja utahensis* (Buddlejaceae). (*1–3:B*; *4:C*)

expected to occur. At present, mode of origin of various instances of vasicentric tracheid presence is speculative, although there is strong circumstantial evidence in a number of cases.

A third type of vasicentric tracheid origin can be hypothesized to be the result of what is termed fiber-tracheid dimorphism here (Fig. 4.2). In this case, in a wood in which fiber-tracheids are characteristically present, what might appear to be fiber-tracheids adjacent to vessels at first glance prove to have slightly to appreciably greater density of pitting than is characteristic of fiber-tracheids. The difference between a fiber-tracheid and a vasicentric tracheid is sometimes very little; for example, in some Proteaceae and Myrtaceae some genera may be said to have true tracheids exclusively (or fiber-tracheids exclusively) while others have fiber-tracheids plus vasicentric tracheids, and careful examination is required to see if cells adjacent to vessels have greater density of pits (or occasionally greater diameter of pits) than cells distal from the vessels. In some species of *Eucalyptus*, there is not much differentiation, whereas differentiation is marked in other species; Metcalfe and Chalk (1950) therefore were conservative in claiming vasicentric tracheids for some eucalypts, but Ingle and Dadswell (1953b) claim vasicentric tracheids for all eucalypts (I believe the latter report is correct). Although one could imagine a phylesis in which a tracheid-bearing ancestor evolved differentiation into vasicentric tracheids plus fiber-tracheids, I believe that the phylesis is more likely to have proceeded from a fiber-tracheid bearing ancestor to descendants in which fiber-tracheids are still present but tracheidlike cells – only a minor modification of the fiber-tracheids – have been formed near vessels. The latter phyletic pathway is suggested because a selective advantage of adding vasicentric tracheids is easy to see, if vasicentric tracheids safeguard the conductive process. The former pathway would have to involve a selective advantage of fiber-tracheids over tracheids while retaining the tracheids at the same time.

Families in which vasicentric tracheids are coupled with fiber-tracheids include (in addition to Myrtaceae and Proteaceae) Dipterocarpaceae, Sapotaceae, and Zygophyllaceae.

In plants of very dry areas, there may be an advantage in multiplication of the number of vasicentric tracheids in a given phylad, with concominant diminution in the proportion of libriform fibers (or fiber-tracheids). A series of instances in which this can be seen are represented by globular cacti (Fig. 4.9, parts 1 and 2). In these, loss of libriform fibers has likely been accelerated by the selective value of a secondary xylem that can contract and expand with availability of water (Mauseth et al. 1995). The vasicentric tracheids all have helical bands, reminding one of primary xylem tracheids. The helical bands are, however, horizontally quite wide, like a helical ramp. The value of wide helices probably relates to the accommodation of changes in volume: horizontally wide helices interfere minimally with expansion and contraction while providing maximal strength; if helices were vertically wide, a smaller portion of the primary wall would be available for expansion and contraction, and the helix would have less flexibility. Other instances of loss of libriform fibers in

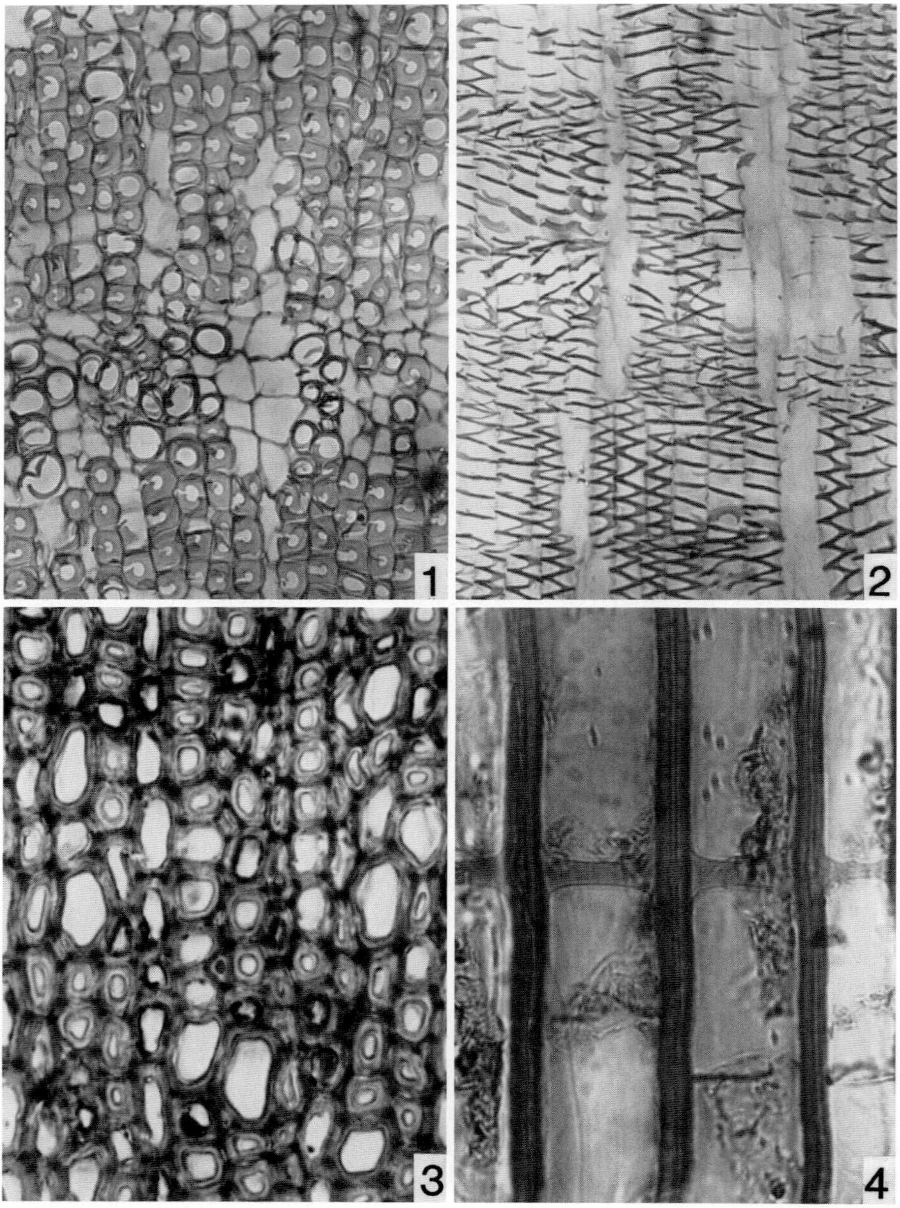

Fig. 4.9. Instances in which vasicentric tracheids are abundant (*1–3*); trabeulae (*4*). *1*, *2* Wide-helix tracheids in secondary xylem of *Coryphantha cornifera* (Cactaceae), a globular cactus. *1* Transection; vessels (appearing to have walls somewhat *dark gray*, band a third of the way from *bottom* of photograph) and wide-helix vasicentric tracheids; *2* radial section; except for parenchyma cells, all cells shown are wide-helix vasicentric tracheids; *3* transection showing vasicentric tracheids abundant; vessels and axial parenchyma cells are less abundant, and no libriform fibers are present; *Loricaria thuyoides* (Asteraceae); *4* radial section showing trabeculae running horizontally (*near center*) through several libriform fibers, *Kohleria elegans* (Gesneriaceae). (*1,2:B*; *3,4:C*)

vasicentric tracheid-bearing phylads can be found in Asteraceae. In *Dimorphotheca cuneata* (Carlquist 1966b, 1985a), *Loricaria thuyoides* (Fig. 4.9, part 3; Carlquist 1961b), and *Tetramolopium humile* (Carlquist 1960a) libriform fibers are very scarce, and vasicentric tracheids are much more common than either vessels or libriform fibers. The probable explanation for these three instances is increase of conductive safety by increasing the proportion of the xylem devoted to the cell type with the greatest conductive safety, vasicentric tracheids. Another series of instances of libriform fiber loss can be seen in the genus *Passiflora* (Carlquist 1985a). Some species of *Passiflora* have tracheids only, whereas other species have both libriform fibers and tracheids (Ayensu and Stern 1964). Because selective value for mechanical strength is low in woody vines, libriform fibers have doubtless been lost concomitantly with increase in number of vasicentric tracheids. This situation can be seen in other woody vines, such as *Clematis* (Fig. 4.7, part 2; fibers are present only in a small strand, lower left, and all other cells in the photograph are vessels, vasicentric tracheids, or parenchyma).

4.12.2 Function of Vasicentric Tracheids

Just as narrow vessels embolize less readily than wide vessels (Ellmore and Ewers 1985; Hargrave et al. 1994), vasicentric tracheids embolize less readily than narrow vessels. If vasicentric tracheids are present, they serve as subsidiary conductive cells in which intact water columns remain. The significance is not merely the persistence of the water columns, but their distribution. Because of adjacence to most or many vessels in a wood, vasicentric tracheids preserve the three-dimensional network of water columns in a stem, not just a portion of this network. Thus, at times of water stress when many vessels might fail, vasicentric tracheids provide water columns that supply all of the foliage as well as stems. Vasicentric tracheids characterize woody dicotyledons with evergreen leaves in areas highly seasonal with respect to water availability (Carlquist 1985b). Many taxa with vasicentric tracheids also tend to group vessels into diagonal aggregations (Sect. 3.4.3.3); these have the effect of uniting all the vessels and their vasicentric tracheids into a continuous network. Failure in any part of such a system could be bypassed, and an ultimate degree of conductive safety (for a vessel-bearing dicotyledon) would be achieved.

4.12.3 Methods for Identifying Vasicentric Tracheids

Vasicentric tracheids might have been identified by more workers and cited in a larger number of families before now if they were easily seen. To determine if vasicentric tracheids are present, one must see if there are imperforate tracheary elements densely provided with bordered pits – tracheids – adjacent to vessels. In a transection, one might, with practice, see whether borders are present or not, and thus see if tracheidlike cells adjacent to vessels are tracheids or not. This is difficult, because bordered pits are much more readily seen in face view in a longitudinal section. By viewing a longitudinal section (a tangential section is usually best), one generally can see whether there are fusiform cells without perforation plates but densely provided with bordered pits, and therefore perhaps tracheids, adjacent to vessels. By viewing a longitudinal section one cannot be sure, however, whether what appear to be tracheids genuinely lack perforation plates or whether perforation plates have been cut away from a particular cell by the sectioning process. One must therefore confirm the imperforate nature of what appear to be tracheids with the aid of a maceration (Fig. 4.8, parts 1 and 2). This is most essential, because small perforation plates can escape notice in a section. Also, vasicentric tracheids may be present but scarce, and a maceration can permit one to determine if that condition is present.

If one has seen appreciable numbers of what appear to be tracheids adjacent to vessels in a longitudinal section as well as in a maceration, vasicentric tracheids can be claimed.

Vascular tracheids can be claimed if one sees what appear to be tracheids in latewood of a longitudinal section and confirms the nature of these cells with macerations. A radial section is best for showing whether the probable tracheids are confined to latewood. If the wood is ring-porous, one can use the vessel diameters as seen in a radial section as a way of localizing tracheids within a growth ring. Yaghmaie and Catling (1984) have macerated earlywood and latewood of *Betula* separately, so they were able to determine that tracheids occurred both in latewood and earlywood and thus are what I term vasicentric tracheids.

Vasicentric tracheids are about the same length as the vessels they accompany, although we have little data on this point. Because narrower imperforate tracheary elements are, in general, longer than wide ones (wide vessel elements), one might expect vasicentric tracheids to be a little longer than the vessel elements they accompany.

Vasicentric tracheids tend to be very similar to the vessel elements they accompany with respect to size and density of pits. This similarity is also evident if helical sculpture is present in vessels. Presence of helical sculpture both in vessels and vasicentric tracheids of a species was reported by Parameswaran and Gomes (1981) in *Ligustrum lucidum* (Oleaceae), and is shown here for *Ceanothus* (Fig. 4.8, part 3) and *Buddleja* (Fig. 4.8, part 4). Helical sculpture may also be seen in vasicentric tracheids of *Arctostaphylos*, *Berberis*, *Trichostema*, and the evergreen species of *Prunus* (Carlquist 1985a).

4.12.4 Systematic and Ecological Occurrence of Vasicentric Tracheids

Metcalfe and Chalk (1950) listed 32 families as having vasicentric tracheids. A later listing (Metcalfe and Chalk 1983) also has 32, although Geraniaceae have been added and Hypericaceae united with Guttiferae (= Clusiaceae in the present book). Of the families in the 1983 list, Staphyleaceae should be omitted, since a study of woods of that family (Carlquist and Hoekman 1985a) failed to reveal vasicentric tracheids. Bonnetiaceae, present on the Metcalfe and Chalk list, probably should be united with Clusiaceae. With those alterations, the number of families recognized by Metcalfe and Chalk (1950) would become 30. As a result of a survey of the southern California flora (Carlquist and Hoekman 1985b), vasicentric tracheids appeared much more prominent than hitherto thought, and proved to characterize prominent large genera (*Arctostaphylos*, *Berberis*, *Ceanothus*) in which they had not hitherto been reported. This led to a search for vasicentric tracheids in other Mediterranean-climate areas of the world, where vasicentric tracheids also prove to be common (Carlquist 1985a).

Systematic affinities proved helpful in discovery of instances of vasicentric tracheid occurrence. For example, discovery of vasicentric tracheids in Berberidaceae led to discovery of them in Papaveraceae and Ranunculaceae as well. Likewise noting presence of vasicentric tracheids in Passifloraceae led me to find them in Erythroxylaceae, Flacourtiaceae, Malesherbiaceae, Turneraceae, and Violaceae. Vasicentric tracheids have undoubtedly evolved polyphyletically in dicotyledons, and their distinctive pathways and appearances (e.g., cacti, *Quercus*) mentioned above are indicative. In addition to Mediterranean-type climates, areas where vasicentric tracheids are exceptionally prominent include deserts and alpine situations (Carlquist and Hoekman 1985b) and dry tropical areas (e.g., Sapotaceae, Zygophyllaceae). Vasicentric tracheids are also unusually well represented in woody vines (Carlquist 1985d), where they may represent a form of conductive safety valuable in these plants, characterized as they are by wide and therefore probably vulnerable vessels.

The list below represents a revision of the listing in my 1985a paper, in which documentation is provided; reports not included in the 1985a paper are documented here with references. The list below contains 80 families; a few additional ones are likely to be found, however. As can be discovered by consulting the 1985a paper, families are listed if one or more genus is known to have vasicentric tracheids; vasicentric tracheids characterize only a few families entirely (e.g., Connaraceae, Dipterocarpaceae, Sapotaceae, Zygophyllaceae).

Acanthaceae
Agdestidaceae (Carlquist 1999d)
Aizoaceae (*Russchia* sp.; new report)
Anacardiaceae
Ancistrocladaceae
Annonaceae (Carlquist 1987c)
Apocynaceae
Araliaceae
Asclepiadaceae
Asteraceae
Austrobaileyaceae (Carlquist 1988c)

Balanitaceae
Berberidaceae
Betulaceae
Bignoniaceae
Boraginaceae
Buddlejaceae
Cactaceae
Calycanthaceae
Capparaceae
Casuarinaceae
Celastraceae
Chenopodiaceae
Clusiaceae
Cneoraceae
Combretaceae
Connaraceae
Cucurbitaceae (*Zanonia*; Carlquist 1985d)
Dioncophyllaceae (Gottwald and Parameswaran 1968)
Dipterocarpaceae
Ericaceae: subfamilies Arbutoideae and Vaccinioideae
Erythroxylaceae
Euphorbiaceae (*Bernardia* cited in Carlquist 1985a in supplemental note; *Kleinodendron riosulense* can be added)
Fabaceae
Fagaceae
Flacourtiaceae
Frankeniaceae
Geraniaceae
Goetzeaceae (Carlquist 1988b)
Goupiaceae
Grossulariaceae
Lamiaceae
Lauraceae
Leitneriaceae (*Leitneria floridana*; Carlquist 1987c)
Linaceae
Loganiaceae
Loranthaceae (*Nuytsia floribunda*; new report)
Malesherbiaceae
Malpighiaceae
Malvaceae
Melastomataceae (Carlquist 1987c)
Moraceae (Carlquist 1987c)
Myrtaceae (especially Australian genera; Ingle and Dadswell 1953b)
Ochnaceae
Olacaceae
Oleaceae
Opiliaceae (Koek-Noorman and van Rijkevorsel 1983)
Papaveraceae
Passifloraceae
Penthoraceae (Haskins and Hayden 1987)
Pittosporaceae
Plumbaginaceae (Carlquist and Boggs 1996)
Polygalaceae
Polygonaceae (Carlquist 1985d)
Proteaceae
Quiinaceae
Ranunculaceae
Rhamnaceae
Rosaceae
Rutaceae
Santalaceae
Sapindaceae
Sapotaceae
Scrophulariaceae
Setchellanthaceae (Carlquist and Miller 1999)
Simaroubaceae
Solanaceae
Stilbaceae (Carlquist 1986c)
Stylobasiaceae
Thymeleaceae
Tremandraceae
Turneraceae
Ulmaceae
Verbenaceae
Violaceae
Zygophyllaceae

Hess (1946a) reports vasicentric tracheids in *Cyrillopsis* (Cyrillaceae). I have not seen wood of this genus, but other Cyrillaceae have true tracheids.

4.13 Reaction Wood

4.13.1 Definition

Reaction wood as commonly defined (e.g., IAWA Committee on Nomenclature 1964) includes tension wood, found on the upper sides of branches in dicotyledons, and compression wood, found on the lower sides of branches of conifers. Both of these have the effect of maintaining the position of a trunk or branch (Wardrop 1955) despite increasing weight or other factors tending to change their positions. Reaction wood may be one of the mechanisms by which a tree attains its characteristic architecture (Fisher and Stevenson 1981).

Tension wood may be recognized at the microscopic level by the presence of "gelatinous fibers" (Fig. 4.10, parts 1–4). "Gelatinous fibers" may be either tracheids, fiber-tracheids, or libriform fibers in which shrinkage patterns appear in walls in prepared slide. These shrinkage patterns relate to absence of lignification in particular layers, and the term "gelatinous" is misleading in that it tends to evoke the image of a mucilaginous substance.

4.13.2 Compression Wood

Although conifers and other gymnosperms (taxads, *Ginkgo*) have compression wood, Höster and Liese (1966) stress not the taxonomic group but the cellular characteristics. Timell (1983) states that "trees and shrubs with a xylem consisting largely of longitudinal tracheids tend to develop compression wood, regardless of whether they are gymnosperms or angiosperms."

One might expect that if compression wood were to be found in angiosperms, it might be in more primitive groups, on account of the fact that vesselless seed plants are characterized by it. Results have been contradictory in the Winteraceae. Dadswell and Wardop (1949) claimed compression wood in the branches in *Tasmannia lanceolata* (= *Drimys aromatica*), but Kucera and Philipson (1977) found neither compression nor tension wood in *Drimys winteri*. Timell (1983), reviewing literature on *Pseudowintera colorata* by Kucera and Philipson (1978) and by Meylan (1981), concludes that xylem on lower sides of the branches in this species cannot be regarded as compression wood.

Buxus stems have been reported to have distinctive wood on the lower sides of branches, and this has been interpreted as compression wood (Höster and

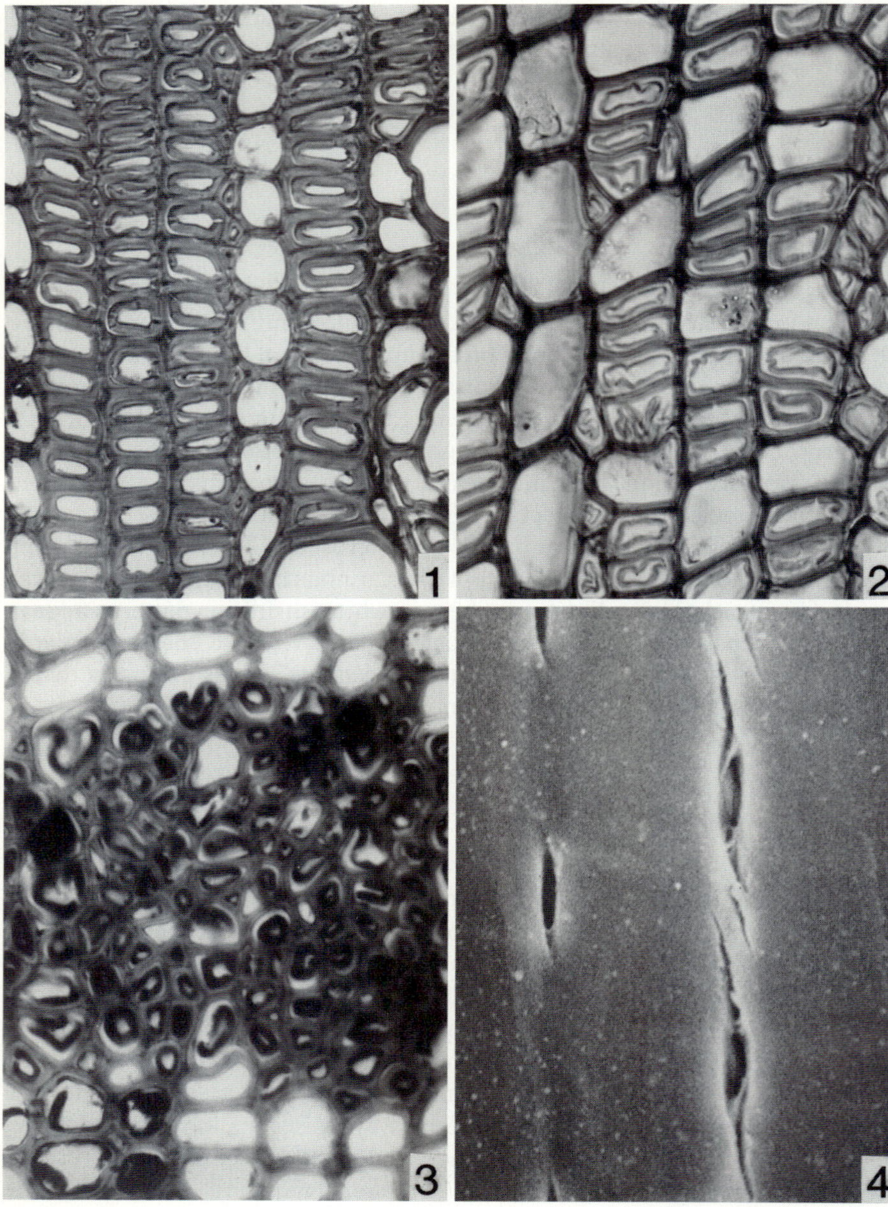

Fig. 4.10. Reaction wood (gelatinous fibers) as seen in transections (*1–3*) and in an SEM view of a longitudinal section (*4*). *1* Wall shrinkage in libriform fibers, *Chamaesyce remyi* (Euphorbiaceae); *2* shrinkage of thin portion of secondary wall in libriform fibers, *Euphorbia piscatoria* (Euphorbiaceae); *3* darker staining inner portion of secondary wall of libriform fibers shows shrinkage, *Maclura pomifera* (Moraceae); *4* inner surface of tracheid; splits extending from pit apertures represent gelatinous nature of wall, *Drimys confertifolia* (Winteraceae). (*1–3*:*C*; *4*:*G*)

Liese 1966). However, the review by Timell (1983) indicates these cells do not fulfill some important criteria for recognition of compression wood.

4.13.3 Tension Wood

4.13.3.1 Characteristics

Tension wood can typically be recognized by the occurrence of a gelatinous layer in the secondary wall, a layer which produces shrinkage patterns as well as distinctive staining patterns (Fig. 4.10, parts 1–4). This layer, typically called a G-layer, may substitute for none, or for one, two, or three of the usual secondary wall layers (S_1, S_2, S_3). The G-layer is composed of virtually nonlignified cellulose, the microfibrils of which are oriented nearly parallel to the longitudinal axis of the cell (Norberg and Meier 1966). Tension wood cells are typically longer than non-tension wood cells in a species (Dadswell and Wardrop 1949).

4.13.3.2 Organographic Occurrence of Tension Wood

Patel (1964) and Höster and Liese (1966) found that tension wood was absent in roots of some taxa in which tension wood is common in stems; in other species, little difference between stems and roots was noted. Patel (1964) found that in roots, distribution of tension wood fibers was not related to orientation of the organ with respect to gravity (upper side vs. lower side). Tension wood occurs in aerial roots of *Ficus* and *Cecropia* (Fisher 1982), and contraction of these roots in relation to climbing can be shown to be based on action of the tension wood. Tension wood does not occur in most buttresses of tropical trees, only some, and it cannot explain the action of buttresses, therefore (Fisher 1982).

4.13.3.3 Systematic Occurrence of Tension Wood

Höster and Liese (1966) have surveyed dicotyledons at large for presence of tension wood, and they present a listing of species studied.

The listing by Höster and Liese (1966) attempts to cover all major orders of dicotyledons, and the data they present show that some families and orders are notably poor in tension wood, whereas others have tension wood abundantly or occasionally. The more typically herbaceous families in the Höster and Liese survey seem mostly to lack tension wood. Onagraceae form a notable exception (Carlquist 1975b, 1987i), as do the succulent species of *Euphorbia* (Fig. 4.10, parts 1 and 2; Carlquist 1970b). This latter instance is curious because of the pervasiveness of the gelatinous fibers, often throughout large portions of wood samples. The weight of succulent stems may be the cause whereby tension wood forms almost constantly because of the constant balancing and counterbalancing of these waterladen and therefore relatively weighty stems. Obviously wood anatomists could profitably examine woods more carefully with respect to tension wood occurrence.

The literature on chemistry and ultrastructure of tension wood is extensive but is not considered here because very little of this literature relates to comparative data.

4.14 Trabeculae

Trabeculae ("trabeculae of Sanio") may occur in vessel elements (Sect. 3.15) as well as in imperforate tracheary elements (Fig. 4.9, part 4) of dicotyledon wood. In either case, a trabecula is a strand of secondary wall material extending radially across a cell lumen, and as shown in Fig. 4.9 (part 4), trabeculae may form a series in a series of cells derived successively from a particular cambial initial. The center of trabeculae consists of material like the middle lamella (or primary wall) of woody cells (Keith 1971), and trabeculae may be responses to some kind of cambial damage. They have no systematic significance.

4.15 Contents of Imperforate Tracheary Elements

4.15.1 Tyloses

Gottwald (1972) has reviewed instances of tyloses in imperforate tracheary elements of dicotyledons. Gottwald was able to confirm the existence of tyloses in imperforate tracheary elements only in Magnoliaceae, in which they could be seen in 22 species of seven genera.

4.15.2 Crystals

Crystals and other phenomena relating to idioblastic cells are reviewed in Chapter 7. However, in terms of imperforate tracheary elements, crystals are relatively infrequent. In Flacourtiaceae, James and Ingle (1956) figure cyrstalliferous fibers in *Flacourtia inermis*, and report them also in the genera *Homalium* and *Scolopia*. R.B. Miller (1975) confirms crystalliferous fibers in the latter two genera. Prismatic crystals have been reported in libriform fibers in Onagraceae in two species of *Gaura*; they are so large in libriform fibers of *Hauya* that they may be called styloids (Fig. 7.6, part 2; Carlquist 1975b). In Punicaceae, in the same order as Onagraceae (Myrtales), crystalliferous fibers are also found (Bridgewater and Baas 1978). In Penaeaceae, also a myrtalean family, crystals occur in chambered axial parenchyma (Carlquist and DeBuhr 1977); because imperforate tracheary elements in Penaeaceae are tracheids, crystals are to be expected in some cell type other than imperforate tracheary elements. Crystalliferous fibers may also be found in *Akocanthera venenata* of the Apocynaceae (Fig. 7.3, part 3). *Pittosporum phillyraeoides* of the Pittosporaceae (Fig. 7.3, part 1), *Eremophila polyclada* of the Myoporaceae (Fig. 7.3, part 2), and *Lactoris* (Carlquist 1990b) have crystals of various sizes in libriform fibers.

4.15.3 Other Substances

Starch is commonly found in imperforate tracheary elements in the case of living or septate fibers, as noted above (Sect. 4.7). Ingle and Dadswell (1953a) report latex in libriform fibers of *Alstonia* and other Apocynaceae.

4.16 Excluded Types of Imperforate Tracheary Elements

Jayme and Azzola (1964) report tracheids with perforation plates at one end in *Fagus sylvatica*; these must be termed vessel elements, despite their narrowness (they may be the vessel elements that terminate a vessel). Fibriform vessel elements have been reported by Lemesle (1956a) as "fibres areolées conductrices" in two genera of Rubiaceae. In *Manettia* of Rubiaceae, Lemesle (1956a) terms these cells "tracheïdes ouvertes." Similarly, fibriform vessel elements in *Corokia* and *Griselinia* are regarded by Patel (1973) as a kind of "fibre . . . with perforation." Despite the fact that fibriform vessel elements may have tracheidlike characteristics, I see no alternative but to define these, as well as other instances (Lemesle 1946), as kinds of vessel elements rather than kinds of imperforate tracheary elements.

Lemesle and Duchaigne (1955a,b) have claimed "pseudotracheids" in Degeneriaceae. In Bruniaceae, similar cells were claimed as a kind of tracheid by Lemesle (1956b). I have studied wood of both families (Carlquist 1978a, 1989e) and conclude that what Lemesle was describing may be simply strands of axial parenchyma cells. Axial parenchyma strands may bear bordered pits, just as ray cells often do, but mere presence of bordered pits does not qualify such cells as tracheids.

5 Axial Parenchyma

5.1 Definition, Means of Observation

Axial parenchyma consists of axially elongate cells or (more commonly) strands of cells, alive at maturity, derived from fusiform cambial initials. Axial parenchyma cells are usually thinner walled than the imperforate tracheary elements with which they are associated. Each cell in a strand of axial parenchyma is usually surrounded by a secondary wall. This definition excludes living fibers, which have contents living at maturity but which tend to have thicker walls than axial parenchyma; if septate, living fibers are subdivided only by a primary wall (cf. Fig. 4.6, parts 3 and 4).

Axial parenchyma distributions are invariably illustrated in transection (Figs. 5.1–5.7). However, one can usually easily distinguish axial parenchyma from other cell types (most notably imperforate tracheary elements) in radial sections. In a radial section, one can easily discern the vertically-oriented strands of cells (e.g., Fig. 5.8, part 3, left and right) and contrast them with imperforate tracheary elements (not subdivided or, if septate, not with each cell surrounded by its own secondary wall) and ray cells (in horizontally oriented sheets of cells). Once one has recognized axial parenchyma cells in a radial section, one can apply the appearances discerned to what one sees in transections and tangential sections.

In illustrating axial parenchyma distributions, authors typically find transections of woods in which axial parenchyma is much thinner walled than the imperforate tracheary elements, providing an easily seen visual contrast (e.g., Figs. 5.2, part 1, 5.3, part 1). However, in some species, wall thickness of axial parenchyma is similar to that of imperforate tracheary elements. In addition to using what one learns about axial parenchyma in a radial section as a way of aiding recognition in a transection of a given species, one may look for cross walls (horizontal walls) in axial parenchyma seen in transections. Only a small proportion of the axial parenchyma cells will show such cross walls in a transection, but those that do can be used as indicators of what axial parenchyma cells look like in transection (diameter, wall thickness) and that information can then be applied to recognition of all axial parenchyma cells in a transection.

Although the cells within a septate fiber may occasionally each be surrounded by a secondary wall, and while a cell of a strand of axial parenchyma may occasionally be septate, these exceptions are so scarce that the distinction mentioned above between a septate fiber and a strand of parenchyma provides few difficulties and has been endorsed by various authors (e.g., Milanez 1938, 1939; Chattaway 1939).

5.2 Types of Axial Parenchyma Distribution

5.2.1 Classification

The traditional classification of axial parenchyma types has been challenged by Braun and Wolkinger (1970); they base their distinctions mostly on the type of tracheary element with which parenchyma is in contact: vessels, tracheids, fiber-tracheids, or libriform fibers. Contact with vessels is usually associated with pits to axial parenchyma larger than those which contact the imperforate tracheary elements. Braun and Wolkinger also take into account contact among parenchyma cells. However, as noted by Metcalfe and Chalk (1983, p. 20), "paratracheal" in the Braun and Wolkinger system equates to diffuse in the traditional system, "paratracheal contact" equates to those cells actually in contact with vessels, and "interfiber" to at least some cells in banded types. Braun and Wolkinger are basically drawing a contrast between parenchyma that contacts vessels (parenchyma may release sugars into vessels; Sauter 1966a,b) and parenchyma that stores photosynthates and/or translocates the photosynthates to other cells. This concept, based as it is on physiological considerations, is worthy of recognition, but the contact-isolation idea may be added to, and need not supplant, the traditional system of axial parenchyma classification.

The traditional system of axial parenchyma classification is based on whether axial parenchyma is present or absent, and whether it is apotracheal (distributed without relation to vessels) or paratracheal (distributed around vessels). Types of apotracheal and paratracheal parenchyma are based upon distribution and relative abundance of parenchyma as seen in transection. The types recognized in the present book include absent, diffuse, diffuse-in-aggregates ("diffuse-aggregate" of some authors), apotracheal banded, marginal, ray-adjacent, vasicentric scanty, vasicentric abundant, abaxial, aliform, confluent, and intervascular; the types "intervascular," "pervasive," and "ray-adjacent" are also proposed for consideration.

5.2.2 Evolutionary Status of Axial Parenchyma Types

The commoner types of axial parenchyma have been compared by Kribs (1937) to types of scalariform perforation plates and length of vessel elements. The results are essentially the same, so the latter table of Kribs is reproduced here (Table 5.1).

Table 5.1. Vessel element length compared to parenchyma types. (Kribs 1937)

Axial parenchyma type	No. of genera	Vessel element length (μm)
Diffuse	104	920
Absent	57	780
Diffuse-in-aggregates	144	650
Vasicentric scanty	61	600
Apotracheal banded narrow	71	510
Terminal (= marginal p.p.)	62	440
Apotracheal banded wide	41	420
Vasicentric abundant	242	310

The results of Kribs' (1937) statistical tabulations cannot be expressed in a simple linear phyletic series of axial parenchyma types. Rather, a complex picture emerges (Fig. 5.1) in which some of the types have been derived polyphyletically. However, certain tendencies are evident: greater degrees of aggregation of cells (vasicentric abundant vs. vasicentric scanty, for example) represent specialization, and greater degrees of grouping with respect to vessels (as opposed to random distribution).

These trends have been interpreted (Carlquist 1975a) as representing development of massive and efficient vertical conduits for photosynthate translocation, paralleled by similar specialization in rays.

5.2.3 Absence of Axial Parenchyma

The parenchyma condition "absent" as designated by Kribs (1937) is often overlooked, but it has several kinds of significance. Most commonly absence or sparsity of axial parenchyma tends to be related to presence of nucleated fibers, which functionally substitute for axial parenchyma. This is certainly not always true, and one can cite exceptions such as *Umbellularia californica* (Lauraceae), in which septate fibers are present in addition to starch-containing vasicentric parenchyma.

If one notes the relatively great length of vessel elements associated with absence of parenchyma in the above table, one would conclude that absence of parenchyma is more primitive than presence of axial parenchyma. While this is valid in a statistical sense, it is certainly by no means always true, and exceptions can be found at both ends of the phylogenetic scale: for example, axial parenchyma does occur in the vesselless family Winteraceae (Bailey 1944a; Carlquist 1981a, 1982a, 1983a), and absence of parenchyma may be found in such a specialized genus as *Fuchsia* of the Onagraceae (Carlquist 1975b).

Absence of parenchyma may be related to limited size of stem, and can also be related to raylessness. If, in rayless species, the duration of the woody

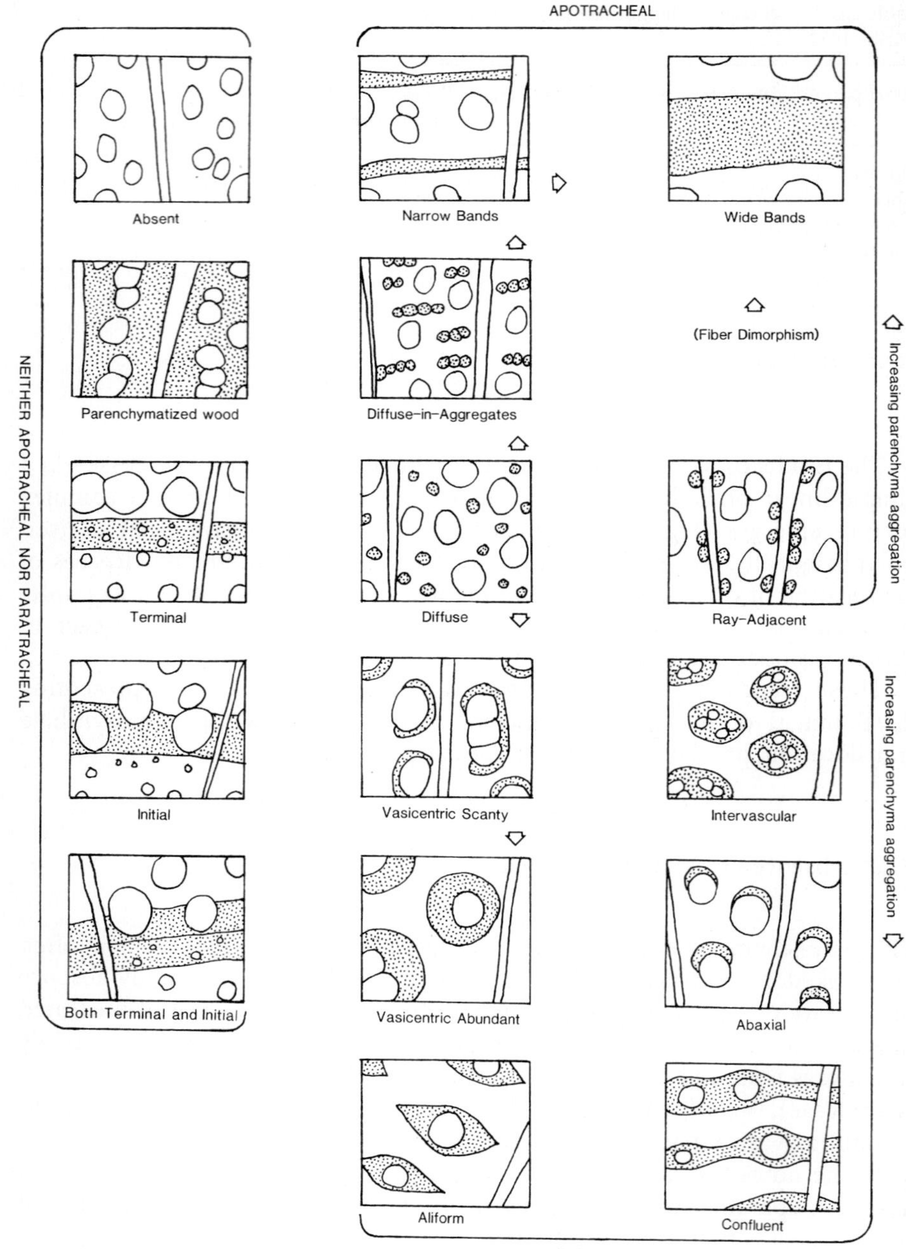

Fig. 5.1. Types of axial parenchyma, as seen in diagrammatic transections, and their probable phylogenetic interrelationships (*vertical lines* rays, *circles* vessels, *stippled areas* parenchyma, *wide horizontal line* limit of growth ring). Diffuse is considered to be the starting point for the other parenchyma types; banded parenchyma may be formed by aggregation of pre-existing parenchyma or from fiber dimorphism (the latter is much less common, and occurs in a limited number of families)

cylinder is so brief that radial translocation of photosynthates is of minimal selective value, that may be true of axial parenchyma also. Raylessness may represent not so much extinction of parenchyma as maximization of mechanical strength (imperforate tracheary elements) to the exclusion of rays. If rays are absent, the interconnecting system, axial parenchyma, also tends to be absent. In *Stylidium*, which is rayless (Carlquist 1981f), axial parenchyma is absent; in the stems studied, woody cylinder diameter was limited. These conditions also occur in *Viviania* (Carlquist 1985f).

5.2.4 Diffuse Parenchyma

Of the parenchyma types other than absent, diffuse easily qualifies as the most primitive according to the Kribs (1937) data reproduced above. Diffuse axial parenchyma can be defined as axial parenchyma strands scattered randomly within the axial (fascicular) secondary xylem (Fig. 5.2). Diffuse parenchyma may be sparse (Fig. 5.2, part 1) or moderately common (Fig. 5.2, part 2). Where diffuse axial parenchyma cells are more abundant, random distributions of axial parenchyma cells inevitably result in small groupings here and there. For example, in Fig. 5.2 (part 3), one sees some cells in contact with vessels, but parenchyma in that figure still qualifies as diffuse. In Fig. 5.2, part 4 we see an example in which axial parenchyma cells are abundant, and one tends to see occasional tangential lines of several parenchyma cells in contact here and there. This example, too, still qualifies as an example of diffuse.

There is no doubt that diffuse parenchyma occurs commonly in woods with tracheids, and this might lead one to suspect that it represents a way of dispersing parenchyma cells among actively conductive cells. However, there are woods with tracheids in which axial parenchyma is not diffuse; in such cases, one finds alternative possibilities potentially related to physiology: for example, rays with upright cells possibly equivalent to axial parenchyma cells are so densely scattered throughout a wood that they may have an effect similar to diffuse axial parenchyma in forming extensive contacts between tracheids and parenchyma. One must also remember that axial parenchyma cells isolated entirely from each other by dead water-conducting cells (tracheids, for example) are not likely because in three dimensions, diffuse parenchyma (as well as, more obviously, other axial parenchyma types) probably represents a network of axial parenchyma cells in contact with each other and with ray cells.

5.2.5 Diffuse-in-Aggregates Parenchyma

Also known as "aggregated diffuse" (Hess 1950) and "diffuse-aggregate" (Kribs 1937), the IAWA Committee on Nomenclature (1964) designates as

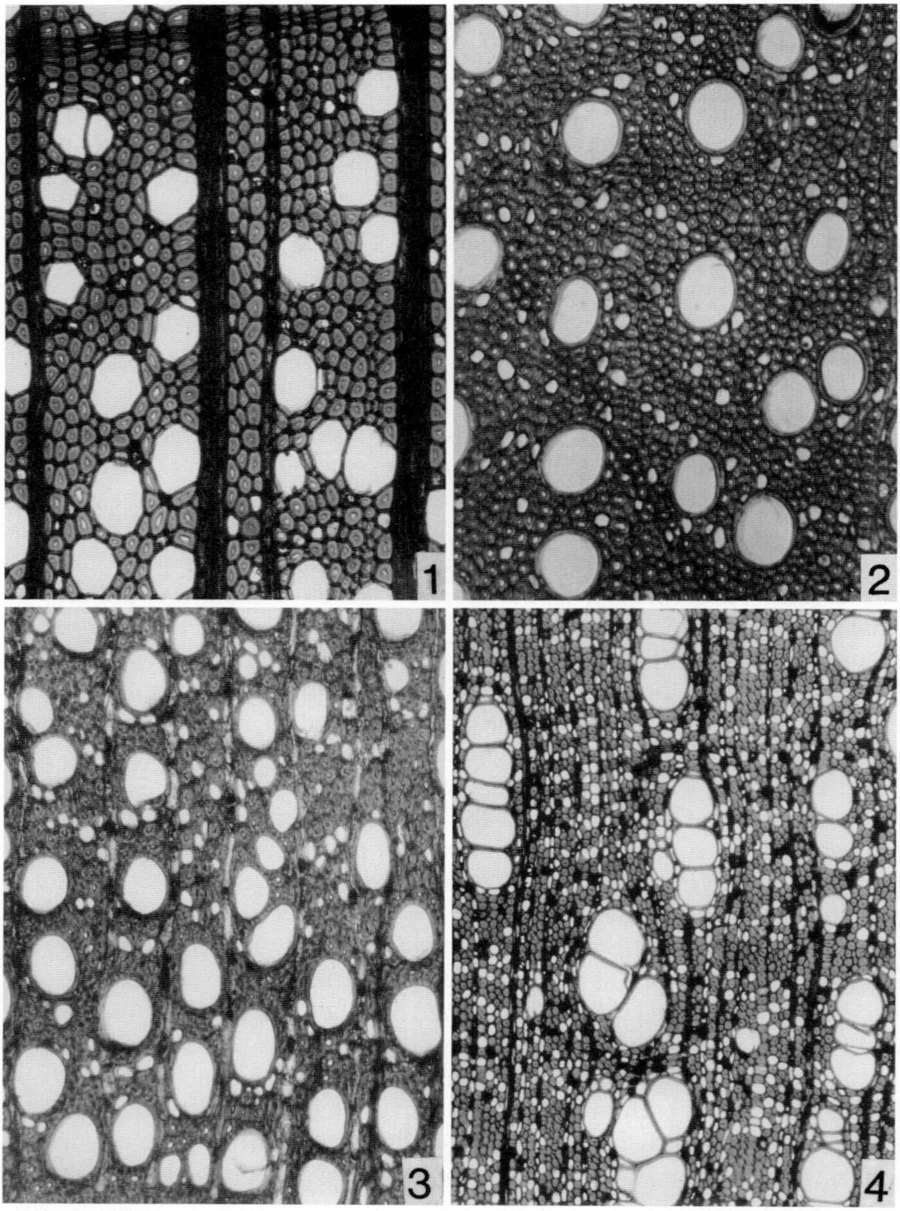

Fig. 5.2. Examples of diffuse axial parenchyma, as seen in transections. *1* Sparse axial parenchyma, *Amelanchier arborea* (Rosaceae); *2* moderately sparse axial parenchyma, *Agonandra* sp. (Opiliaceae); *3* diffuse parenchyma showing small degrees of random aggregation, *Leptolaena bernieri* (Sarcolaenaceae); *4* diffuse parenchyma transitional to diffuse-in-aggregates, *Dipholis salicifolia* (Sapotaceae). (*1–3:B*; *4:A*)

diffuse-in-aggregates those axial parenchyma cells grouped into tangential lines (but typically not forming lines more than one cell wide radially; Fig. 5.3, part 1). Metcalfe and Chalk (1983, their Plate 4c) illustrate tangential clusters of a few cells each as exemplifying diffuse-in-aggregates. "Reticulate" parenchyma combines apotracheal-banded and diffuse-in-aggregates, and is not a maximally useful term. Record (1944b) provided a key to American woods based on this term.

According to the Kribs (1937) data reproduced above, diffuse-in-aggregates represents a considerable advance over diffuse parenchyma in terms of associated vessel element lengths. In terms of functional significance, this advance is plausible, because diffuse-in-aggregates forms a bridging between axial and radial parenchyma systems, by virtue of intersecting rays at frequent points. There are fewer and more extensive contacts between the two systems than with diffuse. Thus, diffuse-in-aggregates provides appreciably greater ease in conduction of photosynthates from radial into axial parenchyma and the reverse.

5.2.6 Apotracheal Banded Parenchyma

Apotracheal banded parenchyma consists of bands two cells wide (at least in some places) to many cells wide. Bands two to three cells wide (Fig. 5.3, parts 2 and 3) would qualify as narrow-banded apotracheal parenchyma, whereas more than three cells in width qualifies a wood as having wide-banded apotracheal axial parenchyma (Fig. 5.3, part 4). Presence of numerous bands has been termed "concentric" parenchyma by Jane (1956), who cited this type in *Lophira*. The term "scalariform" has been applied to successive bands of this sort when seen in transection (Wagenführ 1961), but the vagueness of such terms makes them less useful than the Kribs terms more widely in use. Apotracheal banded parenchyma is sometimes not easy to distinguish from tangentially-oriented paratracheal bands such as confluent (see below). In confluent, one can find that all bands intersect vessels if one traces them tangentially (across rays, in some cases). In apotracheal banded, bands are distributed without reference to vessels; vessels may occur in some of these bands, but no relationship between vessels and bands is evident (e.g., Fig. 5.7, part 1). Possibly where there is a phyletic increase in extent and abundance of the bands, an essentially apotracheal pattern may be derived from a paratracheal pattern. Because paratracheal patterns are common in Fabaceae, one suspects that such a pattern as represented in Fig. 5.7 (part 1) may represent such an instance. One has no alternative but to designate banded types as apotracheal or confluent on the basis of topographic appearances rather than guessing about possible phyletic interconnection as a method of selecting terminology.

Narrow-banded can be considered to precede wide-banded apotracheal parenchyma from the data of Kribs (1937) cited above. The wider the bands,

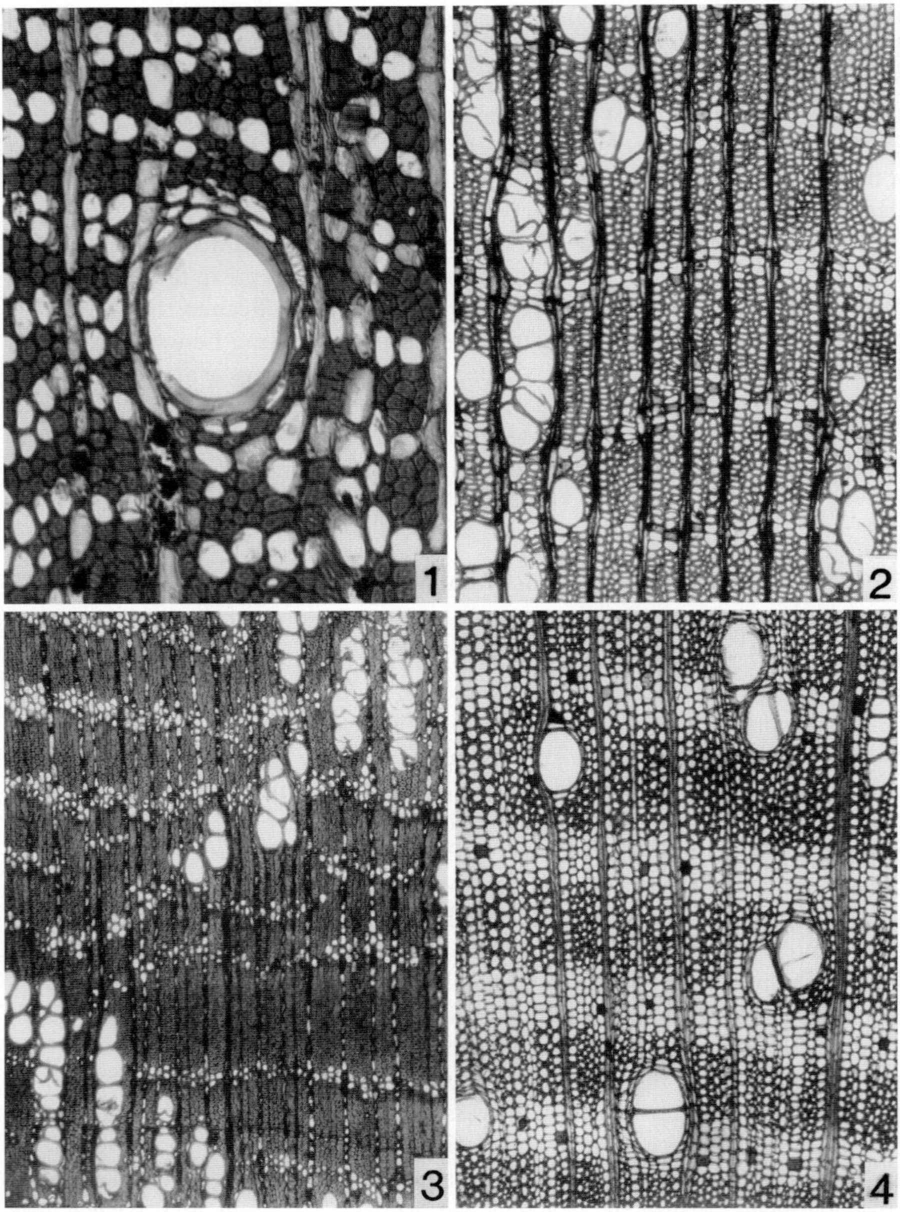

Fig. 5.3. Apotracheal axial parenchyma types: diffuse-in-aggregates (*1*) and banded types (*2–4*). *1* Small transection portion that shows diffuse-in-aggregates, *Scytopetalum klaineanum* (Scytopetalaceae); *2* narrow-banded parenchyma, *Dumohra africana* (Sapotaceae); *3* narrow-banded parenchyma, bands of various thicknesses, *Bumelia persimilis* (Sapotaceae); *4* wide-banded parenchyma, *Ficus rubiginosa* (Moraceae). (*1:B*; *2–4:A*)

the more massive the contacts between axial and radial parenchyma systems, and presumably the more efficient the photosynthate conduction between the two systems. Wide-banded parenchyma characterizes certain families such as Moraceae (Fig. 5.3, part 4; Koek-Noorman et al. 1984a,b,c), Bombacaceae, and certain Fabaceae (Fig. 5.7, part 1).

Fiber dimorphism (see Sect. 4.9) results in bands of tissue that are much like axial parenchyma, and the reader is referred to that discussion. Wood anatomists seem able to distinguish between banded parenchyma that represents a grouping of parenchyma cells that has evolved by means of aggregation, and bands that have resulted from fiber dimorphism (Carlquist 1980a). The presence of degrees of intermediacy between monomorphic and dimorphic fibers is the prime source of evidence, and in fiber dimorphism, only a scattering of cells become subdivided into strands – most retain an undivided status. Also, bands due to fiber dimorphism occur in a few groups in families in which apotracheal banded does not seem to be a widespread or basic type (e.g., in Asteraceae, axial parenchyma is mostly paratracheal).

5.2.7 Marginal Parenchyma

This term has been devised by Hess (1950) to include two types of parenchyma that occur with respect to growth rings: initial, which occurs at the beginning of growth rings, and terminal, which is present as a band at the close of a growth ring (Fig. 2.5, parts 1–4). Marginal parenchyma can also include instances in which both terminal and initial parenchyma are present (Fig. 2.11, parts 3, 4). Terminal parenchyma is much more common than initial, and "terminal" had been used by some workers as a synonym for marginal parenchyma (IAWA Committee on Nomenclature 1964). The term "boundary parenchyma" is used for marginal parenchyma by some authors (e.g., Meylan and Butterfield 1978a). In Kribs (1937), only terminal parenchyma is considered, but that is very likely because instances of initial parenchyma are so infrequent he could not meaningfully incorporate them into his statistical compilations. Initial parenchyma has been reported in *Cedrela odorata*, *Tectona grandis* (Fig. 2.10, part 4), *Terminalia tomentosa*, and other species (Jane 1934; Chowdhury 1964). The types of marginal parenchyma have been used for recognition of a type of growth ring, type 12 (Sect. 2.4.12).

The place of terminal parenchyma in the table of Kribs reproduced above shows that terminal parenchyma is found in relatively specialized woods. However, that should not be interpreted in terms of a progressive phylesis through various other types. Rather, a wood with a capability to form any parenchyma type can produce parenchyma at the close of a growth ring: in the *Hibiscus* shown in Fig. 5.6 (parts 1 and 2), terminal parenchyma is present in addition to both diffuse and vasicentric parenchyma. Thus, a wood in virtually any stage of phylogenetic specialization might form terminal parenchyma. The high degree of specialization in Kribs' table may relate to the fact that temperate woods that have well-marked growth rings are, as a group, more specialized in more numerous features than an equivalent sample of wet tropical species.

5.2.8 Ray-Adjacent Parenchyma

This type was noted by Hess (1950) in *Tetramerista*. It also occurs in some species of *Staphylea* (Carlquist and Hoekman 1985a) and in Basellaceae (Carlquist 1999e). Ray-adjacent parenchyma occurs if the majority of cells in a basically diffuse pattern tend to be clustered along the margins of rays. Ray-adjacent axial parenchyma should not be confused with the sheathing cells of rays (erect cells on the multiseriate portions of multiseriate rays). By checking transections against radial sections, one should be able to report ray-adjacent parenchyma with certainty.

5.2.9 Vasicentric Scanty and Vasicentric Abundant Parenchyma

With these types, we begin descriptions of paratracheal types of axial parenchyma. I am using vasicentric scanty and vasicentric abundant in the sense of Kribs (1937), and thereby departing somewhat from the slightly more ambiguous usage of the IAWA Committee on Nomenclature (1964), in which an incomplete sheath of axial parenchyma cells around a vessel is termed scanty paratracheal and contrasted with vasicentric, which thereby implicitly would include only abundant axial parenchyma adjacent to vessels. In the present book, "paratracheal" covers all types of axial parenchyma distributed with reference to vessels, and the term vasicentric (including both scanty and abundant types) is used to denote axial parenchyma distributed adjacent to vessels (or vessel groups) with cells approximately of equal abundance on all sides of the vessel. Figure 5.4 (part 1) illustrates vasicentric scanty parenchyma, the cells of which form an incomplete sheath rarely more than one cell wide around vessels and vessel groups. In Fig. 5.4 (part 2) is an example that corresponds to the description of vasicentric abundant parenchyma: a sheath of cells at least two cells wide forming a complete sheath around a vessel. A similar condition is shown for the same species, but with parenchyma around a pair of vessels not in contact, in Fig. 5.4 (part 3). Examples that show a cylindrical sheath equally thick in all directions around a vessel, as shown in Fig. 5.4 (part 2), are not as common in dicotyledons as paratracheal types that show the parenchyma sheath thicker or wider in one direction than another.

Kribs' (1937) table, reproduced above, shows that vasicentric scanty parenchyma is found in woods of approximately the same degree of primitiveness as diffuse-in-aggregates, whereas vasicentric abundant ranks as highly specialized. This sequence corresponds to the same interpretation as

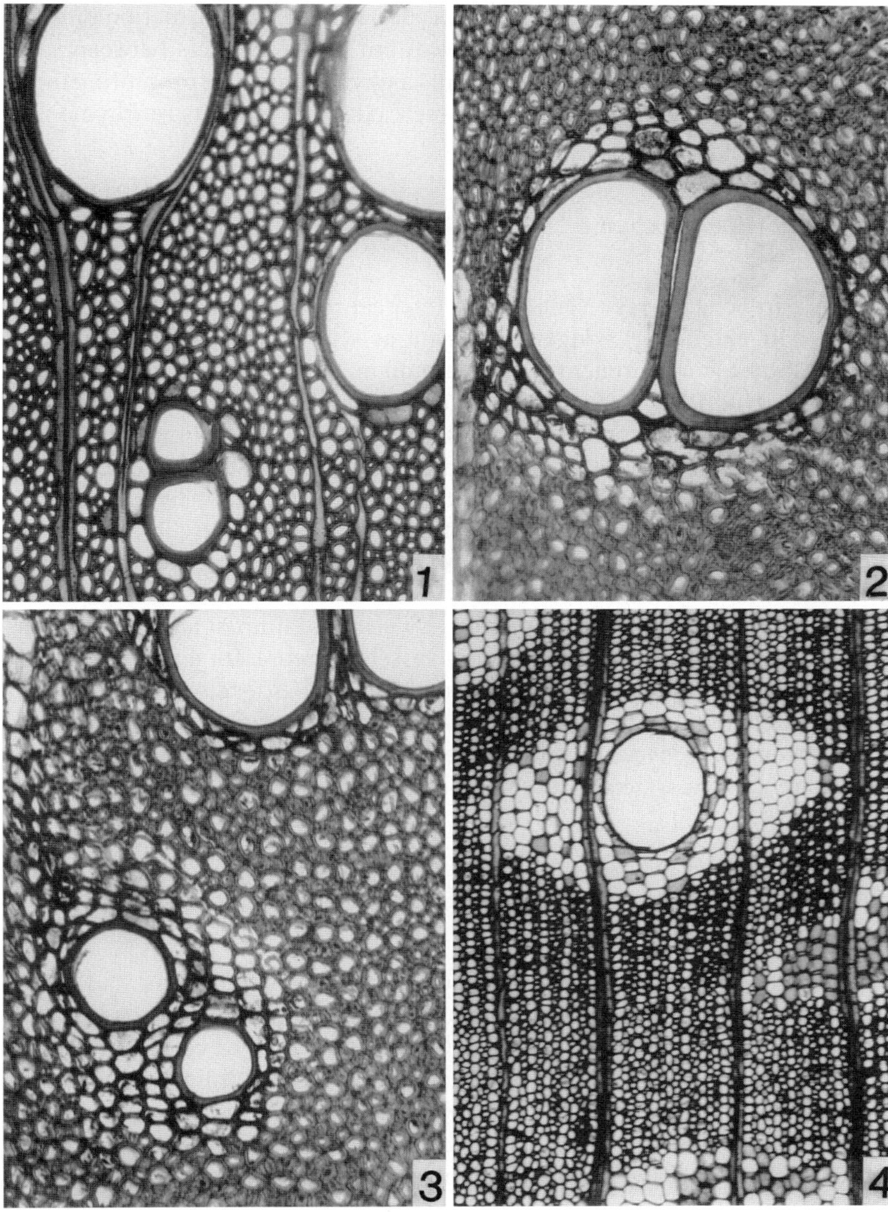

Fig. 5.4. Paratracheal axial parenchyma types. *1* Vasicentric scanty, *Fraxinus quadrangulata* (Oleaceae); *2*, *3* vasicentric abundant, *Tamarix articulata* (Tamaricaceae); *2* axial parenchyma surrounds an adjacent pair of vessels; *3* axial parenchyma surrounds a pair of vessels separated by parenchyma; *4* aliform, *Amburana cearensis* (Fabaceae). (*1–3:B*; *4:A*)

enunciated for apotracheal parenchyma types, namely, that formation of more massive parenchyma sheets forming fewer but larger contacts between axial and radial parenchyma systems represents a phyletic advance promoting more efficient photosynthate conduction between axial and radial parenchyma.

5.2.10 Aliform Parenchyma

Aliform parenchyma represents a paratracheal distribution in which lateral (tangentially-extending) wings are present as seen in a transection (Fig. 5.4, part 4). Aliform axial parenchyma is common in certain tropical legumes, such as *Inga* (Cozzo 1950), and also in certain Moraceae, such as *Brosimum* and *Mesogyne* (Koek-Noorman et al. 1984b). However, illustrations of aliform parenchyma in various works on wood anatomy are misleading in that they tend to imply that a large number of woods may have parenchyma that conforms closely to the ideal illustration of aliform, as shown in Fig. 5.4 (part 3). In fact, in woods that have aliform parenchyma, transitions to confluent and vasicentric abundant types are common.

Aliform parenchyma might not seem, by virtue of limited tangential extensions of the parenchyma sheath around vessels, to represent a special adaptation to conductive efficiency. However, the lateral extensions ("wings") of aliform parenchyma, when seen in three dimensions, do represent a device to increase contact between a paratracheal axial parenchyma system and the radial parenchyma system represented by rays (Carlquist 1985e).

5.2.11 Confluent Parenchyma

Confluent parenchyma is a paratracheal type in which vessels and vessel groups are located in tangential bands of parenchyma (Fig. 5.5, parts 1 and 2). The bands may be indefinite in extent, but can be traced to vessels by following them tangentially in a wood transection. The term "confluent" implies an interconnection, and in fact, in many instances confluent is transitional between an aliform condition and a condition in which bands of indefinite tangential extent occur. The "aliform" component of such a transition is shown by the tendency of the parenchyma sheath to be radially thicker near a vessel, and to taper to a thinner band tangentially on either side of the vessel. Transitional instances are termed aliform-confluent; this condition was cited by Jane (1956) for *Dalbergia latifolia*; it characterizes *Mesogyne insignis* (Koek-Noorman et al. 1984a), *Bosqueiopsis gilletii*, and seven species of *Brosimum* (Koek-Noorman et al. 1984b). In Kribs' (1937) survey, confluent parenchyma was not recognized. Confluent parenchyma can resemble apotracheal banded occurrences (see Sect. 5.2.6).

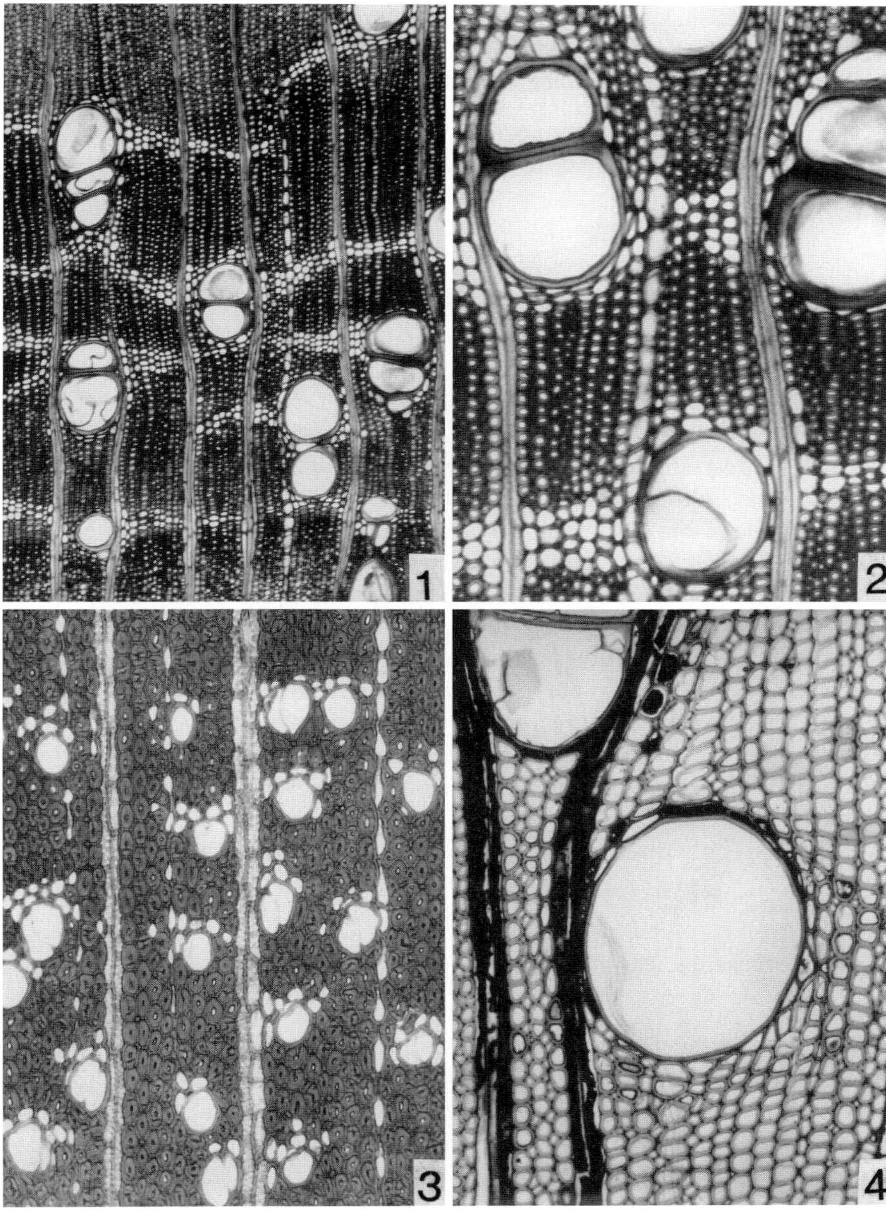

Fig. 5.5. Paratracheal axial parenchyma types. *1, 2 Brosimum uleanum* (Moraceae); *1* confluent, showing that all vessels are included within parenchyma bands; *2* confluent, showing that width of bands is greater closer to vessels than further away from them; *3* abaxial, *Notobuxus natalensis* (Buxaceae); *4* abaxial (parenchyma cells abaxial to vessel, *center*, have dark-staining contents), *Canarium rufum* (Burseraceae). (*1:A*; *2–4:B*)

Confluent parenchyma may be found in *Andira insignis* (Jane 1956) and in Gyrostemonaceae (Carlquist 1978b). It represents a very effective interconnection between axial parenchyma and rays, because all vessels are included in the system.

5.2.12 Abaxial Parenchyma

Paratracheal parenchyma in which the cells are more abundant on abaxial sides of vessels has also been termed "unilaterally paratracheal" parenchyma (Hess 1946d; Jane 1956). There may be only a few cells abaxial to a vessel, as in *Notobuxus* (Fig. 5.5, part 3) and other Buxaceae (Carlquist 1982d) or *Illicium* (Carlquist 1982c) or several cell layers, as in *Canarium* (Fig. 5.5, part 4). Van Vliet (1979) illustrates abaxial parenchyma in *Terminalia bialata* and *Pteleopsis hylodendron*, although he designates it merely as "paratracheal." There is a tendency toward abaxial parenchyma in a number of genera in Canellaceae (Wilson 1960), Clusiaceae, and Lauraceae (Stern 1954; Richter 1981). A key to genera of New World woods in which abaxial parenchyma occurs is offered by Hess (1946d).

No functional explanation for the conformation of abaxial parenchyma has as yet been offered. Abaxial parenchyma is not included in the survey of Kribs (1937). It occurs in many more families with relatively primitive wood, such as Buxaceae, Canellaceae, Clusiaceae, Dichapetalaceae, Ericaceae, Humiriaceae, Icacinaceae, Illiciaceae, Lauraceae, Rhizophoraceae, and Theaceae, than in families with more specialized wood, such as Apocynaceae, Combretaceae, and Fabaceae (only three genera of Fabaceae listed by Hess 1946d).

5.2.13 Intervascular Parenchyma

The term "intervascular parenchyma" is coined here to refer to occurrences in which vessels, although grouped, are often separated from each other by parenchyma cells. This condition, shown here for *Kalanchoe* (Fig. 5.7, part 4), is found in other genera of Crassulaceae with fibrous wood, such as *Aeonium*.

5.2.14 Combinations of Axial Parenchyma Types

Works on comparative anatomy of dicotyledons, by stressing particular types of axial parenchyma, may lead one to think that a given wood characteristically has a single parenchyma type. In fact, some woods simultaneously have two or three axial parenchyma types. The *Hibiscus* shown here (Fig. 5.6, parts 1 and 2)

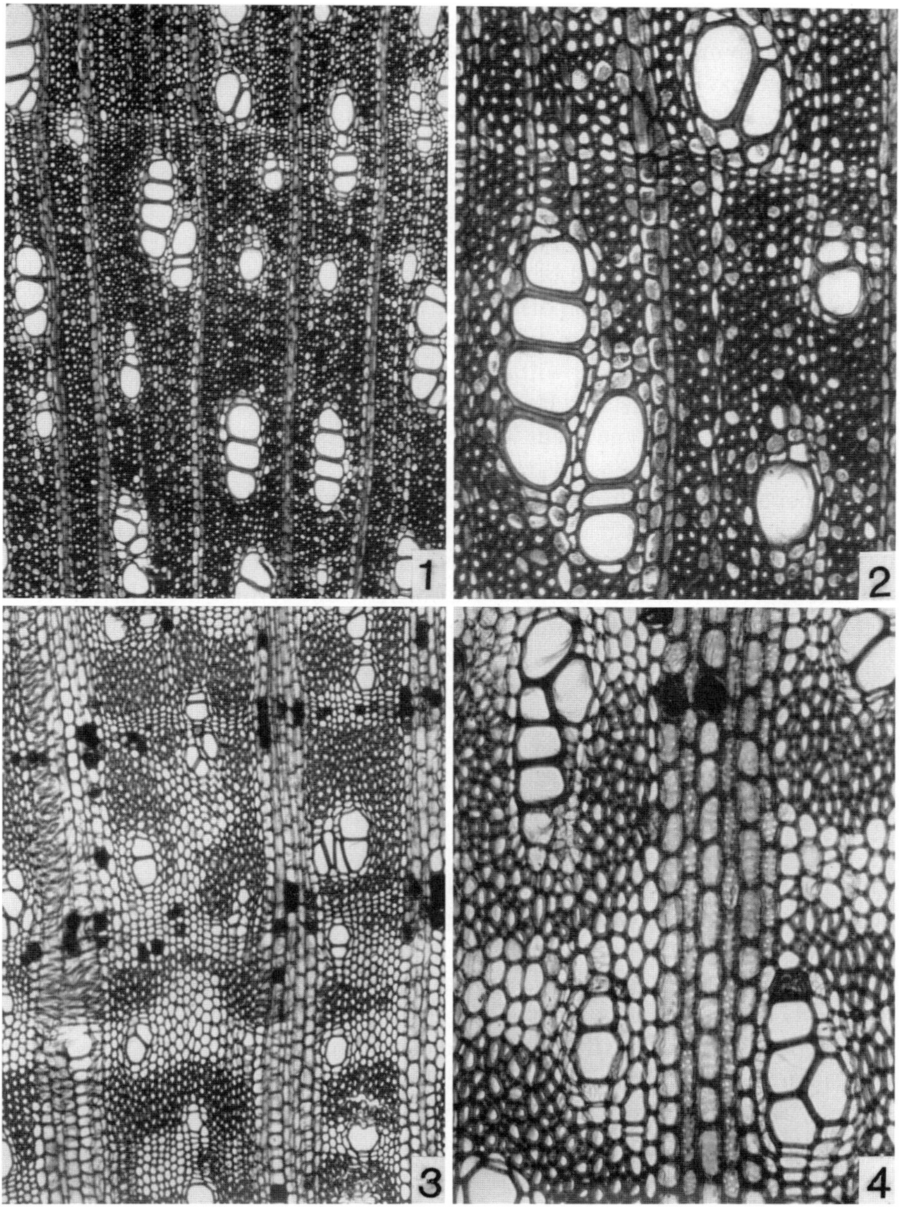

Fig. 5.6. Combinations of axial parenchyma types. *1, 2 Hibiscus* sp. (Carlquist 6088; Malvaceae); *1* a single terminal parenchyma band is shown, *near top*; *2* diffuse axial parenchyma cells and vasicentric scanty parenchyma can be distinguished from fibers; terminal parenchyma band portion *near top of photograph*; *3, 4 Corynocarpus laevigata* (Corynocarpaceae); *3* bands and patches of axial parenchyma are present; *4* various amounts of parenchyma are present adjacent to or surrounding vessels or vessel groups. (*1,3:A*; *2,4:B*)

has diffuse, vasicentric (intermediate between scanty and abundant), and terminal parenchyma. In *Corynocarpus laevigata* (Fig. 5.6, parts 3 and 4) one sees parenchyma that takes banded forms, but diagonal as well as tangential directions are evident. Although the bands do not seem all distributed with relation to vessels, vessels or vessel groups mostly contact parenchyma: *Corynocarpus* furnishes an irregular combination of apotracheal banded, vasicentric scanty, and confluent types. *Paratrophis* has a combination of vasicentric and marginal parenchyma (Patel and Bowles 1978b). In *Apeiba* (Tiliaceae) wood has vasicentric, diffuse, and apotracheal banded types (den Outer and Schutz 1981); the banded parenchyma is further distinctive by virtue of marked radial elongation of the cells. In Staphyleaceae, diffuse, diffuse-in-aggregates, and ray-adjacent parenchyma types may occur within a single species, or only one or two of these may be present (Carlquist and Hoekman 1985a). Other examples in which more than one parenchyma type occur within a single species are offered by Milanez (1944) and Hess (1950).

Another aspect of multiple types of parenchyma within a single species is the occurrence of inconstancy of parenchyma types within a species. An example of this is offered in *Dalbergia latifolia* by Patel and Shah (1980). Inconstancy of this sort suggests that one should be cautious in designating parenchyma types as taxonomic characters if limited quantities of material are at hand.

Nevertheless, there appear to be some examples, as in *Cneorum* (Carlquist 1987g) in which species differ from each other by means of parenchyma types.

5.2.15 Other Axial Parenchyma Types

Koek-Noorman et al. (1984b) reported a tendency for parenchyma bands – basically portions of a confluent pattern – to run diagonally and interconnect with each other in *Trilepsium madagscariense* (Koek-Noorman et al. 1984b).

Hess (1950) proposed a number of additions to axial parenchyma typology. Perhaps because of the intricacy of the distinctions and the fact that the proposed types do not seem strongly tied to taxonomic or functional distinctions, the system of Hess has not met with acceptance. Rather than coining terms for variants, use of the basic types given above plus descriptive adjectives seems the simplest procedure.

5.3 Pervasive Axial Parenchyma

This term is used here to denote phyletic replacement of imperforate tracheary elements, such as libriform fibers, with axial parenchyma. Because earlier studies in wood anatomy dealt so very largely with arboreal species, instances of such extensive parenchymatization of the axial secondary xylem omit this type, cited as "parenchymatized wood" in Fig. 5.1 (left, near top).

Axial parenchyma bulks large in secondary xylem of some succulents. The phenomenon of parenchymatization can be demonstrated well in succulent groups in which relatively woody as well as highly parenchymatous woods are both present. For example in Cactaceae subfamily Cactoideae (Gibson 1973) and in *Opuntia* species with cylindrical stems (Gibson 1977a) as well as those with flattened stems (Gibson 1977b), both "fibrous" and "nonfibrous" woods are present, depending on the genus and species. The more "fibrous" woods in cacti are from arborescent species, as one might expect.

Many plants with stems succulent in appearance may have relatively little axial parenchyma in the secondary xylem itself, but may have large volumes of parenchyma in pith and cortex. This occurs in some succulents one might have thought could have parenchymatous secondary xylem, such as *Aeonium* (Crassulaceae), in which wood is rayless and composed of libriform fibers; axial parenchyma is present only around vessels. However, in most Crassulaceae, such as *Crassula arborea* (Fig. 5.7, part 3), imperforate tracheary elements are entirely absent and fascicular (nonray) xylem consists of vessels scattered in a background of axial parenchyma. Pervasive axial parenchyma may be found in such succulent plants as globular cereoid cacti (Fig. 4.9, parts 1 and 2), *Brighamia* of the lobelioid Campanulaceae (Carlquist 1962a, 1970b), *Carica* of the Caricaceae (Carlquist 1962a, 1998d; Fisher 1980), and in the roots of *Nastanthus* (Carlquist and Devore 1998) and *Silene* (Carlquist 1995c).

Structures obviously related to storage, especially those formed underground, also have nonfibrous wood (e.g., *Pterocactus tuberosus*; Gibson 1978c). In unusual cases, one can find fibrous wood in some shoots, nonfibrous wood in other shoots of a single plant (Gibson 1978a). Such an instance is described by Mauseth and Plemons-Rodriguez (1997), who note that vessel characters are the same in either type of wood; one would expect this because the entirety of a cactus, with its predominance of parenchyma, is probably more significant than localized areas of fibers.

A detailed listing of "avenues for the origin of succulence" in stems and roots of the leaf-bearing cacti has been summarized by Gibson (1975, p. 85) from the work of Bailey; some of these apply only to cacti, as far as is presently known. The pereskioid cacti, in fact, offer interesting examples of modes of structure that are typical of semisucculent trees. Large rays in the pereskioid cacti are related to succulence, but occasional bands of axial parenchyma also clearly are; these tend to occur in outer portions of the secondary xylem (Bailey 1962, 1964) or at somewhat irregular intervals.

Notably wide bands of axial parenchyma characterize semisucculent trees such as *Erythrina* (Fig. 5.7, part 1; Cumbie 1960). Wideness of bands is probably related to relative degree of succulence in trees of Moraceae (Koek-Noorman et al. 1984a,b); such bands are seen in the genus *Cecropia* (Bonsen and ter Welle 1983). The trees *Adansonia* and *Chorisia* of the Bombacaceae have wide apotracheal bands much like those of *Erythrina* (Metcalfe and Chalk 1950; Fisher 1981). The genus *Apeiba* of the Tiliaceae is a good example of wide parenchyma bands (den Outer and Schutz 1981); the cells of these bands

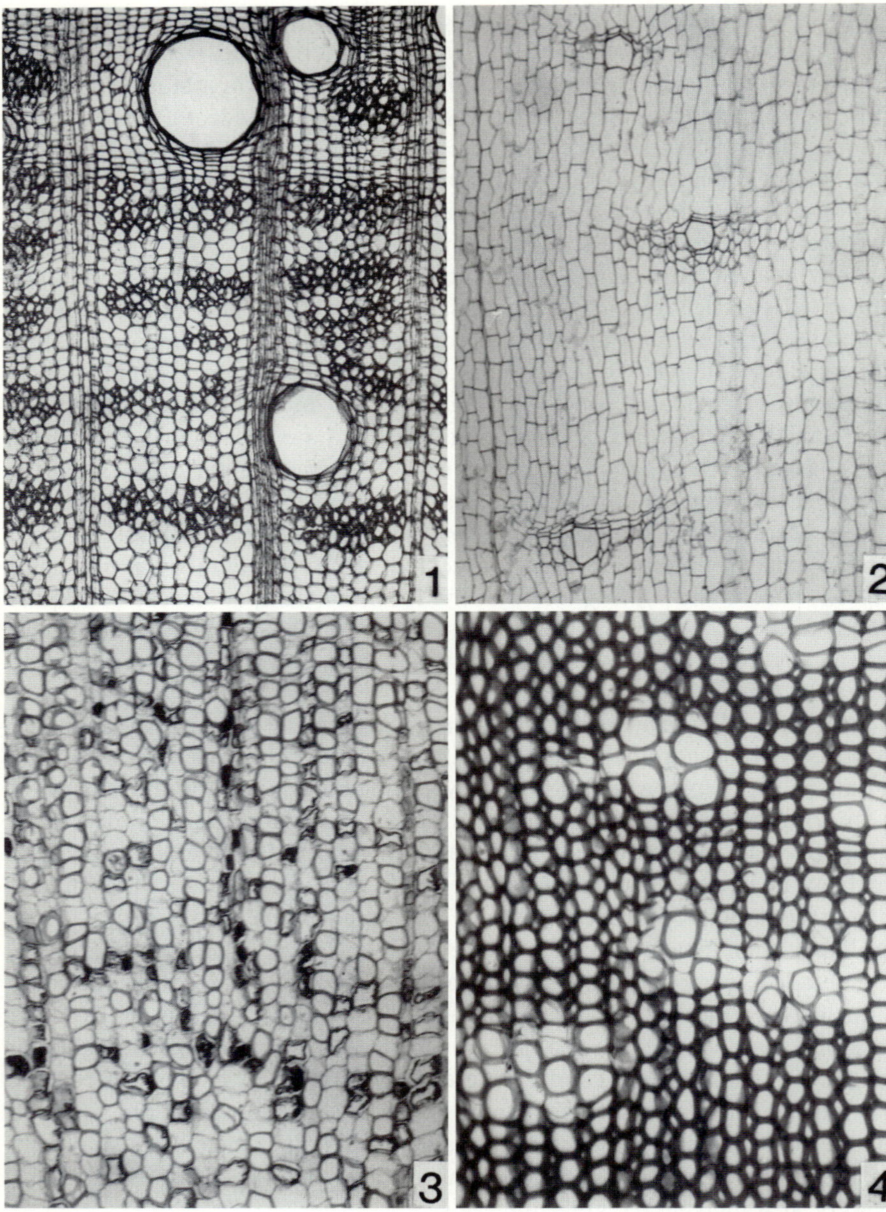

Fig. 5.7. Unusual axial parenchyma types. *1* Bands numerous, wide, and although many bands do not appear disposed with reference to vessels, all vessels do occur in parenchyma bands: *Erythrina constantiana* (Fabaceae); *2* axial parenchyma bands exceptionally wide, perhaps best designated as apotracheal because vessels are sheathed by winglike groupings of libriform fibers *Aeschynomene elaphroxylon* (Fabaceae); *3* secondary xylem consisting wholly of vessels and parenchyma, an example of parenchymatization of the axial xylem: *Crassula arborea* (Crassulaceae); *4* intervascular parenchyma (parenchyma cells scattered among, as well as beside, vessels in the vessel groups), *Kalanchoe beharensis* (Crassulaceae). (*1–3:A*; *4:B*)

experience marked radial elongation. Cell size of the axial parenchyma cells mentioned in this paragraph tends to be appreciably larger than sizes found in nonsucculent woody plants. *Adansonia*, *Chorisia*, and *Erythrina* are drought-deciduous trees that flower during the dry season; the prominent axial parenchyma bands are doubtless related to these events.

Unusually large bands of axial parenchyma occur in woods of some plants that are neither trees nor stem succulents but are, rather, shrubs of dry localities. Metcalfe and Chalk (1950) called attention to this phenomenon in *Alyssum saxatile* and *A. spinosum* (Brassicaceae). In Scrophulariaceae, this phenomenon was reported by Michener (1983) as "fiber walls not lignified the second year" in *Mimulus* (*Diplacus*) *clevelandii*. Such bands occur in Asteraceae in a number of Heliantheae, such as *Argyroxiphium caliginis*, *Dubautia microcephala*, *D. menziesii*, *Hemizonia minthornii*, and *Perymenium strigillosum* (Carlquist 1958a, 1997a, 1998e,f). In Polemoniaceae, these bands may be seen in *Acanthogilia gloriosa* and *Ipomopsis aggregata* (Carlquist et al. 1984). In Hydrophyllaceae, one finds these bands in perennial species of *Phacelia* (Carlquist and Eckhart 1984). These bands may contain water or starch, or both, and might represent a drought-survival mechanism or way of preparing for renewal of conduction and growth at the end of a dry season, or both.

5.4 Parenchyma Proliferation

This term is applied here to a phenomenon characteristic of succulent species of *Fouquieria*. This process is considered along with cambial variants (see Sect. 8.4.5) because cell division allied to cambial action is involved.

5.5 Histological Details of Axial Parenchyma

5.5.1 Nature of Strands

Axial parenchyma is typically subdivided into strands. Strands may be long, with ten or more cells, in families that have long fusiform cambial initials, such as Staphyleaceae (Carlquist and Hoekman 1985a) and Scytopetalaceae (Carlquist 1987h). The majority of dicotyledon families have strands of two to seven cells. Kribs gives data in two of his tables that is relevant; combining these figures, one finds that in woods with diffuse axial parenchyma and with a vessel element length of 920 μm (which we may assume is about the length of a fusiform cambial initial), the average length of a parenchyma strand cell is 154 μm, and therefore there are, in such woods, about six cells on the average.

In woods with vasicentric abundant parenchyma, vessel element length is 310 µm, and parenchyma strand cell length averages 88 µm; therefore, the average strand ought to be between three and four cells long. A few families, other than those that show fiber dimorphism (see Sect. 4.8), typically have undivided axial parenchyma strands; Misodendraceae (Carlquist 1985d) is an example. In instances of fiber dimorphism, the majority of the axial parenchyma cells may be undivided, often with a few cells divided once or twice. The distinction between septate libriform fibers and strand parenchyma is discussed above (Sect. 5.1).

5.5.2 Pitting

The distinction between contact and isolation pits in axial parenchyma has been discussed above (Sect. 5.2.1). A further source of diversity is the presence of bordered pits on axial parenchyma cells. This is discussed in connection with cell types not considered to be tracheids (Sect. 4.16). Bordered pits have been illustrated on axial parenchyma cells of *Nepenthes* (Carlquist 1981b) and *Bubbia* (Carlquist 1983b), and have been reported in *Populus* (Czaninski 1977) and Lardizabalaceae (Carlquist 1984f). Bordered pits are very likely much more widespread on axial parenchyma cells than reports presently indicate. One must stress in this connection that borders are very difficult to see if bordered pits on axial parenchyma cells are observed in face view. Borders are much more easily seen in sectional view, and presence of deposits of amorphous substances in axial parenchyma cells may fill the sectioned pits and outline the borders.

Castro (1985) has studied pit membranes of axial parenchyma cells adjacent to vessels in Lauraceae; pectocellulosic excrescences are present on the parenchyma side of the pit membranes in this family, and tests have assured that these are not artifacts. Lörcher et al. (1987) consider such axial parenchyma cells in Valerianaceae to be transfer cells, perhaps functioning in selective transfer of solutes.

5.5.3 Sclerenchymatous Axial Parenchyma

Axial parenchyma cells that develop very thick walls and therefore may be classified as sclereids have been reported by Boureau (1957) in Apocynaceae (*Aspidosperma aquatica*), Bignoniaceae (*Couralia loxophora*), Euphorbiaceae (*Chaetocarpus, Cunuria, Elateriospermum*), Fabaceae (*Clathrotropis, Martiodendron excelsa*), Linaceae (*Hebepetalum*), and Rutaceae (*Flindersia ifflaiana*). Metcalfe and Chalk (1950) figure sclerified axial parenchyma cells for *Elateriospermum tapos*.

Sclerified axial parenchyma cells may also be seen in the case of many instances of chambered crystals in which the crystal becomes "encapsulated"; in this process, a secondary wall is secreted around the crystal as well as on the cell wall. If this process proceeds far enough, the entire lumen of the crystal-bearing cell becomes filled with secondary wall material (Fig. 7.5, part 2).

5.5.4 Starch

Axial parenchyma is rich in starch in certain species, particularly those in which starch also occurs in ray cells. Starch grains in axial parenchyma are shown here with light microscopy (Fig. 5.8, part 1) and scanning electron microscopy (Fig. 5.8, part 2).

5.5.5 Crystals in Axial Parenchyma

Crystals in wood are covered in detail in Chapter 7. Crystals are mentioned here because they may be related to subdivision of a parenchyma strand into "chambered crystals" – a slight misnomer, because the strand cell is subdivided into rectangular (Fig. 5.8, part 3; left and right) and square cells, each containing a rhomboidal crystal (sometimes with additional minute crystals). As if to cite an exception that proves the rule, Mennega (1973) describes in *Cedrelinga* strand cells subdivided like those that normally would contain chambered crystals yet are lacking in crystals.

In addition to strand cells, undivided fusiform axial parenchyma cells may contain crystals; these may be seen in *Frankenia palmeri* of Frankeniaceae (new report), and are illustrated in Chapter 7 for *Passerina vulgaris* (Fig. 7.8, parts 3–5).

5.5.6 Vessel-Associated Cells and Osmotic Phenomena

Czaninski (1977) finds that axial parenchyma cells in contact with a vessel may be subdivided into three categories: transfer cells, which chiefly occur in primary xylem (see Gunning et al. 1968; Gunning and Pate 1974), contact cells, and vessel-associated cells. The concept of contact cells has been discussed above (Sect. 5.2.1). Czaninski (1977) calls attention to vessel-associated cells as cells that resist attack and accumulate polysaccharides and phenolic substances when attacked, emptying these into vessels. Transfer of sugars into vessels from paratracheal axial parenchyma, thus raising the osmotic

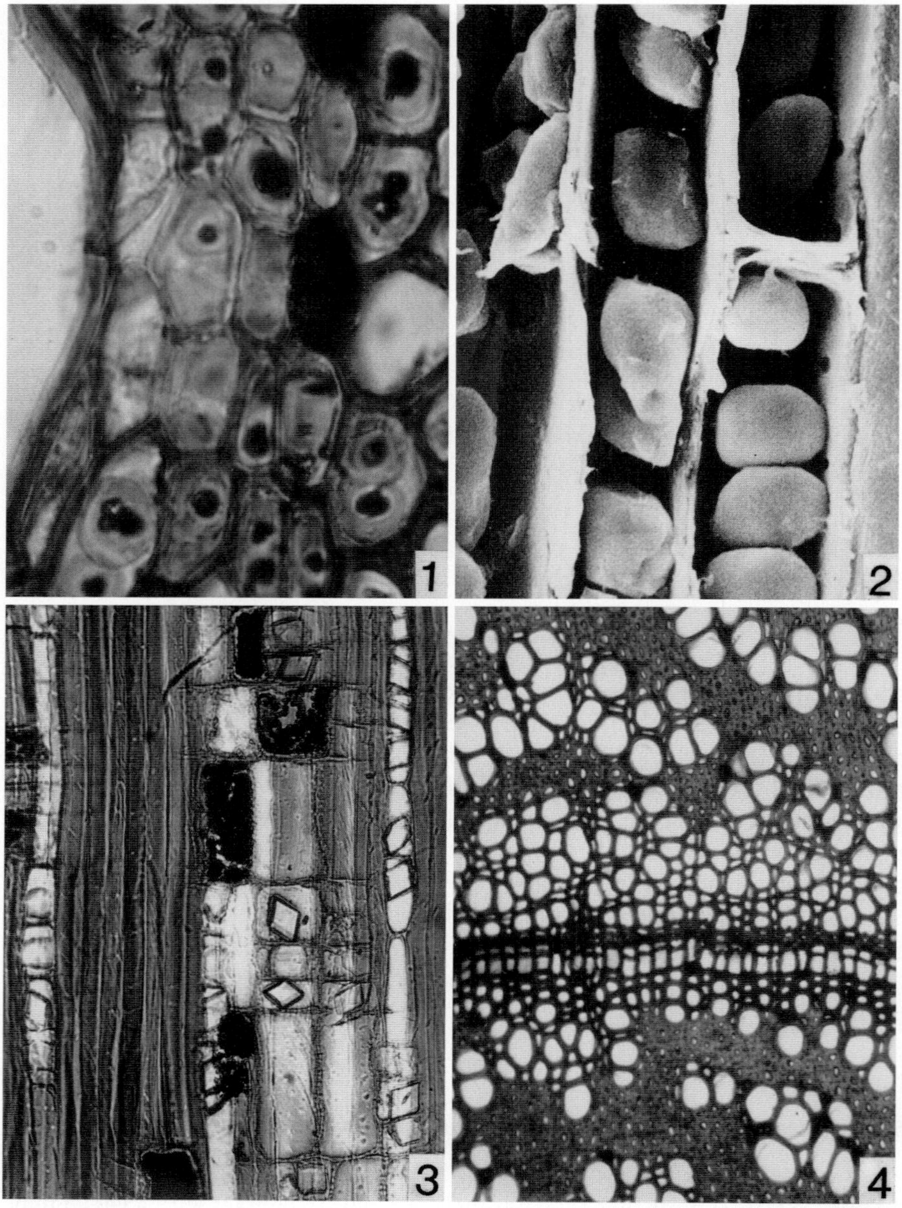

Fig. 5.8. Axial parenchyma cell types, details. *1* Starch grains (*dark spot in center of each grain* is air space) in parenchyma adjacent to vessel (*left*), wood transection, *Cecropia latifolia* (Cecropiaceae or Moraceae); *2* starch grains in axial parenchyma cells, tangential section, SEM photomicrograph, *Cercidium microphyllum* (Fabaceae); *3* radial section showing crystals in both axial parenchyma (*left*, *right*) and ray parenchyma, *Balanops pancheri* (Balanopaceae); *4* interxylary cork band (considered here a specialized type of terminal axial parenchyma) just *below center*, *Artemisia tridentata* (Asteraceae). (*1:C*; *2:H*; *3,4:B*)

potential of sap in vessels and enhancing conduction, has been stressed by Braun (1984) as a function of axial parenchyma. More familiar for plant anatomists is the concept that vessel-associated cells generate tyloses when vessels are deactivated.

5.5.7 Intercellular Spaces

Disjunctive parenchyma, the phenomenon in which parenchyma cells become pulled apart during maturation and remain in contact by means of tubular processes on the cells, is usually found (it is a rare phenomenon) in ray cells. It also may be found in axial parenchyma cells occasionally.

5.5.8 Interxylary Cork

The phenomenon of interxylary cork is not easily referable to any secondary xylem category. It is related to growth ring activity, but even in genera such as *Artemisia*, in which interxylary cork is conspicuous (Fig. 5.8, part 4), it occurs in only a few species (Moss 1936, 1940; Diettert 1938; Moss and Gorham 1953). Suberization of ray cells may also occur in relation to splitting of wood into segments, a phenomenon characteristic of desert shrubs and some portions of herbaceous and "woody herb" dicotyledons (Ginzburg 1963; Waisel et al. 1972; see Jones 1984 for a review). Interxylary cork in the restricted sense does not include suberization of rays, but rather applies to a layer of suberized cells laid down by the cambium at the end of a growth ring. Interxylary cork in this sense may justifiably be regarded as suberized axial parenchyma, hence its inclusion at this point. Interxylary cork has been found in Asteraceae (*Artemisia*, *Crepis*, *Taraxacum*), Boraginaceae (*Mertensia*, *Lithospermum*), Geraniaceae (*Geranium viscosissimum*), Onagraceae (*Chamaenerion*, *Gaura*, *Oenothera*), Polemoniaceae (*Polemonium*), Polygonaceae (*Polygonum glaucum*), and Ranunculaceae (*Aconitum*, *Delphinium*) according to the references cited above as well as Kingsley (1911) and Metcalfe and Chalk (1983, p. 214).

5.5.9 Axial Parenchyma Idioblasts and Secretory Canals

This topic is discussed in greater detail in Chapter 7. However, for convenience of reference to axial parenchyma, mention should be made here of oil cells in axial parenchyma of Lauraceae (Record 1944e; Richter 1981) and Canellaceae (Wilson 1960) and of mucilage in parenchymalike thin-walled fibers of

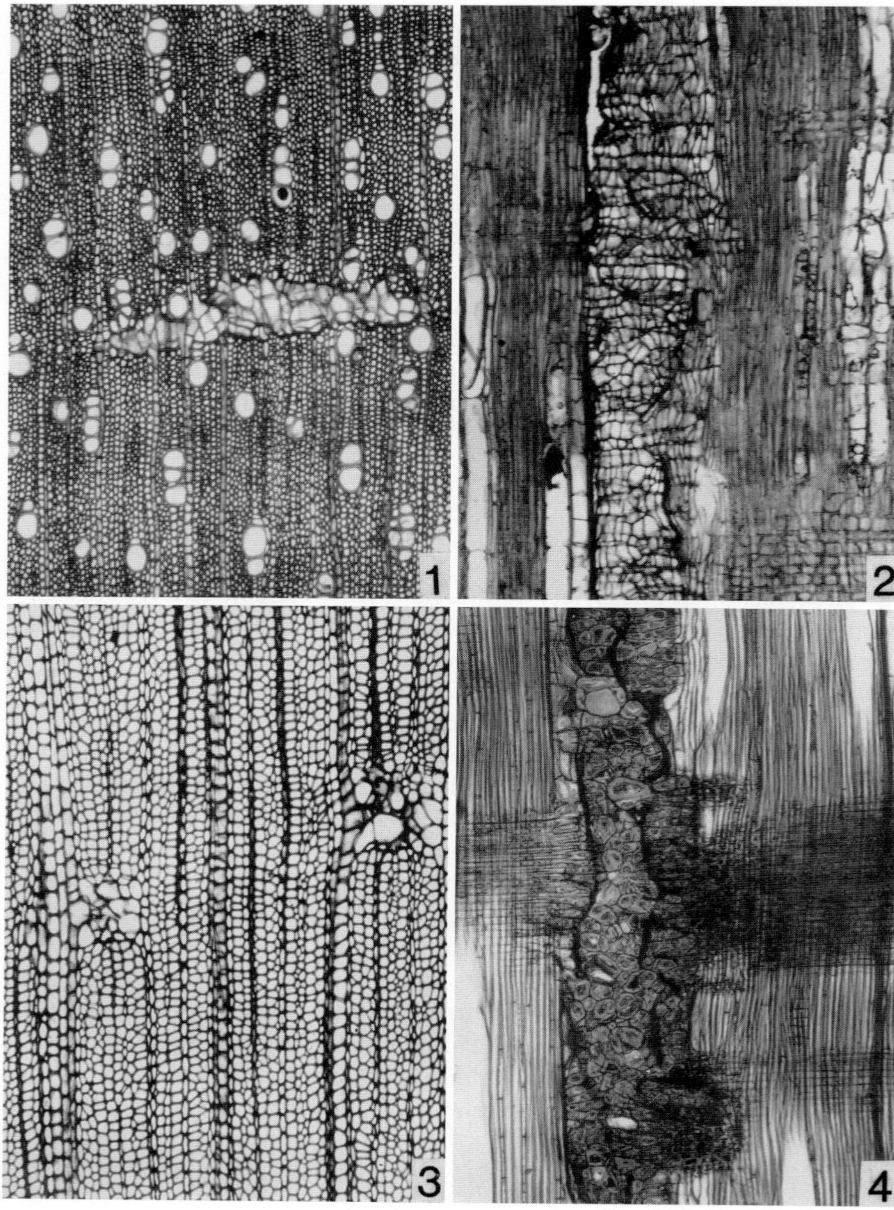

Fig. 5.9. "Pith flecks" (callus tissue intercalated in secondary xylem). *1, 2 Himantochilus marginatus* (Acanthaceae); *1* band-shaped callus tissue as seen in transection; *2* as seen in radial section, callus tissue forms an extensive band; *3* two strands of callus tissue, transection of wood, *Tasmannia purpurea* (Winteraceae); *4* sclerosed callus tissue as seen in radial section, *Betula nigra* (Betulaceae). (*1–4:A*)

Trimenia (Carlquist 1984e). Axial parenchyma in Dipterocarpaceae and Fabaceae contains secretory canals (Record 1944a).

5.5.10 Interxylary Phloem

Although by definition phloem is not the concern of this chapter, one must remember that interxylary phloem (formed from a single cambium) is sometimes sheathed by thick cylinders of axial parenchyma, as in Onagraceae (Carlquist 1975b) and Acanthaceae (Carlquist and Zona 1988a).

5.5.11 Helical Sculpture in Axial Parenchyma

Although helical sculpture is characteristic of vessels and imperforate tracheary elements in certain families, it has been reported in the axial parenchyma cells of Chrysobalanaceae (ter Welle 1975), Ancistrocladaceae and Dioncophyllaceae (Gottwald and Parameswaran 1968).

5.6 "Pith Flecks": Callus Cells Within Wood

The misleading term "pith fleck" (inferring a resemblance to pith parenchyma) has been applied to bands of large parenchyma cells, irregular in orientation, found in various woods (Fig. 5.9, parts 1–3). These bands or pockets are the result of trauma in which the cambium has been injured locally; callus forms, presumably by proliferation of parenchyma cells adjacent to the cambial injury, and soon orderly divisions characteristic of the cambium are restored, so that the callus (which is what a pith fleck represents) is enclosed by normal wood on all sides. The size of the callus pocket presumably corresponds to the size of the injury. Cold and drought are probable sources of injury. Illustrations of such callus have been offered by Stern (1954), Dickison (1972), and Stahel (1972). The callus cells may become sclerosed, as illustrated here for *Betula nigra* (Fig. 5.9, part 4).

6 Rays

6.1 Terminology, Definitions

Xylem rays (wood rays) are sheets of parenchyma extending radially inward from the cambium. In most woody plants, a few primary rays are extended by the cambium as secondary rays, while numerous other secondary rays are added by the cambium. In some herbs or woody herblike plants, many of the primary rays may be extended by the cambium with few modifications and only a small number of new wood rays added by the cambium as growth proceeds.

Rays are ordinarily subdivided into uniseriate, those rays that are only one cell wide at most (as seen in tangential section), and multiseriate rays, which are two or more cells wide at the widest point. Multiseriate rays may have uniseriate wings of varying lengths. The terms "biseriate" and "pluriseriate," although not mentioned by the IAWA Committee on Nomenclature (1964), are sometimes used for rays that are two cells wide and more than two cells wide, respectively.

The height of a ray is considered to be the vertical distance from tip to tip as seen in tangential section; should the ray narrow to a single cell in width and widen again, the height is the entire distance. Ray width is the horizontal axis of the ray as seen in a tangential section; the width of a ray is usually measured at its widest point. Uniseriate rays may be only a single cell in height.

Ray cells that are vertically longer than wide are said to be upright (erect). Cells that are radially longer than wide are termed procumbent. Cells that are about as tall as wide are termed square. These characteristics are defined on the basis of cells as seen in a radial section, and can be applied securely only in a radial section. However, the reader will note that these terms are often used when referring to tangential sections, as in Figs. 6.2–6.10 here. One can do this only by viewing a radial section and applying the information one obtains there about upright, square, and procumbent cells to what one sees in a tangential section. A ray cell taller than wide as seen in a tangential section may, in fact, be square or procumbent rather than upright when one sees it in a radial section – or the reverse (a tile cell is a good example of the reverse). Upright cells can be so tall that in a tangential section they simulate a libriform fiber; in a radial section, the upright cells will be more obviously parts of rays, and will have blunt rather than tapered ends, so that they do not look like fibers, and with this knowledge, an observer can more readily distinguish upright ray cells from libriform fibers in a tangential section.

If a ray has only procumbent cells, it is said to be homogeneous; the term homocellular is used by some authors for this condition. If a ray has square and/or upright cells in addition to procumbent cells, it is said to be heterogeneous (heterocellular). The tendency to use the terms homocellular and heterocellular represents an effort to refer to cellular homogeneity or heterogeneity of rays without using the terms "homogeneous" and "heterogeneous" because these latter terms have become well known and therefore virtually synonymous with the special types Kribs (1935) designated (see Sect. 6.3).

6.2 Ray Dimensions

6.2.1 Ray Height and Width

Papers on comparative wood anatomy commonly present quantitative data on ray height and width. These data are usually presented in the form of means. Heights of multiseriate rays and of uniseriate rays are presented separately. Width of multiseriate rays is measured at the widest point in a ray and may be expressed in microns or in terms of cell number.

Ray height and ray width are often used in comparative work, but one must remember that they are much altered in most species during ontogeny of the xylem cylinder (Fig. 6.1). One may be comparing relative ages of stems when one is comparing species, depending on the nature of the samples. Given comparable material from adult stems, particular species can have characteristic ray dimensions.

6.2.2 Ray Frequency

Ray frequency is calculated by counting the number of rays that intersect an imaginary tangentially running line across a transection. Metcalfe and Chalk (1983) give a normal distribution curve for ray number in dicotyledons. The mean value is 9 rays/mm.

6.2.3 Ray Volume

The area of ray tissue on a tangential section can be estimated in various ways; one simple method has been offered by Gregory (1981). Ray volumes in a survey of various dicotyledons range from 6.1 to 28% of the tangential section (G.E. French in Panshin and de Zeeuw 1980, p. 181). Data are also offered by

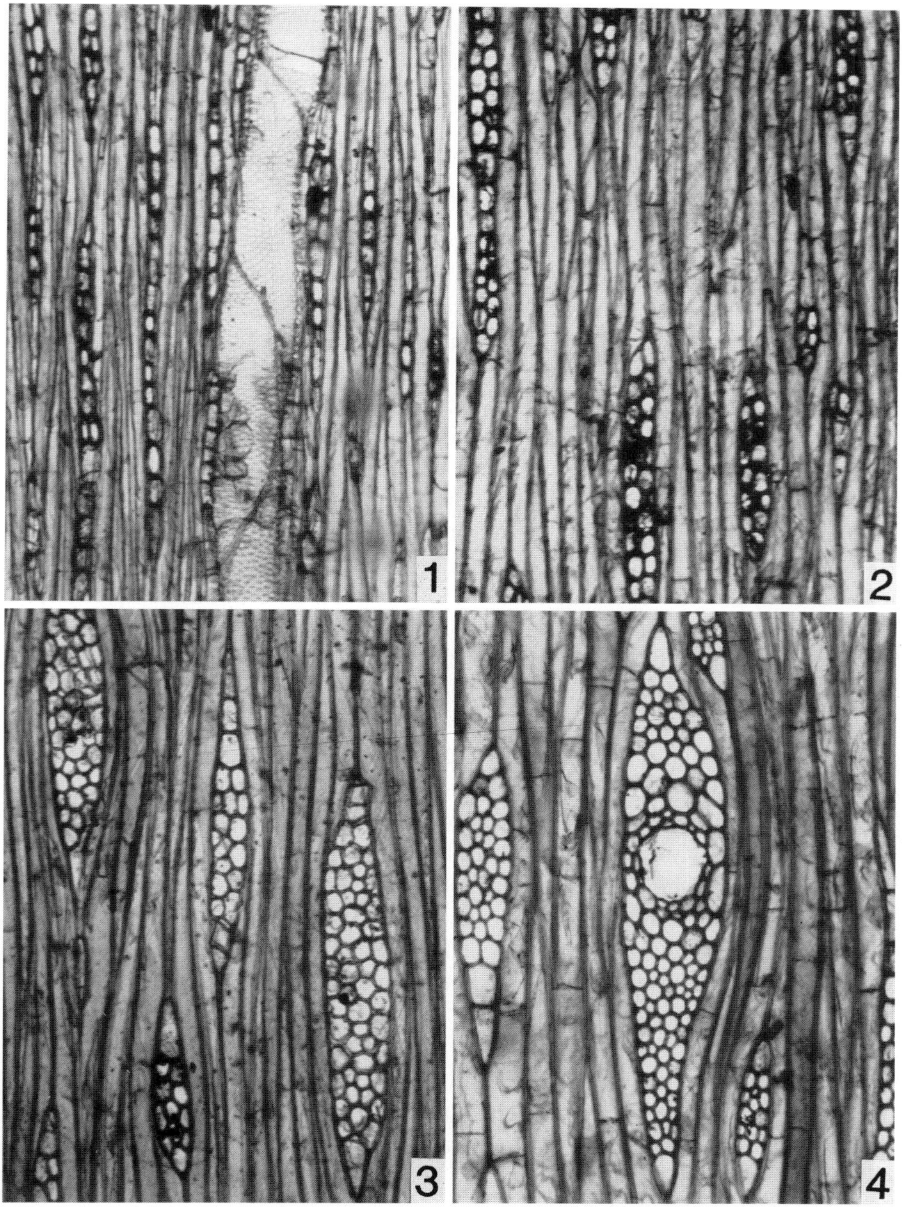

Fig. 6.1. Ontogenetic stages in ray development of *Bursera simaruba* (Burseraceae), shown by tangential sections cut from wood of a single stem. *1* Section from near pith, most rays uniseriate; *2* section cut about 3 mm from pith, most rays biseriate; *3* section cut about 6 mm from pith, rays 2–5 cells wide; *4* section near cambium (5 cm from pith), rays wide multiseriate, secretory cavity present in widest ray. Note that ray cells become vertically shorter during ray ontogeny. (*1–4:B*)

Ghouse and Yunus (1973, 1974a,b), who find ray volumes may run to 40%. Variation among species with respect to ray volume is reported by Petrić and Šćukanec (1975).

6.2.4 Value of Quantitative Measurements

One can cite examples more easily of limitations rather than of decisive value in measurement of ray features, because of particular circumstances: interdependence of wood features, effect of ontogeny, and effect of external environment come into play. Increase in ray width without corresponding increase in ray cell number with polyploidy has been reported in *Parthenium* by Swamy and Govindarajalu (1957). Within a wood sample, new rays appear to be initiated in proportion to increase in stem circumference, and such constancy in ray frequency and volume has been reported by Myer (1922), Linnemann (1953), Sarkany et al. (1957a,b), Taylor (1966, 1969), Wagenführ (1967), and Cumbie (1983). Taylor (1969) found no relationship between ray volume and ray height within the stem, and Erak (1971) reported no relationship between altitude and ray height or frequency. In *Ilex*, Baas (1973) reported no taxonomic correlations between ray size or frequency, and that finding could easily be multiplied may times, although Vander Wyk and Canright (1956) did find ray width to be characteristic of species.

There is a strong tendency for rays to change in dimensions with ontogeny. For example, Barghoorn (1941a) showed increase in ray width within a single stem in *Bursera simaruba*. In this species (Fig. 6.1, parts 1–4), rays begin as uniseriates, and division in ray initials converts rays to multiseriates over time (see Sect. 6.3.1). This tendency probably characterizes the majority of dicotyledons. There tend to be both increase and decrease in ray height. Šćukanec and Petrić (1977) showed differences among individual beech trees with respect to ray volume and ray slenderness; they find ray volume and ray width correlated.

Ferreirinha (1965) has noted a statistical correlation between ray height and vessel element length within a species. This has been confirmed by Giraud (1980) in *Entandophragma*, van den Oever et al. (1981) in *Symplocos*, and Carlquist and Hoekman (1985a) in Staphyleaceae. The reason for this correlation, which may seem unlikely, is that new rays are derived from subdivision of fusiform cambial initials. Vessel elements are approximately the same length as the fusiform initials from which they were derived, and are thus good indicators of probable fusiform cambial initial length in any given species. Ferreirinha (1965) has also reported correlation between fiber length and ray width. This probably happens because both fusiform cambial initial length and ray width tend to increase with age.

Chalk (1955) has found that wood samples with narrow rays (e.g., rays all uniseriate) have a smaller proportion of the wood devoted to ray tissue, whereas samples with wider rays have a higher proportion of the wood present as ray tissue. The functional significance of wider rays needs to be clarified. Certainly where stem and lignotuber wood can be compared species by species, one finds wider rays, often with larger cells, occupying a larger volume of the wood in lignotubers. This can be shown in Bruniaceae (Carlquist 1978a), Geissolomataceae (Carlquist 1975c), and Penaeaceae (Carlquist and DeBuhr 1977). Novruzova (1972) claims a higher percentage of ray tissue in xerophytic species compared with those in mesic sites. The reason for this, if her observation is of general applicability, has not been determined. Perhaps an increase in water storage tissue with xeric habitat is involved.

Although no definite examples are cited here, clearly some woods have few, wide rays (e.g., *Peltogyne*; Fig. 6.6, part 3), whereas others have numerous narrow rays (e.g., *Acnistus*; Fig. 6.7, part 1), and this feature is useful in wood identification.

6.3 Ray Types; Problems in Definition

Although some authors have attempted alternative systems or terms, the ray types offered by Kribs (1935), with the ontogenetic dimension developed by Barghoorn (1940, 1941a,b), have been accepted by most authors. Therefore, these types are adopted here, although in an expanded form (Fig. 6.2) that takes into account protracted or permanent juvenilism in ray structure (Carlquist 1962a), tendencies not taken into account by Kribs (1935). Also taken into account here, although not numbered among the types recognized as a group, are instances of aggregate rays as explained by Eames (1910), Bailey (1911, 1912), and Moseley (1948).

We must take into account ontogenetic changes in the cambium, because without understanding these we do not understand the resultant array of ray types and cannot use them accurately in a comparative way. These ontogenetic changes include the following, based on the Barghoorn papers with additional observations as noted.

1. Fusiform cambial initials may subdivide to form ray initials; if these lie alongside a group of ray initials, the size of the group of ray initials (and the size of the ray derived from them) is enlarged. In this way a ray that begins as uniseriate can be converted to a multiseriate ray. A ray can also be converted to a taller ray if the added ray initials extend above or below the pre-existing ray.
2. Fusiform cambial initials not in contact with ray initials can subdivide, yielding a strand of ray initials In this case, a uniseriate ray can be initiated.
3. Ray initials may elongate to form fusiform cambial initials. This process may subdivide a group of ray initials into two or more groups, leading to "breakup" of a ray. This occurs commonly in Sterculiaceae (Chattaway

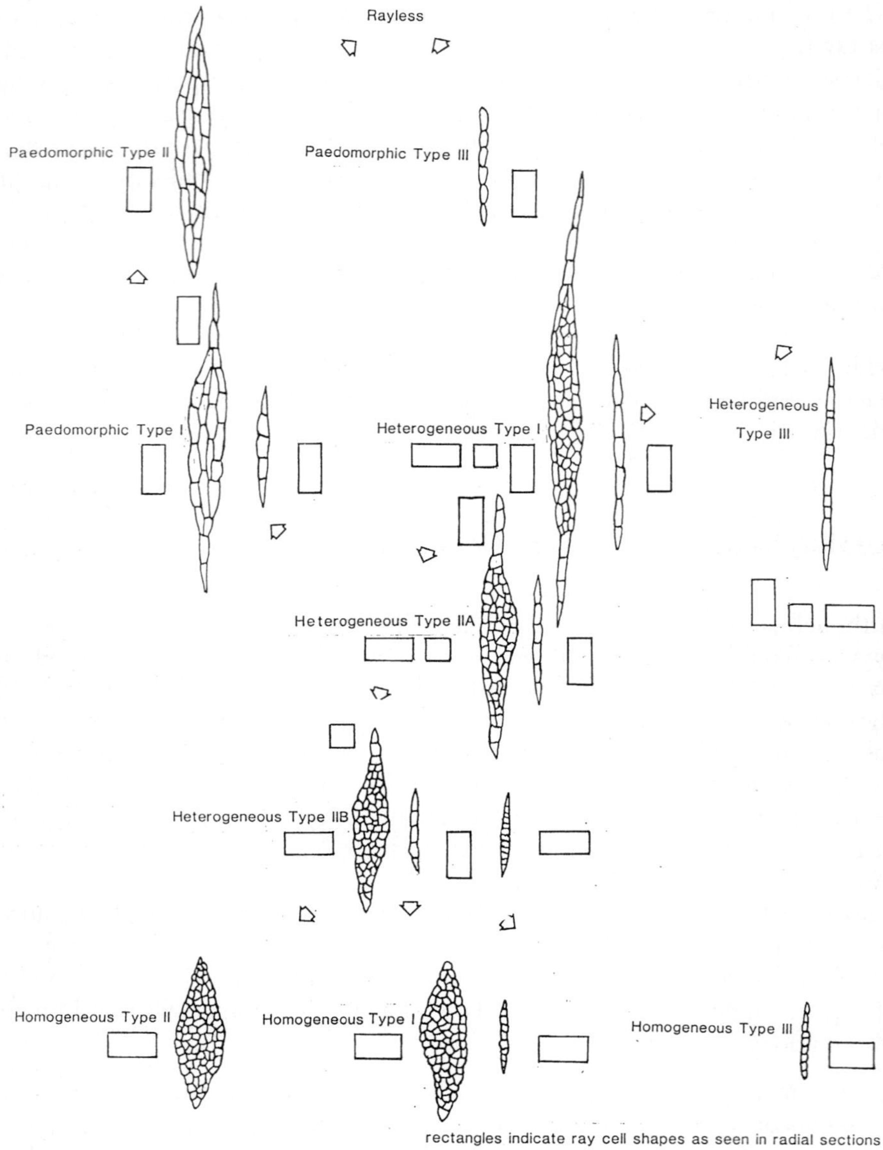

Fig. 6.2. Phylogenetic interrelationships of ray types. The seven ray types of Kribs (1935) are recognized (Heterogeneous I, IIA, IIB, III, Homogeneous I, II, III), and derivation from Heterogeneous type I is indicated. In addition, paedomorphic ray types are proposed, and derivation of these indicated. The phenomenon of raylessness is included in this scheme

1933a) and was reported in Fagales by Jost (1891), Ziglstra (1909), and Moseley (1948) and in other families by Krawczyszyn (1971, 1972).
4. Ray initials may subdivide horizontally, resulting in increased percentage of procumbent cells, so that there is a shift from heterogeneous (heterocellular) to homogeneous (homocellular).
5. Ray initials may subdivide in a longitudinal radial direction, increasing the width of the group of ray initials and resulting in production of wider multiseriate rays (Braun 1955).
6. Any given ray initial or fusiform initial may be lost from the cambium.

Note should be taken that changes 1 and 3 have opposing results. Both are held by Barghoorn (1941a) to occur characteristically. If both were about equally frequent, ray height and width would not change appreciably. As noted above, ray height does tend to bear a statistical correlation with fusiform cambial initial length in any given species, and this would not be true if ray elongation or ray breakup were frequent and in different frequencies, respectively. Change in ray height is mostly effected by derivation, over a long period of time, of most rays from events of subdivision of fusiform cambial initials. Fusiform cambial initials tend to elongate over time (Bailey and Tupper 1918), and ray height tends to increase proportionately.

Ray breakup does occur in woody species, however, where large primary rays are involved. Secondary rays that represented unaltered continuations of primary rays are virtually nonexistent at the periphery of older stems of woody species, showing that ray breakup is active. In species in which cambial activity is sluggish or stem structure is typically herbaceous, or both, limited ray breakup will occur (Carlquist 1962a). For example, Begoniaceae (Carlquist 1985b), Piperaceae, and nonsucculent cacti (Gibson 1977a, 1987b) show minimal breakup of rays.

Subdivision of ray initials in a vertical plane, or in a plane more nearly vertical than horizontal, leads to widening of rays ontogenetically; the data of Barghoorn (1941a) and others show that this does, in fact, occur, and the storied cells at margins of rays of *Lactoris* (Carlquist 1990b) are evidence of this. Rays wider in outer parts of a stem than in inner parts of a stem are characteristic of most dicotyledons. Subdivision of ray initials in a more nearly horizontal plane appears generally not to increase a ray in height, as one might expect, but to lead to increasing numbers of procumbent cells. Both of these tendencies are shown in Barghoorn's (1941a) illustrations of *Bursera simaruba* (similar illustrations here; Fig. 6.1, parts 1–4). One might imagine that a mechanism for restricting subdivision of ray initials only to a horizontal plane, or only to a vertical plane, would be virtually impossible to achieve; therefore, one would expect that rays would become increasingly wider and increasingly composed of procumbent cells with age. Both alterations are, in fact, realized simultaneously in many if not the vast majority of dicotyledon species.

The ontogenetic change cited as 2 above can have two consequences: increase in number of rays in woods that have rays already; and development of rays in woods that begin rayless. Apparently subdivision of fusiform initials into files of ray initials is the most frequent form of adding rays to a stem, judging from the instances Barghoorn (1941a) describes and by the fact that it is an effective and rapid way for production of rays in stems that begin rayless.

The various changes cited above might be expected to vary in degree from species to species, so that species in a genus could actually differ in these respects. In fact, that proves to be true. Paedomorphosis (Carlquist 1962a) is essentially the process of slowing ontogenetic change in the cambium. Thus a wood with paedomorphosis will tend to show (1) less breakup of the primary ray patterns as secondary growth proceeds, so that the primary ray pattern persists indefinitely or little altered into the secondary xylem; (2) minimal horizontal subdivision of ray initials, so that rays consist more predominantly of upright cells, with fewer procumbent cells than would occur in a plant with a more active cambium or, often, no procumbent cells; (3) less increase in width of rays owing to vertical subdivisions of ray initials; (4) less conversion of uniseriate rays to multiseriate rays due to fewer vertical divisions in the ray initials; and (5) origin of few uniseriate rays by virtue of failure of fusiform cambial initials to subdivide into strands of ray initials. All of these tendencies do occur; one can see them, for example, in the rosette-tree species of *Senecio* and one can thereby distinguish wood of these species of *Senecio* from wood of those that have more typically woody stems (Carlquist 1962b).

Barghoorn (1941a) also cited changes that relate to relative presence of uniseriate as compared to multiseriate rays in a given wood:

A In a species with both multiseriate and uniseriate rays in outer secondary xylem, secondary xylem near the pith may have uniseriate rays only. This condition is shown in a number of families such as Illiciaceae (Bailey and Nast 1948; Carlquist 1982c; Fig. 6.5, part 3) and Buxaceae (Carlquist 1982d).
B In a species with only multiseriate rays in the outer secondary xylem, secondary xylem near the pith may have both multiseriate and uniseriate rays.
C In a species with only multiseriate rays in the outer secondary xylem, secondary xylem near the pith may have only uniseriate rays.

These three processes all show the tendency for rays to become wider during secondary growth. Type B conforms to the configuration expected close to the pith in primitive woods (Barghoorn 1940), whereas types A and C are considered specialized in lacking multiseriate rays near the pith. The shift from B to either A or B patterns of early-formed secondary xylem may not be a major change, however.

6.4 The Kribs Ray Types

Kribs (1935) designated six types (Fig. 6.2; also Figs. 6.3–6.10), with two subtypes of one (Heterogeneous II). The primary distinction is between heterogeneous and homogeneous. In this typology, cells approximately square as seen in radial section are counted as upright cells. The second distinction in the Kribs system is based upon whether both multiseriate and uniseriate rays are present in a wood, or whether a wood has multiseriate rays exclusively or uniseriate rays exclusively. A third criterion involves the length of uniseriate wings on multiseriate rays. Inherent in the Kribs classification is not merely convenience, but the fact that when classified in this fashion, the types show statistical linkages to features known to be indicative of wood phylesis: length of vessel elements, for example. Bande and Prakash (1984) have reworked the Kribs terminology in a way that is confusing (apparently, the original Homogeneous II of Kribs is omitted). Their suggestion apparently derives from a recension by Kribs (1950), who noted that Homogeneous I grades into Homogeneous II. Indeed, in the two species figured here as representing Homogeneous II (Fig. 6.6, parts 3 and 4), a few uniseriate rays are present. However, because the uniseriate rays are few, I recognize them as Homogeneous type II, and believe that this is justified in contrasting these woods with Homogeneous I woods in which uniseriate rays are abundant.

The reader is reminded that reliable decisions as to whether procumbent or upright cells are present can only be made on the basis of radial sections, but this information is then applied to what one sees in a tangential section.

The following descriptions of Kribs's six types are offered; wording has been slightly altered from Kribs (1935) in the interests of clarity and conformity to the IAWA Committee on Nomenclature (1964) recommendations.

Heterogeneous Type I. Uniseriate rays are usually very high, numerous, and composed of large upright cells unlike the cells of the multiseriate portion of the multiseriate rays. Multiseriate rays usually have very long uniseriate wings, composed of upright cells like those of the uniseriate rays. The cells of the multiseriate portion of the multiseriate ray are oval as seen in tangential section; in radial section, they prove to be both procumbent and upright (Fig. 6.3).

The examples of Heterogeneous I, shown in Fig. 6.3, show diversity. The multiseriate rays are wider in *Exospermum* (Fig. 6.3, part 1), which is appropriate in view of ontogeny, because the sample was taken from the periphery of a large trunk. In the wood of the other vesselless dicotyledon, *Tetracentron* (Fig. 6.3, part 2), cells range between square and procumbent as seen in radial section, although the rays as seen in tangential section might appear to be predominantly procumbent. *Tetracentron* represents a variant of Heterogeneous I in which cells are much less elongate vertically than is usual in type I. The reverse is true in *Illicium* (Fig. 6.3, part 3), in which cells of the multiseriate portions of multiseriate rays may be mostly upright (multiseriate ray at left) to perhaps half upright (multiseriate ray near center). In *Grubbia* (Fig. 6.3, part 4), uniseriate wings are of various lengths, some only one or two cells long, some much longer, indicating a transition to Heterogeneous II. This last example shows that as one might expect, ray types form a continuum, and one

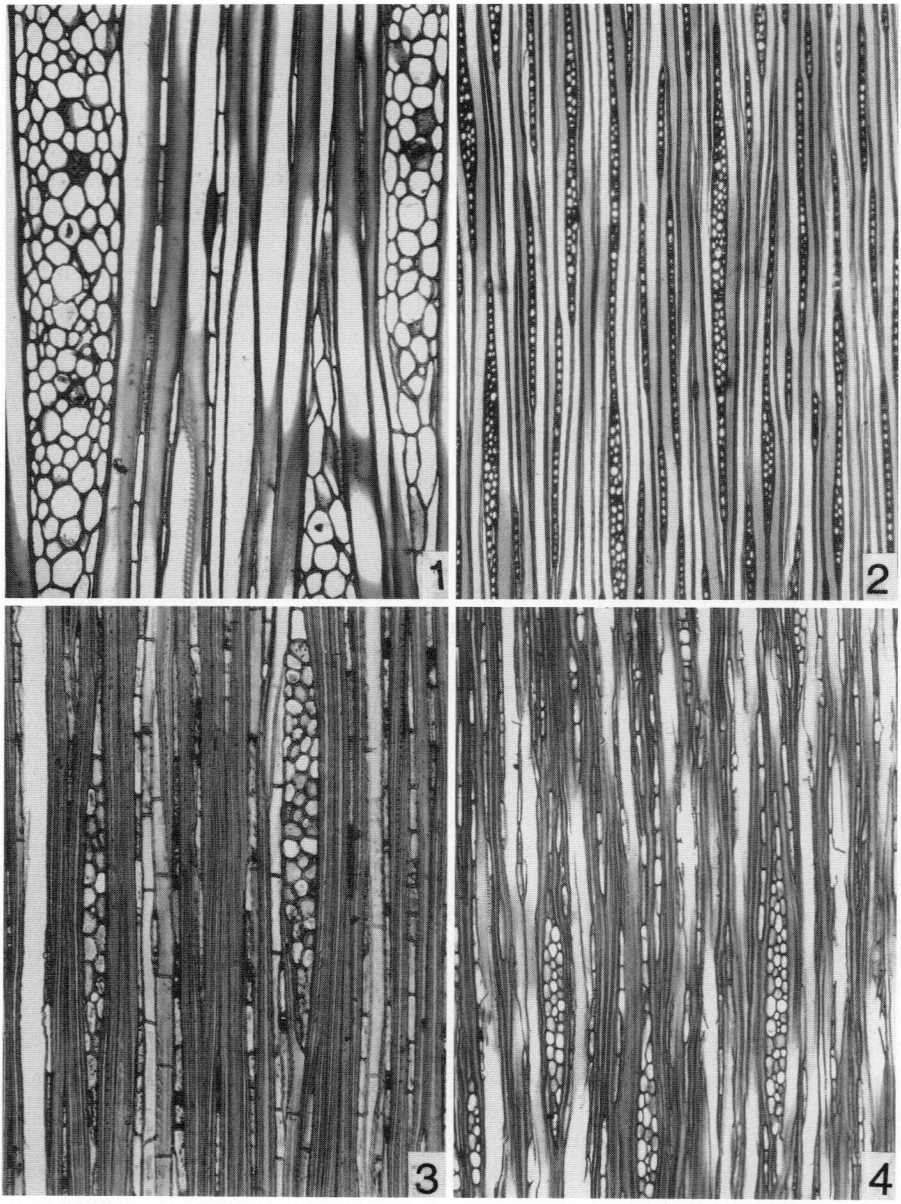

Fig. 6.3. Heterogeneous type I rays, shown from tangential sections. *1* Multiseriate rays wide, *Exospermum stipitatum* (Winteraceae); *2* cells approximately square (as seen in radial section) instead of erect cells present, but rays still qualify as Heterogeneous type I; *Tetracentron sinense* (Tetracentraceae); *3* a few upright cells present in multiseriate portion of multiseriate rays, *Illicium ridleyanum* (Illiciaceae); *4* upright cells relatively abundant in multiseriate portions of multiseriate rays, *Grubbia rosmarinifolia* (Grubbiaceae). (*1–4:A*)

must assign them to the Kribs types based on the majority of features, and some exceptions are permissible.

Heterogeneous Type II. Uniseriate ray cells are usually less markedly upright (than in Heterogeneous type I), and are composed of cells unlike those of multiseriate portions of multiseriate rays (which are mostly procumbent) (Fig. 6.4).

Heterogeneous Type IIA. Uniseriate rays are composed of upright cells. Multiseriate portions of multiseriate rays are composed mostly of procumbent cells; the multiseriate rays bear moderately short uniseriate wings composed of upright cells (Fig. 6.4, parts 1 and 2).

The two examples shown are similar in appearance despite belonging to widely separate families. Both show obvious uniseriate wings which are, however, shorter than the wings of Heterogeneous I rays.

A distinctive variant referable to Heterogeneous type IIA is shown in Fig. 6.7 (part 3), *Iryanthera junius*. In this species, one sees that rays are not more than two cells wide. The biseriate portions of the rays are composed of procumbent cells not much shorter vertically than the cells of uniseriate wings or of uniseriate rays; in these latter two instances, the cells are nearly square to slightly upright in shape.

Heterogeneous Type IIB. Uniseriate rays are of two types, both usually present on any given section: some are composed of upright cells, some are composed of procumbent cells. Multiseriate portions of multiseriate rays are composed mostly of procumbent cells. Uniseriate wings on the multiseriate rays are very short – rarely more than two cells, often only a single cell terminating the ray; wing cells are upright; if several are present, often square.

In Fig. 6.4, both parts 3 and 4 show presence of the two types of uniseriate rays. In Fig. 6.4, part 3 differs from part 4 in having somewhat more numerous upright cells, and in this respect is transitional to paedomorphic type I.

One can express the differences between type IIA and type IIB as representing a greater tendency toward procumbent cells in IIB in the case of uniseriate rays. There is also a shortening of the wings on multiseriate rays.

Heterogeneous Type III. Rays are all uniseriate. Both procumbent and upright cells are present in the rays (Fig. 6.5, parts 1–4).

The four examples in Fig. 6.5 differ in proportion of procumbent to upright cells. Procumbent cells are least abundant in *Homalanthus* (Fig. 6.5, part 2), slightly more abundant in the *Aleurites* (Fig. 6.5, part 1), intermediate in *Pentaphylax* (Fig. 6.5, part 3), and are most abundant in *Dacryodes* (Fig. 6.5, part 4). The *Pentaphylax* section shown represents a section of earlier-formed secondary xylem; biseriate rays and then multiseriate rays appear as secondary growth proceeds in this species (Carlquist 1984g).

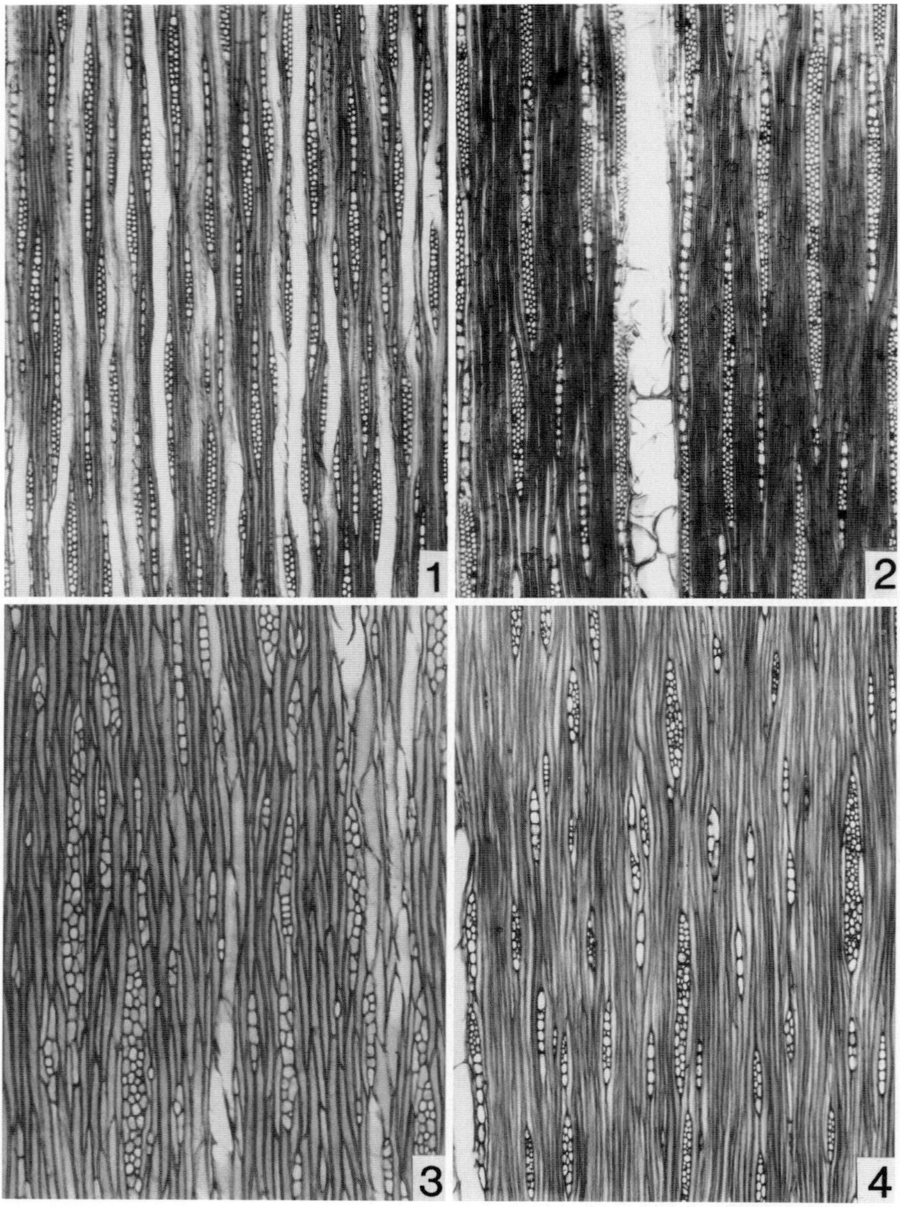

Fig. 6.4. Heterogeneous type IIA (*1, 2*) and type IIB (*3, 4*) rays, shown in tangential sections. *1* Uniseriate wings on multiseriate rays of variable height (transitional to Heterogeneous I), *Parrotia persica* (Hamamelidaceae); *2* uniseriate wings mostly absent on multiseriate rays, *Brosimum uleanum* (Moraceae); *3* upright cells rather common in multiseriate portion of multiseriate rays, *Symbolanthus anomalus* (Gentianaceae); *4* procumbent cells characterize multiseriate portions of multiseriate rays, *Baccharis cassinifolia* (Asteraceae). (*1–4:A*)

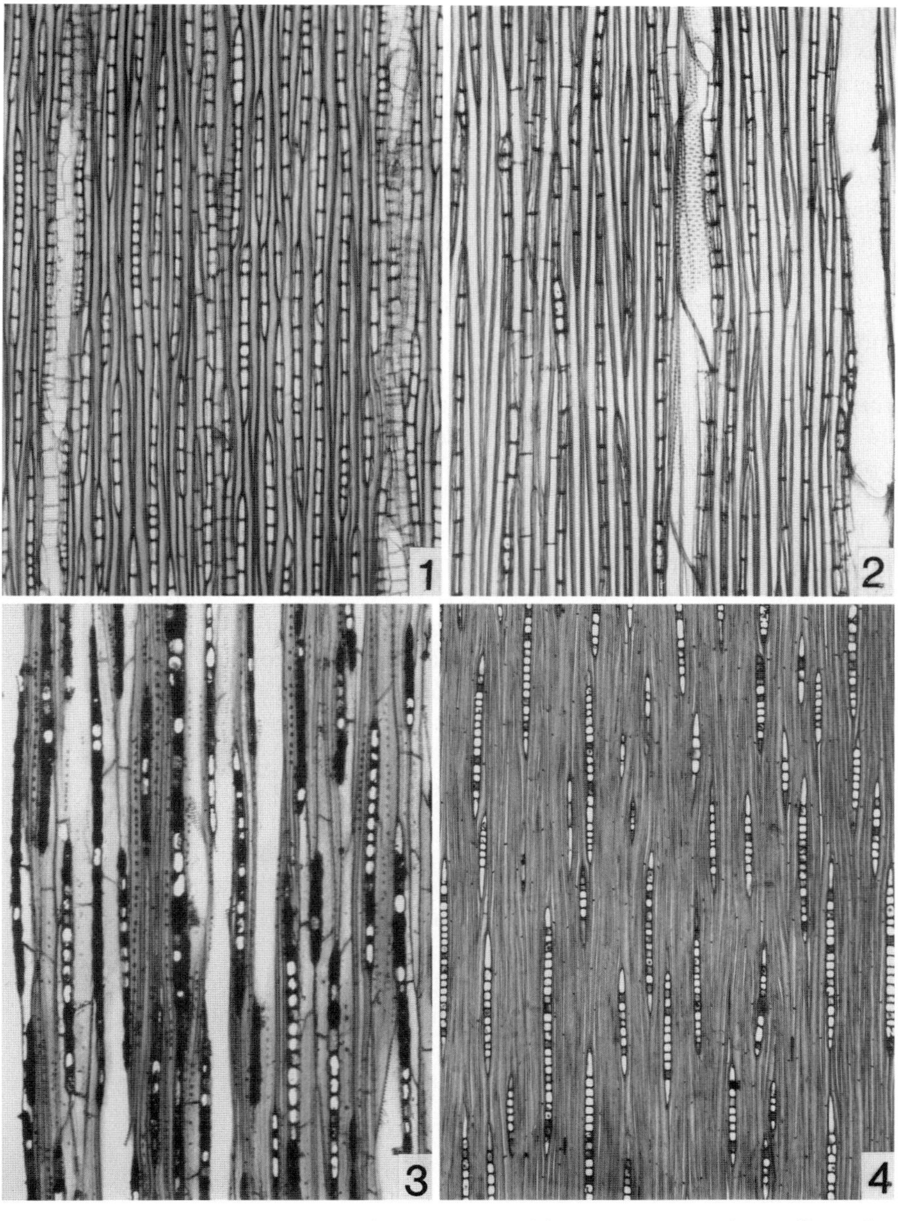

Fig. 6.5. Heterogeneous type III rays, shown in tangential section. *1* Procumbent cells moderately common, *Aleurites fordii* (Euphorbiaceae); *2* procumbent cells relatively few, *Homalanthus populifolia* (Euphorbiaceae); *3* upright cells less common than procumbent cells, *Pentaphylax euryoides* (Pentaphylacaceae); *4* procumbent cells more common than upright cells, *Dacryodes rugosa* (Burseraceae). (*1–4:A*)

The presence of an occasional biseriate ray in a wood that otherwise has only uniseriate heterogeneous rays would not disqualify that wood from being referred to Heterogeneous type III.

Homogeneous Type I. Uniseriate rays are short, because they are composed of procumbent cells. Multiseriate rays are also composed wholly of procumbent cells; wings are rarely present, and rays typically terminate above and below in a file (as seen radially) of either procumbent or square cells.

Uniseriate rays are more common in Fig. 6.6, part 1, less common in Fig. 6.6, part 2. Development of marked procumbency of ray cells thus precedes loss of uniseriate rays.

Homogeneous Type II. Uniseriate rays are scarce to absent. Multiseriate rays are composed entirely of procumbent cells; wings are rarely present, and rays typically terminate above and below in a procumbent cell (as seen in tangential sections; Fig. 6.6, parts 3 and 4).

The scarcity or absence of uniseriate rays in Homogeneous type II represents an evolutionary advance over Homogeneous type I. The tables by Kribs (1935) show that Homogeneous type II is associated with phyletic levels of vessels appreciably more specialized than those associated with Homogeneous type I, and thus his data would argue against abandonment of type II, despite the desire for mutually exclusive definitions. The two examples illustrated both show a small number of uniseriate rays.

Homogeneous Type III. Only uniseriate rays are present. All ray cells are procumbent.

The rays shown in Fig. 6.7 (part 1) are shorter than those of Fig. 6.7 (part 2), showing that even within such a uniform ray configuration one can find quantitative differences that could aid in identification or, within a single genus, in taxonomic distinction.

Kribs (1935) compared the six ray types to end wall morphology of vessel elements and to length of vessel elements. Two of Kribs' tables are reproduced here (Tables 6.1, 6.2),

Kribs' statistical correlations demonstrate that in a primitive wood, the following conditions prevail:

1. Both multiseriate and uniseriate rays are present.
2. Both types of ray are high (of marked vertical length).
3. Multiseriate rays are heterogeneous (heterocellular), uniseriate rays are composed of upright cells.
4. Multiseriate rays have long uniseriate wings.

With phylogenetic specialization, the following changes occur:

1. Either multiseriate or uniseriate rays are lost.
2. There is a tendency for loss of heterogeneity in ray cells: upright cells are lost.

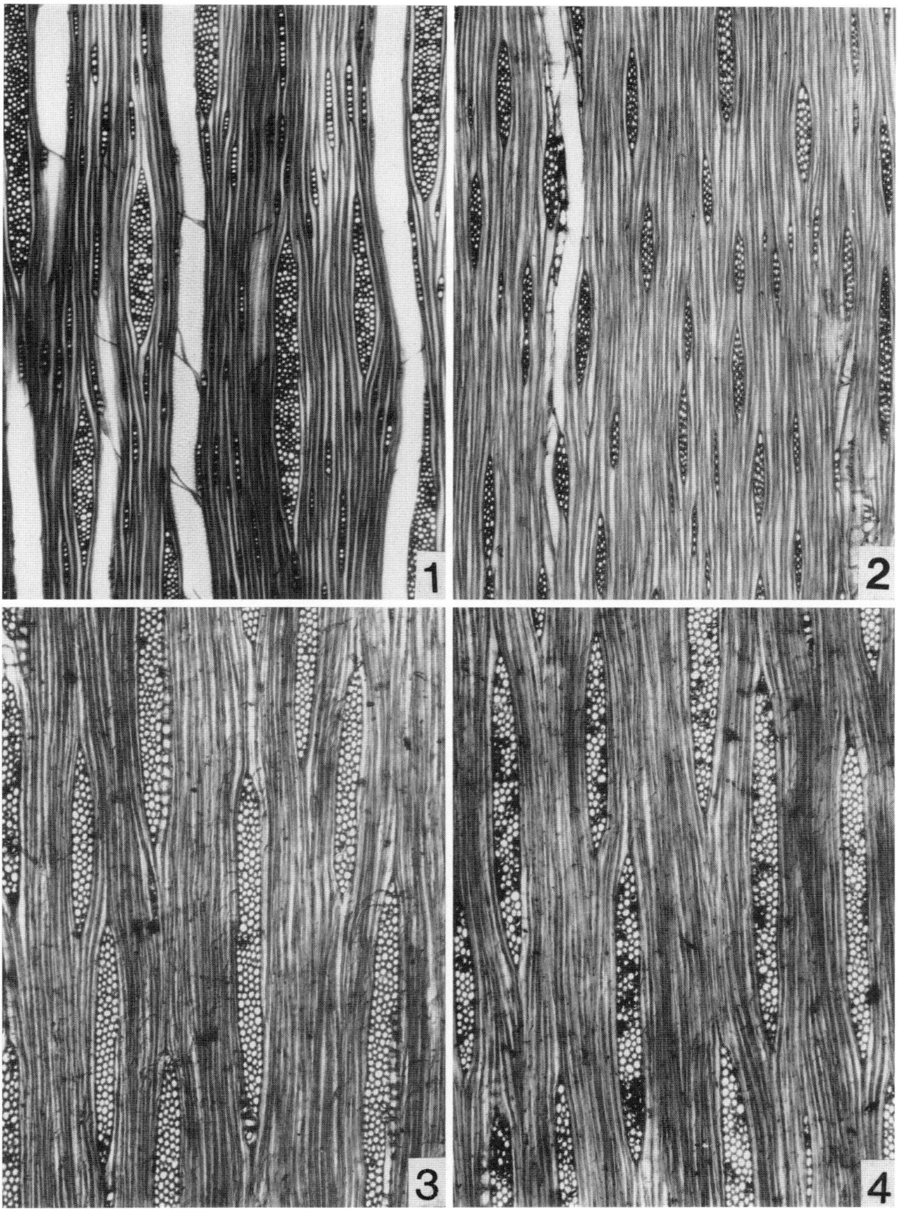

Fig. 6.6. Homogeneous type I (*1*, *2*) and homogeneous type II (*3*, *4*) rays, as shown in tangential sections. *1* Uniseriate rays common, *Acer saccharum* (Aceraceae); *2* uniseriate rays less common, *Phellodendron sachalinense* (Rutaceae); *3* chambered crystals adjacent to multiseriate rays, *Peltogyne parviflora* (Fabaceae); *4* ray cells contain dark-staining compounds, *Peltogyne maranhaensis*. (*1–4*:*A*)

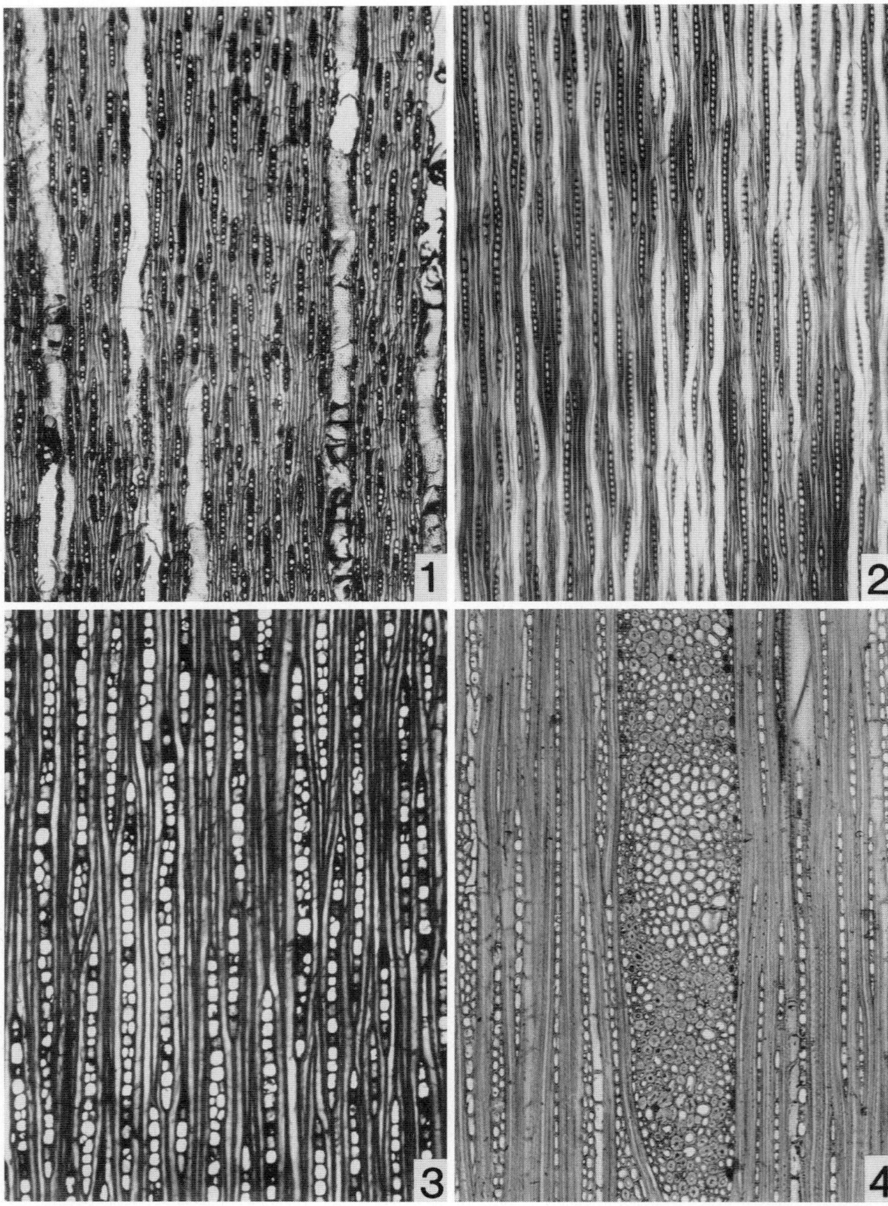

Fig. 6.7. Homogeneous type III rays (*1*, *2*) and variant ray types, shown in tangential sections. *1* Rays short, *Acnistus parviflorus* (Solanaceae); *2* rays taller, *Euonymus japonicus* (Celastraceae); *3* multiseriate rays never more than two cells wide, *Iryanthera junius* (Myristicaceae); *4* rays dimorphic (wider rays markedly wider than other multiseriate rays), *Emmotum holosericeum* (Icacinaceae). (*1–4:A*)

Table 6.1. Total percentage of ray types in each vessel type. (Kribs 1935)

Type of vessel element	Rays (%)	
	Heterogeneous	Homogeneous
Scalariform I	100.00	00.00
Scalariform II	84.37	15.63
Scalariform + simple	86.56	13.44
Simple but oblique	89.63	10.37
Simple-oblique + simple-transverse	44.58	55.42
Simple-transverse	20.90	79.10

Table 6.2. Mean vessel element length for each ray type. (Kribs 1935)

Type of ray	Number of genera	Mean vessel element length (μm)
Heterogeneous type I	210	810
Heterogeneous type III	18	640
Heterogeneous type II	227	580
Homogeneous type I	131	520
Homogeneous type III	41	380
Homogeneous type II	123	350

3. Multiseriate rays become vertically shorter and less numerous.
4. Uniseriate wings on multiseriate rays are reduced, ultimately to a single cell.

Functional reasons for these specializations have been advanced (Carlquist 1975a). Ray cells are involved in storage and radial transport of photosynthates (Braun 1970, 1984) and in radial translocation of ions (Lauchli 1972). We may also assume that axial parenchyma represents a vertical system for translocation of these substances. Therefore, comparing primitive and specialized woods, we see a shift in contact between the two systems.

Primitive woods, typified by Heterogeneous type I rays and diffuse axial parenchyma, feature numerous points of contact between the two systems. Diffuse axial parenchyma consists of parenchyma strands isolated from each other, and heterogeneous rays are relatively narrow. The uniseriate rays and uniseriate wings on multiseriate rays are composed of upright cells. These upright cells are not ideal for radial conduction, but they probably serve as bridges between a radial conduction system, represented by the procumbent cells in the multiseriate portions of multiseriate rays, and a vertical conduction system, represented by the axial parenchyma. The points of contact between the two systems are numerous and limited in extent (e.g., a single diffuse parenchyma cell may contact a single upright cell on the wing of a multiseriate ray).

Specialized woods have, by contrast, fewer but more massive points of contact between rays and axial parenchyma. Axial parenchyma is aggregated into sheets (apotracheal banded) or cylinders around vessels (paratracheal). These sheets or cylinders contact the rays, which tend to be fewer in number and less tall than in primitive woods. Specialized rays tend to consist wholly of procumbent cells; because of their radial elongation, procumbent cells are a cell type very efficient at radial conduction (fewer cross-walls per radial mm). Axial parenchyma cells are composed wholly of vertically elongate cells. Thus rays and axial parenchyma are optimal for radial and vertical conduction, respectively. The efficiency of the two systems requires certainty of contact between them, and points of contact that are large and suited to interconnection between large numbers of cells. Although fewer points of contact between the two systems occur, compared to primitive woods, the points of contact are more massive (e.g., a sheet of apotracheal banded parenchyma intersecting a large multiseriate ray).

6.5 Paedomorphic Ray Types

Barghoorn (1941a) shows samples of rays that begin with a predominance of upright cells but change to a predominance of procumbent cells because of horizontal subdivision of ray initials. However, this transition does not occur in all species: ray initial subdivision, as well as other cambial changes, may be delayed or occur only to a limited extent. Thus, multiseriate rays composed predominantly of erect cells may occur for the life of a plant. Kribs (1935) does not mention such rays – which might have provided a source of confusion in his exposition of ray evolution. Barghoorn (1940, 1941a,b), although he has encountered a few woods with such rays (e.g., *Bocconia*), does not note the implications of this phenomenon. The Kribs sample of wood collections probably did not have many examples of what are termed paedomorphic rays here. The Kribs (1935) study was based on a group of woods from I.W. Bailey's slide collection at Harvard University. Although that collection was diverse, it stressed typically woody plants, particularly trees, and also is slightly biased in favor of more primitive woods, compared to their occurrence in the world flora at large. Few species in categories such as herbs, woody herbs, annuals, succulents, and rosette trees were included in the Kribs (1935) study; because the Bailey collection included only a few of these plants, with their notably paedomorphic rays, it does not apply to them. Study of species in these less woody growth form categories led to the theory of paedomorphosis in dicotyledonous woods (Carlquist 1962a). If we recognize the Kribs ray types, we must also recognize types for species that have permanently juvenilistic rays. This was hinted in an earlier schematization of rays (Carlquist 1961a), but is formalized in a more comprehensive manner here (Fig. 6.2). The new types proposed here are as follows.

Paedomorphic Type I. Both multiseriate and uniseriate rays are present. Uniseriate rays consist of upright cells only. Multiseriate rays are composed predominantly or entirely of upright cells; if procumbent cells are present, they will be in the multiseriate portion of multiseriate rays (Fig. 6.8, parts 1–4).

The rays in this category may have a predominance of wide multiseriate rays with some uniseriate rays (Fig. 6.8, parts 1 and 2), or a predominance of narrow multiseriate rays with numerous uniseriate rays (Fig. 6.8, parts 3 and 4). The examples shown in Fig. 6.8 are noteworthy in showing that between both the upper pair, which simulate each other, and the lower pair, which also do, there is no close degree of phylogenetic relationship. The upper pair, *Trematolobelia macrostachys* and *Senecio mannii*, can both be described as rosette trees, whereas the lower pair could both be described as woody herbs, with stems that do not become large.

Rays corresponding to Paedomorphic type I occur in a scattering of Asteraceae, such as *Verbesina* (Carlquist 1957b, 1958a), *Baccharis* (Carlquist 1958b, 1960a), and *Brachylaena* (Carlquist 1961b). They also can be found in most Onagraceae (Carlquist 1975b), most Goodeniaceae (Carlquist 1969b), most Chloranthaceae (Carlquist 1990a, 1992a,b), and some of the species of *Euphorbia* studied (Carlquist 1970a).

Paedomorphic Type II. Uniseriate rays are absent or quite rare. The multseriate rays consist wholly or predominantly of upright cells; procumbent cells, if present, occur in the multseriate portions of the multiseriate rays (Fig. 6.9, parts 1–3).

Of the examples illustrated here, neither *Cyanea coriacea* (Fig. 6.9, parts 1 and 2) nor *Begonia parviflora* (Fig. 6.9, part 3) have any procumbent cells in rays. *Cyanea coriacea* has markedly vertically elongate ray cells, some of which are even horizontally subdivided (Fig. 6.9, part 2). *Cyanea coriacea* is a rosette shrub with moderately succulent stems (cortex nonfibrous, relatively thick), whereas *Begonia parviflora* has the canelike habit of a large bamboo (no major lateral branches).

Rays corresponding to Paedomorphic type II were illustrated by Barghoorn (1941b) for *Geranium tridens* and *Ardisia brackenridgei*. Other families in which Paedomorphic type II rays occur include some Urticaceae (Bonsen and ter Welle 1984), virtually all of the lobelioid Campanulaceae (Carlquist 1970b), *Bocconia* of the Papaveraceae (Cumbie 1983), some Aristolochiaceae (Carlquist 1993), *Anemopsis* of the Saururaceae (Carlquist et al. 1995), and numerous Asteraceae, such as most of the tribes Cynareae (Carlquist 1958c, 1965b), Vernonieae (Carlquist 1964b), and Eupatorieae (Carlquist 1965a). Rays of *Lactoris* are also of this type, but include storied cells; evidently rays of this species widen by vertical divisions in ray initials (Carlquist 1990b).

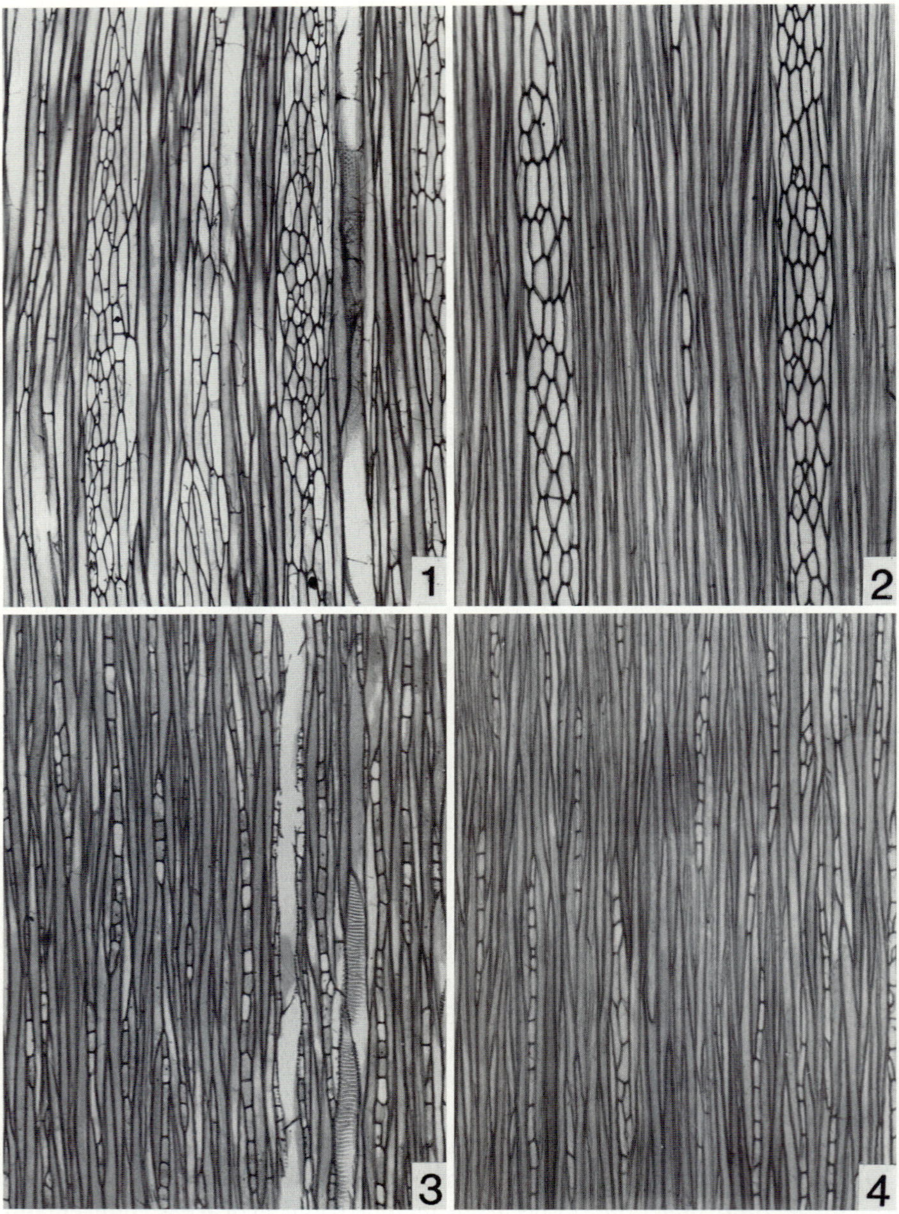

Fig. 6.8. Paedomorphic type I rays, as shown in tangential section. *1* A scattering of procumbent cells present in multiseriate rays, but most ray cells are upright (transitional to heterogeneous IIA), *Trematolobelia macrostachys* subsp. *kauaiensis* (Campanulaceae); *2* uniseriate rays not common, *Senecio mannii* (Asteraceae); *3* multiseriate rays rarely more than two cells wide, *Fuchsia decidua* (Onagraceae); *4* rays biseriate or uniseriate, cells all upright, *Chelonanthus chelonoides* (Gentianaceae). (*1–4:A*)

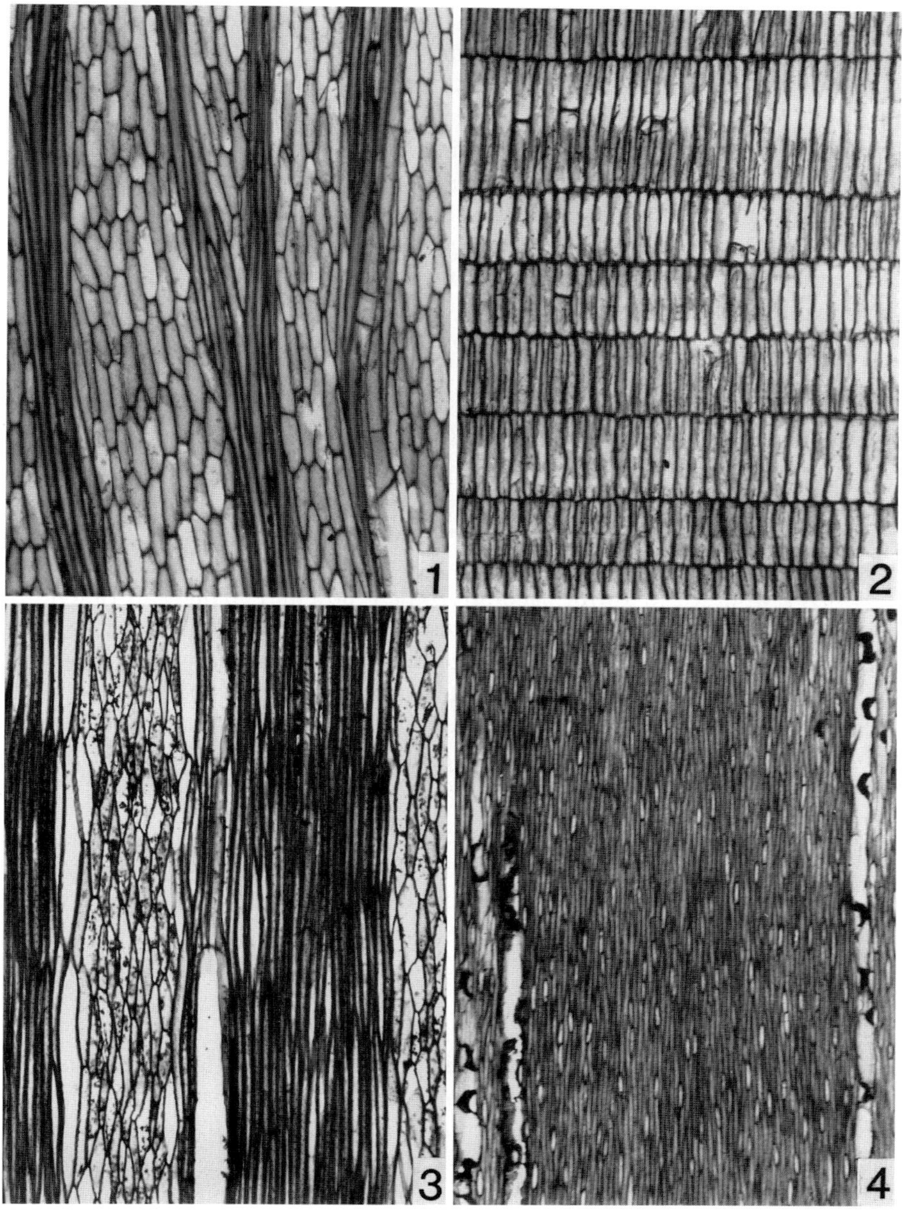

Fig. 6.9. Paedomorphic type II (*1–3*) and type III (*4*) rays, seen in tangential (*1*, *3*, *4*) and radial (*2*) sections. *1*, *2 Cyanea coriacea* (Campanulaceae); *1* tangential section; rays wide; *2* radial section; some cells horizontally subdivided; *3* upright cells at ray margins simulating libriform fibers, *Begonia parviflora* (Begoniaceae); *4* most rays a single cell tall, a few two cells tall, *Setchellanthus caeruleus* (Setchellanthaceae). (*1–4:A*)

Paedomorphic Type III. All rays (or virtually all rays) are uniseriate. Ray cells are upright (Fig. 6.9, part 4, Fig. 6.10, parts 1–4).

Rays representing this type are illustrated here for *Dracophyllum acerosum* (Fig. 6.10, part 1); note a single biseriate ray in Fig. 6.10 (part 2) and *Myrothamnus flabellifolia* (Fig. 6.10, part 3; Carlquist 1976a). An unusual condition referable to Paedomorphic type III is shown here for *Setchellanthus caeruleus* (Fig. 6.9, part 4), in which all rays are uniseriate and only a single cell tall. Also unusual is the condition shown by *Empleuridium juniperinum* (Fig. 6.10, part 4; Goldblatt et al. 1985), in which the ray cells are so tall that they resemble imperforate tracheary elements.

Barghoorn (1941b) illustrated rays of this type for *Corema conradii* (Empetraceae), and such rays characterize all known Empetraceae (Carlquist 1989a). This example, when added to others cited above, suggests that Paedomorphic type III rays can often be found in small woody shrubs. However, this ray type is not restricted to plants with this growth form.

Just as one can find a shift ontogenetically from one of the Kribs types to another (e.g., in *Illicium* or *Pentaphylax* secondary xylem begins with Heterogeneous type III but changes to Heterogeneous type I), so there are examples in which a wood begins with Paedomorphic type I and changes to Heterogeneous type IIB. If paedomorphic ray types were confined only to earlier formed xylem, one could say that recognition of these types is unnecessary. However, they are proposed here precisely because in many species, especially those with growth forms other than typically shrubby or arboreal, rays recognized here as one of the paedomorphic types exist for the entire duration of the secondary xylem.

6.6 Raylessness

6.6.1 Significance of Raylessness

Raylessness could with some justification be considered a fourth paedomorphic ray type; one can see how vertically very elongate upright ray cells can simulate ray cells in *Empleuridium juniperinum* (Fig. 6.10, part 4). Another transitional situation is shown in Fig. 6.11 (part 1) for *Begonia peruviana*: a large ray (left), representing an extension with little alteration of a primary ray, has what could be termed libriform fibers wider that those typical of fascicular areas (e.g., Fig. 3.11, part 1, right), but the difference is one of degree, not of kind. One could term this an example of paedomorphic type II rays or one could, with equal justification, call it a case of incipient raylessness.

Raylessness does represent a form of juvenilism in that there is a delay of horizontal subdivisions in areas of ray initials; there may be virtually no such divisions for the duration of cambial activity, in fact. This interpretation has

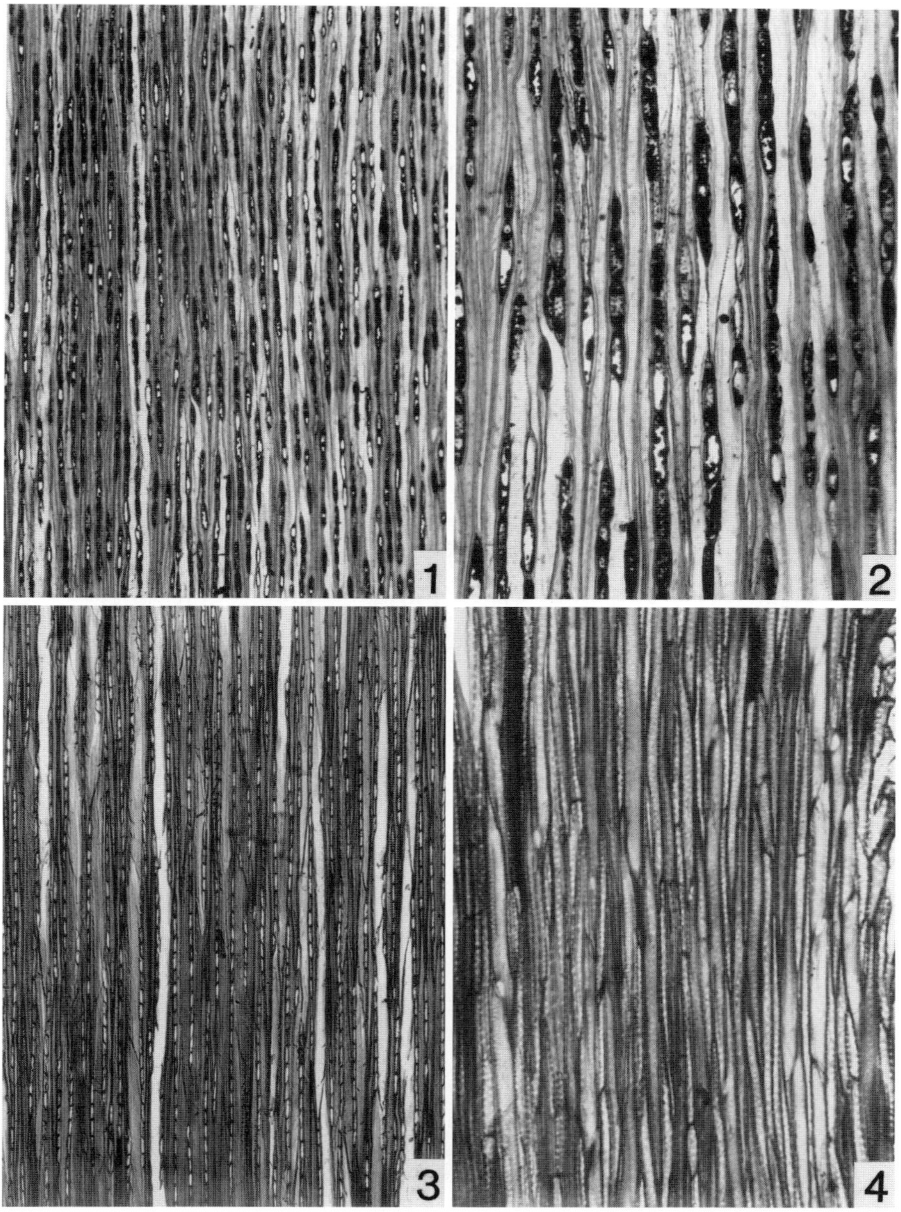

6.10. Paedomorphic type III rays, shown in tangential section. *1, 2 Dracophyllum acerosum* (Epacridaceae); *1* larger portion of wood area to show small size of ray cells; *2* ray cells upright, but not close to libriform fibers in length; *3* rays relatively tall, *Myrothamnus flabellifolia* (Myrothamnaceae); *4* ray cells very tall, difficult to distinguish from imperforate tracheary elements, *Empleuridium juniperinum* (Celastraceae). (*1,3:A*; *2,4:B*)

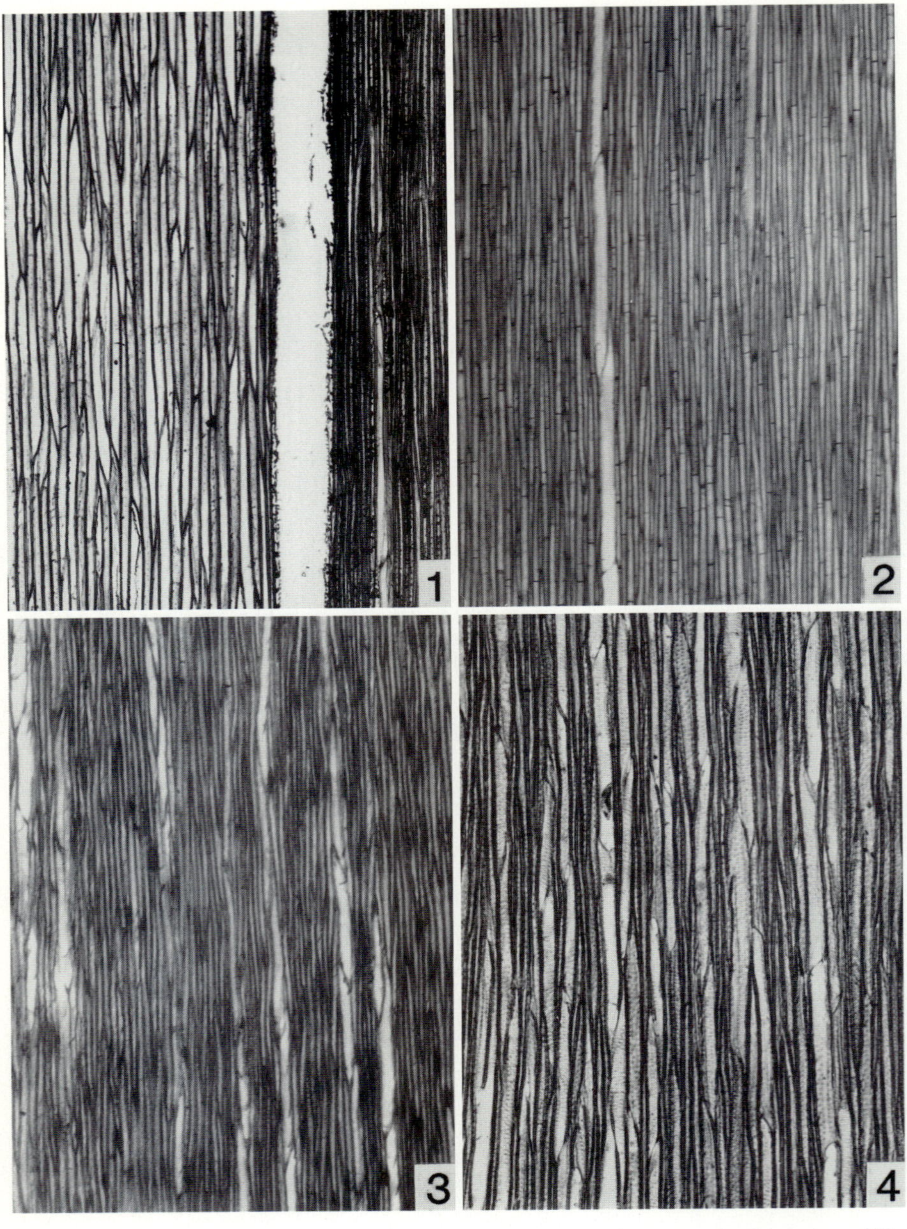

Fig. 6.11. Rayless woods, shown in tangential section. *1* Potential ray area (less) consists of libriform fibers different from ordinary libriform fibers (such as those at right), but a rayless condition can still be claimed, *Begonia peruviana* (Begoniaceae); *2* libriform fibers septate, *Beloperone crenata* (Acanthaceae); *3* libriform fibers storied, *Viola chamissoniana* (Violaceae); *4* narrow vessels very abundant, *Viviania laxa* (Geraniaceae). (*1–4:A*)

been offered earlier (Carlquist 1962a). Raylessness can occur only in a species that is relatively specialized in having short fusiform cambial initials. In such a species, length of fusiform cambial initials will simulate vertical length of ray initials, and failure or delay in subdivision in the ray initials then results in identity in length between fusiform cambial initials and ray initials. To be sure, an additional series of changes besides those involving length must occur: the cells derived from areas of ray initials must mature into fusiform cells with pointed tips rather than upright cells with blunt tips, and cell walls of such cells must simulate those of imperforate tracheary elements in wall thickness and pitting (Carlquist 1970d).

In some woods that begin rayless, horizontal subdivision does occur to the extent that rays do appear. This behavior has been observed in various species, such as *Geranium tridens* (Barghoorn 1941b), *Plantago arborescens* and *P. webbii* (Carlquist 1970b), all species of *Leptodactylon* (Carlquist et al. 1984), *Santolina* (Metcalfe and Chalk 1950; Carlquist 1966b), *Petalonyx* (Carlquist 1984d), *Ixanthus* (Carlquist 1984c), some species of *Phacelia* (Carlquist and Eckhart 1984), and at least the majority of species of *Cyrtandra* (Carlquist and Hoekman 1986a). *Lactoris* (Carlquist 1964b) represents a kind of near-raylessness, like that figured for *Begonia peruviana* (Fig. 6.11, part 1), in which libriform fibers derived from potential ray areas are not like those derived from fascicular areas, although they are libriform fibers.

Raylessness might not be expected in vines, since woody vines tend to feature large ray areas. However, raylessness has been reported in *Cobaea* of the Polemoniaceae (Carlquist et al. 1984) and one species of *Thunbergia* of the Acanthaceae (Carlquist and Zona 1988a). Both of these seem to have stems of relatively short duration, so that a selective value for rays may be minimal. In both of these species, fascicular xylem contains much parenchyma, so that there is, in effect, a substitute for ray parenchyma present.

Barghoorn (1941b) noted that in instances of raylessness, there is "a tendency toward the herbaceous habit of growth."

In a later study (Carlquist 1970b), I stressed the correlation with herbaceous habit, but in a different direction from that implied by Barghoorn: instances of raylessness, viewed in relation to phylesis within their respective families, appear to represent secondary woodiness. For example, *Viola* on the Hawaiian Islands has species that accumulate an appreciable amount of secondary xylem, but are permanently rayless (Carlquist 1974); if one views the ontogenetic development of these species, one finds that they begin with prostrate stems like those familiar in herbaceous species of *Viola*, but then the stems bend upwards like the stems of a shrub (Carlquist 1970d).

Raylessness appears to represent a phylesis in which there is a recent selection for increase in mechanical strength (Carlquist 1975a; Carlquist et al. 1984). Raylessness may represent a way of increasing mechanical strength in a stem of short duration, in which the selective value for radially-oriented parenchyma is minimal. In *Aeonium* (Crassulaceae), wood is permanently rayless. The stems of *Aeonium*, however, have limited duration; although they grow upwards for several years, eventually they fall over and reroot and thus the basal parts of the stem eventually die and the plant is perpetuated by the more distal stem portions. Under these circumstances, the secondary xylem cylinder never becomes thick enough so that development of rays would become selectively advantageous.

Raylessness occurs mostly in relatively specialized phylads of dicotyledons, and libriform fibers are present in most of these. Consequently, one would not expect a difference in pitting between ray cells and libriform fibers: both have simple pits. However, if fiberlike cells are formed in potential ray areas in a rayless species in a phylad with more primitive (tracheid-bearing) wood, would they have simple pits whereas fiberlike cells derived from fascicular areas would have bordered pits? This theoretical possibility appears not to be realized. For example, in *Hebe salicifolia*, no rays are present, but all imperforate tracheary elements have bordered pits, regardless of their point of origin on the cambium (Meylan and Butterfield 1978a).

The work of Pizzolato (1982) shows an apparent chemical mechanism for achievement of raylessness: in *Glycine*, application of 2-4DB resulted in production of raylessness.

6.6.2 Systematic Distribution of Raylessness

The following list has been assembled to record as many reports as possible of raylessness, either raylessness that persists for the duration of a stem or instances in which wood is rayless at first and then develops rays. More instances are likely to be found. Because prior lists have been very incomplete, documentation with references is offered here. I am excluding Barghoorn's (1941b) claim that *Corema* (Epacridaceae), *Staavia* (Bruniaceae), and *Tetratheca* have rayless secondary xylem; all of these begin with Paedomorphic type III rays and continue these rays throughout the duration of the stem.

Acanthaceae: *Beloperone* (Fig. 6.11, part 2), *Diapedium*, *Jacobinia*, and one species of *Thunbergia* (Carlquist and Zona 1988a)
Aizoaceae: most species (Metcalfe and Chalk 1950)
Alseuosmiaceae: *Alseuosmia* (Paliwal and Srivastava 1969; Dickison 1986)
Amaranthaceae: *Bosea*, *Nototrichium*, *Pfaffia* (Metcalfe and Chalk 1950)
Apiaceae: *Pimpinella dendrotragium* (new report)
Asteraceae: *Artemisia* (first year rayless in *A. abrotanum* and *A. rothrockii*, new report), *Lasthenia macrantha* (Carlquist 1959), *Stoebe* (Adamson 1934).
Begoniaceae: *Begonia peruviana* (Fig. 6.11, part 1; Carlquist 1985b)
Chenopodiaceae: rays "typically absent" in the family but present in *Camphorosma* and *Echinopsilon* (Metcalfe and Chalk 1950)
Cistaceae: *Lechea* (Boureau 1957)
Crassulaceae: *Aeonium* (as *Sempervivum*; Barghoorn 1941b)

Frankeniaceae: *Frankenia* (Barghoorn 1941b; Whalen 1987)
Gentianaceae: *Ixanthus* (Carlquist 1984c)
Geraniaceae: *Geranium* (Barghoorn 1941b); *Balbisia*, *Monsonia*, and *Wendtia* (Boureau 1957); *Viviania* (Carlquist 1985f; Fig. 6.11, part 4)
Gesneriaceae: *Besleria* (Barghoorn 1941b), *Chirita*, *Cyrtandra* (Carlquist and Hoekman 1986a)
Halophytaceae: *Halophytum* (Gibson 1978d)
Hydrophyllaceae: *Phacelia* (Carlquist and Eckhart 1984)
Loasaceae: *Loasa*, *Mentzelia*, *Petalonyx* (Carlquist 1984d)
Misodendraceae: *Misodendron gayanum* and *M. recurvum* (Carlquist 1985c)
Nyctaginaceae: most species (Metcalfe and Chalk 1950)
Pentaphragmataceae: some specimens (Carlquist 1997a)
Plantaginaceae: *Plantago* (Carlquist 1970b)
Polemoniaceae: *Cobaea*, *Huthia*, *Eriastrum*, *Leptodactylon* (Carlquist et al. 1984)
Polygonaceae: *Chorizanthe* (Boureau 1957), *Polygonum* (Cumbie 1969)
Primulaceae: *Lysimachia* (Carlquist 1974)
Saururaceae: *Houttouynia*, *Saururus* (Carlquist et al. 1995)
Scrophulariaceae: some species of *Alonsoa*, *Calceolaria*, *Castilleja*, and *Penstemon* (Boureau 1957); *Hebe* (Meylan and Butterfield 1978a)
Simmondsiaceae: *Simmondsia* (Bailey 1980)
Stylidiaceae: *Stylidium* (Carlquist 1981f)
Violaceae: *Viola* (Carlquist 1974; Fig. 6.11, part 3)
Vivianiaceae: see Geraniaceae

The examples illustrated here show raylessness in woods that have a range of features: sparse vessels and septate fibers (Fig. 6.11, part 2), storied libriform fibers (Fig. 6.11, part 3), and abundant vessels (Fig. 6.11, part 4).

6.7 Aggregate Rays

The term aggregate ray has been used to describe clustering of uniseriate or multiseriate rays into larger units (IAWA Committee on Nomenclature 1964). Although libriform fibers may separate the smaller rays within the aggregate ray, vessels are not formed within the confines of a compound ray (if a vessel were present, two compound rays would be said to be present). These rays commonly occur in, and therefore can in a sense be defined by, the fagalean families Betulaceae (Fig. 6.12, parts 1 and 2) and Fagaceae (Fig. 6.12, parts 3 and 4) and the family Casuarinaceae (Moseley 1948), which is probably close to Fagales. They are also reported in Ericaceae, Platanaceae, and Rosaceae (Bailey 1912) as well as in two genera of Euphorbiaceae (*Necepsia* and *Pseudoagrostistachys*; Metcalfe and Chalk 1950) and in *Cryptocarya* of the Lauraceae (Dadswell and Eckersley 1940; Richter 1981). Aggregate rays have been claimed

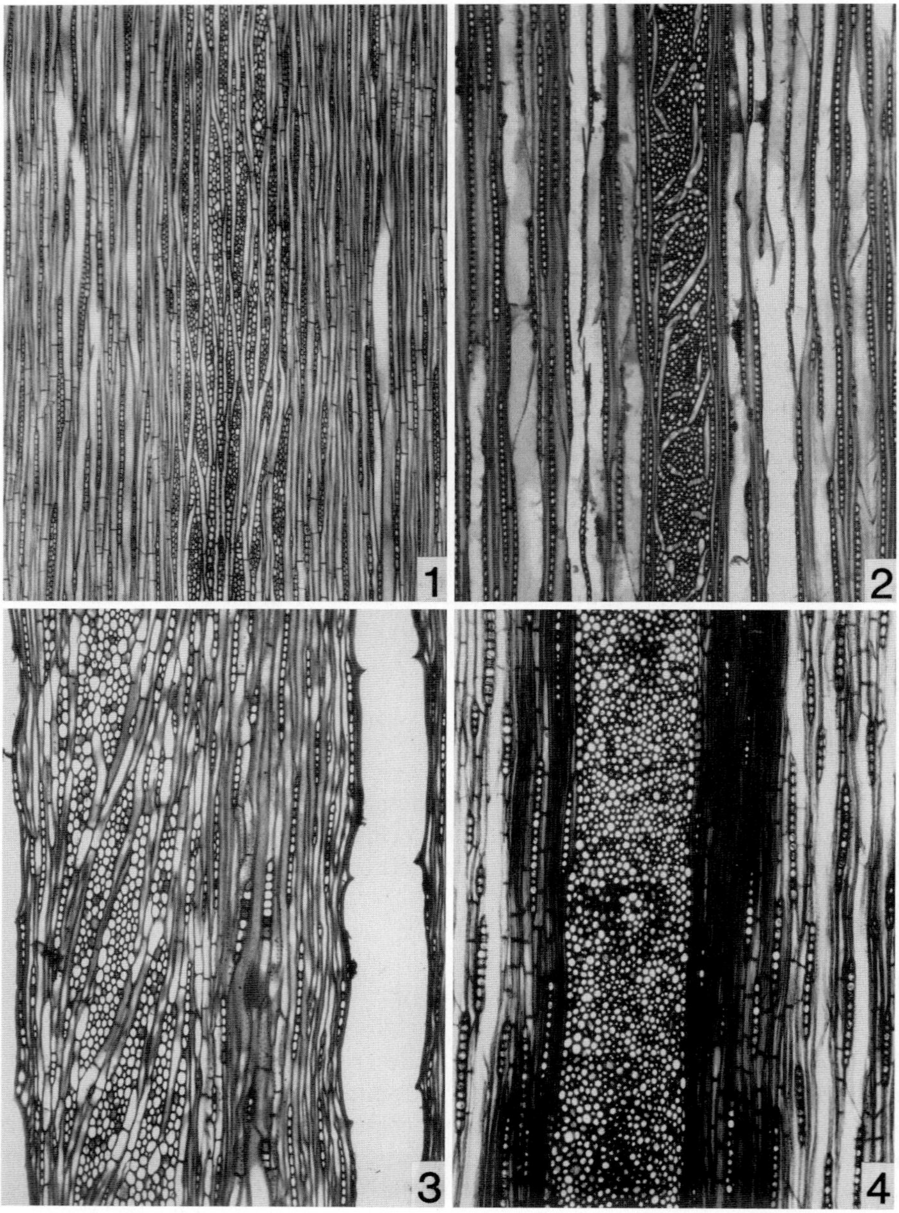

Fig. 6.12. Aggregate rays, from tangential sections. *1* Multiseriate rays in aggregate ray area separated by imperforate tracheary elements, *Carpinus caroliniana* (Betulaceae); *2* aggregate ray well defined but still containing some imperforate tracheary elements, *Alnus rhombifolia* (Betulaceae); *3* multiseriate rays in aggregate ray area, but rays otherwise uniseriate, *Lithocarpus diversiflora* (Fagaceae); *4* aggregate ray devoid of imperforate tracheary elements, *Quercus primus* (Fagaceae). (*1–4*:*A*)

by Ginzburg (1963) for *Artemisia* (Asteraceae), but I have been unable to confirm this. Typically they begin as groupings of rays, and in some taxa (*Quercus* spp.; Fig. 6.12, part 4) may, through elimination of fusiform cambial elements in the corresponding area of the cambium, become large rays devoid of libriform fibers and quite unlike the smaller rays in such a wood. Bailey (1911, 1912) and Eames (1910) have shown that aggregate rays may originate in leaf gap areas of the stele in Fagales. The idea that aggregate rays are, on the contrary, large multiseriate rays in the process of disintegration was claimed (Bailey and Sinnott 1914; Barghoorn 1941a), but the reverse, or synthetic view of aggregate ray ontogeny, is now accepted (Bailey 1912; Jeffrey 1917; Moseley 1948). To be sure, one can occasionally see some breakup of aggregate rays in large stems where large aggregate rays are present (Moseley 1948, p. 257), but breakup of large rays occurs regularly in many dicotyledons (Barghoorn 1941a), and this must not be equated with the idea of conversion of an aggregate ray into its component (often biseriate or triseriate or even uniseriate) ray units. The idea that the aggregate ray represents a primitive feature in dicotyledons was once current, but with a shift away from the concept of Amentiferae as primitive, the aggregate ray is now regarded as a specialization.

In earlier literature, the term aggregate ray was restricted to instances in which uniseriate or narrow multiseriate rays could be seen in groupings, separated by libriform fibers (e.g., Fig. 6.12, parts 1–3), whereas "compound ray" was used to denote the ontogenetic endpoint in which libriform fibers are lost, as in *Quercus* (Fig. 6.12, part 4). The term compound ray has now fallen into disuse (Metcalfe and Chalk 1983, p. 22). These authors note that in some woods, a continuum exists between aggregate rays in which imperforate tracheary elements subdivided an aggregate ray and the rays in which no such imperforate tracheary elements occur, so that a division between the two types would be quite arbitrary. For a review of this situation, see Eeckhout (1951).

Philipson et al. (1971) have shown that the cambial initials that give rise to aggregate rays are not all sharply divided between fusiform initials and typical ray initials, but include cells intermediate in shape; this is demonstrated in terms of mature wood cells by the fossil oak illustrated by Eames (1910). Other details of aggregate ray occurrence are given by Bailey (1912).

As seen in transection, the xylem cylinder of a young stem with some secondary growth is typically indented (namely, where the contour of the pith bulges outward) where aggregate ray formation begins. Where the trunk has been injured, aggregate rays may be absent and the cambium reverts to production of uniseriate rays. Species of Fagales differ in the degree to which aggregate rays become converted into homogeneous masses lacking intercalated libriform fibers (compare the four photographs in Fig. 6.12).

Investigation is needed to see if the large rays of Proteaceae and Rosaceae – families alleged to be related to Fagales in some systems – are at all related to aggregate rays of Fagales or represent an independent line of evolution.

6.8 Ray Dimorphism

In some wood anatomy literature one sees the phrase "rays of two distinct sizes." This phenomenon is noteworthy because in most woody dicotyledons there is no sharp distinction between uniseriate, biseriate, narrow multiseriate, and wider multiseriate rays – these form a continuum even though one may be able to categorize portions of it (e.g., uniseriate rays). Some woods have little variation (e.g., rays all uniseriate). However, some woods show a sharp break between wide, tall multiseriate rays and uniseriate rays. This is visible, for example, in *Lardizabala*, *Sinofranchetia*, and other Lardizabalaceae (Carlquist 1984f).

The divergence between any two ray classes is indicative in this family of the fact that large multiseriate rays are continuations of primary rays, whereas the uniseriate rays are later in origin, formed in the fascicular part of the cambium by means of subdivision of fusiform cambial initials. Wood of this nature can be found in other scandent taxa. The prominence of extensions of primary rays in secondary xylem of vines and lianas has been stressed by various authors (Schenck 1893; Haberlandt 1914; Carlquist 1975a, 1985f; Sieber and Kucera 1980) as a means for enhancing flexibility of vines. In this interpretation, fascicular portions of xylem could twist, like strands in a cable, and the soft parenchyma would yield, minimizing fracture of the fascicular portions.

Examples of dimorphic rays may be found in nonvining species, as shown here for *Emmotum holosericeum* (Fig. 6.7, part 4). One can cite many genera of Malvales (Metcalfe and Chalk 1950), *Trochodendron* (Bailey and Nast 1945), and, less conspicuously, Winteraceae (Bailey 1944a; see Fig. 6.3, part 1).

6.9 Interconnected Rays

I am coining the term "interconnected rays" to denote instances in which one can see multiseriate rays that narrow to uniseriate portions, then widen out again; several uniseriate portions may interconnect multiseriate portions in some examples (Fig. 6.13, part 1; note long ray at left). This phenomenon was termed "*zusammengesetzt*" by Moll and Janssonius (1906–1936), and that term was accepted by Metcalfe and Chalk (1983, p. 24). However, this German word is used for a different phenomenon, that of the aggregate ray: "*zusammengesetzter Markstrahl*" according to the German version of the IAWA Committee on Nomenclature (1964). Therefore a new term is needed, and I am offering the term interconnected ray because of the appearance that several multiseriate rays are interconnected by uniseriate portions. Metcalfe and Chalk (1950) list interconnected rays for a number of Euphorbiaceae: *Claoxylon cuneatus*, *Gavarretia terminalis*, *Grossera vignei*, *Hevea discolor*, *Macaranga barteri*, *Mallotus acuminatus*, and *Neoscortechinia arborea*.

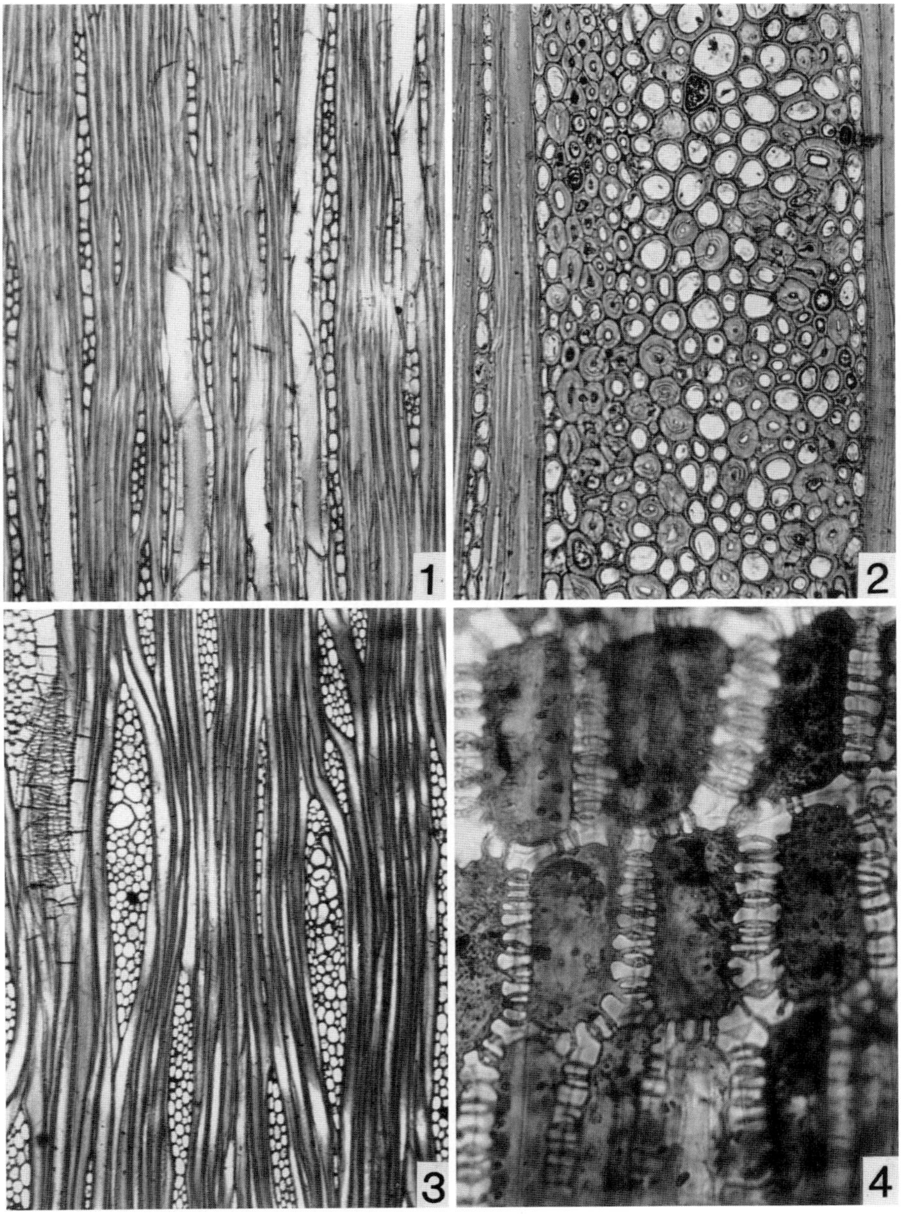

Fig. 6.13. Details of ray histology, from tangential (*1–3*) and radial (*4*) sections. *1* Interconnected rays (multiseriate ray portions interconnected by uniseriate segments), *Bontia daphnoides* (Myoporaceae); *2* sclereids of various wall thicknesses intermixed with ordinary ray cells, *Emmotum holosericeum* (Icacinaceae); *3* libriform fibers running radially present in multiseriate rays, *Cecropia sciadophylla* var. *decurrens* (Moraceae or Cecropiaceae); *4* sectional view of ray cell walls shows borders present on many pits, *Vaccinium arboreum* (Ericaceae). (*1,2:B*; *3:A*; *4:C*)

6.10 Ray Histology

6.10.1 Tile Cells

Originally described by Moll and Janssonius (1909–1936), tile cells were unfortunately defined too narrowly by the IAWA Committee on Nomenclature (1964 and preceding glossaries). That definition is what is now called the *Durio* type of tile cell (Fig. 6.14, parts 1–4), and reads as follows: "a special type of apparently empty upright cell of apparently the same height as the procumbent ray cells and occurring in indeterminate horizontal series usually interspersed among the procumbent cells." Other instances of tile cells have come to light in which the erect cells are not nearly so short either vertically or radially, and are, in fact, much taller than the procumbent cells (Fig. 6.15, parts 1, 3, and 4) or only a little taller (Fig. 6.15, part 2). The instances of tile cell occurrence where the tile cells are much taller than the procumbent cells in a ray have been termed the *Pterospermum* type. Chattaway (1933b) has offered a broader definition of tile cells so as to include both the *Durio* and the *Pterospermum* types: "a special type of erect cell, without visible contents, occurring in radial series, much narrower radially than the procumbent cells of the ray and interspersed among them." Manchester and Miller (1978) have reviewed this situation and find an intermediate type (tile cells only a little taller than the procumbent cells) in *Guazuma*; these may correspond to the example shown in Fig. 6.15 (part 2) here. Manchester (1980) has described a fossil genus, *Chattawaya*, in which unusually large tile cells are present.

Tile cells are limited to four families, and occur as follows (list modified from Chattaway 1933a,b).

1. *Durio* type: Bombacaceae (genera of Durioneae – *Boschia, Coelostegia, Cullenia, Durio, Neesia*), Malvaceae (*Kydia*), Sterculiaceae (*Guazuma, Kleinhovia, Leptonychia, Scaphopetalum*), Tiliaceae (*Columbia, Luehea, Lueheopsis, Mollia, Mortoniodendron* p.p., *Vinticena*).
2. *Pterospermum* type: Bombacaceae (*Montezuma, Ochroma*), Malvaceae (genera of Ureneae – *Malvaviscus, Pavonia, Urena* – but probably also *Hibiscus mutabilis*; Webber 1934), Sterculiaceae (*Pterospermum, Melochia* p.p.), Tiliaceae (*Belotia, Duboscia, Grewia, Mortoniodendron* p.p., *Trichospermum*, but possibly also *Sparmannia* and *Triumfetta*, in which the rays are composed mostly of erect cells with only a few procumbent cells).
3. Intermediate: *Reevesia* and *Triplochiton* of the Sterculiaceae according to Metcalfe and Chalk (1950); Manchester and Miller (1978) cite *Guazuma*.

My material of *Triplochiton* (Fig. 6.15, part 1) follows the *Pterospermum* type. The two types definitely do represent extremes that are easily recognizable, but a continuum in shape and size is present (Metcalfe and Chalk 1983, p. 14), and one should not be surprised if a particular species does not fit readily into one category or the other, or if it seems to show both types in

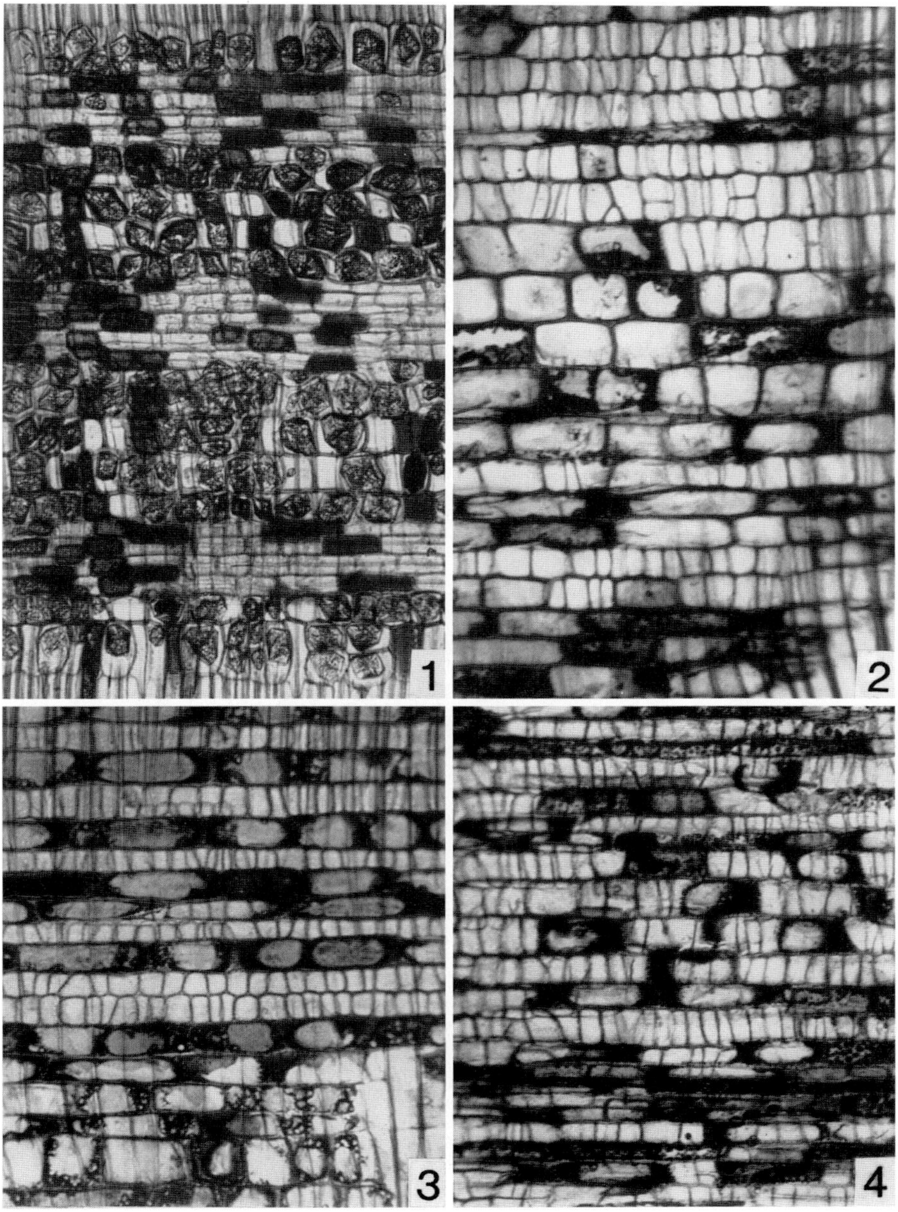

Fig. 6.14. Division of labor in ray cells as shown in tile cells (*2–4*) and ray cells other than tile cells (*1*), in radial section. *1* Crystals occur in upright cells, dark-staining deposits more common in procumbent cells, *Pleiogynium solandri* (Anacardiaceae); *2* tile cells vertically as tall as procumbent cells, *Lueheopsis rugosa* (Tiliaceae); *3, 4 Mollia lepidota* (Tiliaceae); *3* tile cells vertically shorter than procumbent cells or about the same; *4* tile cells vertically a little taller to about the same in vertical height as procumbent cells. (*1–4:B*)

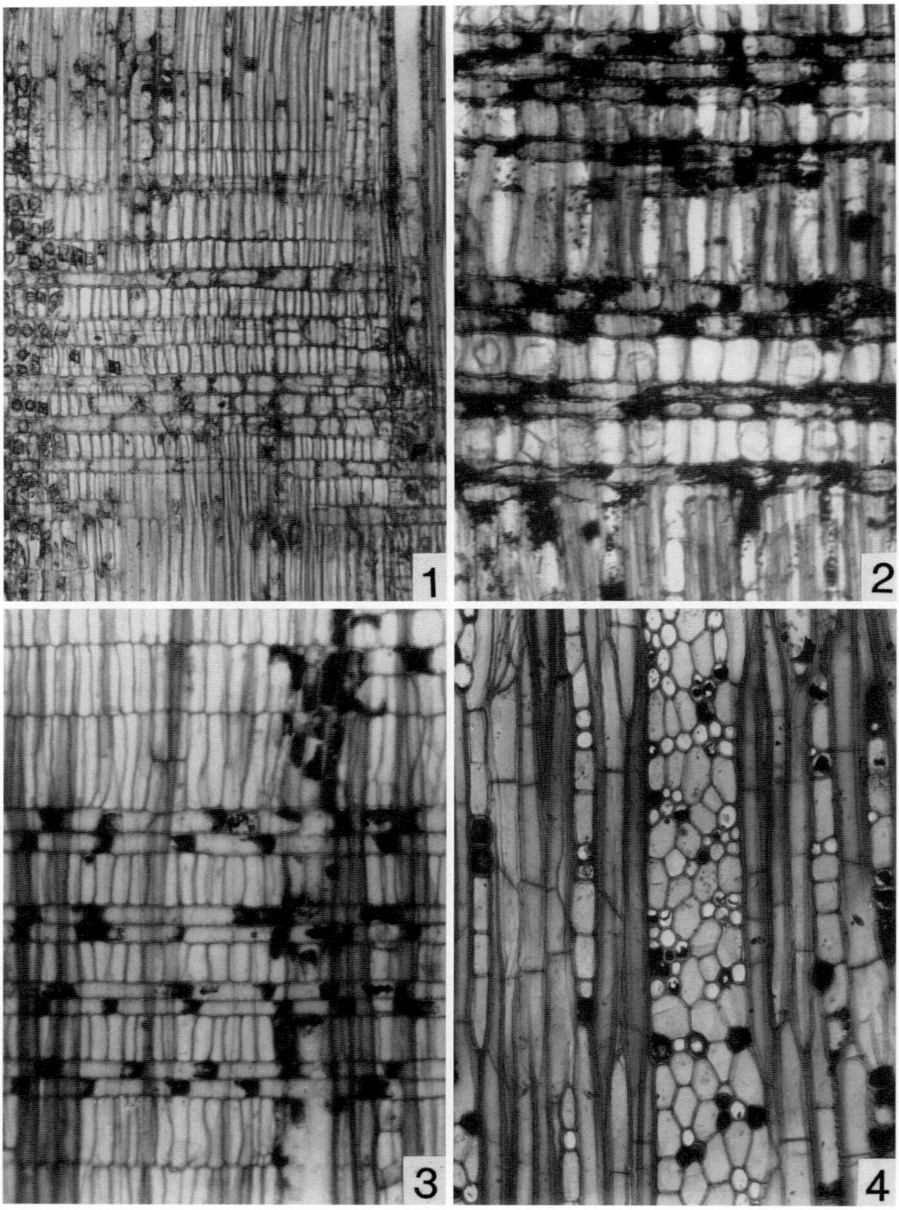

Fig. 6.15. Tile cells in Sterculiaceae, in radial (*1–3*) and tangential (*4*) sections. *1* Tile cells vertically a little taller than procumbent cells, *Triplochiton scleroxylon*; *2* tile cells somewhat taller vertically than procumbent cells and containing crystals, *Pterospermum diversifolium*; *3, 4 Pterospermum acerifolium*; *3* tile cells about twice as tall vertically as procumbent cells; *4* narrow diameter and dark contents identify procumbent cells in a tangential section, gray color of tile cells caused by tangential walls in face view. (*1:A*; *2–5:B*)

different locations. Because upright ray cells tend to become shorter as a stem grows, one might expect that the tallness of tile cells might decrease ontogenetically, and studies are needed in this respect.

Although tile cells are defined in terms of what one sees in radial sections, the ray cell dimorphism that tile cells represent can also be seen in a tangential section (Fig. 6.15, part 4)

The idea that tile cells are lacking in contents is contradicted by the examples shown in Fig. 6.15 (parts 1 and 2). In the *Triplochiton* (Fig. 6.15, part 1), crystals are formed in tile cells (and axial parenchyma cells) on a seasonal basis. In *Pterospermum diversifolium* (Fig. 6.15, part 2), crystals are formed in tile cell, and dark-staining deposits occur in other cells.

The mention of contents in tile cells brings to mind other instances of ray histology that follow similar patterns. One tends to have the idea from the anatomical literature that tile cells are a unique, isolated phenomenon. For this reason, an example much like the tile-cell phenomenon is illustrated here (Fig. 6.14, part 1). In this example, *Pleiogynium solandri*, upright ray cells tend to contain crystals whereas procumbent cells tend to contain accumulations of amorphous dark-staining compounds. Dark-staining compounds of this sort characterize procumbent cells in many species that have tile cells (Figs. 6.14, parts 3 and 4, 6.15, parts 1–4). While these compounds occur in a few of the upright cells in *Pleiogynium solandri*, only that minor exception and the fact that upright ray cells in that species are not quite so narrow as in instances of tile cell occurrence prevents one for claiming tile cells in *Pleiogynium*. This example shows the probable interpretation of tile cells – and other instances of differentiation of cells within a ray: functional differentiation of some kind, either for conduction or for accumulation of some kind of crystal or other compound probably functional in herbivore deterrence.

6.10.2 Perforated Ray Cells

Perforated ray cells are secondary xylem cells derived from ray initials but with perforation plates and lateral wall pitting like those of vessels. One can see the perforation plates of ray cells in face view in radial sections (Fig. 6.16, parts 1, 3–5). However, in a tangential section one can see that perforated ray cells connect a vessel on one side of a ray with a vessel on the opposite side of that ray, as shown here in Fig. 6.16 (part 2), and as illustrated by Styer and Stern (1979a,b) and Botosso and Vidal Gomes (1982).

A monographic treatment of perforated ray cells was offered by Chalk and Chattaway (1933). An excellent SEM study of assorted perforated ray cells has been presented (Nagai et al. 1994). Some of the subsequent contributions in which perforated ray cells are mentioned or figured include Carlquist (1960a, 1962b, 1978b, 1983c, 1987d), Stern (1967), Koek-Noorman (1970, 1972), Styer and Stern (1979a,b), McLean and Richardson (1973), Koek-Noorman and

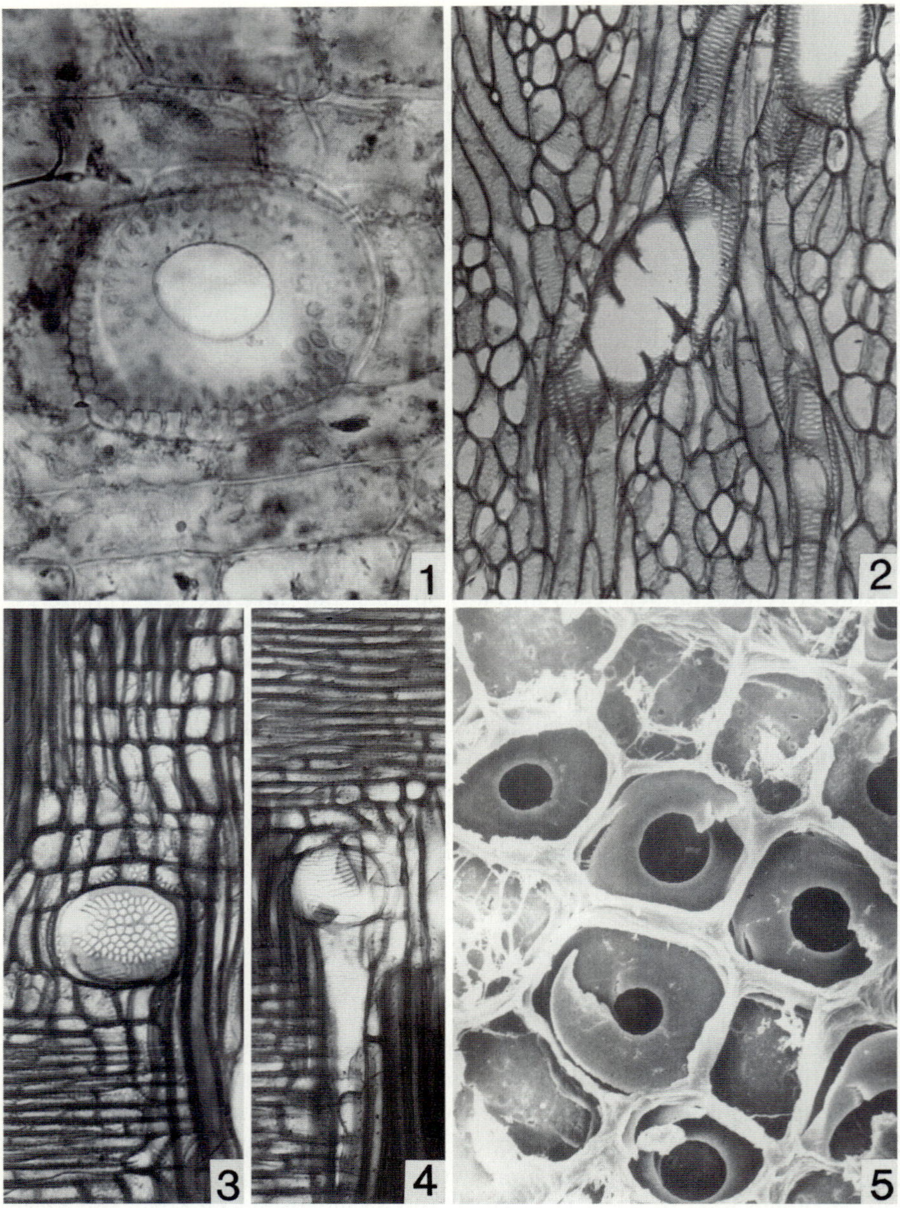

Fig. 6.16. Perforated ray cells in radial (*1*, *3–4*) and tangential (*2*) sections; ray idioblasts (*5*). *1* Perforated ray cell, *Batis maritima* (Bataceae); *2* perforated ray cells in sectional view, *Lobelia tortuosa* (Campanulaceae); *3–4* perforated ray cells, *Staphylea bumalda* (Staphyleaceae); *3* perforated ray cell with multiperforate perforation plate; *4* perforated ray cell with scalariform perforation plates, larger than neighboring ray cells; *5* transection of ray area of wood, showing wide-helix imperforate idioblasts, *Anacampseros marlothii* (Portulacaceae). (*1,5:C*; *2–4:B*; *5:F*)

Hogeweg (1974), van Vliet (1976b), den Outer and van Veenendaal (1980), Mennega (1980), Nazma et al. (1981), Botosso and Vidal Gomes (1982), Carlquist et al. (1983), Giraud (1983), Dayal et al. (1984), Rao et al. (1984), and Rudall (1985). Botosso and Vidal Gomes (1982) and van Vliet (1976b) emphasize series of perforated ray cells that run radially rather than tangentially, and term these "radial vessels."

Perforated ray cells seem sometimes to be found in woods in which breakup of larger rays into smaller segments is occurring actively. However, there is no reason to believe that this is always or even usually true. Perforated ray cells can occur in very narrow as well as very wide rays. They may be found in relatively herbaceous species, such as *Batis maritima* (Fig. 6.16, part 1) or *Lobelia tortuosa* (Fig. 6.16, part 2), or they may occur in woody species, such as *Staphylea bumalda* (Fig. 6.16, parts 3–4). If a species characteristically has simple perforation plates in vessels, the perforated ray cells will be simple also (Fig. 6.16, parts 1 and 2). If a species has scalariform perforation plates in vessels, perforated ray cells will have scalariform perforation plates (Fig. 6.16, parts 4 and 5) or some modification of a scalariform type (Figs. 3.10, part 1, 6.16, part 3). Perforated ray cells usually have bordered pits on lateral walls, but the pit size may be reduced compared to that in typical vessels (Carlquist 1960a).

Although perforated ray cells are a phenomenon that has drawn attention of various workers, there is no reason to believe that they are a characteristic feature of particular taxonomic groups. For this reason, no list of families in which perforated ray cells occur is offered here. They probably will be found in many families and genera other than those in which they have currently been reported.

6.10.3 Radial Vascular Tissue in Rays

The radial vascular tissue described by Chattaway (1948) in rays of *Banksia* and *Dryandra* is of a nature somewhat different from that of the perforated ray cells described above. The radial vascular tissue is a group of vessels that run radially for indefinite distances, rather than single vessels that cross rays in a tangential direction. The radial vascular strands are always connected with the vertical vascular tissue of the secondary xylem, but not with the innermost secondary xylem; they originate in a ray at some point after secondary growth has begun. Phloem does not accompany the vessels in the instances described by Chattaway (1948). The vascular tissues in rays of *Alluaudia* described by den Outer and Veenendaal (1980) are apparently similar to the situation described by Chattaway.

6.10.4 Radial Imperforate Tracheary Elements in Rays

I have observed an instance in *Cecropia* (Fig. 6.13, part 3) in which ray cells appear to be markedly dimorphic in size. Examination of radial sections reveals that this is not some kind of idioblastic formation, but rather is a case of inclusion within a ray of radially oriented libriform fibers, like those outside of the rays in this species.

6.10.5 Sclereids

Although there have been no special listings or other forms of attention called to occurrences of sclereids in rays, this phenomenon can be seen occasionally. Illustrated here is an instance in which groups of sclereids are intermixed with thinner walled ray cells (Fig. 6.13, part 2). In some cases, an entire ray may be composed of sclereids. An interesting instance of this sort was described in *Misodendron gayanum*, in which the fascicular xylem consists of vessels plus axial parenchyma, whereas the ray cells are like short fibers (Carlquist 1985c).

6.10.6 Ray Cell Contents: Crystals and Silica

Crystals in woods (Sect. 7.1) and silica in wood (Sect. 7.4) are covered elsewhere because they occur not only in rays, but in axial parenchyma and even fibers. In describing crystals in a wood, however, one may wish to enter material about crystals in connection with information on rays if crystals occur only in rays. Worthy of mention in the present context is the tendency of upright ray cells to be subdivided horizontally into two or more cells, each of which bears a single rhomboidal crystal. Ray cells subdivided into a series of crystal-bearing cells in rays have been figured for a wide range of families (e.g. Stern 1967; Carlquist 1978a, 1980b). Subdivision of procumbent ray cells into crystal-bearing cells has also been recorded, as in *Wrightia* (Ingle and Dadswell 1953a). Ray cells may be altered in shape by inclusion of an unusually large crystal (e.g., *Hauya*; Carlquist 1975b). These features are mentioned in this chapter because subdivision of ray cells may occur without presence of crystals (although the cells produced by subdivision are usually not so small as those produced when crystals are present). Also, unusually large ray cells do not necessarily connote crystal presence: they may be oil cells or mucilage cells, for example.

6.10.7 Oil Cells and Mucilage Cells

Although the contents of oil cells and mucilage cells may seem at first glance rather different, the taxonomic and organographic occurrence of these cells is very similar, and consequently instances in which oil cells and mucilage cells have been reported in woods have been reported as a single list, rather than two, by Metcalfe and Chalk (1950, p. 1354). Baas and Gregory (1985), who list systematic occurrence of these cell types, concur in the idea that these two kinds of cells may be alternative manifestations of a single basic type. Oil and mucilage cells are covered in greater detail in Chapter 7 because they occur in axial parenchyma as well as in ray parenchyma.

6.10.8 Tanniniferous Tubules

Tanniniferous tubules, a structure reported by Garratt (1933) in the rays of Myristicaceae, are interpreted here as cellular in nature and are covered in detail in Chapter 7.

6.10.9 Laticifers

Laticifers occur in rays of several families of dicotyledons. Information on mode of occurrence and on systematic distribution of laticifers is included in Chapter 7.

6.10.10 Disjunct Ray Cells

Ray cells that appear to have pulled apart with each other prior to maturation, and which therefore are interconnected to each other by fingerlike processes or other restricted points of contact, are known as disjunctive ray cells. They have been reported by Cozzo and Ragonese (1959) in *Eucalyptus* and Ragonese (1977) in Myrtaceae, and by Heimsch (1942), Webber (1945), and Cozzo (1951) in Sapotaceae.

6.10.11 Wall Characteristics and Pitting of Ray Cells

Ray cells range from thin-walled to thick-walled in dicotyledons; they are so often lignified that mention of instances in which ray cells are nonlignified is warranted when descriptions are prepared. In some genera, rays may contain both cells with lignified walls and some with nonlignified walls (*Akebia*; Carlquist 1984f). Notably thick-walled ray cells characterize species in some families, such as Stilbaceae (Carlquist 1987c).

Pitting among ray cells may be simple or bordered. The latter condition is far more common that one would realize from the literature. The reason is that when ray cells have thick walls, observing borders on pits when one observes pits in face view is difficult. However, if one looks at radial sections and notes whether pits on tangentially-oriented walls have borders as seen in sectional view, one can often observe that borders are present (Fig. 6.13, part 4). Wood anatomists are not accustomed to looking at pits in sectional view, so that bordered pits are rarely mentioned. A figure showing bordered pits in ray cells of *Casuarina* (Moseley 1948, p. 257) and the report of bordered pits in ray cells of *Metrosideros* (Sastrapadja and Lamoureux 1969) demonstrated that these pits may be present more widely than had been suspected. I have now figured them in a number of families (Carlquist 1975c, 1977a, 1978a,b, 1980b, 1981b, 1983b, 1989a, 1999c), but these represent only a small fraction of the families in which they could be found if sought in the above-mentioned manner.

There seems no evident correlation of pit border presence in ray cells with phylogenetic position of a taxon within the dicotyledons. The presence of borders especially on tangentially-oriented walls suggests retention of large pit membrane area on those walls as an adaptation to maximization of flow of photosynthates in a radial direction. Borders are not present exclusively on tangentially-oriented walls of rays cells, they may be present on other walls as well, but there can be little doubt that they are more commonly present on the tangentially-oriented walls. Flow of photosynthates in ray cells is a matter related to considerations Braun (1970) has discussed.

Braun (1970) has discussed his idea on differentiation among ray cells on the basis of pitting. These ideas were earlier presented in a series of papers (Braun 1967a,b, 1968a,b). Braun finds that ray cells can be divided into contact cells, with large pits, densely placed, facing vessel elements (or tracheids, in gymnosperms and primitive angiosperms), isolation cells, with very small and sparse pits facing imperforate tracheary elements of the fascicular xylem and each other, and intermediary or "mediating" cells. Braun envisages ray organization as changing with phylesis. The earlier stage ("contact stage") features contact cells only, or, in multiseriate rays, contact cells in the central portions of multiseriate rays. Intermediate stages of advance ("contact-isolation stage"), in Braun's interpretation, feature introduction of isolation cells as sheathing cells on the sides of multiseriate rays, or in the central portions of uniseriate rays. The last stage ("isolation stage") features presence of isolation cells exclusively. This last stage corresponds to homogeneous (homocellular) rays. Examples of these types are given by Braun (1967b, 1968a).

One must take Braun's ideas in conjunction with our understanding of progressive evolutionary change of rays based on Kribs' constructs. Mention was made above in connection with the Kribs ray types of a phyletic tendency for shift from numerous small points of contact between axial and radial wood parenchyma to larger but fewer points of contact between the two systems. With increasing specialization, the rays become increasingly specialized for radial conduction of photosynthates, the axial parenchyma increasingly specialized for vertical conduction; intermediate types of cells (e.g., upright cells in uniseriate rays and uniseriate rays) are progressively extinguished in this phylesis. In primitive rays, the upright cells have capabilities for vertical conduction as well as radial conduction.

One could note that in paedomorphic rays, with their predominance of upright cells, capability for vertical conduction is maintained. Paedomorphic stems are stems with relatively small accumulations of secondary xylem, so that de-emphasis of radial conduction mechanisms in comparison to stems of woodier plants is understandable.

Raylessness represents the ultimate loss of radial conduction capability, and occurs only in stems that have relatively small radial thickness; if rayless stems do experience addition of secondary xylem to an appreciable extent, they tend to develop rays (which are usually paedomorphic rays).

In considering photosynthate conduction, one must account for the function performed by that conduction. Parenchyma stores photosynthates, often in the form of starch. Such storage may relate to sudden events of growth, flowering, and fruit production. One portion of such sudden growth, that of rapid leafing out in springtime in winter-deciduous trees, has been related to hydrolysis of starch in axial and ray parenchyma by Sauter (1966a,b; Sauter et al. 1973). The release of sugars into vessels in this event is achieved by the "contact cells," which have large pits that face vessels and therefore seem suited to this activity. The fact that contact cells are more prominently pitted in roots than in stems (Braun 1984) is logical in this connection, given the tendency of roots in temperate woody plants to serve to a large extent for photosynthate storage.

Van Vliet (1976a) has reported unilaterally compound pits on vessel-ray contacts in Rhizophoraceae, and notes that these correlate with tribal limits in this family. Observation of vessel-ray pitting of this sort, and of ray pitting in general, is to be encouraged, not merely because it is of probable physiological significance, but because it has been neglected or not described in detail in some groups.

6.10.12 Intercellular Space Systems in Rays

Various workers have assumed that the intercellular space system of rays functions as a gas exchange system, although this assumption is now being reviewed (Back 1969). Although the possibility that the intercellular space system may serve for water transportation has been entertained by a number of authors (see Kucera 1985a), little or no evidence to support that idea has been presented. One can find pit fields facing intercellular spaces in ray cells, but this is not evidence that that cells are actually absorbing water from the intercellular spaces; demonstration of water in intercellular spaces of rays also is not evidence that conduction is occurring. Zimmermann's (1983) idea that intercellular spaces in wood (and most of these spaces are in rays) could serve for water storage seems illogical. Mathematical calculations presented by Zimmermann show that, in fact, it would be difficult to force water into intercellular spaces or extract water from them.

The presentation of alternative hypotheses such as these does have the merit of promoting observations on the intercellular space system of rays. An intercellular air space system, until evidence otherwise is presented, seems essential because ray cells have indefinite viability and therefore have a requirement for gas exchange to support respiration.

Information of every conceivable sort about morphology of the intercellular space system of rays in wood has been assembled by Kucera (1985a,b,c,d), and the reader should use those papers as an index to pertinent literature on this topic.

6.10.13 Secretory Canals in Rays

In a sense, secretory canals in rays represent a specialized portion of the intercellular space system in rays. Because the accumulations of secondary plant products in rays are allied in function to contents of idioblasts (e.g., laticifers) in rays, secretory canals in rays ("radial canals"), as well as axial secretory canals associated with axial parenchyma ("vertical canals") are discussed in Chapter 7.

6.10.14 Wide-Band Idioblasts in Rays

The wide-band tracheids in woods of globular cacti have been described by various authors (Sects. 4.9.1, 4.9.2). However, isolated idioblasts of this description occur in rays of *Anacampseros* of the Portulacaceae (Fig. 3.16, part 4; Carlquist 1998b). These idioblasts doubtless serve for water storage; the helices would prevent collapse of the cells when shrunken.

6.11 Systematic Distribution of Ray Types

One could wish for a systematic listing of families according to the Kribs (1935) ray system. Such lists have not been attempted, to my knowledge, and one can imagine several reasons why. Some distinctions within the Kribs system are arbitrary, because his types represent an evolutionary continuum. Also, there is ontogenetic alteration – woods that begin as Heterogeneous type III and end as Heterogeneous type I are easy to find, and examples of other transitions could be mentioned. The two lists that follow represent categories that are relatively easy to define. The first list includes families in which one or more genera have the ray type indicated, and no implication should be taken that the families are necessarily predominantly or exclusively of the ray type indicated; in the second list, the genera are specified except for small families in which the entire family may be assumed to qualify.

Very wide rays (about six cells wide or more, including aggregate rays and rays representing extensions of primary rays; list modified from Yatsenko-Khmelevsky 1954)

Actinidiaceae
Akaniaceae
Annonaceae
Apiaceae
Aquifoliaceae
Araliaceae
Aristolochiaceae
Asclepiaceae
Asteraceae (*Artemisia*)
Bataceae
Begoniaceae
Berberidaceae
Betulaceae
Bombacaceae
Burseraceae
Cactaceae
Campanulaceae (including Lobelioideae)
Caricaceae
Casuarinaceae
Chloranthaceae
Clusiaceae
Cochlospermaceae
Coriariaceae
Corylaceae
Cyrillaceae
Datiscaceae
Dilleniaceae
Elaeocarpaceae
Ericaceae (including Vacciniaceae)
Euphorbiaceae
Fabaceae (some)
Fagaceae
Flacourtiaceae
Fouquieriaceae
Garryaceae
Greyiaceae
Grossulariaceae
Gyrostemonaceae
Hippocrateaceae
Icacinaceae
Juglandaceae (some species of *Juglans*)
Lamiaceae
Lardizabalaceae
Lauraceae
Lecythidaceae
Loasaceae
Loganiaceae
Loranthaceae
Malvaceae
Marcgraviaceae
Meliaceae
Misodendraceae
Monimiaceae
Moraceae
Myricaceae
Myrsinaceae
Myrtaceae (*Gomidesia*)
Papaveraceae
Passifloraceae
Piperaceae
Platanaceae
Polygonaceae

Proteaceae
Ranunculaceae (*Clematis*)
Rhamnaceae
Rhizophoraceae
Rosaceae
Rubiaceae
Sabiaceae
Santalaceae
Scrophulariaceae
Scytopetalaceae
Simarubaceae
Staphyleaceae
Sterculiaceae
Tamaricaceae
Theaceae
Theophrastaceae
Thymeleaceae
Tiliaceae
Trochodendraceae
Ulmaceae
Verbenaceae
Violaceae
Viscaceae
Vitaceae
Vochysiaceae
Winteraceae

Rays uniseriate or nearly so (list modified from Record 1944d).

Acanthaceae (*Beloperone*)
Anacardiaceae (*Conocladia*)
Apocynaceae (*Aspidosperma*, *Cameraria*, *Geissospermum*, *Himatanthus*, *Plumeriopsis*)
Betulaceae (*Alnus*)
Bignoniaceae (*Enallagma*, *Jacaranda*, *Tabebuia*)
Brunelliaceae
Burseraceae (*Tetragastris*)
Capparaceae (*Capparis*, *Steriphoma*)
Caryocaraceae
Celastraceae (*Euonymus*)
Chrysobalanaceae
Clusiaceae (*Calophyllum*, *Caraipa*, *Haloclathra*)
Cneoraceae
Columelliaceae
Combretaceae (*Anogeissus*, *Buchenavia*, *Combretum*, *Conocarpus*, *Laguncularia*, *Lumnitziera*, *Terminalia*)
Connaraceae (*Cnestidium*, *Connarus*, *Rourea*)
Diclidantheraceae
Dilleniaceae (*Doliocarpus*)
Duckeodendraceae
Ebenaceae (*Diospyros*)
Empetraceae
Euphorbiaceae (*Anomocalyx*, *Euphorbia*, *Hippomane*, *Manihot*, *Nealchornea*, *Pera*)
Fabaceae (*Brownea*, *Dalbergia*, *Elizabetta*, *Havardia*, *Heterostemon*)
Fagaceae (*Castanea*, *Castanopsis*, *Chrysolepis*)
Gonystylaceae
Hamamelidaceae (*Bucklandia*, *Fothergilla*, *Hamamelis*, *Parrotiopsis*)
Hippocastanaceae
Hippocrateaceae (*Salacia*)
Humiriaceae
Hydrangeaceae (*Fendlera*)
Leitneriaceae
Loganiaceae (*Antonia*, *Bonyunia*, *Potalia*)
Lythraceae (*Grislea*)
Melastomataceae (Melastomoideae, e.g., *Mouriri*)
Meliaceae (*Aglaia*, *Amoora*, *Aphanomixis*, *Cabralea*, *Cipadessa*, *Dysoxylum*, *Guarea*, *Trichilia*, etc.)
Myristicaceae (*Dialyanthera*)
Myrothamnaceae
Myrtaceae (*Eucalyptus* p.p., *Melaleuca*, *Pleurocalyptus*, *Tristania* p.p., *Xanthostemon*)
Nyctaginaceae (*Neea*, *Pisonia*, *Torrubia*)

Olacaceae (*Ptychopetalum*)
Oleaceae (*Menodora*)
Polygonaceae (*Coccoloba*, *Eriogonum*, *Ruprechtia*)
Rubiaceae (*Amyris*, *Cephalanthus*, *Gleasonia*, *Henriquezia*, *Kotchubaea*, *Platycarpum*, *Tocoyena*)
Rutaceae (*Choisya*)
Salicaceae (*Populus*, *Salix*)
Sapindaceae (*Cupania*, *Cupaniopsis*, *Pseudima*, *Toulicia*)
Sapotaceae (*Chromolucuma*, *Lucuma*, *Madhuca*, *Payenia*, *Pouteria*)
Scrophulariaceae (*Scoparia*)
Setchellanthaceae
Simaroubaceae (*Quassia*)
Simmondsiaceae
Solanaceae (*Cyphomandra*, *Fabiana*, *Lycium*)
Styracaceae (*Pamphilia*)
Surianaceae
Theaceae (*Franklandia*, *Gordonia*, *Pelliciera*)
Zygophyllaceae (*Porlieria*)

7 Cell Contents, Secretory Structures

7.1 Crystals

7.1.1 Observation and Description

When mentioning occurrence of crystals in woods of dicotyledons, wood anatomists usually specify degree of abundance, crystal shapes and sizes, and organographic distribution. By "organographic distribution," one means whether crystals are present in rays, axial parenchyma, or imperforate tracheary elements, whether they are in all cells or only part of them in any of those tissues, and whether crystals are distributed in all parts of a plant and evenly distributed throughout a growth ring. Wood samples are often not adequate to provide all of the data one would ideally like in this respect.

The chemical composition of crystals is often not specified. Unless there is a contrary indication, workers assume that calcium oxalate is present. This assumption is justified in many but not all cases. Calcium oxalate survives most microtechnical treatments, and its lack of solubility in contrast with that of other crystalline materials offers an inferential means of identification often used in practice although not often specified.

Crystals, particularly calcium oxalate crystals, are usually seen readily by means of the light microscope because of their refractiveness, although in some preparations they are not obvious. The use of polarized light greatly enhances the visibility of crystals by presenting crystals bright against a dark background (Figs. 7.1, part 3, 7.7, part 5); iridescent patterns are sometimes formed (Fig. 7.7, part 1), and sometimes entire crystals turn particular colors under polarized light (Fig. 7.1, part 2; dark-appearing crystals are showing this phenomenon). Illustrations using partially polarized light (Fig. 7.1, parts 2 and 3) as opposed to fully polarized light (Fig. 7.1, part 1) can be recommended because the polarized light produces a bright appearance, highlighting the crystal, while still showing the background and thus revealing the context of the crystals better. Most of the illustrations in the present account do not use polarized light so as to simulate what the reader is likely to see without polarization.

Caution should be used in relying on polarized light for detection of crystals. Some polarizing devices are more effective than others. Also, some preparations fail to show the degree of brightness of others despite equal content

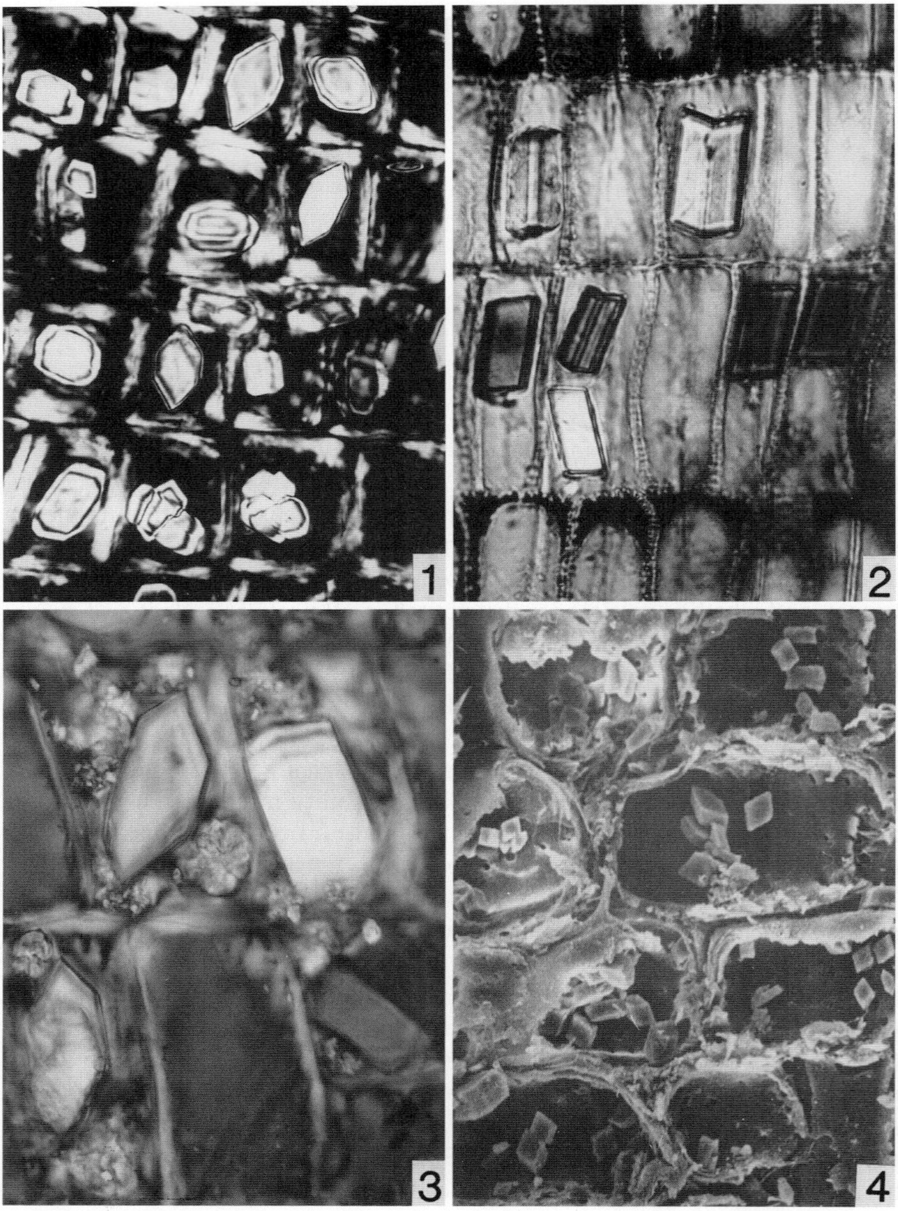

Fig. 7.1. Rhomboidal (prismatic) crystals in ray cells of radial sections, seen with polarized light (*1–3*) and with SEM (*4*). *1* Fully polarized light, *Scaevola spinescens* (Goodeniaceae); *2* partially polarized light; some of the mirror image-shaped crystals become colored in the polarized light; *Ludwigia octovalvis* (Onagraceae); *3* large crystals plus an assemblage of smaller crystals or crystal fragments, *Raspalia globosa* (Bruniaceae); *4* numerous small crystals alike in size in each ray cell, *Graptophyllum insularum* (Acanthaceae). (*1–3*:*C*; *4*:*E*)

of calcium oxalate crystals. Note should also be taken that spurious crystals, representing precipitation of calcium or other salts during microtechnical processing, may occur as artifacts. Certain fungi, as a byproduct of metabolic activities, deposit crystals in wood (Muhammad and Micko 1984), and these offer sources of error for the comparative anatomist interested in describing crystals formed by the wood cells.

Scanning electron microscopy (SEM) can be recommended for demonstration of crystal presence (Figs. 7.1, part 4, 7.2, 7.7, part 4, 7.8, parts 1 and 2, 7.9, parts 4 and 5, 7.10, parts 1–4). Many examples could be cited, such as the atlas of Meylan and Butterfield (1978a). Unless a cell is "opened" for SEM inspection by sectioning, the crystals it contains will not be visible during SEM inspection. Therefore, most workers will probably want to use light microscopy for screening, then turn to SEM if available for rendering of particular crystal occurrences. Because sectioned material of wood is used for SEM studies, crystals seen by SEM in cells cut open by the sectioning are often fractured by the sectioning (Fig. 7.10, part 5), and one must search extensively to find those relatively few crystals that are exposed but intact. In the relatively few instances in which crystals are numerous per cell (Fig. 7.1, part 4), this is not applicable.

7.1.2 Calcium Oxalate Crystals

Calcium oxalate crystals, although ordinarily identified by pragmatic means of appearance and solubility, can be identified with greater certainty by chemical tests (Pobeguin 1943; Metcalfe and Chalk 1983, p. 87). Note should also be taken of the crystallographic monograph of Frey (1929), who takes into account both the monohydrate and trihydrate forms. The same author (Frey-Wyssling 1981) has updated this survey.

Chattaway (1955, 1956) monographed systematic occurrence of calcium oxalate crystals in wood. The term "crystal," with a modifying word (e.g., "acicular crystal") is common in wood literature, and one may generally assume that the author means calcium oxalate crystals, although exceptions do occur. Very few instances have been subjected to microchemical testing because the assumption is so widespread that calcium oxalate is by far the predominant compound present in crystalline form.

7.1.2.1 Rhomboidal Crystals; Subdivision of Crystal-Bearing Cells; Encapsulated Crystals

Rhomboidal crystals ("prismatic" of some authors) are by far the most common type of calcium oxalate crystal in woods, as the list below suggests. If one compares listings of rhomboidal crystals in wood with listings for those

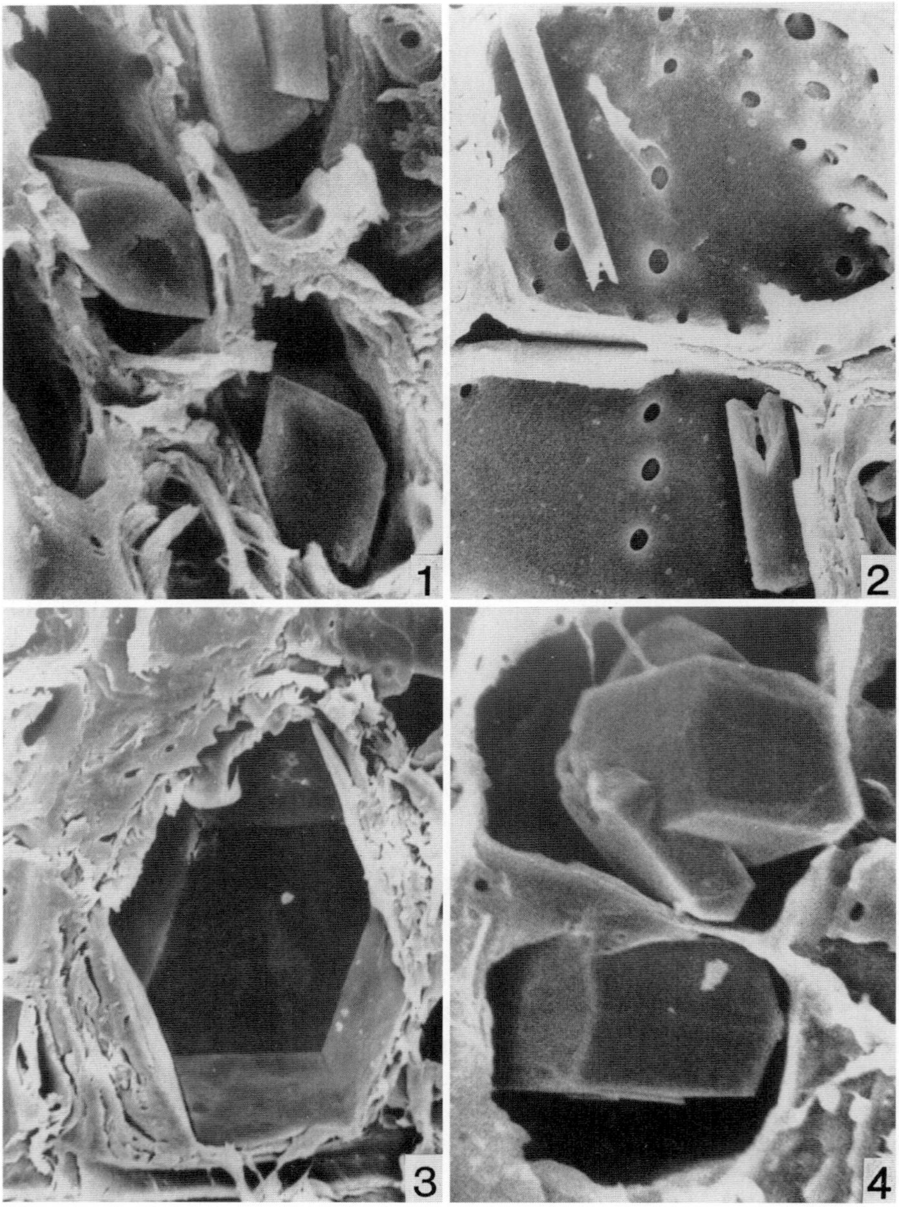

Fig. 7.2. SEM photographs of crystals in ray cells from tangential (*1*, *3*) and radial (*2*, *4*) sections. *1* Hexagonal crystals borne singly or (*above*) in pairs, *Scaevola spinescens* (Goodeniaceae); *2* elongate crystals that show mirror-image construction of halves, *Montinia caryophyllacea* (Montiniaceae); *3* large crystal, encapsulated within a sclereid, *Pittosporum phyllareoides* (Pittosporaceae); *4* pair of cells containing, *below*, a single crystal, and, *above*, a compound crystal, *Kokia rockii* (Malvaceae). (*1*,*3*,*4*:*F*; *2*:*G*)

crystals in all parts of the plant, one finds that rhomboidal crystals occur in wood in many fewer taxa.

The term "solitary crystal" is often used for a rhomboidal crystal found singly in a cell not altered in some way in relation to containing a crystal. Solitary crystals may be seen in Fig. 7.1, parts 1 and 2 (two crystals in one cell shown;Figs. 7.2 parts 1–4, 7.5 parts 3 and 4).

The term solitary crystal may also be used for instances in which one crystal per cell occurs but in which there has been subdivision of a parent cell into units containing one crystal each (with nearby non-crystal-bearing cells not subdivided). Crystals solitary in such subdivided cells are shown here in Figs. 7.3, parts 1–3, 7.4, parts 4 and 5, 7.5, parts1 and 2, 7.6, parts 1–3.

Upright ray cells subdivided horizontally into a crystal-containing series of cells are found in certain species. These are shown here in Figs. 7.3, parts 1 and 2, 7.6, parts 1 and 3. Procumbent or square cells subdivided into crystal-bearing cells may occasionally be found. Some families characteristically have upright ray cells subdivided into crystal-bearing series of cells, whereas others do not (see listing below).

Axial parenchyma cells may be subdivided into cells each of which contains one crystal, or one large crystal plus smaller ones. In some families (notably Scytopetalaceae; Carlquist 1987h), some cells of an axial parenchyma strand are so subdivided whereas others are not; this condition is shown here in Fig. 7.4 (part 4). In other groups, all or most of particular axial parenchyma strands are subdivided, whereas other axial parenchyma strands are not at all subdivided (Figs. 7.3, part 3, 7.4, part 5, 7.6, part 2); this is particularly conspicuous in the genus *Averrhoa* (Fig. 7.5, parts 1 and 2), in which there is vasicentric axial parenchyma not subdivided into crystal-bearing cells plus diffusely distributed axial parenchyma cells all of which are subdivided into crystal-bearing cells. Subdivided crystal-bearing axial parenchyma cells are often described as "chambered crystals." The term chambered crystal is useful, but awkward in that one may wish to refer to the cells containing crystals and their distribution rather than to the crystals.

Although rhomboidal crystals are most frequently found in axial and ray parenchyma, there are also instances in which "crystalliferous fibers" are found. Examples of crystalliferous fibers are shown here in Figs. 7.4, part 1 (left), 7.4, parts 2 and 3, and 7.17, part 2. "Crystalliferous fibers" in the usage followed here denotes crystals in imperforate tracheary elements essentially like other imperforate tracheary elements in a wood in having thick lignified walls and in not being subdivided into a series of crystal-bearing cells (in which case "chambered crystals" would be said to be present).

Rhomboidal crystals may be formed characteristically as one large crystal plus many smaller ones (Figs. 7.1, part 3, 7.3, part 4, 7.4, part 1) or even a large one with intermediate ones (Fig. 7.2, part 4). These smaller crystals are often termed "crystal fragments," but that designation is probably not correct in implying that the smaller crystals are irregular in shape. Very likely, study by SEM will reveal that smaller crystals, not adequately viewed by light

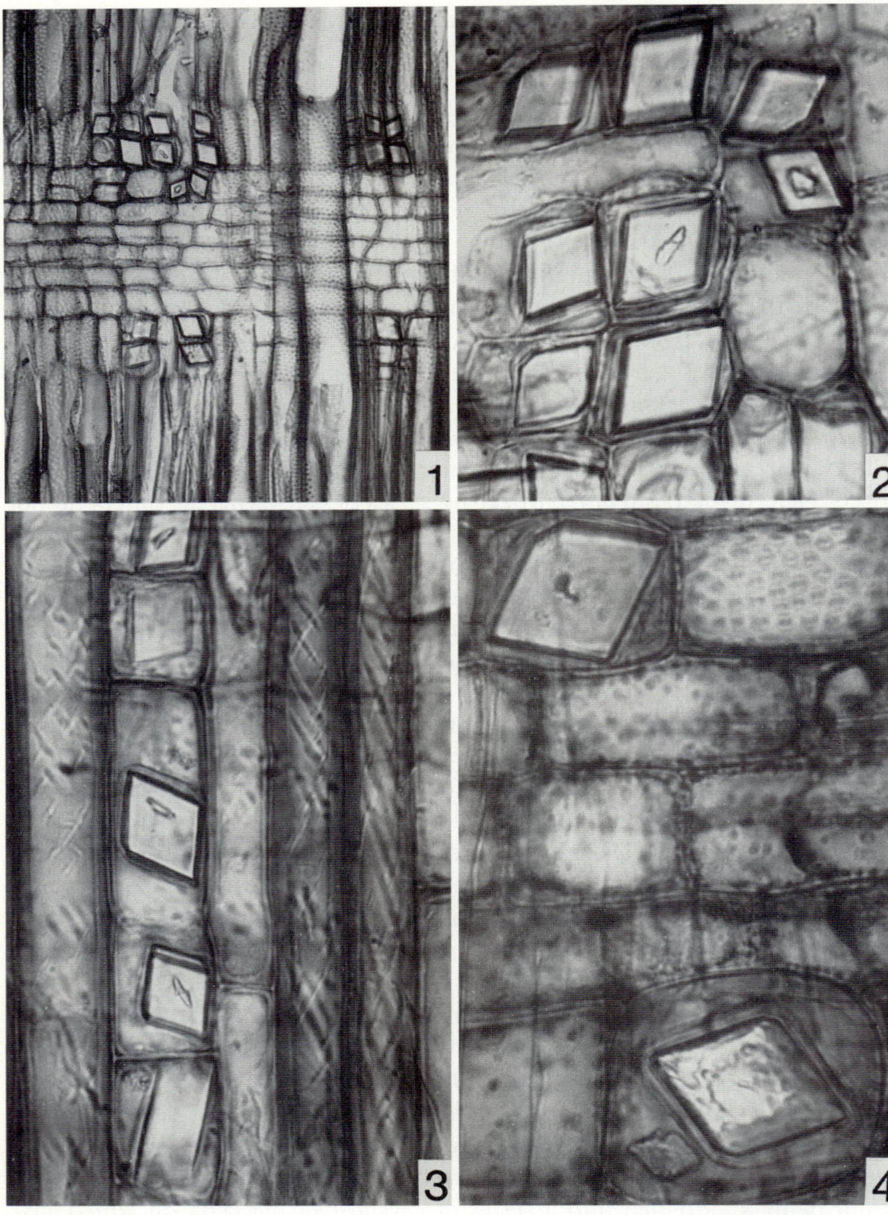

Fig. 7.3. Rhomboidal crystrals in ray cells (*1, 2, 4*) and axial parenchyma cells (*3*) of *Pittosporum* (Pittosporaceae). *1–3 P. lifuense*; *1* crystals mostly in subdivided marginal cells of rays; *2* rhomboidal crystals, some of which have spaces in centers (enlarged portion of 1); *3* crystals in ray cells of various length, a condition intermediate between chambered and nonchambered; *4* crystals in procumbent ray cells; lower crystal surrounded by a layer of secondary wall material (encapsulated), *P. napaliense*. (*1:B*; *2–4:C*)

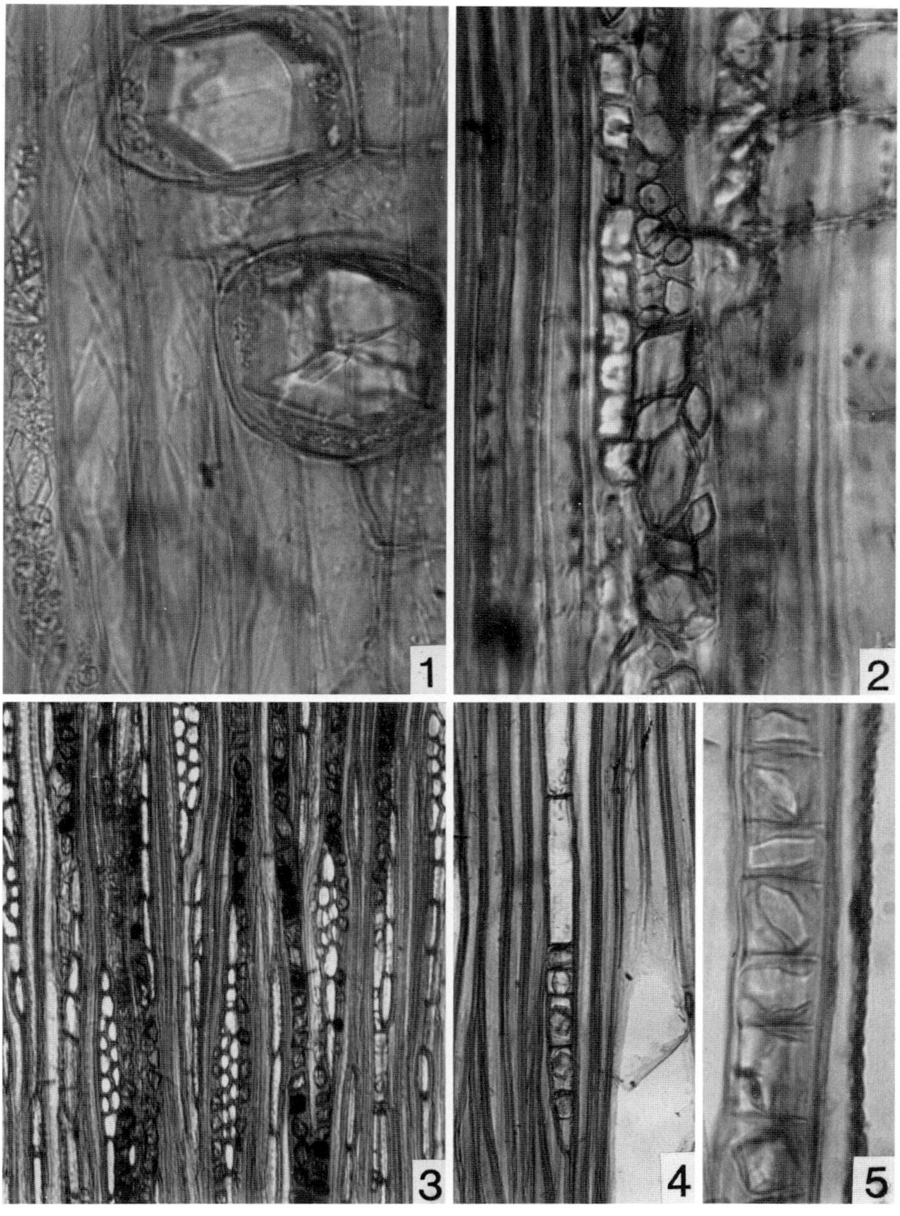

Fig. 7.4. Types of rhomboidal crystal occurrence. *1* Crystals of various sizes in libriform fiber (*extreme left*), large plus small crystals in two ray cells, *right*, *Pittosporum phillyraeoides* (Pittosporaceae); *2* hexagonal crystals of various sizes in fiber-tracheids, *Eremophila polyclada* (Myoporaceae); *3* hexagonal crystals of uniform size in some fibers, *Akocanthera venenata* (Apocynaceae); *4* axial parenchyma strand in which crystal-bearing cells are subdivided ("chambered crystals"), but noncrystalliferous cells are not subdivided, *Schizomeria ornata* (Cunoniaceae); *5* chambered crystals of axial parenchyma, *Aeschynomene elaphroxylon* (Fabaceae). (*1,2,5:C*; *3,4:B*)

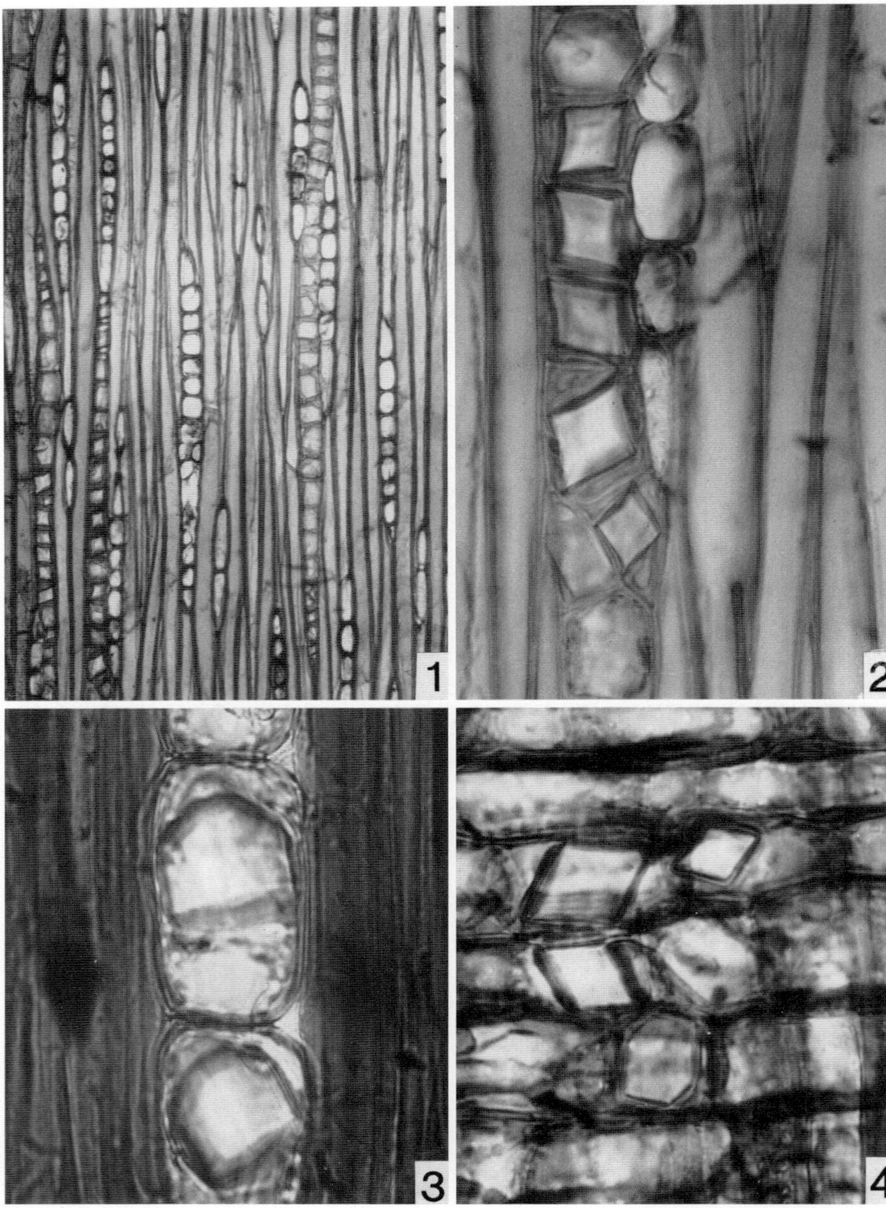

Fig. 7.5. Chambered (*1, 2*) and nonchambered (*3, 4*) occurrence of rhomboidal crystals. *1, 2 Averrhoa bilimbi* (Oxalidaceae), tangential sections; *1* long strands of chambered crystals are adjacent to some rays; all axial parenchyma strands are subdivided into crystal-bearing cells, no non-crystal-bearing strands occur in the diffuse parenchyma; *2* portion of an axial parenchyma strand showing rhomboidal crystals, one each in squarish cells; *3, 4 Rhopalocarpus lucidus* (Sphaerosepalaceae); *3* portion of axial parenchyma strands from radial section; strand cells contain one crystal each but are not subdivided into squarish cells; *4* rhomboidal crystals in procumbent ray cells; no ray cells are subdivided into crystal-bearing cells. (*1*:*B*; *2–4*:*C*)

microscopy, are in fact merely smaller rhomboidal crystals. Instances of crystalliferous fibers frequently show a mixture of large plus small crystals (Figs. 7.4, parts1 and 2. 7.17, part 2), although a few cases have only crystals of a single size class (Fig. 7.4, part 3).

If very small crystals are formed in large numbers in a single cell, "crystal sand" is said to be present. Crystal sand is discussed below (Sect. 7.2.1.7). However, small crystals formed in smaller number (see Fig. 7.1, part 3) may be covered under the heading of rhomboidal crystals.

Shapes of rhomboidal crystals differ. Most frequently in face view they have four sides, corresponding to the geometric definition of rhomboid (Figs. 7.1, part 4, 7.3, parts 1–4, 7.5, parts2 and 4). In other instances, a hexagonal outline in face view is apparent (Figs. 7.1, part 1, 7.2, part 1, 7.4, part 1). The crystal of Fig. 7.2 (part 3) has more than the usual number of faces. In a few cases, crystals are formed in what may be called a "mirror image" fashion (Carlquist 1997b; Figs. 7.1, part 2, 7.2, parts 2 and 4). Mention should be made of these various manifestations, which are of systematic value. Further discussion of rhomboidal crystal types will be found in the monograph of Frey (1929). Some crystals have slightly rounded angles rather than sharp ones. Another item of structural importance is the occurrence of an apparent lacuna in the center of a crystal (Figs. 7.2, part 2, 7.3, parts 1–4).

Cells may secrete wall material around crystals. If such wall material is very thin, it is likely to be overlooked in light microscopy. The term "encapsulated crystal" describes enclosure within a wall covering the crystal. In some instances, deposition of wall material is so extensive that the entire lumen of the cell between the primary wall and the crystal is filled with secondary wall material (Figs. 7.2, part 3, 7.5, part 2). Parameswaran and Richter (1984) have produced excellent electron micrographs of encapsulated crystals in sectional view. This could be termed a kind of sclereid.

The following list of instances in which rhomboidal crystals occur represents a reworking and expansion of data from Chattaway (1955, 1956). Abbreviations are as follows: A = axial parenchyma; F = fibers (imperforate tracheary elements); R = rays; c = encapsulated crystal or sclerosed crystal-bearing cells; e = enlarged cells bearing crystals; s = subdivided ("chambered") cells; v = variable in size (mixed large and small crystals within a single cell); inclusion of abbreviation within parentheses = infrequent occurrence. Abbreviations apply to the genus or genera they follow. Many new records remain to be found. References are not cited except in occurrences of special interest, such as *Misodendron* subgenus *Angelopogon* (Misodendraceae), in which libriform fibers are restricted to strands paired beside latewood portions of each fascicular area, and cells bearing rhomboidal crystals surround these strands (Fig. 7.5, part 4).

Acanthaceae: *Beloperone*, *Graptophyllum* RvAs
Aceraceae (RAs)
Aextoxicaceae

Akaniaceae (R)
Alangiaceae: *Alangium* Rs(A)
Anacardiaceae: 33 genera R; *Pleiogynium* (Figs. 6.14, part 1, 7.16, part 4), *Rhus*, *Semecarpus*, *Spondias* (A)
Annonaceae: *Goniothalamus*, *Tetrastemma* (R)
Apiaceae: various genera (R); *Eryngium bupleuroides* A
Apocynaceae: 22 genera R; some also A; F in *Akocanthera* (Fig. 7.3, part 3)
Aquifoliaceae: *Ilex* Rs(A)
Araliaceae: *Brassaiopsis*, *Dendropanax*, *Didymopanax*, *Pseudopanax* R
Asclepiadaceae: *Leptadenia*
Asteraceae: *Athanasia*, *Ericameria*, *Gutierrezia*, *Proustia*, *Tetradymia* R; *Olearia* Rv; *Eriocephalus* F
Avicenniaceae: *Avicennia* R(A)
Balanopaceae: *Balanops* RsAs
Berberidaceae: *Berberis*, *Mahonia* R
Bignoniaceae: *Couralia*, *Diplanthera*, *Dolichandrone*, *Oroxylon*, *Stereospermum* R
Bombacaceae: 10 genera, mostly R but A in a few cases
Boraginaceae: *Boureria*, *Cordia* R; *Cordia* (A)
Brassicaceae: *Descurainia* Rv
Bruniaceae: *Linconia*, *Titmannia* R; *Audouinia*, *Nebelia* Rs; *Berzelia* RAcs (Fig. 7.5, part 2); *Brunia* (Fig. 7.5, part 3), *Lonchostoma*, *Staavia* RsAs; *Raspalia* RsvAs
Buddlejaceae: *Buddleja* Rv (Carlquist 1997b; Fig. 7.7, part 1)
Burseraceae: 11 genera mostly R but A in a few cases
Canellaceae: *Canella*, *Capsicodendron*, *Cinnamosma*, *Pleodendron*, *Warburgia* R
Capparaceae: *Atamisquea*, *Capparis*, *Morisonia*, *Steriphoma* R
Caryocaraceae: *Anthodiscus*, *Caryocar* Rs
Casuarinaceae: *Casuarina* RAs
Celastraceae: 12 genera – some R, some As (probably none with both)
Chenopodiaceae: *Kochia* Rv; 9 other genera Rv (Fahn et al. 1986); also As (3 genera), Fs (2 genera)
Chloanthaceae: *Pityrodia* (R)
Cistaceae: *Cistus* (R; Fahn et al. 1986)
Clusiaceae: 13 genera As; *Pentaphalangium* also R
Cneoraceae: *Cneorum* RA, As (but in different species; Carlquist 1987g)
Combretaceae: *Anogeissus* R(A); *Combretum* RvA; *Conocarpus* Rv(A); *Guiera*, *Macropteranthes* (As); *Pteleopsis*, *Terminalia* (RsAs)
Connaraceae: *Connarus*, *Cnestidium*, *Cnestis*, *Ellipanthus*, *Jollydora*, *Manotes* As
Convolvulaceae: *Ipomoea* (A)R
Coriariaceae: *Coriaria* R
Cornaceae: *Griselinia*, *Marlea*, *Mastixia* R; *Curtisia* Rv
Corylaceae: *Carpinus* Res, *Ostrya* Re
Corynocarpaceae: *Corynocarpus* R

Cunoniaceae: 14 genera As
Daphniphyllaceae: *Daphniphyllum* (As)
Davidsoniaceae: *Davidsonia* As
Dichapetalaceae: *Chailletia, Dichapetalum, Gonypetalum, Tapura* R
Dipsacaceae: *Scabiosa* (R)
Dipterocarpaceae: *Balanocarpus, Vateria* R; *Cotylelobium, Dryobalanops* As; *Hopea* RA, *Parashorea* RsAs; *Pentacme* Rs; *Shorea, Vatica* RAs
Ebenaceae: *Diospyros, Maba* RAs
Elaeocarpaceae: *Aceratium* Rsv; *Aristotelia, Crinodendron* RsA; *Elaeocarpus* RsA; *Sloanea* Rv
Epacridaceae: *Acrotriche, Leucopogon* As, *Monotoca* Asc, *Trochocarpa* Asc
Ericaceae: *Oxydendrum* As; *Vaccinium* R
Erythroxylaceae: *Erythroxylum* As
Escalloniaceae: *Quintinia* As
Euphorbiaceae: many genera R; many genera As; many genera RAs
Fabaceae: many genera As (Asc in about a third of these), R also in some: *Cynometra, Daniella, Guilandina, Indigofera, Maniltoa, Pseudosindora, Robinia, Sesbania, Sophora*
Fagaceae: *Castanea, Chrysolepis, Nothofagus, Shiia* As; *Lithocarpus, Pasania, Quercus* RAs; *Fagus* Rs
Flacourtiaceae: many genera R, a few also A (*Homalium, Pangium, Ryparosa, Trichadenia*)
Geissolomataceae: *Geissoloma* RvAv
Gesneriaceae: *Cyrtandra, Rhytidophyllum* R
Gonystylaceae: *Gonystylus* Rv(A)
Goodeniaceae: *Coopernookia, Scaevola* R
Goupiaceae: *Goupia* R
Grubbiaceae: *Grubbia* Rv(A)
Hamamelidaceae: *Altingia, Distylium, Ostrearia* R; *Symingtonia* Rs; *Rhodoleia* RAs
Himantandraceae: *Galbulimima* RAsc
Hippocastanaceae: *Billia* As
Hippocrateaceae: *Salacia* Rs
Humiriaceae: *Saccoglottis* As
Icacinaceae: *Apodytes, Citronella, Gonocaryum* Rs; *Calatola, Platea, Tylecarpus, Urandra* R; *Medusanthera, Pennantia* (R); *Rhyticaryum* RAv; *Lisianthera* (RsA)
Juglandaceae: *Engelhardtia* A; *Hicoria* Ae; *Juglans* Ase
Julianiaceae: *Juliania, Orthopterygium* R
Lactoridaceae: *Lactoris* F (Carlquist 1990b)
Lecythidaceae: *Barringtonia* Rv; *Bertholletia, Lecythis* Asc; *Cariniana, Chydenanthus, Couroupita* Asc; *Planchonia* (RAs)
Linaceae: *Ctenolophon* RsAs; *Hebepetalum* Rs; *Hugonia, Lepidobotrys, Orthocosmus* As
Loganiaceae: *Strychnos* RAsv

Loranthaceae: *Amyema*, *Psittacanthus* R; *Loranthus* Rc
Lythraceae: *Lagerstroemia* As
Malesherbiaceae: *Malesherbia* (R)
Malpighiaceae: *Bunchosia* R(A)
Malvaceae: 10 genera R (Rs in *Kokia*; Fig. 7.2, part 3); A also in *Hibiscus*, *Lagunaria*, *Papuadendron*, *Thespesia*
Marcgraviaceae: *Norantea* Rs
Meliaceae: most genera AS; uncommonly R in *Carapa*, *Cedrela*, *Khaya*, *Soymida*, *Toona*, *Vavaea*, *Xylocarpus*
Melianthaceae: *Bersama*
Misodendraceae: *Misodendron* subgenus *Angelopogon* (Fig. 7.6, part 4), around strands of fibers (Carlquist 1985c)
Moraceae: RA in at least 10 genera; A only in a few (*Antiaris*, *Maclura*, *Parartocarpus*)
Moringaceae: *Moringa* RA
Myoporaceae: *Eremophila* R (Fv in one species; Fig. 7.4, part 2)
Myristicaceae: *Horsfieldia*, *Iryanthera* R
Myrsinaceae: *Aegiceras*, *Discocalyx*, *Rapanea*, *Suttonia* R; *Conomorpha*, *Myrsine* (R)
Myrtaceae: As in 20 genera; Asv in many of these
Nyssaceae: *Nyssa* As
Ochnaceae: *Cespedesia*, *Lophira* AS; *Elvasia* AsR; *Ochna* Rec; *Ouratea* Rc; *Testulea* R
Octoknemataceae: *Octoknema* R
Olacaceae: As in most genera; R also in *Ochanostachys*, *Ongokea*, *Strombosia*
Oleaceae: *Chionanthus* A; *Fraxinus*, *Ligustrum*, *Schrebera* Rv
Onagraceae: *Gaura* (Fv)
Oxalidaceae: *Averrhoa* (Fig. 7.4, parts 1 and 2), *Sarcotheca* As
Pedaliaceae: *Rogeria*, *Uncarina* R
Pentaphylacaceae: *Pentaphylax* (A)
Phytolaccaceae: *Gallesia*, *Seguiera* R
Pittosporaceae: *Billardiera, Marianthus* R; *Bursaria*, *Citriobatus* RS, *Pittosporum* RsAsc (Fig. 7.3, parts 1–4)
Platanaceae: *Platanus* R
Polygalaceae: *Carpolobia*, *Phlebotaenia* As
Polygonaceae: *Coccoloba* AFs; *Podopteris*, *Ruprechtia* Fs; *Muehlenbeckia* Rs; Neomillspaughia A; *Triplaris* AsFs
Proteaceae: *Hakea*, *Roupala* R; *Stenocarpus* Rs
Punicaceae: *Punica* Fs (Bridgewater and Baas 1978)
Quiinaceae: *Touroulia* As
Rhamnaceae: 21 genera R; *Maesopsis* As
Rhizophoraceae: *Anopyxis* As; *Bruguiera*, *Cassipourea* R; *Carallia*, *Combretocarpus* RAs; *Ceriops*, *Rhizophora* RA
Rosacaeae: numerous genera As
Rubiaceae: many genera R (sometimes c); Koek-Noorman (1969)

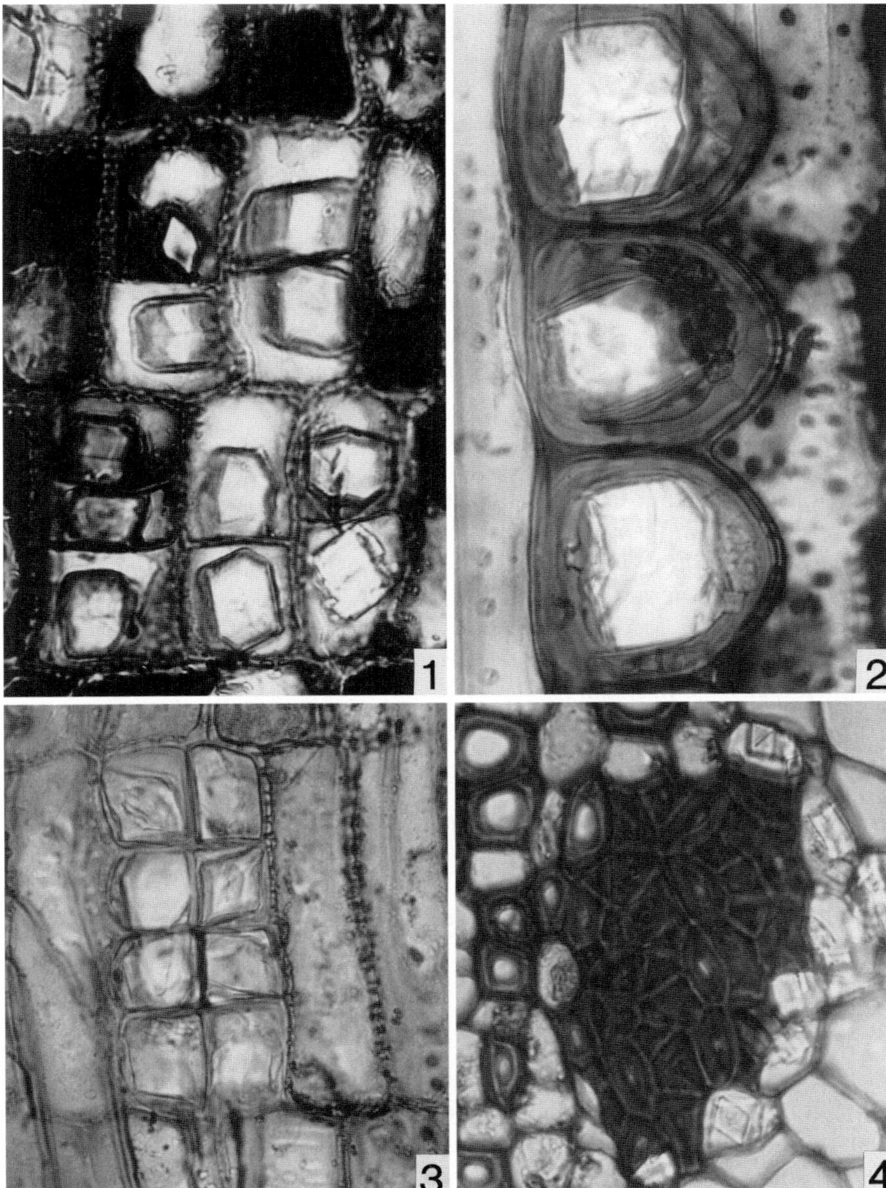

Fig. 7.6. Rhomboidal crystal occurrence in subdivided cells, radial sections (*1–3*) and transection (*1*). *1* Upright ray cells subdivided into pairs of crystal-bearing cells, *Huertea cubensis* (Staphyleaceae); *2* chambered crystals of axial parenchyma; the sclerosed cell wall merges with secondary wall material encapsulating the crystals, *Berzelia ecklonii* (Bruniaceae); *3* upright ray cells subdivided into groups of four crystal-bearing cells, *Brunia stokoei* (Bruniaceae); *4* crystal-bearing cells occur only at periphery of fiber strands in wood of *Misodendron brachystachyum* (Misodendraceae), see also Fig. 2.13, parts 3 and 4. (*1–4:C*)

Rutaceae: 41 genera As; *Feronia, Flindersia, Phellodendron* AsR; *Calodendrum, Micromelum* R
Salvadoraceae: *Salvadora* R
Santalaceae: *Acanthosyris, Cervantesia* R; *Exocarpus, Santalum* RAs
Sapindaceae: 47 genera As; *Ellatostachys, Euphoria, Pometia* RAs; *Litchi* R
Sapotaceae: 15 genera As; *Chrysophyllum, Mimusops* RAs
Scytopetalaceae: *Rhaptopetalum, Scytopetalum* As
Simaroubaceae: 6 genera R; *Ailanthus* RAs; *Cadellia, Irvingia, Klainodoxa* As
Solanaceae: *Grabowskya* Re (Carlquist 1992h)
Sonneratiaceae: *Duabanga* Av; *Sonneratia* R
Sphaerosepalaceae: *Rhopalocarpus* AR (Figs. 7.4, part 3, 7.5, part 4)
Staphyleaceae: *Huertea* RsFs (Fig. 7.6, part 1); *Turpinia* Rs
Sterculiaceae: 22 genera R; RAs in many of these
Styracaceae: *Styrax* RAs
Symplocaceae: *Symplocos*
Tamaricaceae: *Tamarix* R (Fahn et al. 1986)
Theaceae: *Gordonia* RAs; *Schima* As
Theophrastaceae: *Jacquinia* As
Thymeleaceae: *Phaleria*
Tiliaceae: 13 genera R; 5 genera RA; *Diplodiscus, Schoutenia* RAs
Tremandraceae: *Platytheca* RAs; *Tetratheca* Rv; *Tremandra* R(s)
Trigoniaceae: *Trigoniastrum* As
Ulmaceae: *Apelocera, Aphanathe, Phyllostylon* R; *Celtis* R(RsA); *Zelkova* Re
Urticaceae: *Pipturus, Urtica* R
Verbenaceae: *Vitex* R
Violaceae: *Alsodeia* and some other genera R
Viscaceae: *Viscum* R (Fahn et al. 1986)
Vochysiaceae: *Erisma, Qualea* As
Zygophyllaceae: *Bulnesia, Guaiacum, Larrea, Nitraria, Porlieria* As

7.1.2.2 Elongate or Rodlike Crystals

These terms can be used to denote crystals that are more than twice as long as wide but not narrow and needlelike (more than ten times as long as wide: acicular crystals and raphides). The term "styloid" has evidently been used for this category of crystals by Metcalfe and Chalk (1983, p. 217). Styloid may be used to refer to an unusually large elongate crystal, such as that shown in Fig. 7.7 (part 2) or Fig. 7.7(part 3). The listing of families with styloids assembled by Metcalfe and Chalk (1983) does not separate those families with styloids in woods from those families that have styloids elsewhere in the plant body but not in wood. The list of families with elongate crystals below is therefore based on the listing of Chattaway (1955), with modifications where appropriate. Abbreviations are as follows: A, in axial parenchyma; F, in libriform fibers; R, in rays. Other conventions are as above for rhomboidal crystals.

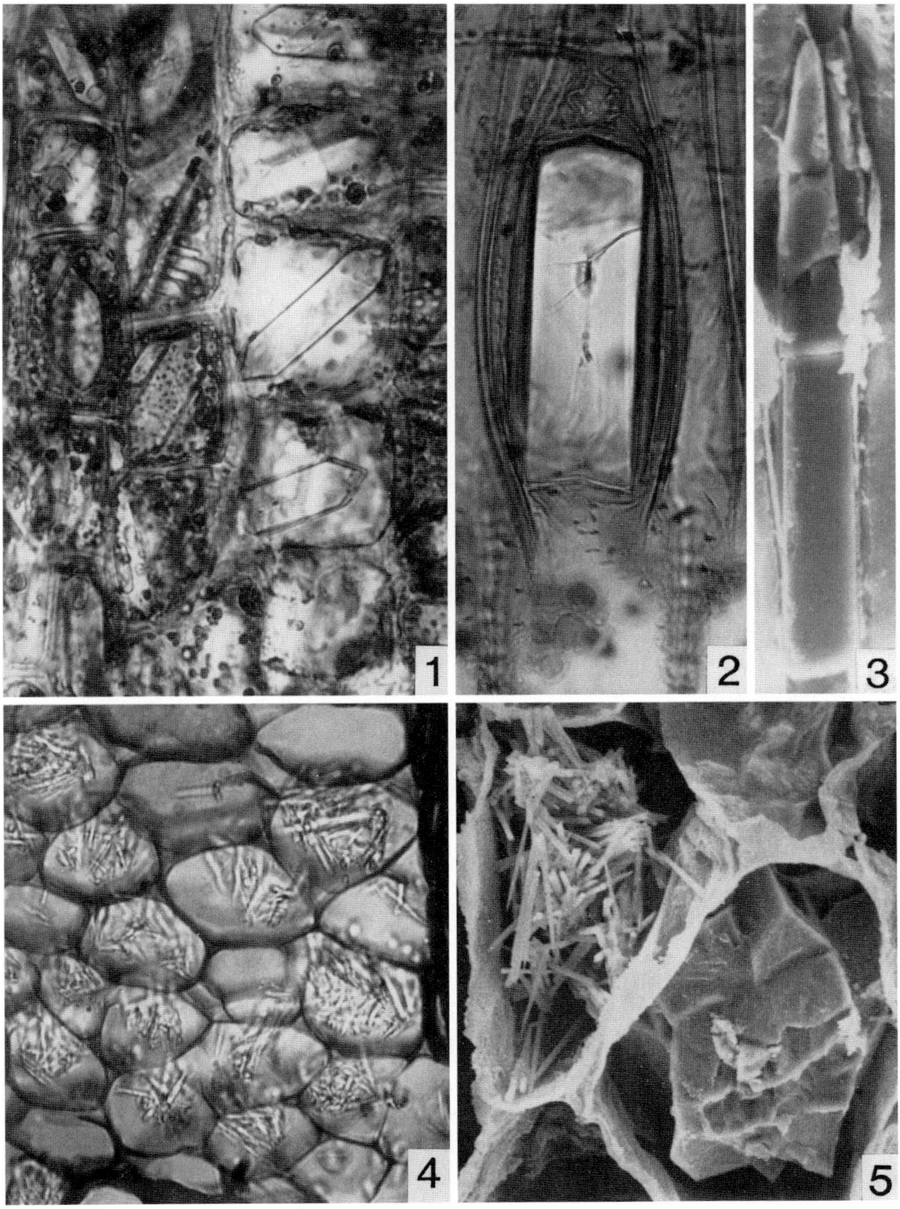

Fig. 7.7. Elongate (*1, 2*) and acicular (*3, 4*) crystals. *1* Rodlike crystals in ray cells of radial section, *Buddleja utahensis* (Buddlejaceae); *2* styloid (plus smaller crystals, *above*) in an enlarged fiber, radial section, *Hauya elegans* (Onagraceae); *3* portion of styloid (breaks evident) from idioblast in axial xylem of tangential section, *Trichostigma octandrum* (Phytolaccaceae); *4* acicular crystals in the abundant parenchyma that sheaths interxylary phloem strands, transection, *Thunbergia alata* (Acanthaceae); *5* acicular crystals (*upper left*) plus a druse, radial section, *Tetrastigma voinierianum* (Vitaceae). (*1,2,4:C; 3,5:E*)

Apocynaceae: *Rejoua* RF
Asteraceae: *Proustia*
Avicenniaceae: *Avicennia* RA
Bignoniaceae: *Radermachera* R
Buddlejaceae: *Buddleja* R (Fig. 7.6, part 1); *Peltanthera* (Carlquist 1997b)
Celastraceae: *Siphonodon* R
Combretaceae: *Anogeissus* RS; *Terminalia* A
Elaeocarpaceae: *Sloanea* R
Euphorbiaceae: *Glochidion* R
Lauraceae: *Cryptocarya* RA; *Dehaasia, Licaria, Lindera, Litsea* R
Melastomataceae: *Henriettea* R (ter Welle and Mennega 1977); *Dichaetanthera* A (parenchyma sheathing interxylary phloem; van Vliet 1981)
Monimiaceae: *Matthaea, Tetrasynandra* R
Montiniaceae: (Fig. 7.2, part 2; Carlquist 1989d)
Myrsinaceae: *Aegiceras* R
Myrtaceae: *Psidium* R; *Rhodamnia* R
Oleaceae: *Ligustrum, Olea* R, *Notelea* RA
Onagraceae: *Hauya* F (Fig. 7.7, part 2); the mirror image crystals of *Ludwigia* (Fig. 7.1, part 2) are sometimes rhomboidal, sometimes elongate
Phytolaccaceae: *Gallesia, Seguiera, Trichostigma* (Fig. 7.7, part 3)
Proteaceae: *Hakea* R
Rosaceae: *Prunus* R
Rubiaceae: *Hymenodictyon* RA; *Ixora*
Thymeleaceae: *Pimelea* RA
Urticaceae: *Dendrocnide* A
Verbenaceae: *Premna, Vitex* R

7.1.2.3 Acicular Crystals

Acicular crystals are defined here, much as in Metcalfe and Chalk (1983), as slender crystals, more than ten times as long as wide, which are not arranged in a tight packet surrounded by a gelatinous sheath (Fig. 7.7, parts 1–5). The latter characteristics are regarded here as those of raphides.

The following list of families in which acicular crystals occur is an expansion of the list of Chattaway (1955). Although Metcalfe and Chalk (1983, p. 217) give a list of families containing acicular crystals, their list is not limited to occurrence in wood. Obviously, from the list below, only a few families do have acicular crystals in woods; two have been added to Chattaway's (1955) listing, and more are to be expected as wood studies progress.

Acanthaceae: *Mendoncia, Thunbergia* A (acicular crystals occur in parenchyma sheathing interxylary phloem; Carlquist and Zona 1987; Fig. 7.7, part 4).
Lauraceae: *Actinodaphne, Cryptocarya, Iteadaphne* R

Verbenaceae: *Gmelina, Premna* R
Vitaceae: *Tetrastigma* R (Fig. 7.7, part 4; new report, but acicular crystals are to be expected in other genera)

7.1.2.4 Raphides

Raphides are defined as crystals more than ten times as long as wide, formed oriented parallel to each other in a tight packet surrounded by a gelatinous sheath (Fig. 7.8, parts 1–5). These features are illustrated here by means of SEM (Fig. 7.8, parts 1 and 2). The listing of families with raphides in wood here is an expansion of the list of Chattaway (1955), which in turn is based on the account of Hess (1936). Raphides typically occur in cells large in comparison to neighboring cells. Cells bearing raphides are therefore idioblastic; raphide-bearing cells may be borne in pairs, as they characteristically are in some Onagraceae (Fig. 7.8, part 4; Carlquist 1975b). Abbreviations are as follows: A = axial parenchyma; F = libriform fibers; R = rays.

Actinidiaceae: *Saurauia* A
Aizoaceae: *Carpobrotus, Russchia* R (new report)
Dilleniaceae: *Curatella, Davilla, Doliocarpus* RA; *Doliocarpus, Hibbertia* A; *Tetracera, Wormia*
Greyiaceae: *Greyia*
Marcgraviaceae: *Marcgravia, Souroubea*
Melastomataceae: *Bredia* R (van Vliet 1981)
Nyctaginaceae: *Bougainvillea, Calpidia, Colignonia, Commicarpus, Neea, Pisonia, Torrubia* A
Onagraceae: *Gaura, Hauya, Lopezia* (Fig. 7.8, parts 4 and 5), *Ludwigia, Oenothera* (Fig. 7.7, part 3), *Xylonagra* A
Phytolaccaceae: *Phytolacca*
Rubiaceae: *Coelospermum, Morinda* A; *Calycodendron, Calycosia, Cosmibuena, Coussarea, Craterispermum, Faramea, Gillespiea, Hamelia, Hillia, Mapouria, Pagamea, Plectronia, Psychotria, Straussia* R (see Koek-Noorman 1969)
Rutaceae: *Raputia*
Tetrameristicaceae: *Tetramerista* R
Urticaceae: *Dendrocnide* R (Bonsen and ter Welle 1984)
Vitaceae: *Tetrastigma, Vitis* R

7.1.2.5 Crystal Sand

Crystal sand is the formation of numerous small rhomboidal crystals per cell. These crystals are often so small that they do not appear as crystals, but rather as a granular content (Fig. 7.9, parts 1–5). When examined by SEM (Fig. 7.9, parts 4 and 5), the nature of crystal sand becomes readily evident. Polarized

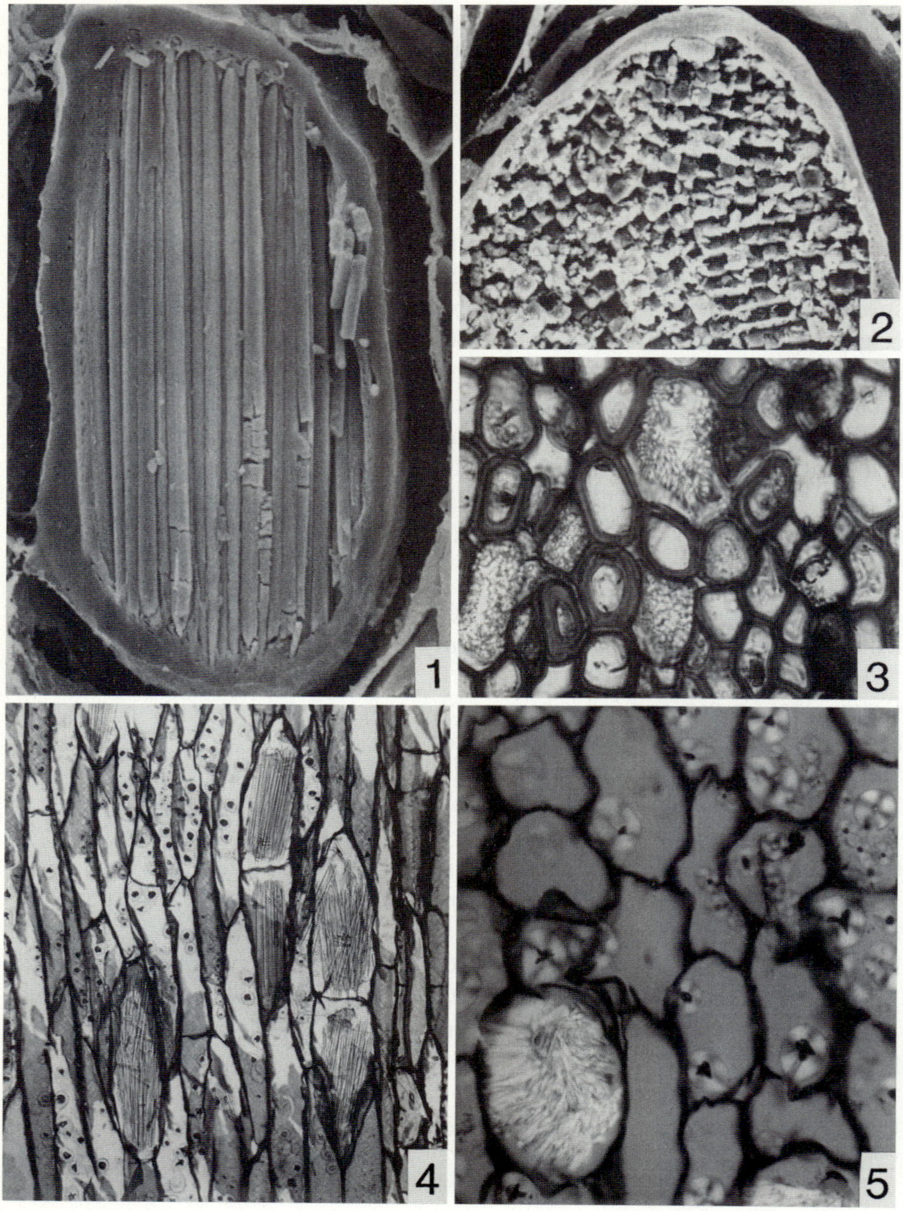

Fig. 7.8. Raphides in woods. *1, 2 Tetrastigma voinierianum* (Vitaceae), radial section, SEM photomicrographs; *1* raphides in longisection, surrounded by dried gelatinous sheath; *2* portion of transection of packet of raphides and associated gelatinous sheath; *3* raphides in transection, wood transection *Oenothera linifolia* (Onagraceae); *4, 5 Lopezia lopezioides* (Onagraceae), axial parenchyma associated with interxylary phloem; *4* raphides in longisection, raphide-bearing cells paired, nonpolarized light; *5* raphides in transection (*lower left*), starch grains, partially polarized light. (*1,2:E*; *3,5:C*; *4:B*)

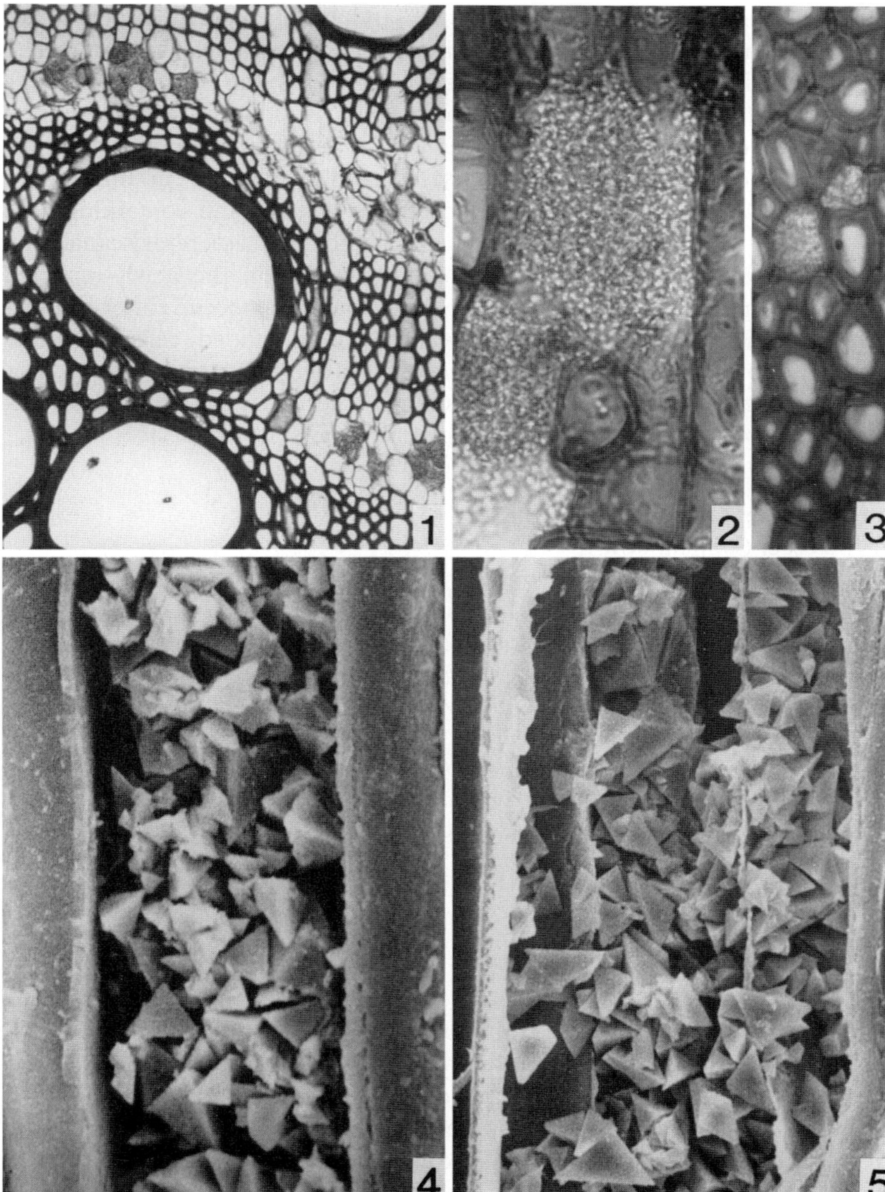

Fig. 7.9. Occurrence of crystal sand in woods. *1* Crystal sand (*gray*) in axial parenchyma associated with interxylary phloem, transection of *Craterosiphon scandens* (Thymeleaceae); *2* crystal sand in ray cells, walls between ray cells break down, *Nolana crassulifolia* (Nolanaceae); *3–5 Passerina vulgaris* (Thymeleaceae); *3* two crystal-sand-bearing cells (*center*), transection; *4 5* SEM photomicrographs of crystal sand idioblasts (note tetrahedral form of crystals) from longisections; *4* thick-walled fiber containing crystal sand; *5* thin-walled fibriform parenchyma cells containing crystal sand. (*1,3:B*; *2:C*; *4,5:H*)

light ought to be a reliable test for crystal sand, but in practice birefringence does not always occur. Because of the granular nature of crystal sand, it can be confused with other kinds of cell contents. Very likely, the report by Metcalfe and Chalk (1983, p. 224) of radial canals in wood for Solanaceae and Thymeleaceae is based on appearances that have since proved to be crystal sand and are so reported below. Crystal sand may occur in thin-walled parenchyma cells in which cell walls between crystal-sand-containing cells may break down, yielding the appearance of a laticifer or a secretory canal (Fig. 7.9, part 2). Attention is called here to the fact that in the newly-reported instance of crystal sand occurrence in *Passerina*, it may occur either in libriform fibers (Fig. 7.9, part 4) or in thin-walled (but unidivided) cells that therefore should probably be called axial parenchyma (Fig. 7.9, parts 3 and 5).

The listing below is modified from the list of Chattaway (1955). Abbreviations are: A = axial parenchyma, F = libriform fibers, R = rays.

Amaranthaceae: *Bosea* A
Boraginaceae: *Cordia*, *Patagonula* AR
Icacinaceae: *Gomphandra* R
Lauraceae: *Actinodaphne*, *Lindera* R
Nolanaceae: RA (Fig. 7.9, part 2; Carlquist 1987a)
Rubiaceae: *Adina*, *Anthocephalus*, *Calycophyllum*, *Diplospora*, *Hodgkinsonia*, *Mastixiodendron*, *Neonauclea*, *Randia*, *Timonius* R
Sapotaceae: *Bumelia*, *Chrysophyllum*, *Mastichodendron*, *Palaquium*, *Pouteria*, *Sideroxylon* A
Solanaceae: *Grabowskya* A (Carlquist 1992h); *Lycium* A (Fahn et al. 1986); *Nicotiana* A (Carlquist 1987a)
Thymeleaceae: *Craterosiphon* A (Fig. 7.9, part 1); *Passerina* AF (Fig. 7.9, parts 3–5; new report).

7.1.2.6 Druses

Druses are compound crystals, sometimes termed "cluster crystals" (Fig. 7.10, parts 1–5). Although generally approximately spherical, druses may be cuboidal if they grow so as to fill up a cuboidal space, as in Penaeaceae (Carlquist and DeBuhr 1977). The compound nature of a druse is not evident on surfaces broken open by sectioning – the adjacent crystals are thoroughly fused (Fig. 7.10, part 5); if only the edge is sectioned, smaller fragments may be produced (Fig. 7.10, part 1). Some druses consist of rhomboidal units much alike in size (Fig. 7.10, part 4), but others contain minute rhomboids as well as larger ones (Fig. 7.10, part 3). Druses generally occur singly per cell, occasionally more than one per cell.

The listing below is based upon Chattaway (1955), but modified by addition of new reports, which are cited with references. Metcalfe and Chalk (1950) did not offer a list of which families have druses (cluster crystals) specifically in

Fig. 7.10. SEM photographs of druses in ray cells, from radial (*1–4*) and tangential (*5*) sections, SEM photomicrographs. *1* Nearly intact druse (some fracturing of surface at *right* because of sectioning), *Ipomoea arborea* (Convolvulaceae); *2* intact druse, *Ipomoea cardiosepala*; *3* druse composed of large and small crystals, *Operculina palmeri* (Convolvulaceae); *4* druse composed of flattened and angular crystals, *Aristolochia sipho* (Aristolochiaceae); *5* druse cut in half, component crystals fused, starch grain portions, *Tetrastigma voinerianum* (Vitaceae). (*1–5:F*)

wood, although their later listing (Metcalfe and Chalk 1983, p. 216) does denote wood occurrence by taking advantage of Chattaway's compilation. Abbreviations are as follows: A = axial parenchyma; e = enlarged cells; R = rays; s = cells subdivided; parentheses = infrequent occurrence.

Anacardiaceae: *Rhus* Rs
Araliaceae: *Hedera* (R; Fahn et al. 1986)
Aristolochiaceae (R; Fig. 7.10, part 4; Carlquist 1993)
Avicenniaceae: *Avicennia* (R; Fahn et al. 1986)
Basellaceae: *Anredera* R (Carlquist 1999e)
Betulaceae: *Betula*
Bombacaceae: *Bernouillia, Ceiba, Eriodendron, Ochroma* R
Bruniaceae: *Audouinia, Berzelia, Raspalia, Thamnea* R (Carlquist 1982a)
Burseraceae: *Haplolobus* (R)
Cactaceae: *Cereus, Dendrocereus, Leptocereus, Pereskia* (and very likely many other genera) R
Caryocaraceae: *Caryocar* Rs
Caryophyllaceae: *Arenaria macradenia* R (Carlquist 1995c)
Clusiaceae: *Bonnetia, Garcinia, Pentaphalangium* RAe
Cochlospermaceae: *Cochlospermum* As
Combretaceae: *Combretum, Guiera, Terminalia* A
Convolvulaceae: *Exogonium* AR; *Ipomoea* R (Fig. 7.10, parts 1–3; Carlquist and Hanson 1991)
Corylaceae: *Carpinus, Ostrya* R
Dipterocarpaceae: *Hopea, Vateria, Vatica* R
Euphorbiaceae: *Acalypha, Macaranga, Mallotus, Phyllanthus, Sapium* R; *Manihot* RsAe
Fabaceae: *Gleditsia* R
Loasaceae: *Mentzelia* (Carlquist 1984d)
Malvaceae: *Althaea, Cienfugosia, Malvaviscus, Paritium, Sphaeralcea, Thurberia* R; *Hibiscus* RAs
Melastomataceae: *Dichaetanthera* RA (van Vliet 1981); tribe Blakeae (less commonly in other Melastomoideae) R (van Vliet et al. 1981)
Meliaceae: *Cedrela* RAs; *Entandophragma* A; *Toona* As
Misodendraceae: *Misodendron* subgenus *Angelopogon* (Carlquist 1985c)
Myrtaceae: *Eugenia* As
Penaeaceae: *Brachysiphon* A; *Saltera* R
Quiinaceae: *Quiina* As
Rhamnaceae: *Ceanothus, Rhamnus* R; *Colubrina* RS
Rhizophoraceae: *Carallia, Crossostylis* R (van Vliet 1976a)
Rosaceae: *Prunus* ReAs; *Pygeum* R
Sonneratiaceae: *Sonneratia* R (Rao et al. 1989)
Sterculiaceae: *Pterocymbium* R, *Sterculia* RAs
Theaceae: *Camellia, Gordonia, Schima* R
Ulmaceae: *Celtis* Rs

Urticaceae: *Boehmeria*, *Forskolea*, *Gyrotaenia*, *Pipturus*, *Urera* R; *Nothocnide*, *Poikilospermum*, *Urera* A (Bonsen and ter Welle 1984)
Viscaceae: *Viscum*
Vitaceae: *Cayratia* R; *Tetrastigma* R (new report; Fig. 7.10, part 5)

7.1.2.7 Sphaerocrystals

The term "sphaerocrystal" has been applied to compound crystals that are essentially like druses except that the surface of the crystal mass is smooth rather than rough with the projecting tips of the component crystals. Sphaerocrystals may be seen in vascular rays of woods of some Fouquieriaceae (new report) and many Cactaceae (Metcalfe and Chalk 1950).

7.1.2.8 Function of Calcium Oxalate Crystals

Experimental demonstration of the effectiveness of calcium oxalate crystals presents almost insoluble problems because the function is probably in defense against herbivores. The number of toxic or irritant substances in woods is enormous (Stewart 1966), and finding a species pair in which the only defensive difference is presence of calcium oxalate crystals is extremely difficult.

The probable defensive nature of calcium oxalate crystals can be hypothesized on several bases. Floras of calcium-rich soils do not show calcium oxalate accumulation to an appreciably greater extent than those of normal soils, so preferential accumulation of calcium is not a function primarily of edaphic factors. The idea, once prevalent, that calcium oxalate is merely a metabolic byproduct and therefore a waste product, has by default been abandoned as many secondary plant products have been demonstrated to have defensive activity, and the essential economy involved in production of these compounds becomes increasingly evident.

Active deposition of calcium oxalate crystals by fungi has been noted repeatedly (Muhammad and Micko 1984, Horner et al. 1985), but that is not necessarily evidence even in fungi of production of a waste substance – it may well deter fungal predators.

The distribution of calcium oxalate crystals in tissues of plants is a very compelling form of circumstantial evidence. One often finds that calcium oxalate crystals, like accumulations of other toxic and irritant substances, are more common closer to the surface of a plant than further toward the interior. The distribution of calcium oxalate crystals in wood shows just such a kind of differential: in all of the crystal categories listed above, a much smaller number of taxa have crystals in wood than have crystals in other parts of a plant. To see calcium oxalate crystals present in phloem or cortical parenchyma of a particular species but absent in wood is a common experience in plant anatomy.

The size and distribution of calcium oxalate crystals suggests that they are effective against chewing insects and molluscs; in addition, acicular crystals and raphides are annoying to the mouth and throat parts of vertebrate herbivores. The sharp edges, hardness, and extremely low solubility of calcium oxalate crystals makes them potentially effective in providing annoyance to mouth parts of insects. The fact that predation occurs in wood (or other vegetative parts) despite presence of crystals is not evidence of ineffectiveness of these crystals. One must remember that armaments of plants are usually rich in redundancy, so that secondary plant products, as well as mechanisms such as spinescence and crystal presence, are involved, and no one item in this armament can be singled out as determining success or failure in deterrence of predation. Also, predatory insects, molluscs, and vertebrates are constantly evolving resistance to various plant defense mechanisms, and presence of some predation does not indicate a defense is ineffective.

7.1.3 Crystals Other Than Calcium Oxalate

Sulfates of calcium or of calcium and potassium (Miller 1978) are probably infrequent, but their relative solubility makes recognition of these relatively difficult. More conscientious workers may report "ghosts" of crystals that have been mostly dissolved by needed microtechnical procedures (e.g., Fahn et al. 1986 in their account of Capparaceae). Miller (1978) used polarized light as a way of distinguishing sulfate crystals (white in polarized light) from calcium oxalate crystals (iridescent or colored with his equipment). Electron microprobe analysis in association with a scanning electron microscope can also be used effectively, and by means of that tool, Miller (1978) demonstrated potassium calcium sulfate in wood of 31 *Capparis* species.

Berberin, a crystalline flavonone, can be found in Ranunculales (Berberidales) but probably is not reported often because of its solubility. Berberin is associated with the yellowish color common in woods of these families.

Other crystalline secondary plant products will doubtless be detected as microtechnical methods advance. Obviously, methods that involve boiling in water or storing in alcohol or acids (even weak organic acids) will result in dissolving of certain classes of compounds. Spot tests for certain chemicals are often used (chiefly by dendrologists rather than wood anatomists) and some of these can reveal presence of certain crystalline materials (although most of these tests are for noncrystalline compounds).

7.2 Cystoliths

A cystolith is a layered primary wall intrusion in a cell; calcium carbonate is deposited in the layers and on the surface of the cystolith, although the calcium carbonate is rarely observed in preparations because of its solubility (Fig. 7.11, parts 1 and 2). Cystoliths are characteristically borne one per cell, and the cell containing a cystolith is termed a lithocyst. Cystoliths occur idioblastically: sometimes singly, but often in groups of two or four, as in Opiliaceae (Koek-Noorman and van Rijkevorsel 1983). For elegant illustrations of cystoliths in wood, the reader is referred to the accounts of ter Welle (1980) and Koek-Noorman and van Rijkevorsel (1983).

Record (1925a, 1927) recorded occurrences of cystoliths and other kinds of calcium carbonate deposits in woods. Lithocysts tend to be larger than other cells of an axial parenchyma strand in *Sparattanthelium* (ter Welle 1980), and larger than other ray cells in Opiliaceae (Koek-Noorman and van Rijkevorsel 1983), but in Acanthaceae, the lithocysts are the same size as or only slightly larger than other adjacent ray cells. The number of families with cystoliths in wood is much smaller than the number of families with cystoliths elsewhere in the vegetative portions of the plant. The following taxa are known to have cystoliths in wood.

Acanthaceae: rays of *Bravaisia* (Fig. 7.11, parts 1 and 2) and *Sanchezia* (Fig. 7.11, part 3; Carlquist and Zona 1988a); rays of *Trichanthera* (ter Welle 1980; Carlquist and Zona 1988a).

Hernandiaceae: axial parenchyma of all species of *Sparattanthelium* (ter Welle 1980).

Opiliaceae: rays of *Cansjera*, *Champereia*, *Lepionurus*, *Meliantha*, *Opilia*, *Rhopalopilia*, and *Urobotrya* (ter Welle 1980; Koek-Noorman and van Rijkevorsel 1983); cystoliths were claimed for *Agonandra* by Record (1925a), but subsequent authors have not found them.

Ulmaceae: *Celtis* (Bonsen and ter Welle 1984)

Urticaceae: rays of *Gyrotaenia* (Bonsen and ter Welle 1984)

7.3 Starch

Starch, although an obvious component of many plant cells, is mentioned at this juncture to emphasize the diversity of appearances starch has in microtechnical preparations (Fig. 7.10, part 5, Fig. 7.11, part 4).

Starch is best preserved in liquid-preserved wood samples. Although liquid preservation (50% ethanol or a solution using 50% ethanol plus other ingredients is useful) is not employed often because long-term maintenance of liquid-preserved collections is difficult, for particular studies it is definitely to be recommended. In sections made from liquid-preserved preparations, starch grains may be made more visible by staining with iodine (in a potassium iodide solution), providing a color ranging from pale purple to deep purple-black. Because this requires an additional step, it is rarely undertaken.

Observation of starch grains in sections made from liquid-preserved preparations is facilitated without staining by means of polarized light (Fig. 7.7,

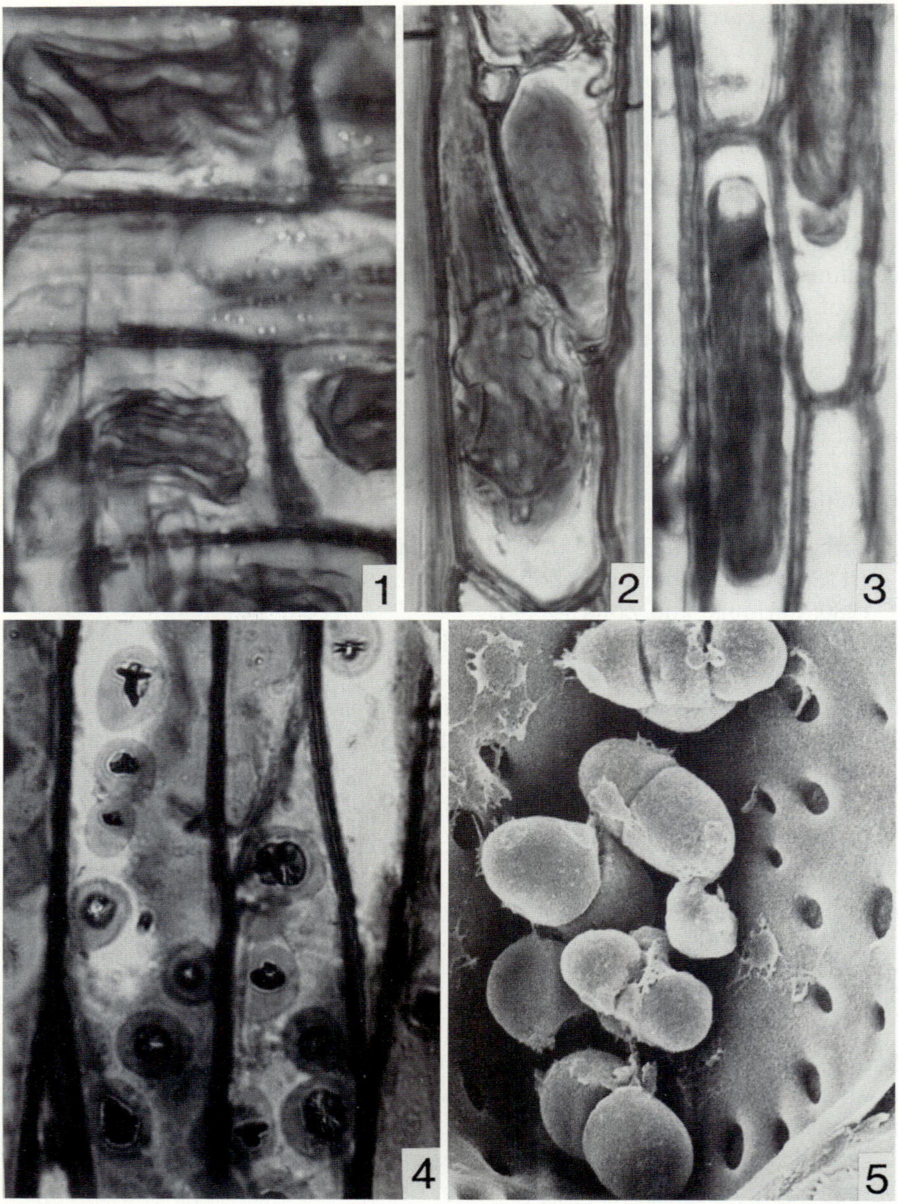

Fig. 7.11. Cystoliths (*1–3*) and starch grains (*4, 5*) in wood. *1* Layered cystolith remnants from ray cells, radial section, *Bravaisia floribunda* (Acanthaceae); *2* cystoliths, tangential section, *Bravaisia integerrima*; *3* cystoliths, tangential section, *Sanchezia decora* (Acanthaceae); *4* starch grains (*blackish air space in centers of grains*), parenchyma associated with interxylary phloem, longisection, *Lopezia lopezioides* (Onagraceae); *5* compound starch grains, SEM photomicrograph of ray cell from radial section, *Sarcandra glabra* (Chloranthaceae). (*1–4:C*; *5:G*)

part 5). Less often mentioned in microtechnical literature is the irregularly shaped air space that appears in starch grains that have been dehydrated, as is always the case on permanent slides (Fig. 5.8, part 1). Because of refractiveness, the air space usually appears black (sometimes reddish when out of focus). The air space results from the withdrawal of water during microtechnical dehydration from the hydrated central portion of a starch grain. This air space is an invariably good indication of starch presence. One can find it even in degraded starch grains that have been much altered by treatment with alkalis (e.g., ethylene diamine).

With scanning electron microscopy, starch grains are revealed superbly. Shown here are examples that are unusual: much elongate starch grains in *Tetrastigma* (Fig. 7.9, part 5) and compound (grouped, mutually compressed) starch grains in *Sarcandra* (Fig. 7.11, part 4). Although both of these examples represent preparations made from liquid-preserved material, starch grains may also be observed in woods that have been dried and then boiled prior to sectioning (e.g., Fig. 5.8, part 2).

The value of observing starch in woods is manifold. One can identify living fibers by means of starch grain presence. The presence of starch indicates not merely storage, but other functions we are only beginning to appreciate (see Braun 1970, 1984; Sauter 1966a,b, for example).

7.4 Silica Bodies and Other Silica Deposits

Silica bodies in woods (Fig. 7.12, parts 1–4, Fig. 7.13, parts 1–4) can sometimes be recognized by their most commonly encountered forms (often spherical, sometimes rough), although silica bodies range from minute (Fig. 7.12, part 1) to medium (Fig. 7.13, parts 2 and 4) to large (Fig. 7.12, parts 3 and 4), and may be elongate, sometimes paired (Fig. 7.12, part 3). Although studies by SEM (Figs. 7.12, part 4, 7.13, parts 1–4) reveal rough surfaces commonly, the degree of roughness varies from little (Figs. 7.12, parts 1 and 2, 7.13, part 1) to conspicuous (Figs. 7.12, parts 3 and 4, 7.13, part 4) and highly porous (Fig. 7.13, part 3), although in sectional view, most silica bodies are solid (Fig. 7.13, part 1, lower right). With light microscopy, silica bodies often appear pale purple, and they exhibit no birefringence or staining ability. They are insoluble in reagents other than hydrofluoric acid. For various details concerning silica occurrence, the reader may wish to consult the large monograph on silica bodies by Netolitzky (1929a).

Various types of silica occurrence in woods of dicotyledons have been reviewed by ter Welle (1976a,b), and these reports were summarized by Metcalfe and Chalk (1983, pp. 219–220). Silica occurrence is of interest to wood anatomists because presence of silica is often of diagnostic significance – only a minority of dicotyledon woods contain visible silica accumulations. Silica bodies can increase in size with age (ter Welle 1976a). Within families, generic

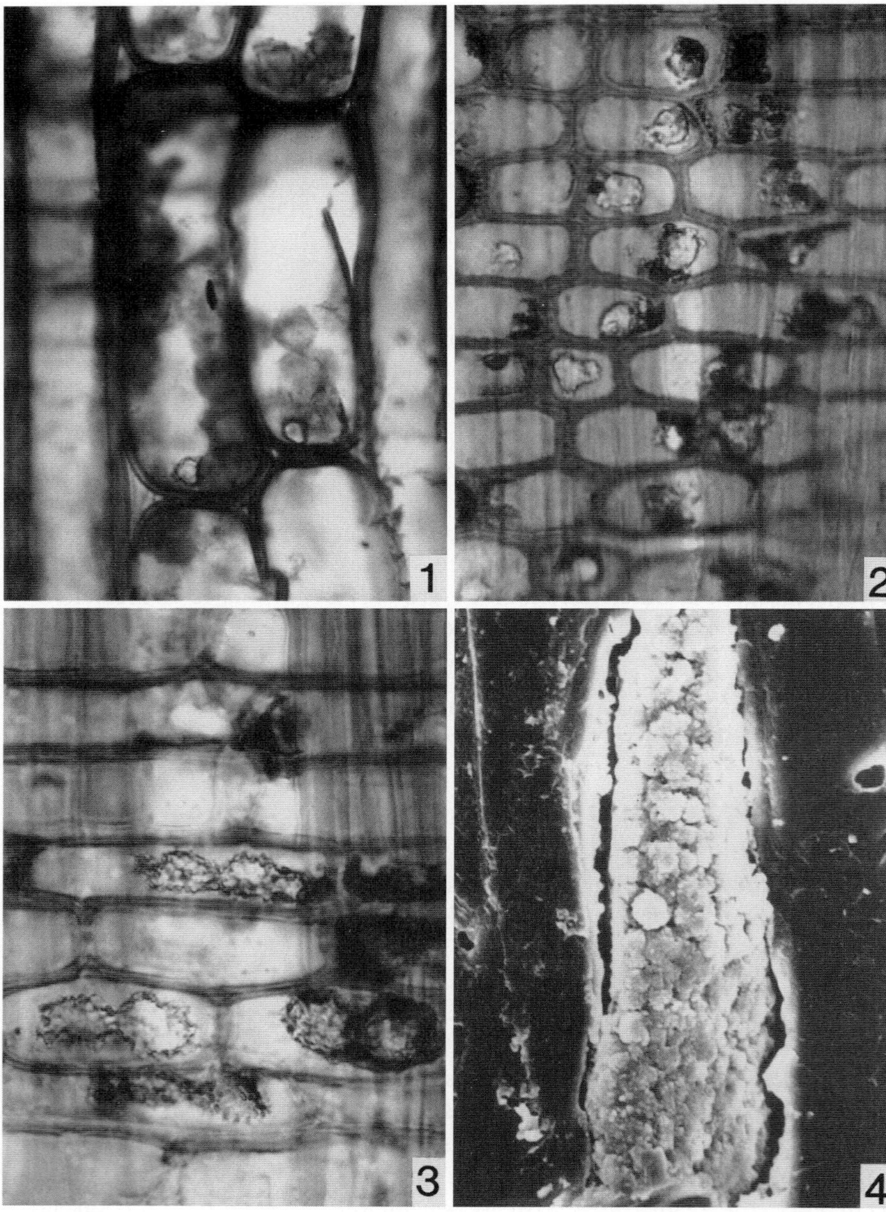

Fig. 7.12. Silica body occurrence in woods, radial (*1–3*) and tangential (*4*) sections. *1* Small silica bodies, at *bottom* in three of the ray cells, *Mimusops heckelii* (Clusiaceae); *2* sphaeroidal silica bodies in ray cells, *Leptolaena bernieri* (Sarcolaenaceae); *3* rough silica bodies, formed in pairs, in ray cells of *Eschweilera odorata* (Lecythidaceae); *4* elongate silica body from idioblast in axial xylem, showing rough surface, *Limonium rumicifolium* (Plumbaginaceae). (*1–3*:*C*; *4*:*D*)

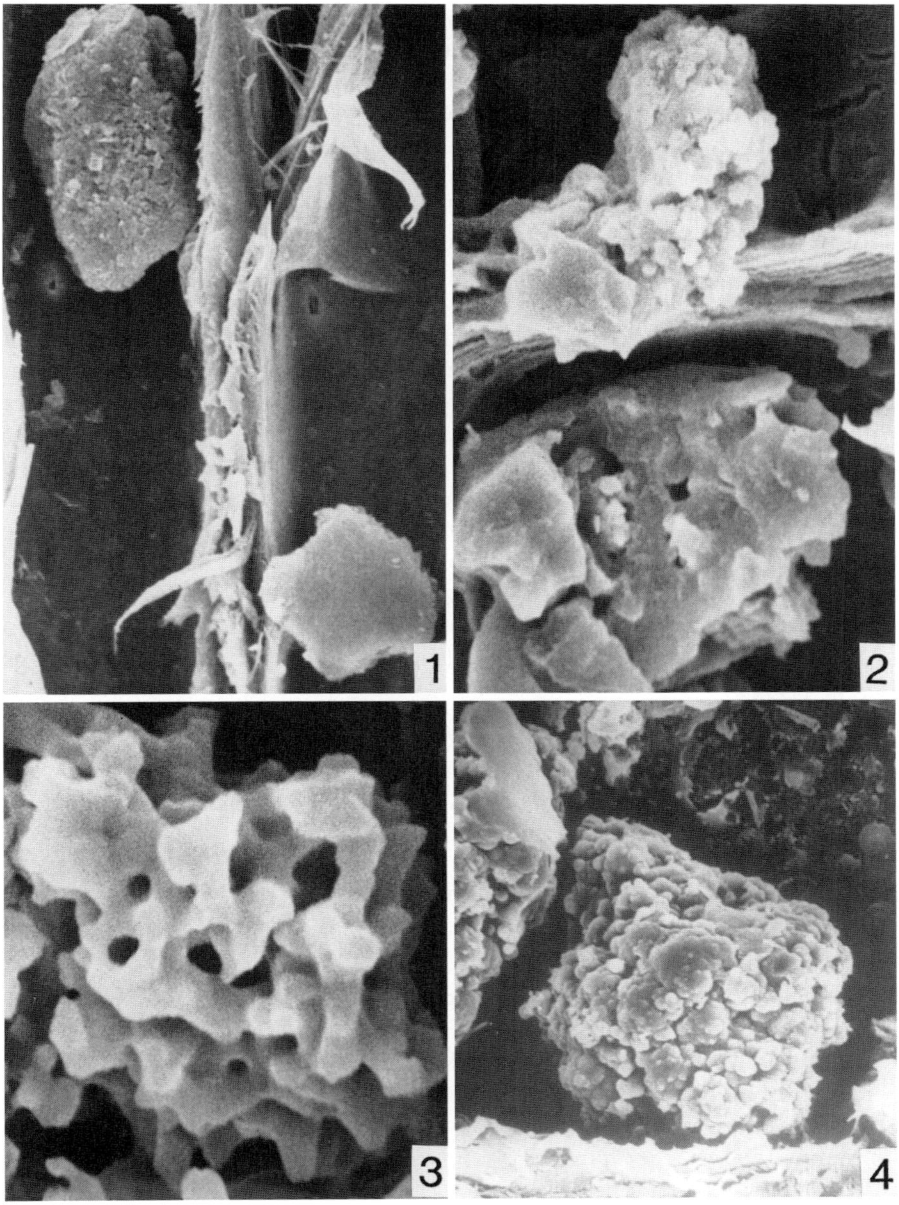

Fig. 7.13. Silica bodies, SEM photomicrographs. *1* Intact silica body (*upper left*), silica body cut in half (*lower right*) from longisection of libriform fibers, *Protium insigne* (Burseraceae); *2, 3* various shapes and textures of silica bodies, *Chrysobalanus icaco* (Chrysobalanaceae), ray cells in radial section; *2* irregularly shaped intact grain (*above*), fractured, somewhat porose grain (*below*); *3* grain with spongy appearance; *4* grain with rough surface from ray cell, *Meliosma herberti* (Sabiaceae). (*1,2,4*:*G*; *3*:*H*)

distribution can be of importance. In Vochysiaceae, silica bodies are plentiful in *Qualea*, but rare in other genera (Quirk 1980). In Rhizophoraceae, only in *Gynotroches* are visible silica accumulations abundant (van Vliet 1976a). Other examples can be cited in Dipterocarpaceae (Gottwald and Parameswaran 1966) and Quiinaceae (Gottwald and Parameswaran 1967). Vitreous silica (amorphous deposits not in spherical bodies) is reported in rays of *Dendrocnide* and *Leucosyke* but not in other genera of Urticaceae (Bonsen and ter Welle 1984). Because of the considerable taxonomic importance of silica occurrence, an updated version of silica occurrence, based on the above compilations but with other reports as cited below, is presented here. References are given for reports not cited by Metcalfe and Chalk (1983). In each family, the genus or genera with silica bodies are followed by "B"; those genera with silica but not in the form of bodies of appreciable size are not followed by a letter. Even those genera with silica bodies do not uniformly contain silica bodies, and the literature should be consulted, because silica bodies can range from very rare in a genus (e.g., *Lindackeria* and *Mayna* of Flacourtiaceae) to very common (e.g., all Chrysobalanaceae). Silica bodies are ordinarily found in ray cells, but may be found in axial parenchyma and libriform fibers (Fig. 7.12, part 1) of Burseraceae, in axial parenchyma of *Coelostegia* (Bombacaceae) and *Michelia* (Magnoliaceae), in libriform fibers and Anacardiaceae (James and Ingle 1956), and in fibriform cells of Plumbaginaceae (Fig. 7.12, part 4; Carlquist and Boggs 1996).

Anacardiaceae: *Anacardium*, *Loxopterygium*, and other genera (see Dong and Baas 1993) B; *Lannea*, *Melanorrhoea*, *Parishia*, *Swintonia*
Ancistrocladaceae: *Ancistrocladus* B (Gottwald and Parameswaran 1966)
Bombacaceae: *Bombax*, *Coelostegia*, *Quaribea* B
Boraginaceae: *Cordia* (Gottwald 1980)
Burseraceae: *Dacryodes*, *Paraprotium*, *Protium* (Fig. 7.12, part 1), *Trattinckia* B; *Canarium*, *Santiria*
Caryocaraceae: *Anthodiscus* B
Chrysobalanaceae: all species examined B (Fig. 7.12, parts 2–4)
Clusiaceae (including Bonnetiaceae): *Archytea*, *Clusia*, *Haploclathra*, *Kielmeyera*, *Odomatopus*, *Tovomita* B; *Mimusops* (new report; Fig. 7.11, part 1) B; *Caloxylon*, *Garcinia*
Combretaceae: B rare (van Vliet 1979; van Vliet and Baas 1984)
Connaraceae: *Agelaea*, *Connarus*, *Pseudoconnarus*, *Rourea* B
Dilleniaceae: *Hibbertia* B (Dickison 1984)
Dioncophyllaceae: *Dioncophyllum* B (Gottwald and Parameswaran 1968)
Dipterocarpaceae: rays of a few species, e.g., *Anisoptera* B
Erythroxylaceae: *Erythroxylum* B
Euphorbiaceae: *Actinostemon*, *Maprounea*, *Micrandra*, *Senefeldera* B; *Antidesma*, *Baccouria*, *Phyllanthus*, *Sapium*, *Uapaca*
Fabaceae: *Apuleia*, *Dialium*, *Dicorynia*, *Sclerolobium*, *Tachigalia* B; new additions: *Diptychandra*, *Distemonanthus* B (Koeppen 1980)

Fagaceae: *Nothofagus* (Patel 1986)
Flacourtiaceae: *Lindackeria, Mayna* B; tyloses of *Hydnocarpus* (James and Ingle 1956)
Hippocrateaceae: *Prionostemma* B (Mennega 1972a)
Humiriaceae: *Saccoglottis* B
Lauraceae: *Cryptocarya, Licaria, Mezilaurus, Ocotea* B
Lecythidaceae: *Allantoma, Cariniana, Corythophora, Couratari, Eschweilera* (Fig. 7.12, part 3), *Holopyxidium, Lecythis* B
Magnoliaceae: *Michelia*, possibly other genera
Meliaceae: *Aphanopyxis, Guarea, Trichilia* B; *Chisocheton*
Menispermaceae: *Anomospermum* B
Moraceae: *Brosimum* B (Koek-Noorman et al. 1984b)
Myrtaceae: 10 genera B (Ingle and Dadswell 1953b)
Olacaceae: *Liriosma* B
Plumbaginaceae: *Limonium* B
Polygonaceae: *Neomillspaughia, Ruprechtia, Symmeria*
Proteaceae: *Euplasa, Panopsis, Roupala*
Quiinaceae: *Lacunaria* B
Rhabdodendraceae: *Rhabdodendron*
Rhizophoraceae: *Gynotroches, Kandelia, Pellocalyx* B
Rubiaceae: *Mitragyna* B
Rutaceae: *Erythrochiton, Galipea* B; *Acronychia, Evodia*
Sabiaceae (Meliosmaceae): *Meliosma* B (Fig. 7.13, part 4)
Sapindaceae: *Toulicia* B
Sapotaceae: genera other than *Manilkara* B
Sarcolaenaceae: *Leptolaena bernieri* (new report; Fig. 7.11, part 2) B
Simaroubaceae: *Simaba* B
Solanaceae: *Acnistus* B (Carlquist 1992h)
Sterculiaceae: *Heritiera, Tarrietia*
Styracaceae: *Styrax* B, may be present or absent in a single species (Dickison and Phend 1985)
Theaceae: *Ternstroemia* B
Theophrastaceae: *Clavija* B
Tiliaceae: *Brownlowia, Luhea* B
Ulmaceae: *Gironniera, Parasiphonia, Ulmus*
Urticaceae: *Dendrocnide* (Bonsen and ter Welle 1984)
Vochysiaceae: *Qualea* B (Quirk 1980)
Winteraceae: *Drimys* B (Carlquist 1988a)

7.5 Laticifers and Latex Deposits

In wood of dicotyledons, laticifers occur chiefly in rays, and the reports listed below are for rays (Fig. 7.14, parts 1–4). In addition, vertically-running laticifers

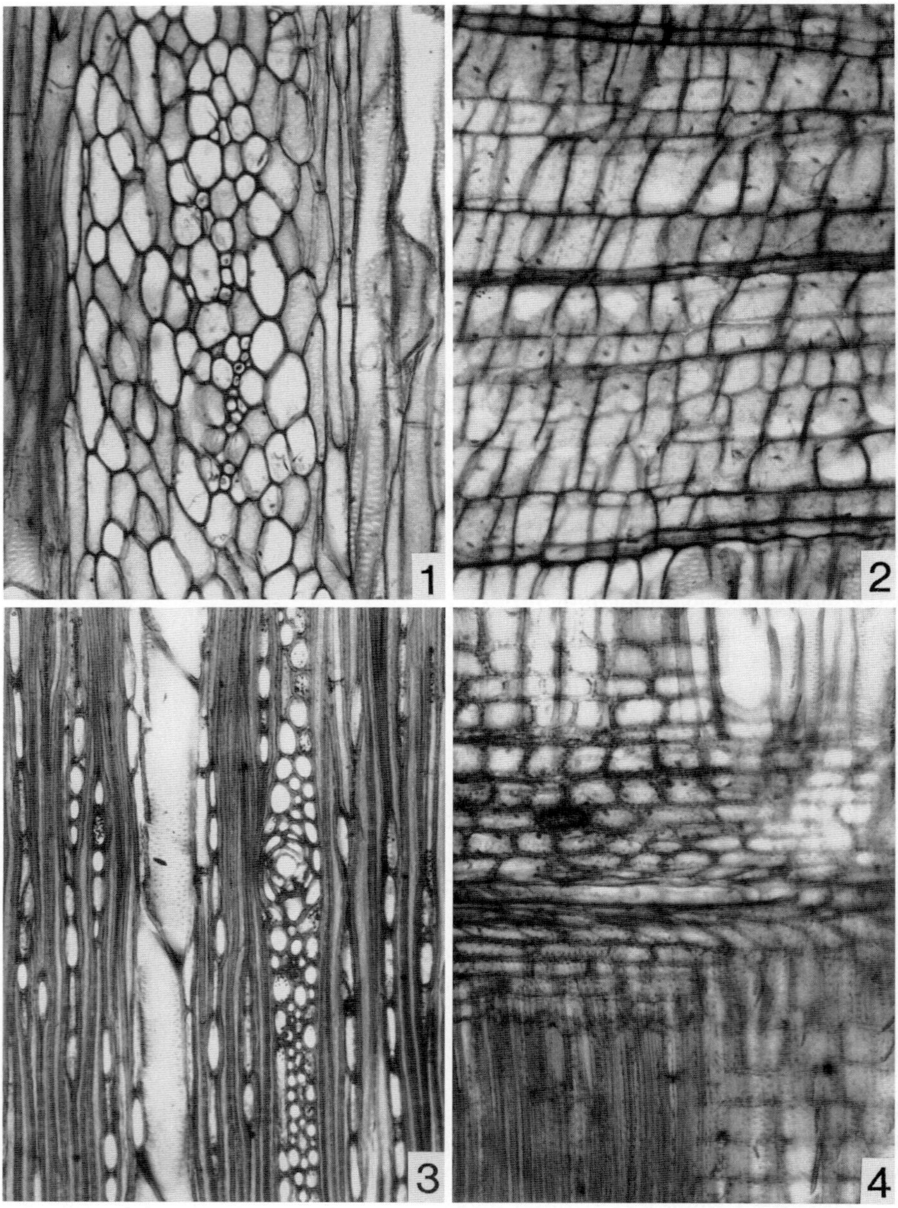

Fig. 7.14. Laticifers in ray cells, tangential (*1, 3*) and radial (*2, 4*) sections. *1, 2* Articulated laticifers in *Lobelia tortuosa* (Campanulaceae); *1* numerous laticifers (small cells) in transection; *2* three laticifers in longisection; *3, 4* nonarticulated laticifers in *Chamaesyce celastroides* (Euphorbiaceae); *3* single laticifer in ray to *left of center*; *4* laticifer in longisectional view, *near center*. (*1–4:B*)

("latex tubes") have been found among the libriform fibers in Moraceae in *Artocarpus* and *Bagassa* (Topper and Koek-Noorman 1980). An attempt has been made in compiling the list below to exclude instances of secretory canals, which are an intercellular phenomenon, and to include only reports of laticifers which are cellular.

Apocynaceae: *Alstonia* (Sidyasa and Baas 1998), *Peschiera, Stemmadenia,* and other genera (Record 1944f; Metcalfe and Chalk 1950; Ingle and Dadswell 1953a)
Asclepiadaceae: *Tacazzea apiculata* (new report)
Campanulaceae, subfamily Lobelioideae: *Lobelia tortuosa* (Fig. 7.14, parts 1 and 2; Carlquist 1969a).
Caricaceae: *Carica* (Carlquist 1998d; Fisher 1980)
Euphorbiaceae: *Croton* (Rudall 1989), *Cunuria* (Record and Hess 1942), *Dalechampia* (Heimsch 1942), *Euphorbia* including chamaesyceoid species (Fig. 7.14, parts 3 and 4; Heimsch 1942; Carlquist 1970a), *Hippomane* (Heimsch 1942), *Neoalchornea, Sebastiania*
Moraceae: *Acanthosphaera, Anonocarpus, Antiaris, Artocarpus, Bagasa, Bosquiea, Brosimopsis, Brosimum, Castilla, Chlorophora, Dorstenia, Ficus, Helianthostylis, Helicostylis, Naucleopsis, Noyera, Ogcodeia, Olmedia, Olmedioperebea, Olmediophaena, Parartocarpus, Perebea, Trophis, Trymatocarpus* (Record and Hess 1943; Metcalfe and Chalk 1950; Koek-Noorman et al. 1984a,b)

Systematic significance is evident from the above distributions. Laticifers are absent from Moraceae subfamily Conocyphoideae (Tippo 1938), as well as from the families close to Moraceae (Cannabaceae and Urticaceae).

7.6 Tanniniferous Tubules

Garratt (1933) coined the term "tanniniferous tubules" for structures he found in rays of wood of Myristicaceae. He found them in all species of the family he examined, but they have not been located in any family other than Myristicaceae. They are more abundant in some species of Myristicaceae than in others. Presence of tanniniferous tubules has been mentioned by Armstrong and Wilson (1980) in *Horsfieldia*. They may contain scattered deposits of dark-staining amorphous materials (Fig. 7.15, parts 1–3) or an entire tubule may be filled with such compounds (Fig. 7.15, part 4). The tubules are smaller in diameter than other ray cells (Fig. 7.15, parts 1–4). As of this date, no authors have expressed an opinion on the nature of these tubules, i.e., whether they represent cellular or intercellular phenomena.

My observations revealed that the tanniniferous tubules do have very thin radially-running primary walls. Tangentially-running walls are not typically evident, but tangentially-oriented walls were seen in cells judged to be

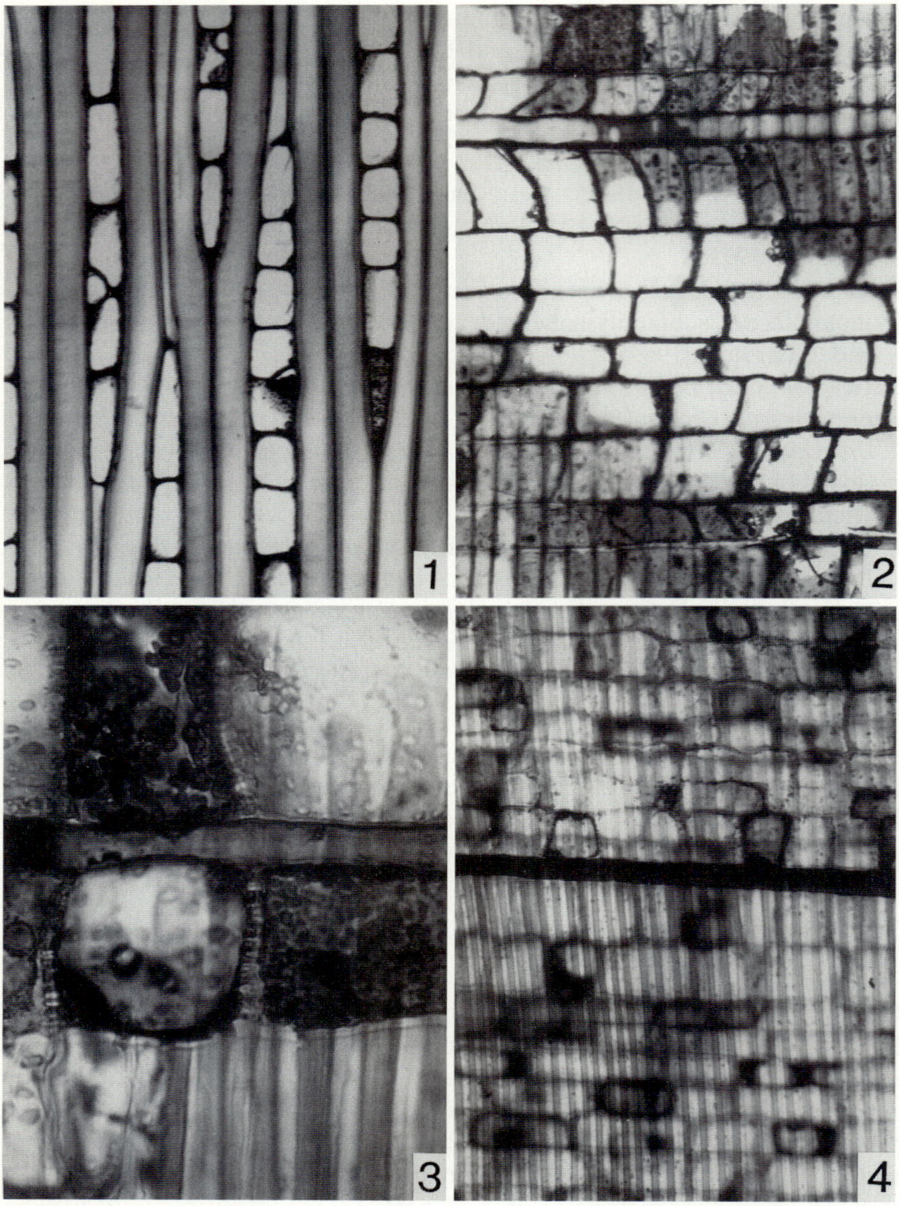

Fig. 7.15. Tanniniferous tubules in rays of Myristicaceae, in tangential (*1*) and radial (*2–4*) sections. *1* Three tanniniferous tubules (triangular cells) in transection, *Dialyanthera otoba*; *2* tubule *near top*, dark-staining contents near its *center*, *Dialyanthera otoba*; *3* tubule *near center*, with minimal contents, although dark-staining deposits can be seen in adjacent ray cells, *Iryanthera laevis*; *4* tubule in longisection, filled with dark-staining contents, *Knema heterophylla*. (*1,2,4:B*; *2:C*)

intermediate in size and morphology between procumbent ray cells and tanniniferous tubules. Walls of the tubules may be wavy in outline where not in contact with ray cells.

Tanniniferous tubules may, in my opinion, be likened to the narrowest of procumbent ray cells. As an hypothesis, I suggest that tanniniferous tubules may represent markedly procumbent ray cells in which tangentially-oriented walls break down and dissolve early in ontogeny, so that an uninterrupted radial cell series, or tubule, forms. Careful ontogenetic studies based on liquid-preserved material and focusing on secondary xylem near the cambium are needed to test this hypothesis.

7.7 Oil and Mucilage Cells

Record (1944f) and Metcalfe and Chalk (1950) have grouped oil-containing cells with cells containing other materials, notably mucilage, and listed them together. These cells are illustrated in Fig. 7.16 (parts 1–4) and Fig. 7.17 (part 1). The reason for this practice is the tendency in some families – especially Annonaceae, Lauraceae, and Myristicaceae, for oil cells to occur in some taxa, but mucilage-containing cells with the same general appearance and in the same locations to occur in related genera and species. The grouping of oil and mucilage-containing cells has been followed by Baas and Gregory (1985), who describe the differences between the two kinds of cells. A systematic listing of mucilage cells was presented by Gregory and Baas (1989). Oil cells have a three-layered wall, with a suberized layer between cellulosic layers; a "cellulosic cupule" is present; in mucilage cells, the wall is wholly cellulosic and no cellulosic cupule is present. Oil cells have plastids, and oils are formed largely within the vacuoles; in mucilage cells, mucilage is deposited between the cell wall and the plasma membrane only; plastids are usually absent but Golgi vesicles are present. Oil cells contain terpenes, fats, and flavonoid aglycones (the terpenes may account for dark-staining residues in microtechnical preparations, so cells with such contents can be tentatively identified as oil cells). Mucilage cells contain mucilaginous polysaccharides. Despite these differences, a basic similarity between the two cell types is certainly conceivable and worthy of further investigation, as suggested by Baas and Gregory (1985). As with other idioblasts, oil and mucilage cells occur less commonly in wood than in other vegetative parts of dicotyledons.

Mucilage and oil cells clearly are of systematic value: all of the families listed below belong to the Annonales (Magnoliales, Ranales of various authors). Although not all the families of that order are listed below, those that are not have oil or mucilage cells elsewhere in the plant body. The fiberlike mucilage idioblasts reported in *Trimenia* of the Trimeniaceae (Carlquist 1984e) are very similar to the "tube cells" in *Ocotea* of the Lauraceae (Richter 1981).

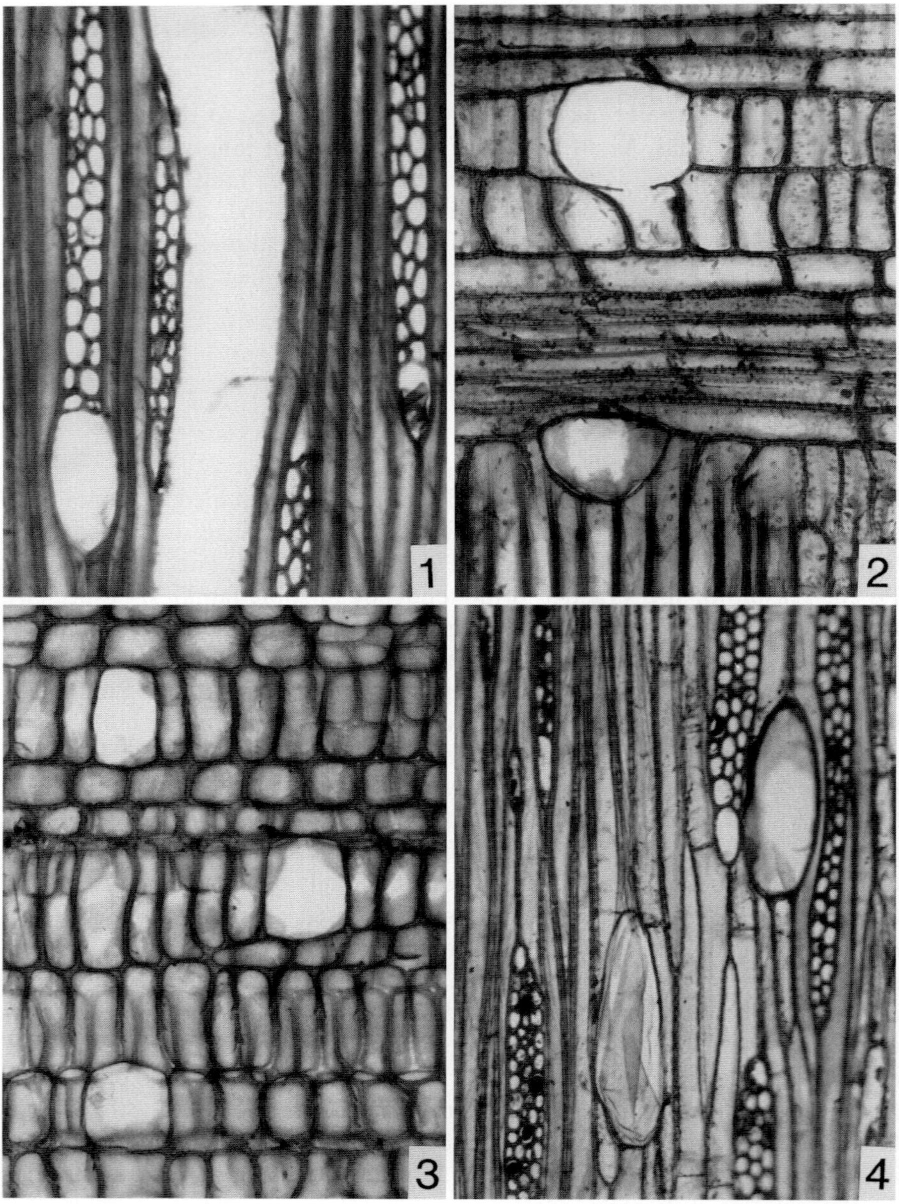

Fig. 7.16. Oil (or mucilage) cells in ray cells (*1–3*) and in axial xylem (*4*). *1, 2 Michelia platyphylla* (Magnoliaceae); *1* oil cell at ray margin, *below*, tangential section; *2* oil cells at ray margins, *above and below*, radial section; *3, 4 Eupomatia laurina*; *3* three ray cells that are slightly larger and paler are oil cells, radial section; *4* two elongate oil cells in a tangential section. (*1–4:B*)

The listing that follows is based on the list of Baas and Gregory (1985), extracting only those taxa in which oil and mucilage cells occur in wood, and altering the list with respect to several taxa (note also the excluded instances following the list). Abbreviations are as follows: A = axial parenchyma; F = fibers or fiberlike cells; M = mucilage cell; O = oil cell; R = ray; parentheses indicate uncommon occurrences. Designation of cells as mucilage cells or oil cells is tentative in Annonaceae, Lauraceae, Myristicaceae, Piperaceae, Trimeniaceae, and perhaps other families, as indicated by the designation M/O, which means that the cells have not yet been identified with certainty (other instances cited as O may not be oil cells or primarily oil-containing cells). References on which the list is based are cited by Baas and Gregory (1985).

Annonaceae: 12 genera OR(A)
Aristolochiaceae: *Apama* M/ORA
Cactaceae M (Gibson 1977a; Gregory and Baas 1989)
Canellaceae: 4 genera ORA
Degeneriaceae: OR (Carlquist 1989e)
Eupomatiaceae: *Eupomatia* OR (Fig. 7.16, part 3; Carlquist 1992d)
Hernandiaceae: *Hernandia* OA(R)
Lauraceae: 150 genera M/O (*Cinnamomum*; Fig. 7.16, part 4)
Magnoliaceae: 6 genera OR (*Michelia*; Fig. 7.16, parts 1 and 2)
Monimiaceae: *Daphnandra*, *Doryphora* O(R); *Hortonia* ORA
Myristicaceae: *Knema*, *Myristica*, *Virola* ORA
Piperaceae: *Piper* M/O(R)
Rutaceae: *Euxylophora* OR
Schisandraceae: *Kadsura*, *Schisandra* ORA (Carlquist 1999c)
Trimeniaceae: *Piptocalyx* M/OR; *Trimenia* MF+OR
Winteraceae: *Bubbia*, *Exospermum* (Fig. 7.17, part 1), *Zygogynum* OR

Some supposed secretory structures of rays of Apocynaceae (Record 1944f; Ingle and Dadswell 1953a) are laticifers as cited above (Sect. 7.5), but other secretory structures in rays of this family need further investigation. The alleged oil or mucilage cells of *Grabowskya* (Solanaceae) listed by Record (1944f) and Baas and Gregory (1985) have been restudied by me and prove to be crystal sand idioblasts (see Sect. 7.1.2.5) as shown by birefringence of contents, like those of other crystal-sand-bearing idioblasts of Solanaceae (Carlquist 1992h) and Nolanaceae (Carlquist 1987a).

7.8 Cells with Amorphous Deposits

Under this heading are considered cells that are neither oil nor mucilage cells but have dark-staining contents (Fig. 6.14, parts 1–4, Fig. 7.17, parts 2 and 3). Authors such as Metcalfe and Chalk (1950) use terms such as "resinous contents" or "gummy contents" to describe these cells which may occur in

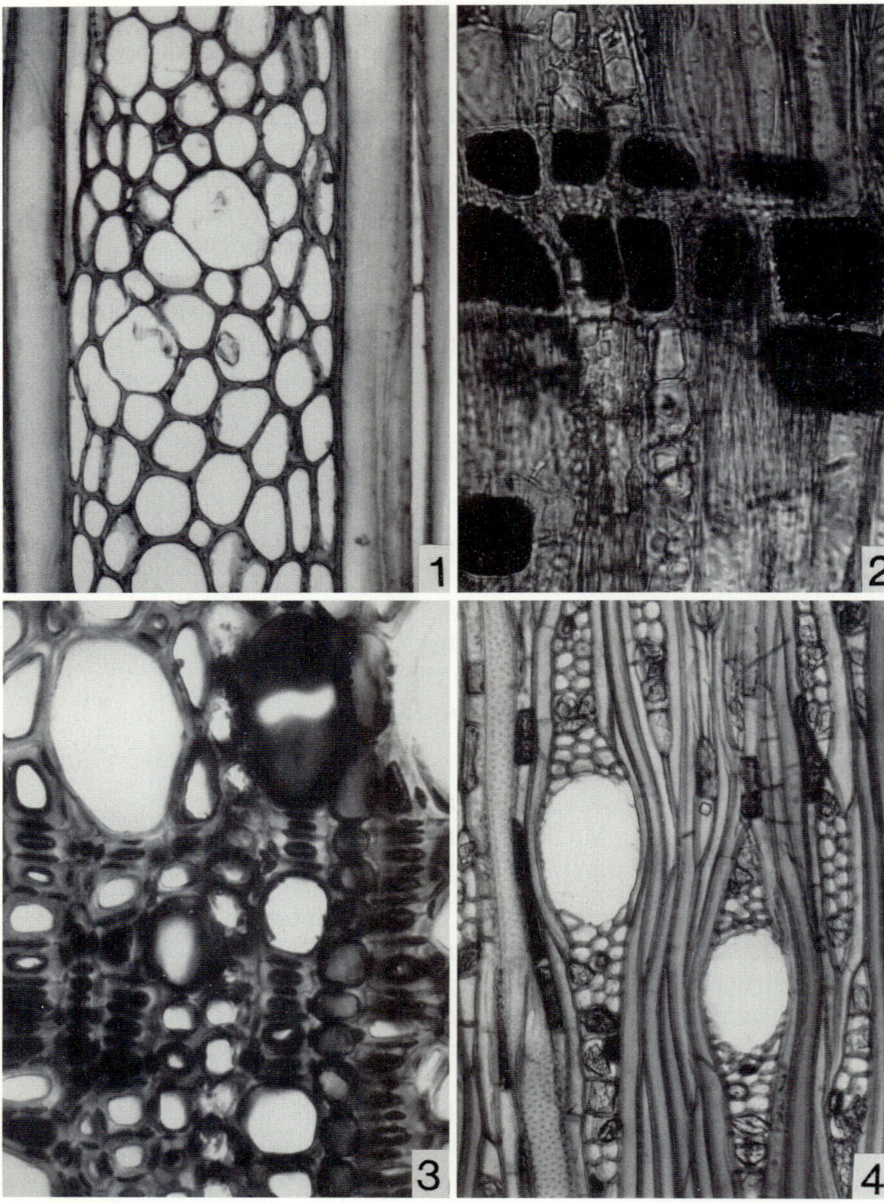

Fig. 7.17. Amorphous accumulations and secretory structures. *1* Two ethereal oil cells (thin-walled, larger than other ray cells) from multiseriate ray of tangential section, *Exospermum stipitatum* (Winteraceae); *2* dark-staining compounds in ray cells, radial section, crystals in libriform fibers behind ray, *Gaura sinuata* (Onagraceae); *3* dark-staining compounds in vessels and tracheids of wood transection, *Arctostaphylos auriculata* (Ericaceae); *4* two secretory canals in rays, tangential section, *Pleiogynium solandri* (Anacardiaceae). (*1*,*4*:*B*; *2*,*4*:*C*)

idioblastic fashion or as sheets of tissue. Grayish granular contents that can be identified with reasonable certainty as tannins or phlobaphenes (oxidized tannins) are also occasionally mentioned in literature on wood anatomy. Secretory structures of all kinds have been reviewed by Fahn (1979), who shows that further work is necessary because plant anatomists have not, with few exceptions, devoted attention to chemical identification of cell contents. Cooperative efforts involving both plant anatomists and plant chemists are likely to yield important advances in our knowledge.

There is probably a large congruence between the anatomical citations mentioned above and what is known in the wood chemistry and dendrology literature as heartwood exudates. This topic has been reviewed by Hillis (1987). These exudates are probably in large measure toxic or irritant compounds (Stewart 1966) that function as chemical defenses against boring insects, fungi, and other wood-destroying organisms, and are secreted in wood cells no longer active in water conduction or photosynthate storage.

Pending use of more precise methods of determining the chemical identity of the various substances found in wood cells, wood anatomists are encouraged to report the existence of such deposits. A vague term such as "amorphous compounds," together with the color and staining reactions of such deposits, is perhaps better than offering an adjective that implies texture and chemical composition (e.g., resinous), which cannot be ascertained easily and which may be misleading.

7.9 Intercellular Canals (Secretory Canals)

Under this heading are discussed large intercellular spaces of indefinite extent filled with a secondary plant product. For a discussion of the intercellular space system of wood other than that represented by secretory canals, see Section 6.12.12.

7.9.1 Radial Canals

Radial canals are secretory canals running radially in ray tissue; they are usually borne singly within rays and occur in the widest portions of multiseriate rays (Fig. 6.1, part 4, Fig. 7.17, part 4, Fig. 7.18, part 1). Radial canals probably contain secondary plant products that function as defense mechanisms against herbivorous insects and vertebrates.

Radial canals have evolved polyphyletically in woods of dicotyledons, but so few times that they are indicative of relationship in some cases. For example, the reader will note that the list below contains several families of the order Sapindales (Anacardiaceae, Burseraceae, Julianiaceae, Sapindaceae). The

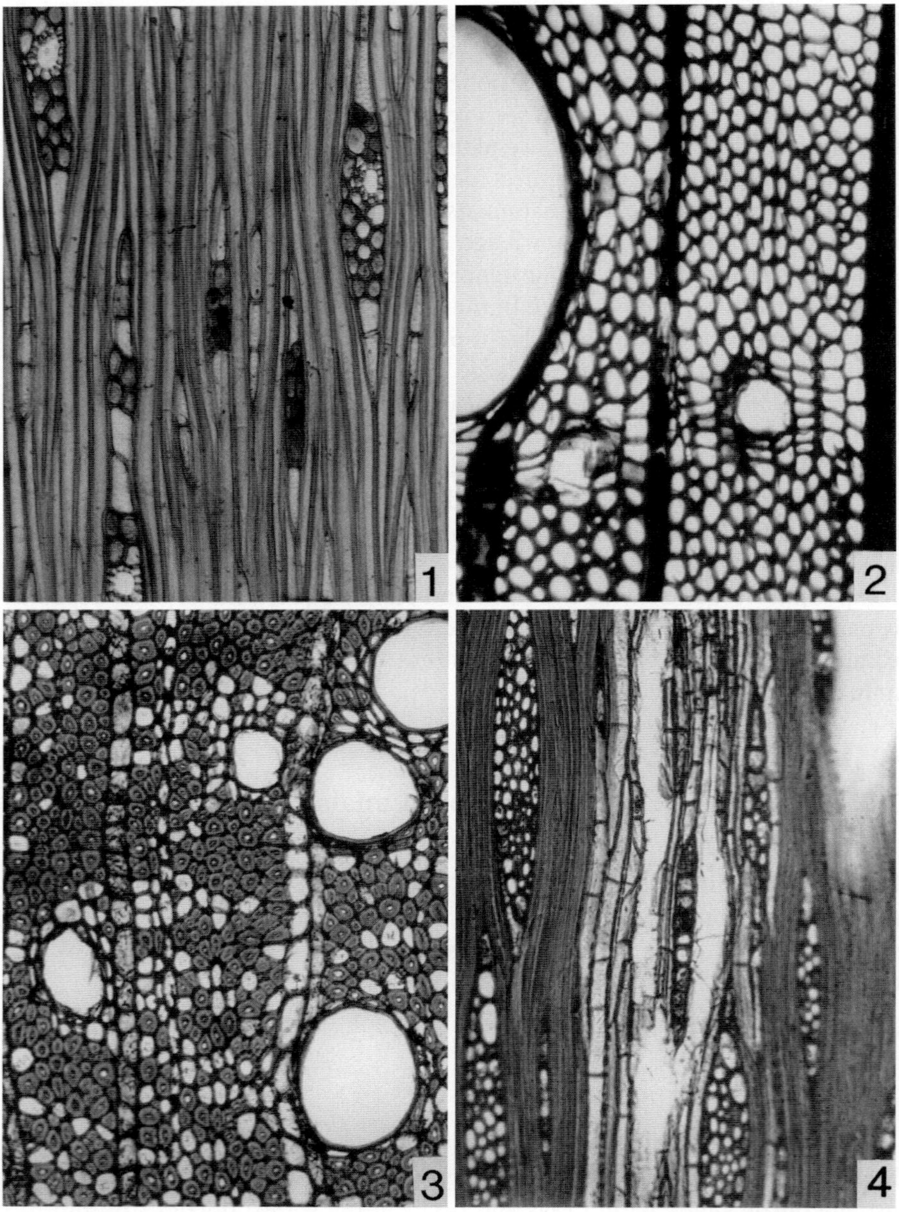

Fig. 7.18. Radial (*1*) and vertical (*2–4*) canals in wood. *1* Secretory canals (*upper left*, *upper right*, *lower left*) in rays, tangential section, *Harpephyllum caffrum* (Anacardiaceae); *2* two vertical canals (*near bottom*, in band of parenchyma); portion of vessel, *upper left*, *Shorea negrosensis* (Dipterocarpaceae); *3*, *4 Vatica mangachapoi* (Dipterocarpaceae); *3* two secretory canals (*smaller circles*), three vessels (*right*) in wood transection; 4 longisection of vertical canal (forking space, running vertically, *center*). (*1–4:B*)

family Fabaceae, formerly thought to be rosalean, is now thought to belong to Sapindales (Rutales), so that radial canals are an evidence of this relationship. Two families universally thought to be very closely related, Apiaceae and Araliaceae, both have radial canals.

The listing below is a product of combining and modifying the listings of Record (1925b, 1944a) and Metcalfe and Chalk (1983, p. 224).

Anacardiaceae: 33 genera (Figs. 7.17, part 4, 7.18, part 1)
Apiaceae: *Eryngium*, *Peucedanum*, *Steganotaenia* (new reports)
Apocynaceae: 23 genera
Araliaceae: *Arthrophyllum*, *Cheirodendron*, *Didymopanax*, *Heptapleura*, *Sciadodendron*
Asteraceae: *Artemisia*, *Chrysothamnus*, *Hymenoclea*
Burseraceae: *Boswellia*, *Bursera* (Fig. 6.1, part 4)
Cactaceae: several genera (Record 1944a)
Clusiaceae: *Mammea*, *Ochrocarpus*, *Rheedia*
Cochlospermaceae: *Cochlospermum*
Crypteroniaceae: *Crypteronia*
Dipterocarpaceae: *Shorea*
Euphorbiaceae: *Alchornea*, *Croton*, *Mabea*, *Pera*, *Sapium*
Fabaceae: *Hardwickia*, *Herminiera*
Hamamelidaceae: *Altingia*, *Liquidambar*
Julianiaceae: *Amphipterygium*, *Juliania*
Loganiaceae: *Anthocleista*
Myrtaceae: *Eugenia*, *Leptospermum*
Rosaceae: *Pygeum*
Rubiaceae: *Hymenodictyon*
Sapindaceae: *Deinbollia*
Ulmaceae: *Gironniera*

The citations of Metcalfe and Chalk (1983, p. 224) for radial canals in Solanaceae and Thymeleaceae are not accepted here. Both of these families have crystal sand idioblasts (Sect. 7.1.2.5); these, when clustered, show breakdown of intervening walls, creating appearances that might have been mistaken for intercellular canals with contents.

7.9.2 Vertical Canals

Metcalfe and Chalk (1950, p. 1353) listed "normal vertical canals" and "traumatic vertical canals." The latter represent instances in which the cambium has been killed in local areas; such injury can occur in many taxa for a variety of reasons, and is not considered worthy of recording in a systematic context here. "Normal" vertical canals represent secretory canals that occur in axial parenchyma and run parallel to it (Fig. 7.18, parts 2–4). Obviously vertical canals can only occur when axial parenchyma strands are massive enough to contain them, and so many families with axial parenchyma types such as diffuse or diffuse-in-aggregates cannot be expected to have vertical canals.

The listing of families with vertical canals is not the same as the listing of families with radial canals, although Anacardiaceae, Dipterocarpaceae, and Fabaceae are on both lists. As with radial canals, the concept of Sapindales (Rutales) in the Thorne (1983) sense is reinforced by the systematic distribution of vertical canals: if Fabaceae and its close relative Connaraceae belong to Sapindales, one can cite two other Sapindalean families that have vertical canals, Anacardiaceae and Simaroubaceae. The listing of families with "normal" vertical canals is based upon compilations of Record (1925b, 1944a) and Heimsch (1942).

Anacardiaceae: *Drimycarpus*
Connaraceae: *Connarus*, *Cnestidium*
Cornaceae: *Mastixis*
Dipterocarpaceae: all of the family except Monotoideae (Fig. 7.18, parts 2 and 3)
Fabaceae: *Copaifera* and six other genera of the subfamily Caesalpinoideae
Simaroubaceae: *Simarouba*

8 Cambial Variants (Anomalous Secondary Growth)

8.1 Classification and Nature of Types

"Anomalous secondary growth" is the term under which have been grouped cambial conformations, cambial products, and cambial numbers which differ from the most common "normal" condition, namely, a single cylindrical cambium that produces phloem externally and xylem internally. The term "variants" is employed here as a way of referring to the less common types; "anomalous" may give the misleading impression of a disorderly action.

Agreement is general that the "normal" cambium is ancestral in dicotyledons, and that from that type, the variant types described below have been derived. Origins of the variant types have very likely been many and, for any given type, polyphyletic, for they represent alternative solutions to production of secondary tissues. This does not rule out the possibility that two related families with the same cambial variant may share an ancestor with that variant.

The basis for cambial variant classification has been largely topographic. However, an ontogenetic element is implicit also, in that different kinds of developmental events lead to the types achieved. A number of studies incorporating ontogeny have been performed, and taking advantage of these appeals as an appropriate procedure in devising terminology. The distinctive histological results of diverse ontogenies permit workers to recognize the nature of particular types. Thus, there is no longer any reason to use such misnomers as "included phloem," or to apply "interxylary phloem" to phloem produced by successive cambia.

The variants reported to date fall in three main categories: (1) successive cambia; (2) a single cambium that yields interxylary phloem as well as xylem internally; (3) cambia that begin as single (or in a few cases, multiple and simultaneous) normal cambia that produce phloem externally and xylem internally and which have or develop a conformation other than cylindrical.

The monograph of Pfeiffer (1926) is basic to the study of cambial variants. In addition to extended descriptive coverage, the literature citation of that monograph is exceptionally extensive. A brief summary of types with useful diagrammatic illustrations has been offered by Chalk and Chattaway (1937). Detailed wood histology of West African lianas (a group rich in cambial variants) is presented by Obaton (1960). Obaton's monograph is notable for considering ontogenetic stages and for showing cellular details as well as diagrams.

Attention is called to the fact that phloem histology is poorly understood in most examples where cambial variants are involved because dried wood samples are studied. This is also true in dried wood samples where thin-walled parenchyma is present. While liquid preservation of wood samples is not logistically easy to manage during field work in areas where woods with cambial variants are abundant (e.g., rain forests where lianas abound), further studies on this topic will hopefully attempt to use liquid-preserved materials to a larger extent. Phloem and parenchyma are not well represented if dried specimens are used, so sections of liquid-preserved specimens are presented here.

8.2 Successive Cambia; the Lateral Meristem and Its Products

I am following the term "successive cambia" as used by Schenck (1893) and Pfeiffer (1926; Figs. 8.1, 8.2, parts 1–4, Fig. 8.3, parts 1 and 2). Metcalfe and Chalk (1983, p. 211) called this phenomenon "concentrically alternating xylem and phloem". They and others used the terms "interxylary phloem" and "included phloem" where successive cambia are involved. However, I restrict the term interxylary phloem to a type where a single cambium (Sect. 8.3) is involved. "Interxylary phloem" ("included" phloem) is a misnomer where successive cambia are involved, as stressed by several workers, because the phloem produced by each cambium is in a normal position relative to the xylem produced by that cambium, and a band of parenchyma or sclerenchyma (conjunctive tissue) separates the vascular tissue bands from each other. Thus, with successive cambia, the phloem is not included in wood or intercalated into wood at all. The term successive cambia is well established and has been used by various authors (Fahn and Zimmermann 1982). However, the phenomena of successive cambia, their products, and the origin of these cambia has given rise to a diversity of interpretations not with respect to the histological plan, which is not in question, but to the origin and perpetuation of a lateral meristem that gives rise to these cambia. With respect to those cellular events, diverse terminology and interpretations have arisen, and a review resolving those controversies, which do not affect recognition of the various cambial variants in any case, is not undertaken here. Rather, a series of simple terms that are easily applicable is used here.

In plants with successive cambia, the formation of the first cylinder of vascular bundles and of the secondary growth by a cambium in that cylinder are the same as in the vascular cylinder of dicotyledons with a single cambium (Fig. 8.1, upper left). In plants with successive cambia, what I am terming a lateral meristem forms in the cortical parenchyma, as noted by Pfeiffer (1926) and others (Fig. 8.1, upper row center). This lateral meristem produces conjunctive tissue and vascular cambia to the inside, and secondary cortex (parenchyma) to the outside (band at top of Fig. 8.1.1, upper right, is meristematic and has given rise to a second series of bundles; the dotted line is the lateral meristem). Each of the successive cambia, as soon as identifiable by pro-

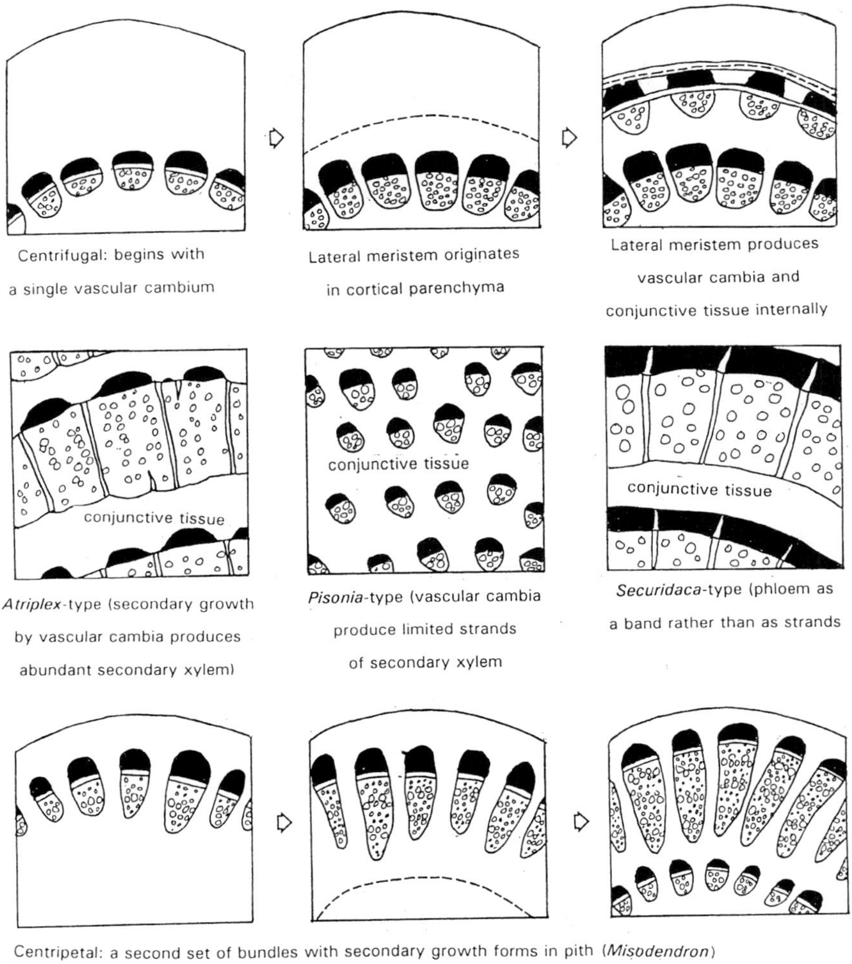

Fig. 8.1. Ontogenetic development (*top row*) and types of successive cambial activity. The products of a lateral meristem are conjunctive tissue and a series of vascular cambia that may produce very little (e.g., *Pisonia*), or much more (*Atriplex*) secondary xylem. *Securidaca* differs in having continuous bands of phloem rather than strands of phloem. Origin of lateral meristem is from cortex (*top, center*). In some species, the first lateral meristem is a single layer that continues to act, producing conjunctive tissue and vascular cambia internally and secondary cortex externally (Fig. 8.2, parts 3 and 4). In others, a diffuse lateral meristem adds to radial thickness of secondary cortex and within that secondary cortex, vascular cambia form (Fig. 8.2, parts 1 and 2). Centripetal successive cambial stages in *Misodendron* (*bottom row*)

duction of secondary phloem to the outside and secondary xylem to the inside, may be termed a vascular cambium, and its products are identical to those of vascular cambia in dicotyledons at large.

These phenomena are shown in photographic form for *Stegnosperma* (Fig. 8.2, parts 1 and 2) and *Barbeuia* (Fig. 8.2, parts 3 and 4). In studies on *Stegnosperma* (Carlquist 1999b), more secondary cortex is produced by divisions

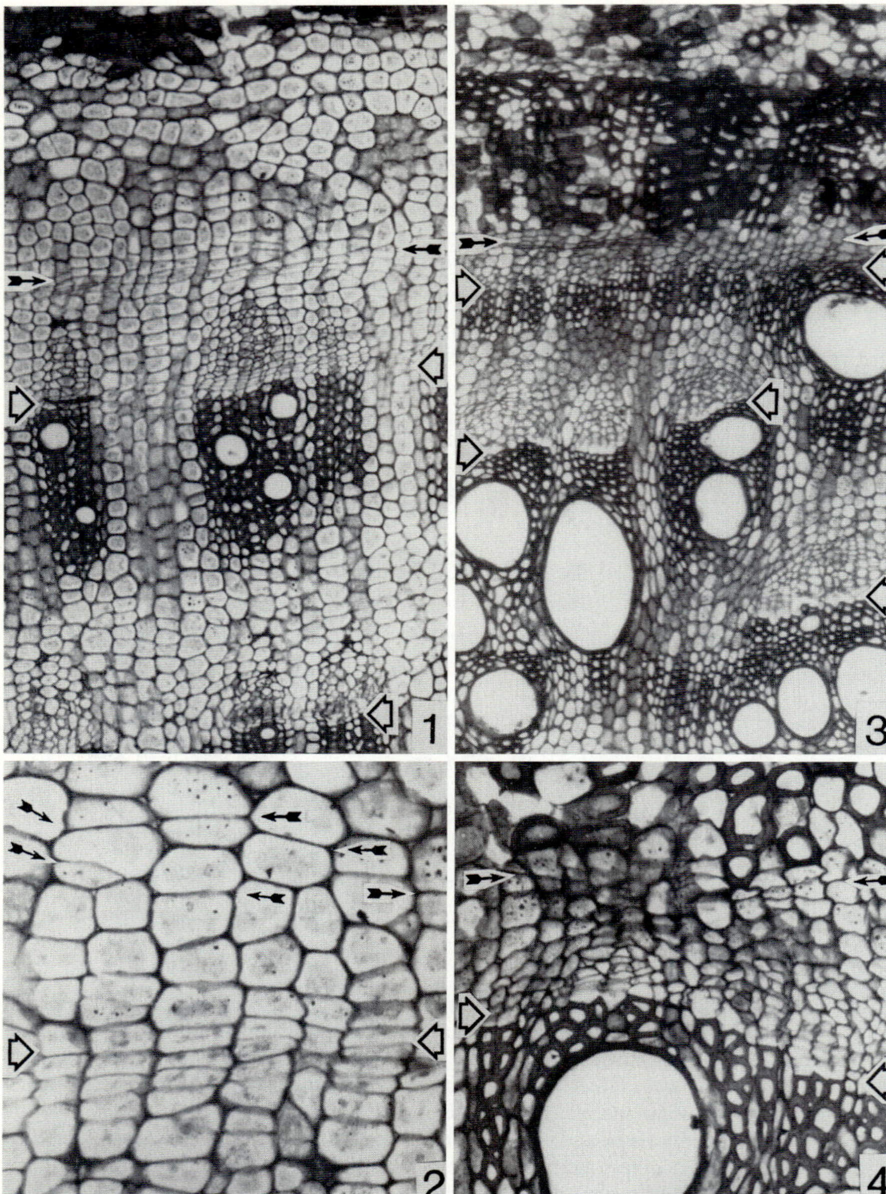

Fig. 8.2. Transections of stems with successive cambia (*1, 3*) with enlargements of areas that show activity of lateral meristems (*2, 4*). *1, 2 Stegnosperma halimifolium*; *1* increment of vascular tissue (*center*; outer portion of a second vascular band at *bottom*); *2* vascular cambium (*bottom*) and divisions producing secondary cortex at *top*; *3, 4 Barbeuia madagascariensis*; *2* primary cortex with patches of sclereids at *top*, below that secondary cortex (radial rows of cells) also containing many sclereids, and, in lower two-thirds of photograph, products of the lateral meristem and the vascular cambia; *3* secondary cortex (*above*) and secondary phloem and xylem produced by a vascular cambium below. *Narrow arrows* denote divisions of lateral meristem, *wide arrows* vascular cambia. (*1,3:A*; *2,4:B*)

in various layers, as shown by the upper set of arrows in Fig. 8.2; a new vascular cambium is producing files of cells below (wider arrows). The occurrence of several layers of cells (as traced radially) acting to produce secondary cortex may be regarded as a "diffuse" lateral meristem in *Stegnosperma*. In *Barbeuia* (Carlquist 1999a), only a single layer acting as a lateral meristem could be identified (wider arrows; Fig. 8.2, part 4; above). The lateral meristem produces sclerenchyma as well as parenchyma externally as secondary cortex, a tissue that does not serve as a site for new lateral meristems (the sclerenchyma patterns would show that); only new vascular cambia (and their products) and conjunctive tissue are produced to the inside. The lateral meristems in plants with successive cambia are evidently bidirectional, and the respective external and internal products of the lateral meristem are as given above.

Unidirectional cambia have been demonstrated only in a genus that does not have successive cambia, *Stylidium*; see Sect. 8.3).

The conjunctive tissue lying between the phloem of one band and the xylem formed by the next band may consist of parenchyma (Fig. 8.2, part 1) or sclerenchyma (Fig. 8.3, part 2). The thickness of xylem formed from each cambium can vary; in some species, such as *Cocculus laurifolius* of the Menispermaceae, several centimeters of secondary xylem may be formed by a particular cambium. Phloem in the successive bands tends to appear as isolated strands (Fig. 8.1; *Atriplex*; Fig. 8.3, part 1; *Simmondsia*) or as a more nearly continuous (extending tangentially) band outside the xylem (Fig. 8.1; *Securidaca*). Although successive cambia are often thought to form concentric rings, as in the beet (*Beta*), arcs of vascular tissue that extend for limited tangential distances are often produced instead. In *Pisonia* (Nyctaginaceae) they are not really bands as such, but vascular strands looking like collateral bundles scattered in a background of conjunctive tissue (Fig. 8.1; center). Not often stressed is the fact that there are vascular interconnections between the products of successive cambia (Zamski 1979; Zamski and Azenkot 1981).

The occurrence of successive cambia is of considerable importance systematically. An interesting series of examples occur in Caryophyllales (Centrospermae). Numerous genera in this order have successive cambia, yet some families of the order (Cactaceae, Didiereaceae, Halophytaceae, Portulacaceae) have only a single vascular cambium per stem (or root). Some genera of the family Phytolaccaceae (sensu lato) have successive cambia, while others (*Hilleria*, *Monococcus*, *Schindleria*, *Trichostigma*) have only a single cambium, despite the large stems formed in some (Carlquist 2000b). Studies in progress are likely to reveal whether successive cambia are ancestral or have been evolved within one or more clades of Caryophyllales.

Record (1933) found that the wood of *Rhabdodendron* has successive cambia. *Rhabdodendron* had hitherto been placed in Rutaceae or Chrysobalanaceae, but Record made a compelling case for exclusion of *Rhabdodendron* from those two families on the basis of the presence of successive cambia.

The presence of successive cambia in *Simmondsia* (Fig. 8.3, part 1) alerted workers to the probability that this genus does not belong in Buxaceae.

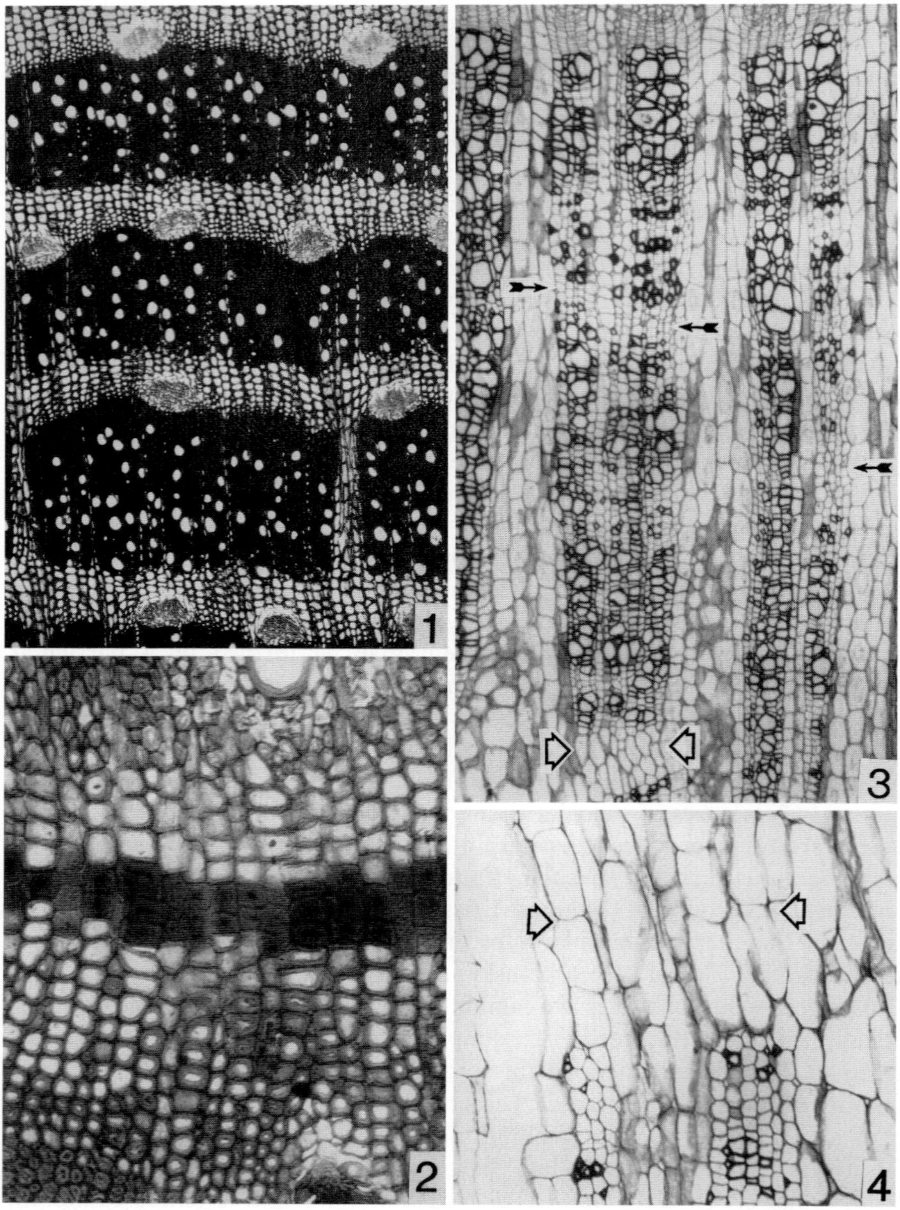

Fig. 8.3. Woods with successive cambia (*1, 2*) and proliferation of axial parenchyma (*3, 4*). *1* Moderate quantities of secondary xylem (*dark*), conspicuous conjunctive parenchyma bands (*light*), phloem strands; *Simmondsia chinensis* (Simmondsiaceae); *2* portion of conjunctive tissue between bands produced by successive cambia, showing band of sclerenchyma; *Avicennia nitida* (Avicenniaceae); *3, 4 Fouquieria (Idria) columnaris* (Fouquieriaceae), transection of lower stem; *3* outside of secondary xylem (portion of secondary phloem at *top*), showing prominent bands of axial parenchyma (*narrow arrows*) and a zone of radially elongate axial parenchyma cells (*wide arrow pair*); *4* area of secondary xylem internal to the zone shown in *3*; the strand of xylem (*below*) has become separated from other portions of the fascicular area to which it belongs by marked radial elongation of axial parenchyma (*wide arrow pair*). (*1,3,4:A*; *2:B*)

Likewise, Verbenaceae become a more easily defined and natural group if one excludes genera with successive cambia. The mangrove genus *Avicennia* (Studholme and Philipson 1966; Zamski 1979) is segregated as Avicenniaceae. A group of shrubby genera are separated as Chloanthaceae, formerly called Dicrastylidaceae (Carlquist 1981c). *Chrysanthemoides monilifera* (Asteraceae) has been excluded from the genus *Osteospermum* now that successive cambia have been revealed in that species (Adamson 1937).

The following list represents an original compilation. Appropriate references are cited, but fewer references are listed for families in which successive cambia have long been known (e.g., Pfeiffer 1926). Some reports are in need of further investigation or confirmation.

Agdestidaceae: *Agdestis* stems and roots (Carlquist 1999d)
Aizoaceae: woodier genera (herbaceous genera have little or no secondary growth)
Amaranthaceae: *Bosea* (Chalk and Chattaway 1937), *Charpentiera*, *Nototrichium* (Carlquist 1974), and other woodier genera
Asteraceae: *Chrysanthemoides* (Adamson 1934)
Avicenniaceae: *Avicennia* (Fig. 8.3, part 2; Studholme and Philipson 1966; Zamski 1979)
Barbeuiaceae: *Barbeuia* (Fig. 8.2, parts 3 and 4; Carlquist 1999a)
Basellaceae: *Anredera* stems and roots (Carlquist 1999e)
Capparaceae: *Boscia* (Adamson 1936), *Cadaba*, *Forchhammeria*, *Maerua*, *Stixis*
Caryophyllaceae: *Spergularia* stem (Solereder 1908), roots of various genera (Carlquist 1995c).
Chenopodiaceae: numerous genera (Fig. 8.3, part 1; Artschwager 1920; Pfeiffer 1926; Fahn and Schchori 1968; Baird and Blackwell 1980; Yarrow and Popham 1981; Zamski and Azenkot 1981; Fahn and Zimmermann 1982; Fahn et al. 1986)
Chloanthaceae (Dicrastylidaceae): all genera (Carlquist 1981c)
Connaraceae: *Rourea pulchella* (Boureau 1957), *Agelaea*, *Connarus*, and *Spiropetalum* (Obaton 1960)
Convolvulaceae: *Argyreia* (Pant and Bhatnagar 1975), *Calonyction*, *Convolvulus*, *Dicranostyles* (Mennega 1969), *Ericybe*, *Hewittia*, *Ipomoea* (Lowell and Lucansky 1990), *Maripa*, *Merremia*, *Porana*, and some other genera (see Carlquist and Hanson)
Cucurbitaceae: *Adenopus*, *Luffa*, *Melothria*, *Mormodica*, *Physedra*, *Sphaerosicyos* (Zimmerman 1922), and roots of genera with tuberous roots like *Marah*
Dilleniaceae: *Doliocarpus* (Chalk and Chattawa 1937)
Dioncophyllaceae: *Dioncophyllum* (Boureau 1957)
Fabaceae: *Derris*, *Koompassia*, *Machaerium*, *Mucuna*, *Pachyrhizus*, *Pueraria*, *Rhynchosia*, *Strongylodon*, *Wisteria* (Boureau 1957)
Frankeniaceae: *Frankenia* (Barghoorn 1941b)

Hippocrateaceae: *Cheiloclinium, Salacia* (Solereder 1908)
Icacinaceae: *Icacina mannii, Raphiostylis* (Obaton 1960)
Loranthaceae: *Nuytsia floribunda* (Pfeiffer 1926)
Menispermaceae: most genera (Solereder 1908; Chalk and Chattaway 1937; Obaton 1960; Carlquist 1996e)
Nyctaginaceae: most genera, including *Boerhaavia, Maheshwari* (1930), *Bougainvillea* (Esau and Cheadle 1969), *Cryptocarpon, Heimerliodendron* (Studholme and Philipson 1966), *Mirabilis* (Mikesell and Popham 1976), *Neea, Pisonia, Torrubia* (Chalk and Chattaway 1937), *Phaeoptilium*
Passifloraceae: *Adenis* (Obaton 1960), *Adesmia* (Ayensu and Stern 1964)
Phytolaccaceae: *Gallesia, Phytolacca* (Wheat 1977; Mikesell 1979; Carlquist 2000b), *Petiveria* (Carlquist 1998a), *Rivina*
Plumbaginaceae: *Aegialitis* (Carlquist and Boggs 1996)
Polygalaceae: *Bredemeyera, Catocoma, Somesperma, Moutabea, Securidaca* (Solereder 1908)
Rhabdodendraceae: *Rhabdodendron* (Record 1933)
Rubiaceae: *Basanacantha* (Boureau 1957), *Chiococca* (Pfeiffer 1926)
Sapindaceae: *Serjania pinnata* (see Sect. 8.4.7)
Simmondsiaceae: *Simmondsia* (Bailey 1980; Carlquist 1982d; Fig. 8.1, part 2)
Stegnospermataceae: *Stegnosperma* (Fig. 8.2, parts 1 and 2; Horak 1981; Carlquist 1999b)
Thymeleaceae: *Wikstroemia chamaejasme* (Solereder 1908)
Vitaceae: *Tetrastigma* (new report)

One might apply a term such as "centripetal successive cambia" to an unusual phenomenon. In *Misodendron*, subgenus *Angelopogon*, a primary cambium forms considerable xylem and phloem, but a second series of bundles with cambial activity forms not outside of the first cylinder, but inside it, in the pith (Behm 1895; Carlquist 1985c). Families in which this kind of activity occurs are listed:

Acanthaceae: *Afromendoncia* (Obaton 1960)
Bignoniaceae: *Campsis, Clytostoma, Tecomaria* (Pfeiffer 1926)
Caryocaraceae: *Caryocar nuciferum* root (Blank 1939)
Caryophyllaceae: *Acanthophyllum* (Pfeiffer 1926), *Dianthus* (Carlquist 1995c).
Misodendraceae: *Misodendron* subgenus *Angelopogon* (Behm 1895; Carlquist 1985c) (Fig. 8.1, bottom)

In some species of Piperaceae, such as *Macropiper excelsum* and *Piper jaborandi* (Metcalfe and Chalk 1950) as well as in *Phytolacca* (Carlquist 2000b), there are pith bundles that experience secondary growth. However, these bundles are not laid down as a result of the development of a second cambium, but are part of the primary stem. Some of the examples in the list above might represent pith bundles (bundles laid down simultaneously with the bundles of the vascular cylinder) rather than later developed bundles interconnected by a cambium.

8.3 Interxylary Phloem Produced by a Single Cambium

The use of the term "interxylary phloem" for instances involving successive cambia has resulted in confusion, and I recommend that it be restricted to instances in which a single cambium forms phloem internally (Fig. 8.4, parts 1–4, Fig. 8.5). The term "foraminate" has been used for what is termed interxylary phloem here. Attention is drawn to the term "int*ra*xylary phloem" (easy to confuse with "int*er*xylary phloem") which denotes phloem formed at the margin between pith and primary xylem. Interestingly, int*ra*xylary phloem may be a sort of "precursor" to int*er*xylary phloem: for example, the order Myrtales characteristically has intraxylary phloem, but only a scattering of species in some myrtalean families have interxylary phloem. Interxylary phloem is easy to distinguish from phloem produced by successive cambia, because in the latter case, strands of vessel-bearing secondary xylem are always paired with strands of secondary phloem, whereas in interxylary phloem, strands of interxylary phloem are intercalated into secondary xylem that has a distribution of vessels not related to the phloem.

Interxylary phloem can take various forms (Fig. 8.5). Strands can be composed of one or a few sieve tubes and little or no phloem, as in *Stylidium*; instances of interxylary phloem of this sort are so inconspicuous that they are difficult to observe. Other reports of slender interxylary phloem strands include *Stoebe* of the Asteraceae (Adamson 1934) and *Ixanthus* of the Gentianaceae (Carlquist 1984c). Large phloem strands thickly sheathed with parenchyma, in contrast, may be seen in *Thunbergia alata* (Fig. 8.4, parts 1 and 2; Carlquist and Zona 1988a). Bands of interxylary phloem wider tangentially than radially characterize many Onagraceae (Carlquist 1975b, 1977b, 1983f, 1987i), such as *Gaura* (Fig. 8.4, parts 3 and 4). In this last-cited example, parenchyma sheathing the phloem is intermediate in thickness, ranging from one to two cells (Fig. 8.4, part 4). Attention is called to the parenchyma associated with interxylary phloem, because it is ordinarily not mentioned, and the reader is left with the impression that interxylary phloem strands consist wholly of sieve tube elements and companion cells. In addition to parenchyma sheathing phloem strands, occasional parenchyma cells may be found intermixed among the sieve tubes in interxylary phloem. In *Combretum* (Combretaceae) and *Erisma* (Vochysiaceae), interxylary phloem strands are borne in the large bands of confluent axial parenchyma (Chalk and Chattaway 1937; Fig. 8.5). In some taxa with interxylary phloem, strands of phloem are round in transection, whereas in others the phloem is formed as tangentially broad bands, as in *Myriocarpa* of the Urticaceae (Chalk and Chattaway 1937). Although sections of dried wood samples easily reveal instances of interxylary phloem, the histology of phloem in sections prepared from such dried samples is often not satisfactory, and hopefully more studies using pickled stem material will be done in the future.

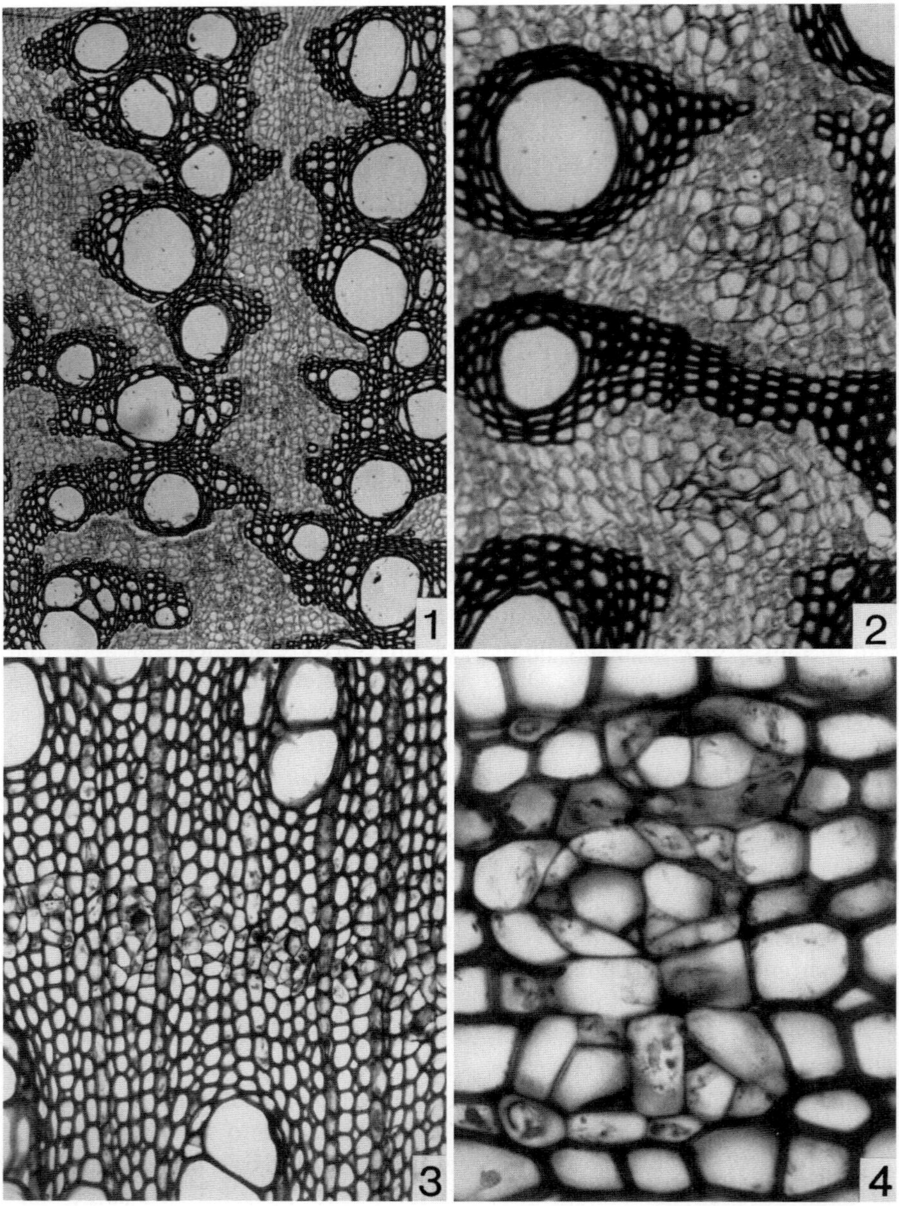

Fig. 8.4. Interxylary phloem, as seen in wood transections. *1, 2 Thunbergia alata* (Acanthaceae); *1* interxylary phloem strands occur in the axial parenchyma patches (*gray*), vessels are sheathed in fibers (*blackish*); *2* portion transection, showing two complete strands of interxylary phloem (cells larger, slightly darker gray) in axial parenchyma (lighter gray, smaller diameter cells); *3, 4 Gaura parviflora* (Onagraceae); *3* band of interxylary phloem, *left to right* across middle of photograph; *4* sieve tubes and associated companion cells, sheathed by a single layer of parenchyma. (*1:A*; *2,4:C*; *3:B*)

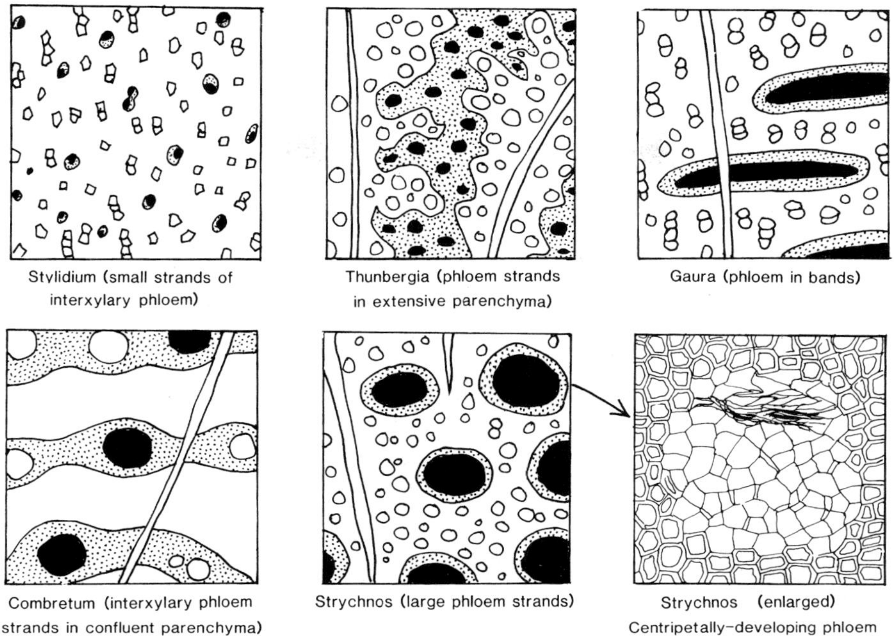

Fig. 8.5. Modes of interxylary phloem occurrence. Phloem strands may be very small, with only one or two sieve tubes (*Stylidium*), or larger; in bands (*Gaura*); in confluent axial parenchyma (*Combretum*); and with older phloem crushed at abaxial tips of phloem strands (*Strychnos*)

An interesting early study on interxylary phloem based on living material grown in greenhouses was that of Scott and Brebner (1889) on *Strychnos* (Loganiaceae). These authors showed that each of the interxylary phloem strands of *Strychnos* stems contain a cambium that produces new sieve tube elements and companion cells toward the outside, so that in each interxylary phloem strand earlier formed sieve tube elements and companion cells are eventually crushed and appear as a band near the outer edge of the strand (Fig. 8.4). This has been termed "centripetal" interxylary phloem, and it is widespread within the genus *Strychnos* (Cockrell 1941). That term may be superfluous, because cambial activity may be more widespread in instances of interxylary phloem than previously thought (but in no case does such a cambium produce any xylem).

The families listed below have interxylary phloem derived from a single cambium. In a few of these, such as roots of *Asclepias* or *Brassica*, one finds a few strands of phloem scattered in a background that consists mostly of parenchyma: these are questionable examples of interxylary phloem. Most of the examples cited, however, represent a woody secondary xylem, produced by a single cambium, in which strands or bands of interxylary phloem are scattered. The list was derived from reports by Pfeiffer (1926) and Metcalfe and Chalk (1950), expanded by more recent reports and corrected for terminology. Some of the reports are in need of further study or correction.

Apocynaceae: *Lyonsia straminea*
Asclepiadaceae: *Asclepias obtusifolia*, *A. syriaca*, *Ceropegia*, *Leptadenia* (Singh 1943)
Asteraceae: *Stoebe* (Adamson 1934)
Brassicaceae: *Brassica*, *Cochlearia*, *Raphanus* (Pfeiffer 1926)
Clusiaceae: roots of *Endodesmia* (Pfeiffer 1926)
Combretaceae: *Calycopteris*, *Combretum*, *Guiera*, *Thiloa* (van Vliet 1979)
Convolvulaceae: *Ipomoea versicolor* (Solereder 1908)
Cucurbitaceae: *Cucurbita*, *Lagenaria*
Euphorbiaceae: *Dalechampia*
Gentianaceae: *Chironia*, *Crawfordia*, *Ixanthus* (Carlquist 1984c), *Orphium*, roots of many others
Icacinaceae: lianoid genera of tribe Sarcostigmateae
Loganiaceae: *Antonia*, *Bonuynia*, *Logania*, *Norrisia*, *Strychnos* (Mennega 1980; van Veenendal and den Outer 1993)
Lythraceae: *Lythrum salicaria* root (Gin 1909)
Malpighiaceae: *Dicella*, *Stigmaphyllon*
Melastomataceae: *Kibessia*, *Lijndenia*, *Memecylon*, *Mouriri*, *Olisbea*, *Pternandra* (Chalk and Chattaway 1937)
Onagraceae: *Epilobium*, *Gaura*, *Lopezia*, *Oenothera*, *Zauschneria* (Carlquist 1975b, 1977b, 1983f, 1987i)
Salvadoraceae: *Azima* (den Outer and van Veenendaal 1981), *Dobera*, *Salvadora*
Solanaceae: roots and rhizomes of *Atropa belladonna*, roots of *Datura stramonium* and *Scolopia carniolica*
Thymeleaceae: *Aquilaria*, 8 other genera
Urticaceae: *Myriocarpa* (Chalk and Chattaway 1937)
Vochysiaceae: *Erisma*, *Erismadelphus*, *Qualea*

8.4 Cambia Normal in Products But Abnormal in Conformation or Dispersion

Under this heading are considered all cambial variants in which the cambium produces phloem to the outside, xylem to the inside, but in which the cambium has a three-dimensional shape other than cylindrical, and may be either single or multiple; if multiple, successive cambia are not involved. The multiplicity of cambia in this category arises either from simultaneous origin in several sites or from fracture of the cambium into several portions.

8.4.1 Stem Flattened in Transection

Lianas that have a bandlike form to their stems are a common sight in tropical rain forests. This form occurs, in the case of a single cambium, by deposition of more xylem in one or two sites of the stem (Fig. 8.6, part 1). The stem can flatten unilaterally if one site receives a greater deposition of secondary xylem, or bilaterally, if two sites receive a greater deposition. Although flattened stems occur in freestanding situations, the flattened stem is often related to the climbing habit of lianas in that the broad face of the flattened axis contacts branches of trees.

Genera with a single cambium and with stems oval in transection are listed below. This list is modified from Pfeiffer (1926), Bhambie (1972), and Metcalfe and Chalk (1983).

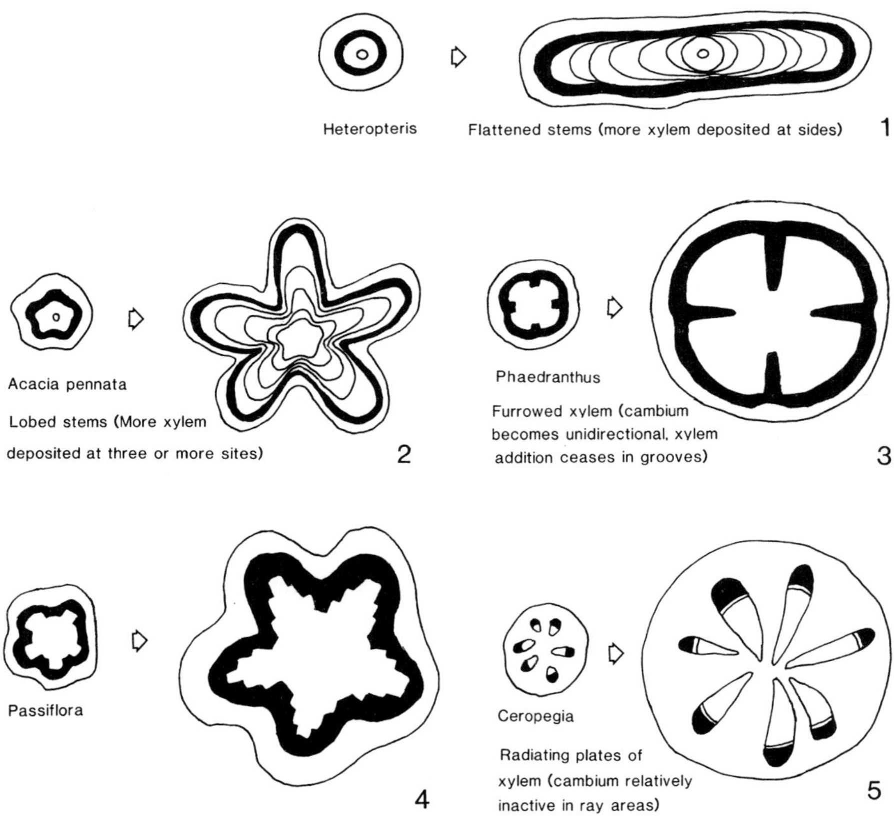

Fig. 8.6. Diagrammatic representations, showing ontogenetic sequences, in cambia that have normal products (phloem to outside, xylem to inside) but abnormal conformation. Cambium produces more xylem in some places than others to achieve flattened or lobed stems. Grooved xylem results from cambium that acts unilaterally (little or no xylem) in grooves; progressively, unilateral activity may expand along margins of grooves, resulting in ledged (zigzag) appearance. Presence of xylem plates separated by rays results from sluggish cambial activity in ray areas

Celastraceae: *Celastrus*, *Euonymus*
Combretaceae: *Quisqualis*
Malpighiaceae: *Heteropteris*
Marcgraviaceae: *Marcgravia*
Moraceae: *Ficus*
Piperaceae: *Piper*
Polygonaceae: *Atraphaxis*, *Coccoloba*
Rubiaceae: *Manettia*, *Sabicea*
Verbenaceae: *Lantana*, *Lippia*, *Petraea*

Examples of lianas with markedly flattened stems but single cambia are listed below (list compiled from Obaton 1960; Metcalfe and Chalk 1983; and other sources).

Apocynaceae: *Landolphia owariensis*
Aristolochiaceae: *Aristolochia*, *Pararistolochia*
Fabaceae: *Abrus*, *Bauhinia* (Basson and Bierhorst 1967), *Milletia*, *Rhynchosia*
Malpighiaceae: *Heteropteris* (Fig. 8.6, part 1)
Polygonaceae: *Coccoloba*
Ulmaceae: *Celtis*
Urticaceae: *Urera obovata*
Vitaceae: *Cissus*

8.4.2 Stems Lobed in Transection

8.4.2.1 Stems Lobed But Not Fragmented

Flattened stems can be found in lianas with successive cambia. This combination is illustrated by Pfeiffer (1926) for *Machaerium* of the Fabaceae (Fig. 8.6, part 2), *Anomospermum* (Menispermaceae), and *Securidaca* (Polygalaceae); to this list I can add a species of *Tetrastigma* (Vitaceae). Several other genera could also be added to this list.

If a single cambium deposits secondary xylem in greater quantities at three or more sites, stems lobed in transection result (Fig. 8.6, part 2). In the following is a partial listing of genera with this characteristic (compiled from various sources):

Celastraceae: *Celastrus*, *Euonymus*
Combretaceae: *Combretum dolichandrone* (Obaton 1960)
Fabaceae: *Acacia pennata* (Fig. 8.6, part 2; Obaton 1960), other genera
Rubiaceae: *Canthium* (Francis 1926), *Mussaenda* (Obaton 1960)
Verbenaceae: *Lantana* (Bhambie 1972)

Stems lobed in transection are not confined to lianas; they characteristically occur in buttressed trees of rain forest areas (Francis 1924).

8.4.2.2 Stems Lobed and Fragmented

Another phenomenon that should be cited in this regard is represented by desert shrubs that show fragmentation of the woody cylinder. The cambium becomes fragmented as the woody sectors separate from each other, so this can indeed be considered a cambial variant. This tendency is seen in such genera as *Artemisia* (Asteraceae), *Eriogonum* (Polygonaceae), and *Larrea* (Zygophyllaceae). For a review on this topic, the reader should consult Ginzburg (1963) and Jones (1984).

Stems markedly lobed but with more than one cambium can include *Mormodica* of the Cucurbitaceae (Pfeiffer 1926) with successive cambia, and *Serjania* of the Sapindaceae (Pfeiffer 1926). Many more examples could be added.

8.4.3 Stems with Xylem Furrowed

This term ("interrupted stem" of Chalk and Chattaway 1937) has a more precise application than the heading might indicate. It is not merely a heightened form of the furrowed stem habit. Rather, as shown by the angular edges on xylem contours in the examples of Fig. 8.5 (parts 3 and 4), species with furrowed xylem show conversion of cambium to unilateral (producing phloem only) or nearly unilateral (only a little secondary xylem compared to the abundant secondary phloem) production at particular sites (Schenck 1893). In the *Phaedranthus* example (Fig. 8.6, part 3), the four sites where furrows form represent early conversion of cambium to unilateral activity, and no further sites for unilateral, cambial activity are developed, so that the xylem cylinder is normal except for the deep grooves at those four points. In some other Bignoniaceae, additional sites intermediate between the original sites convert to unilateral cambium acitivity, so that shallower grooves in addition to the deeper ones are present, as in *Pithoctenium* and *Lundia* of the Bignoniaceae (Pfeiffer 1926). Grooved xylem is so common in scandent Bignoniaceae that Pfeiffer (1926) offers a key to wood of Bignoniaceae based on variations of this habit. In *Doxantha*, Dobbins (1969, 1971) has shown that in the indented areas the cambium functions normally at first but then becomes unidirectional rather suddenly. The sites at which this happens correspond to phyllotaxy and leaf traces (Dobbins 1981). Chemical activity of leaves has been implicated in cessation of bidirectional cambial activity at these locations (Dobbins 1970).

Another variant on the grooved xylem habit is shown by *Passiflora glandulosa* (Fig. 8.6, part 4; Ayensu and Stern 1964). In this example, conversion of cambium to unilateral (or nearly unilateral) activity occurs progressively: it begins at five sites, then additional segments of cambium are converted on the margins of the grooves, so that the grooves become wider and wider.

A listing of families and genera with grooved secondary xylem can be offered on the basis of Pfeiffer (1926) and Obaton (1960); the latter author terms this phenomenon "tiges sillonées."

Bignoniaceae: *Adenocalymma, Anemopaegma, Arrabidaea, Bignonia, Callichlamys, Clytostoma, Cuspidaria, Distictis, Doxantha, Fridericia, Glaziovia, Haplolophium, Lundia, Melloa, Paragonia, Petastoma, Pithoctenium, Phaedranthus, Phryganocydia, Pleonotoma, Pyrostegia, Stizophyllum, Tanaecium*
Convolvulaceae: *Bonamia, Neuropeltis, Prevostia*
Fabaceae: *Centrosema plumieri*
Hippocrateaceae: *Hippocratea, Salacia*
Malpighiaceae: *Triapsis odorata*
Icacinaceae: *Neostachyanthus, Pyrenacantha*
Passifloraceae: *Crossostemma, Passiflora* (Ayensu and Stern 1964)

8.4.4 Xylem in Plates

One may consider that dicotyledons in which fascicular areas are few and are separated by large ray areas that consist of thin-walled parenchyma constitute a kind of anomaly with respect to cambial activity. Pfeiffer (1926, p. 31) does consider examples of this tendency. One can see this clearly in *Misodendron*, subgenus *Angelopogon*, in which cambial activity is much less in ray areas than in fascicular areas, and in which breakup of ray areas is nonexistent. Examples of this tendency can be listed as follows, although depending on one's interpretation, various numbers of additional examples might be cited.

Asteraceae: *Bidens, Mikania* (Pfeiffer 1926)
Menispermaceae: *Cocculus, Menispermum, Tinospora* (Pfeiffer 1926)
Misodendraceae: *Misodendron* subgenus *Angelopogon* (Carlquist 1985c)

The occurrence of nonlignified axial xylem is cited by Metcalfe and Chalk (1950, p. 961) as an anomaly in some species of *Breweria, Convolvulus, Exogonium*, and *Ipomoea* (Convolvulaceae). This should not be listed as a cambial variant. One could cite herbaceous annuals in which this occurs. However, the pertinence in listing examples of nonlignified xylem becomes greater when one is discussing vining genera like *Ceropegia* (Asclepiadaceae), because in woodier members of the families in which these occur (e.g., *Asclepias albicans*), wood is typical in having a ground plan of imperforate tracheary elements with lignified secondary walls. One may consider this as related to the vining habit without considering it a cambial variant. In the same regard, the extension of wide primary rays unaltered into the secondary xylem in vining genera (e.g., *Aristolochia*) is an example of a characteristic of vines (Carlquist 1985d, 1993) without being an example of a cambial variant – it is merely an example of a cambium with a slow pace of cell divisions.

8.4.5 Wood Portions Dispersed or Separated by Divisions in Parenchyma and Parenchyma Expansion

This term is used to describe the behavior of axial parenchyma in stems of the succulent species of *Fouquieria*, such as *F. (Idria) columnaris* (Fig. 8.3, parts 3 and 4). Inward from the cambium, one can see patches of axial xylem (Fig. 8.3, part 3). Further inward, these patches consist of radially elongate cells which separate the fibrous axial xylem into isolated patches. The next effect is to widen the stem and to provide very large parenchyma cells that can serve for water storage, or, when dead, still be the source of a stem widening that offers a kind of support. This behavior has also been reported in *Carica pentagona* (Carlquist 1998d). Such parenchyma proliferation has been included in this chapter rather than under the heading of axial parenchyma because there is not merely radial elongation of cells, but production of more parenchyma by means of tangential divisions in the parenchyma, and these divisions can be considered a kind of cambial action. Differential enlargement of rays is a source of breakup of fibrous wood in *Anredera* of the Basellaceae (Carlquist 1999e).

Divisions of a cambial kind within axial parenchyma can be found in those species in which the stem begins with a cylinder of xylem, but in which segments become separated, and in which each segment has its own cambium. The cambium of each of these segments tends to extend around the segment partially or wholly, to a greater extent than would happen if the segments remained united.

Parenchyma abundance is related to dispersion of wood segments in species and genera that possess this cambial variant. One might possibly call this a group of cambial variants, since one cannot be sure that the mechanism of formation is identical in all examples with dispersed fragments of wood, and some degree of diversity is indicated by the work of Handa (1937). A clue to the mechanism probably more important than others is represented by Obaton's (1960) term "parenchyme de remplissage." This parenchyma, in essence, forms in zones instead of imperforate tracheary elements; these parenchyma zones may show proliferation, separating the wood segments further (Fig. 8.7, part 1). However, parenchyma proliferation is difficult to separate from tendency of cambium to form in this parenchyma, partially or even wholly encircling the fragments and assuring further separation. The kind of parenchyma abundance that leads to fragmentation is shown for *Thunbergia alata* in Fig. 8.4 (parts 1 and 2). *Thunbergia* is closely related to *Mendoncia* and the appearance shown in Fig. 8.4 (part 1) is essentially the same as the pattern shown in early stages of stem development in *Afromendoncia* and *Mendoncia*. The tendency of stems with dispersed wood fragments to have abundant quantities of parenchyma is evident (Schenck 1893; Pfeiffer 1926; Handa 1937, 1938; Wagner 1946). More ontogenetic studies such as those of Handa (1937, 1938) on *Bauhinia* and Obaton (1960) on *Afromendoncia* are needed. Cambia that form in ray-adjacent axial parenchyma produce secondary phloem in *Coccinia* of

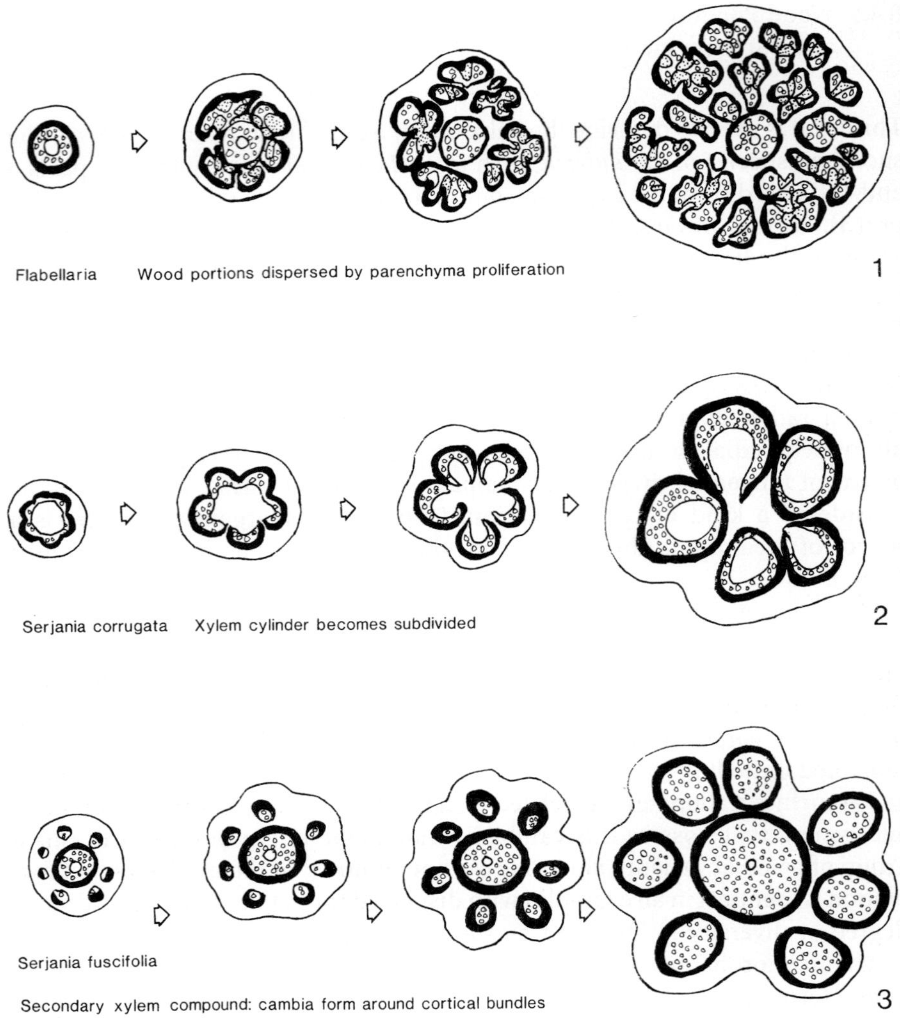

Fig. 8.7. Cambial variants in which the cambium is or becomes separated into several segments. Four stages in development are shown for each species. *Flabellaria* (Malpighiaceae) has origin of xylem segments outside the original cylinder; these segments are expanded and dispersed by parenchyma proliferation. *Serjania corrugata* (Sapindaceae) has a xylem cylinder consisting of arcs (in sectional view), the arcs are converted to cylinders; *Serjania fuscifolia* has cortical bundles, xylem is added to each of these and to the original cylinder

the Cucurbitaceae (Carlquist 1992g) and *Anredera* of the Basellaceae (Carlquist 1999e); these instances do not result in much displacement of xylem strands compared to *Flabellaria*, but are like initial stages in the splitting of xylem strands in *Flabellaria* (Fig. 8.7, part 1).

The listing of examples of wood with dispersed wood fragments below is based on the data accumulated by Pfeiffer (1926) and Obaton (1960).

Acanthaceae: *Afromendoncia, Mendoncia*
Apiaceae: *Azorella selago*
Basellaceae: *Anredera* (Carlquist 1999e)
Apocynaceae: *Tetrapteris*
Caricaceae: *Carica*
Caryophyllaceae: *Acanthophyllum*
Convolvulaceae: *Bonamia* (Carlquist and Hanson 1991), *Merremia, Neuropeltis, Prevostea*
Cucurbitaceae: *Coccinia* (Carlquist 1992g)
Fabaceae: *Bauhinia* spp. (e.g., *B. championi, B. japonica, B. langsdorffiana*), *Kunstleria ridleyi*
Fouquieriaceae: *Fouquieria*
Icacinaceae: *Icacina, Iodes, Phytocrene, Pyrenacantha*
Malpighiaceae: *Banisteria, Flabellaria, Mascagnia, Tetrapteris*
Vitaceae: *Cissus*

8.4.6 Divided Xylem Cylinder

This term may be applied to those species in which the stem has a vascular cylinder that is furrowed as seen in transection (Fig. 8.7, part 2). During secondary growth, each of the arcs becomes surrounded by cambium, and subsequent stages of secondary growth feature addition of secondary tissues to each of the xylem segments independently. This cambial variant, also known as "cleft xylem mass," was originally studied by Radlkofer (1875) in *Urvillea*. This cambial variant is also known from *Serjania corrugata* and closely related species of *Serjania*. Both *Serjania* and *Urvillea* belong to Sapindaceae.

8.4.7 Compound Secondary Xylem

In this type of cambial variant, a normal cylinder of secondary xylem and secondary phloem develops in a stem; in addition, cambia develop around cortical bundles, converting each of these to cylinders of secondary xylem surrounded by secondary phloem (Fig. 8.7, part 3). Because these cortical bundles are pre-existing sites for cambial development unlike the parenchyma providing the sites for cambial development in successive cambia, this is considered an independent cambial variant.

This cambial variant has been identified in 91 of 172 species of *Serjania* and 16 of 122 species of *Paullinia* (both genera of Sapindaceae), by van der Walt et al. (1973). The three-dimensional course of vascular bundles has been studied in one of the species of *Serjania* with this variant, *S. subdentata*, by Johnson and Truscott (1956). A similar analysis, showing secondary growth of the cortical bundles, has been performed by Obaton (1960) on *S. pinnata*, which also has this variant.

Compound xylem masses possibly referable to this variant have been found in subterranean axes of *Gnidia* (Thymeleaceae) by Leandri (1931). Metcalfe and Chalk (1983) cite this example as representing both "compound xylem masses" (p. 61) and "concentrically alternating xylem and phloem" (p. 212), however.

8.5 Combinations of Cambial Variants

Some examples encountered in reviewing this topic do not conform to any single type, because in fact they combine two variants in a single stem (Fig. 8.8). I believe that these should not be referred to a single type, which would only conceal the nature of these examples. A few examples are cited here, but more could easily be added.

Flattened stems can be found in lianas with successive cambia. This combination is illustrated by Pfeiffer (1926) for *Machaerium* of the Fabaceae (Fig. 8.8, part 1), *Anomospermum* (Menispermaceae), and *Securidaca* (Polygalaceae); to this list I can add a species of *Tetrastigma* (Vitaceae). Several other genera could be added to this list.

Afromendoncia and *Mendoncia* (Acanthaceae) illustrate the phenomenon of xylem segments dispersed by parenchyma. However, interxylary phloem, like that for *Thunbergia* (Fig. 8.3, parts 1 and 2) has now been discovered in *Mendoncia* (Carlquist and Zona 1987). Thus, *Mendoncia* (Fig. 8.8, part 2) is an example of both phenomena combined.

Thinouia ventricosa of the Sapindaceae (Fig. 8.8, part 3) is an example in which successive cambia (in a distinctive form of that phenomenon) occur. New cambia originate in parenchyma of the cortex. Each of these cambia produces a cylinder of secondary xylem and secondary phloem. Very likely some mechanism for increase in parenchyma in the cortical parenchyma occurs also. In any case, a stem with markedly lobed outline is produced. Thus, *Thinouia ventricosa* combines successive cambia with lobed stem outline.

Another intriguing example figured by Metcalfe and Chalk (1983, p. 60) is furnished by *Phytocrene macrophylla* of the Icacinaceae. Successive cambia are evident in this species, but in each of the vascular bands, xylem is grooved, indicating occurrence of unilaterally-active cambium in some sites.

8.6 Phylogeny, Habit, and Ecology of Cambial Variants

Cambial variants have originated in only a fraction of dicotyledon taxa (but in two genera of Gnetales, *Gnetum* and *Welwitschia*). What selective value do cambial variants have, how and why have they originated?

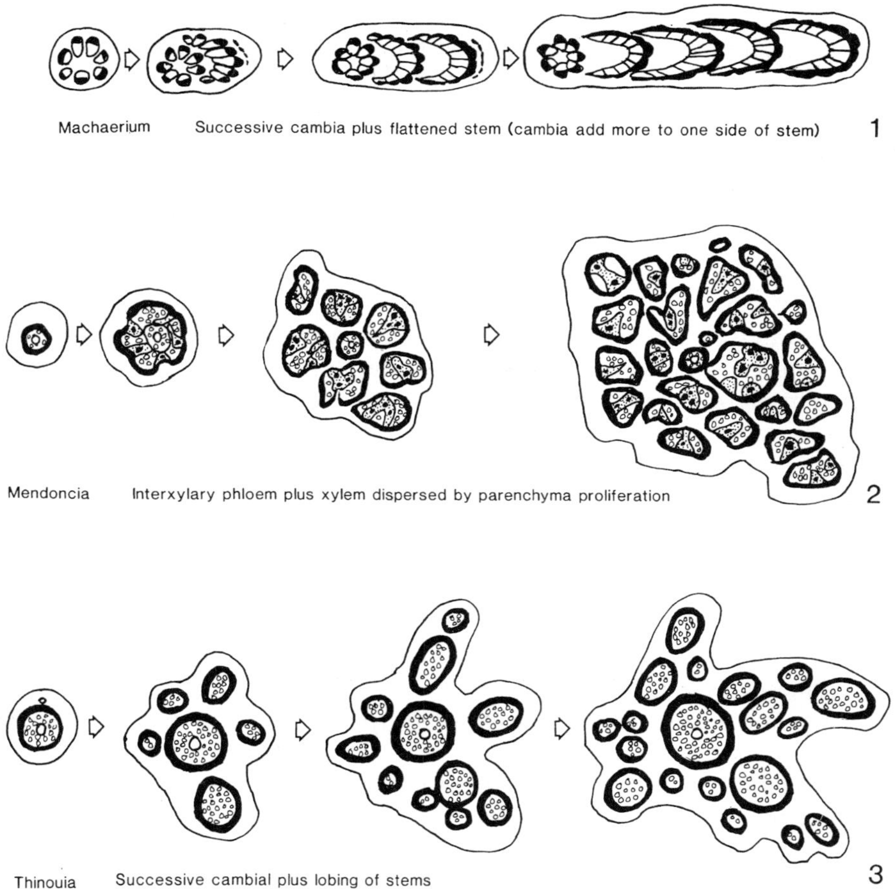

Fig. 8.8. Stems of lianas that show a combination of cambial variants. *Machaerium* (Fabaceae) combines flattened stems with successive cambia. *Mendoncia* (Acanthaceae) has interxylary phloem, plus fragmentation of the cambium into segments. *Thinouia* (Sapindaceae) has a kind of successive cambial activity (origin of strands of vascular tissue in cortex) plus lobed stems

In attempting to correlate cambial variants with habit and ecology, one must first note that while cambial variants are particularly conspicuous in lianas and vines, there are a number of instances in which cambial variants occur in plants of entirely different habits. For example, in the beet (*Beta vulgaris*, Chenopodiaceae), the rings of vascular tissue produced by successive cambia and the primary thickening meristem mechanism may represent a successful way of providing innervation for a storage structure, much as vascular strands are scattered throughout other storage organs without successive cambia, such as potatoes or sweet potatoes.

In Onagraceae, all species have intraxylary phloem, but only a scattering of species in about six genera have interxylary phloem. If one examines the systematic and oganographic distribution of the instances of interxylary phloem occurrence in this family, one finds that the species with interxylary phloem have rather sudden flowering (and therefore fruiting) events. This suggests that the interxylary phloem strands may be channels for translocation of photosynthates to aid rapid formation of flowers and fruits (Carlquist 1975b). The parenchyma associated with the phloem strands may also play a part in this regard, since it can be observed to be rich in starch in these taxa (Figs. 7.8, parts 4 and 5, 7.11, part 4).

Strychnos (Fig. 8.5; see enlarged portion) develops cambia in relation to the interxylary phloem strands, so that new secondary phloem is produced in each of the strands (although no new secondary xylem) and older phloem is crushed in each strand (van Veenendal and den Outer 1993). Possibly, this production of more phloem permits the conductive system of older parts of the stem to remain functional. If so, the phloem could function in storing or mobilizing photosynthates.

The function of grooved stems in relation to tree buttresses is obvious in the trees in which those occur. The function of tree buttresses has been argued, although increasingly the relationship between shallowness of tropical soils and the function of buttresses in adding strength to the shallow root system is commonly stressed by authors in ecology.

The function of splitting in desert shrubs is not entirely clear at the present time (Jones 1984), although those discussing this phenomenon tend to view this habit as a way for the water-conductive system of a shrub to be canalized into separate portions.

The function of the flat stems of lianas seems related to their climbing habit, either when appressed to stems of supporting tree trunks (e.g., *Marcgravia*) or leaning on them (e.g., *Heteropteris*). Obaton (1960, pp. 192–193) lists the cambial variant types in relation to habit categories, and one finds that flattened stems are entirely restricted to lianas, and are the only stem type found in epiphytic lianas.

The abundance of parenchyma in stems of lianas and vines has long been noted, and the function of this parenchyma – or conversely, the dispersed nature of xylem and phloem units within the parenchymatous background as well as the dispersed nature of phloem in various cambial variants have attracted various hypotheses. Among those that can be offered are the following:

1. Parenchyma between wood segments provides greater flexibility for lianas and vines, permitting them to adapt to shifts induced by growth in the trees on which they are supported; the parenchyma, furthermore, is held under this hypothesis to permit torsion of the stems with a minimum of damage to the vessels and sieve tubes.
2. Parenchyma abundance in stems of lianas and vines provides sites for starch storage. This starch storage can supply massive flowering and fruiting when the vine or liana reaches canopy status. Additionally, the starch

storage could function the way it does in some trees such as the sugar maple, *Acer saccharum*, in which starch in parenchyma is hydrolyzed into sugars that are transferred into vessels, where the increase in osmotic pressure generates flow in the vessels.
3. Parenchymatization represents an alternative background tissue, present instead of fibers and sclerenchyma because lianas and vines are not self-supporting.
4. Parenchyma permits repair of the vascular system by formation of new cambia that can replace deactivated conducting cells with new functional cells. This would permit lianas that fall to the ground with collapse of their supporting trees to survive and continue functioning. A corollary of this hypothesis is that redundancy of conducting tissue, when it is scattered throughout a stem, provides a degree of safety that permits noninjured xylem and phloem to continue conduction even though some parts of a stem are injured.

The first of the two above hypotheses can be traced to Schenck (1893), but it has been endorsed up to the present (Carlquist 1975a, 1985d). While it may not explain parenchyma presence by itself, and parenchyma may have manifold functions in stems of scandent plants (just as it has in nonscandent plant stems), the pervasiveness of parenchymatization in lianas when one views plants as a whole is persuasive. Individual kinds of structure are also pertinent. The stems of *Ceropegia*, for example, have only a few plates of xylem and are entirely parenchymatous except for vessels and phloem. These stems twist a great deal, but are short-lived (arising annually from tubers in some species), so that an explanation that involves mechanical considerations seems more plausible than one that involves response to injury or redundancy to continue function despite injury. Likewise, those species that have radiating plates of xylem separated by wide rays where cell division is very sluggish are not ideally designed for regeneration of tissue.

The second hypothesis, that of starch storage to support flowering, fruiting, and conductive activities (Carlquist 1985f), is an outgrowth of the ideas of Sauter (1966a,b) and Braun (1970, 1983, 1984). Although these authors originated their ideas of shunting sugar from parenchyma into vessels in terms of conduction in deciduous temperate trees, they may well apply to a wide range of growth forms. In view of the vulnerability of the large vessels of vines, any mechanism that would safeguard the water columns of these large vessels seems of possible selective value. The idea that parenchyma in lianas supports massive flowering and fruiting events should be investigated. Certainly the stem parenchyma represents the only major location where such storage could occur.

The idea that parenchyma presence in lianas merely represents relaxation of selection for self-supporting stem structure featuring more mechanical tissue has not been seriously expressed by authors to my knowledge. However, it is cited here as one possibility that can be entertained if only to be rejected.

Certainly this idea is very difficult to test experimentally, and is the kind of idea that one would view seriously only if other hypotheses seemed lacking in support.

The idea that parenchymatization of stems of lianas and vines represents a site for regeneration was endorsed by Dobbins and Fisher (1986) and Fisher and Ewers (1989). While these authors do demonstrate that parenchyma is capable of regeneration in response to wounding, one wonders whether that action would be effective in traumatic events such as crash of a liana – such a liana probably could not be restored to a reproductive state (again growing to a canopy position) easily even if its conductive system were intact.

Dobbins and Fisher (1986) do, however, mention redundancy of conductive tissue as of value in lianoid and vining stems, and one can readily agree with this idea – indeed, redundancy of conductive tissue is potentially of selective value in a number of situations. The merit in this concept is that if one supposes that evolution proceeds economically, one is likely to hypothesize that devices which prevent cessation of function are of more value than those which would result in regeneration of the same structures following a hiatus in functioning. Scattering of phloem as well as xylem strands throughout a stem might well help achieve redundancy that would permit continuation of function despite injury.

With respect to the dispersion of phloem strands throughout a stem that is produced by several of the cambial variants mentioned above, another hypothesis is conceivable. One notes that by no means all instances of interxylary phloem or successive cambia occur in scandent dicotyledons, although certainly such cambial variants are more common in scandent forms than they are in nonscandent life forms. The three-dimensionalization of phloem in particular is a point to consider here. Both interxylary phloem produced by a single cambium and the strands of phloem in stems with successive cambia provide a remarkable dispersion of phloem throughout a stem. This dispersion does not seem related to improved regeneration capacity. It might be related to redundancy. However, three-dimensionalization of a phloem system potentially offers an ideal mechanism for distributing photosynthates throughout a stem, transferring them to storage sites and permitting optimal rapidity and/or capacity for retrieval of photosynthates. Because products of successive cambia do form a network with each other, dispersion of phloem throughout a stem to enhance storage or retrieval of photosynthates represents an entirely feasible system. One might even ask why this system has not evolved more often. The monocotyledon stem (even those with a primary thickening meristem) does offer this three-dimensionalization of the photosynthate-conductive system, suggesting that this may be a reason why successive cambia and interxylary phloem have been evolved in dicotyledons. Dicotyledons without either of these mechanisms rely on the network of rays and axial parenchyma within stems (and roots) to achieve photosynthate storage and retrieval, and while these systems and their specializations in organization seem logical responses to these functions, certainly the dispersion of phloem throughout a stem represents, topographically at least, potentially an even better scheme. Three-dimensionalization of both phloem and xylem is potentially advantageous if the phloem and xylem throughout a stem continue to function over a period of years, rather than for only a single year. We know that phloem in palms does function for long periods of time, for example, so that monocots do have this condition. Is most of the conductive tissue in a liana stem functional over a long period of time? If so, the three-dimensionalization of xylem and phloem in dicotyledonous liana stems is correlated with longevity of xylem and phloem as it is in monocotyledon stems with long duration.

9 Cambial Ontogeny (Storying, Paedomorphosis, Other Changes)

9.1 Introduction

We cannot fully understand phenomena such as raylessness, cambial variants ("anomalous secondary growth"), and storying by viewing the end products alone. Even if one can characterize these end products with appropriate terminology, degrees of intermediacy in expression (e.g., storying of all cell types or tissues in a wood vs. storying of some) and the interrelationships between these expressions can only be meaningfully interpreted when placed in an ontogenetic context.

Because wood has rigid walls and is laid down sequentially from cambia, it is an archive we can study by means of serial tangential sections. One can trace changes in ray width by this means, for example. Another means for study of certain of these developmental events is study of radial sections. For example, one can compare length of vessel elements at various points along a radial section and see if vessel elements increase or decrease in length as one goes from the pith outward.

Esau (1965) offers a useful terminological distinction between divisions in the cambium that increase the number of cells in the cambium (multiplicative divisions) and divisions of cambial initials that yield xylem to the outside and phloem to the inside (additive divisions).

9.2 Increase in Circumference of the Cambium and Storying

9.2.1 Mechanism of Fusiform Cambial Initial Increase

Increase in number of cells, the result of multiplicative divisions, is the mechanism by which increase in circumference of the cambium is achieved. The way in which these divisions occur is what is considered here.

We will not consider changes in number of ray initials, because that is of minor importance compared to initiation of new rays; new rays are chiefly derived by subdivision of fusiform initials into ray initials, and thus to understand increase in ray number as well as in number of fusiform cambial initials, we must understand how the number of fusiform cambial initials can be increased in a particular stem or root.

There are some kinds of multiplicative divisions in fusiform cambial initials that do not result in increase in circumference of the cambium. A transverse division of a fusiform cambial initial not followed by vertical displacement of the cells does not increase the circumference of the cambium. Such a division merely results in a shortening of the length of fusiform cambial initials. Such divisions do occur, and explain the shortening over time of xylem cells, such as the shortening of vessel elements observed in species showing paedomorphosis (*Talinum guadalupense*; Fig. 9.2, part 1). Divisions such as these are commonly transverse. They occur in procambium. If they did not occur, late metaxylem tracheary elements would be as long as protoxylem tracheary elements, because metaxylem procambial cells would elongate to exactly the same degree as protoxylem tracheary elements stretch. As the graphs of *Eriobotrya japonica* (Fig. 9.2, part 1) and *Liriodendron tulipifera* (Fig. 9.2, part 2) show, transverse divisions that shorten the length of tracheary elements do occur, and this process also occurs in the vascular cambium.

Some divisions of fusiform cambial initials convert a fusiform cambial initial to a strand of ray initials. These divisions can generally be detected easily because more than two cells usually result from this process. As there are multiplicative divisions of fusiform cambial initials, conversion of fusiform cambial initials to ray initials has the effect of maintaining a balance between number of fusiform cambial initials and number of ray initials (Braun 1955).

The divisions of fusiform cambial initials that lead to increase in cambial circumference range from nearly transverse to oblique and, in few cases, there are lateral divisions cutting off arclike portions of fusiform cambial initials (Cheadle and Esau 1964). The more nearly transverse divisions are more common in woody dicotyledons than the more highly oblique and lateral divisions. Esau (1977) applies the term "pseudotransverse" to those apparently transverse fusiform cambial initial divisions that lead to increase in circumference of the cambium, whereas "transverse" is applied to divisions that subdivide the procambium in the primary body.

Bailey (1923) clarified the way in which all vascular plants with cambia increase circumference by multiplicative divisions. Prior to Bailey's paper, increase in cambial circumference was thought due to radial longitudinal divisions. Radial longitudinal (vertical) divisions do account for multiplicative divisions in storied cambia. However, multiplicative divisions are pseudotransverse divisions in woods with nonstoried cambia. If the cells resulting from pseudotransverse divisions remained in place, no increase in cambial circumference would occur. However, each of the daughter cells of such divisions undergoes intrusive growth. Each of the daughter cells achieves approximately the length of the parent cell by "growing past each other." As seen in a tangential section, two cells, if not side by side, do tend to occupy about twice the lateral space the parent cell occupied.

9.2.2 Manifestations of Storied Structure

At the outset of secondary growth, the cambium in a species that characteristically has storied wood begins in a nonstoried condition. The more numerous the multiplicative divisions, the more rapidly the cambium becomes storied in such a species(Fig. 9.1). If the multiplicative divisions occur slowly, as in succulents and rosette shrubs, the onset of storied structure is slower.

Increase in ray number is held to occur in two ways: (a) breakup of a ray into two portions by invasion of fusiform cambial initials; and (b) subdivision of a fusiform cambial initial into a strand of ray initials (Barghoorn 1941a; Cumbie 1983). If rays originated primarily by (a), there would be few storied rays in a species in which fusiform cambial initials are typically storied. For example, in *Macropiper*, fusiform cambial initials and therefore vessels and libriform fibers are storied, whereas rays are not storied (Patel and Bowles 1980). In citing this example, one must note that the observations by Patel and Bowles are correct, but they are incorrect in regarding *Macropiper* as an example of active ray breakup – exactly the reverse is true. As their illustrations or any preparation of a piperaceous stem show, *Macropiper* shows minimal ray breakup (and thus does demonstrate paedomorphosis, contrary to Patel and Bowles). A situation similar to that in *Macropiper* is seen in *Begonia* (Fig. 6.9, part 3), a genus with growth habit and wood plan similar to that of *Macropiper* (Carlquist 1985b).

If rays form abundantly by subdivision of fusiform cambial initials – alternative (b) above – storied rays are abundant (Fig. 9.1, part 4). Abundant storied rays occur in such families as Fabaceae (Cozzo and Christiani 1950) or Zygophyllaceae (Cozzo 1948; see listing in Sect. 9.2.4 below). If uniseriate rays are abundant, and represent recent derivations from fusiform cambial initials, storied rays would be abundant in a species with a storied cambium. If only multiseriate rays occur in a species with a storied cambium, rays are unlikely to be storied (Fig. 9.1, parts 1–3) – the rays represent events of derivation from fusiform cambial initials that are too infrequent.

Most families in which storying occurs do not have storied rays (Fig. 9.1, parts 1–3). Examples of this are easy to cite, as the listing below in Section 9.2.4 shows. In some of these examples, one can see that vessels and axial parenchyma are storied, whereas libriform fibers are not (e.g., Fig. 9.1, parts 1 and 2). In a species with storied fusiform cambial initials that has nonstoried fibers, one can demonstrate that the libriform fibers are relatively long, and undergo much elongation after they are derived from fusiform cambial initials. Vessel elements and axial parenchyma strands do not elongate very much after derivation from the fusiform cambial initials that generate them, so vessels and axial parenchyma show storying much more than libriform fibers do. One should not conclude, if one sees nonstoried libriform fibers (or fiber-tracheids), that a given species does not have storied fusiform cambial initials. One must sometimes look carefully, therefore, to see if storying is present (and,

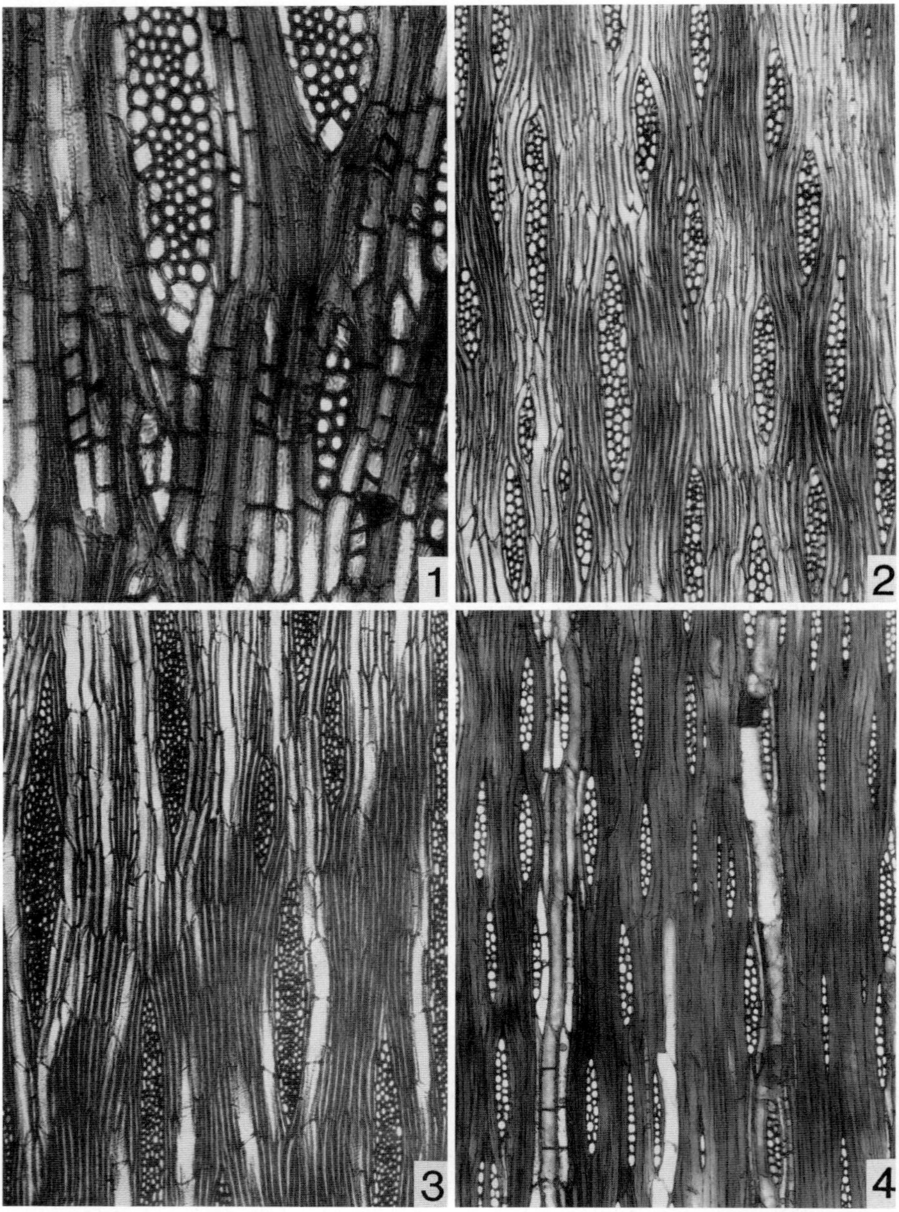

Fig. 9.1. Types of storying, as shown by tangential sections. *1* Axial parenchyma strands storied, *Rhopalocarpus lucidus* (Sphaerosepalaceae); *2* vessels (including narrow vessels that look much like libriform fibers) storied, libriform fibers not storied to as great a degree, *Olearia nummularifolia* (Asteraceae); *3* vessels, libriform fibers, axial parenchyma (in strands of two cells) storied, *Cassinia longifolia* (Asteraceae); *4* rays show storied pattern (vessels and axial parenchyma strands also conform to storied pattern), *Brachylaena merana* (Asteraceae). (*1:B*; *2–4:A*)

although it should not be necessary to say so, one can search for storying only in tangential sections, not radial sections). Species with storied cambia and very short libriform fibers (which thereby elongate relatively little during maturation) often show storying in the libriform fibers (Figs. 6.11, part 3, 9.1, part 3).

The storied pattern in a dicotyledon wood may be relatively stable, once it is achieved. However, Wloch and Zagorska-Marek (1982) find that the storied cambium may be somewhat more dynamic. Hejnowicz and Zagorska-Marek (1974) showed that by displacement of cell tips, a storied pattern may be obliterated.

9.2.3 Evolutionary Significance of Storied Structure

Bailey (1923) claims that storied cambia occur in species with shorter fusiform cambial initials, and that since phyletic advance seems to be related to shortening of fusiform cambial initials (Bailey and Tupper 1918), specialized dicotyledons tend to have storied cambia more commonly than dicotyledons with more numerous primitive features. Although there is no quantitative documentation of this, the listing of Section 9.2.4 will tend to confirm this for those familiar with the phylogenetic status of the various families listed there. As with all other phylogenetic statements, there are exceptions because some families with numerous primitive features will tend to have a few specialized ones (e.g., Lauraceae).

The reason for this phyletic tendency must be somewhat speculative at present. Perhaps formation of a vertical (longitudinal radial) wall forming a multiplicative division is easy to achieve, for physical reasons, in shorter fusiform cambial initials, whereas the same cytokinetic processes in a long fusiform cambial initial would be difficult: the insertion of a cell plate into a narrow cell tip may not be readily achieved. A possible piece of evidence for this idea is found in the data of Cheadle and Esau (1964). In *Liriodendron*, multiplicative divisions (in a nonstoried cambium) are most commonly near-transverse (50.8%), less commonly oblique (39.8%), and rarely lateral – longitudinal but not tip-to-tip (9.6%). Perfect tip-to-tip longitudinal multiplicative divisions were not reported by Cheadle and Esau (1964) in *Liriodendron*, which has relatively long fusiform cambial initials. The fact that strands of axial parenchyma in dicotyledons at large are subdivided by transverse, rarely oblique, walls might be evidence that transverse divisions are more readily achieved than oblique or vertical divisions in vertically elongate cells. *Liriodendron* has long, narrow fusiform cambial initials compared to the short, wide fusiform cambial initials of species with storied cambia.

Intrusive growth of derivatives from fusiform cambial initials takes place readily as the derivatives mature into imperforate tracheary elements. In many families of dicotyledons the imperforate tracheary elements in secondary xylem are several times as long as the fusiform cambial initials from which they are derived. This circumstance suggests that intrusive growth is not at all difficult, so that intrusive growth within the cambium by the fusiform initials represents a more feasible way of achieving increase in cambial circumference than one might have thought intuitively.

Imperforate tracheary elements are longer than vessel elements in storied woods, and they continue to become longer as a stem grows (Davidson 1976; Fig. 9.2, part 2) even though the vessel elements stay the same. One can hypothesize a selective value for longer fibriform elements for mechanical reasons (Wardrop 1951; Wellwood 1962). Therefore, the intrusive growth method of increasing cambial circumference is not merely understandable, it is a part of the intrusive growth in fusiform cambial initials that is continually occurring and which also produces the upswing of the age-on-length curve seen in a typically woody dicotyledon *Liriodendron tulipifera* (Fig. 9.2, part 2). In the graph just cited, vessel element length can be assumed to be the same as fusiform cambial initial length. Note should be taken that fusiform cambial initials do not continue to be longer with increase in stem diameter in *Octomeles*, whereas they do in *Liriodendron*, indicating that intrusiveness of fusiform cambial initials is active in nonstoried cambia, inactive in storied cambia (the length of vessel elements is an accurate indication of length of fusiform cambial initials). Intrusiveness of the imperforate tracheary elements (after they are derived from fusiform cambial initials) occurs in both genera.

In typically woody dicotyledons, transverse divisions that produce shorter fusiform cambial initials are virtually nonexistent (these divisions occur in instances of paedomorphosis, however). Truly transverse divisions (other than those that convert a fusiform cambial initial to a strand of ray initials) would disturb or even prevent a storied pattern. Transverse divisions do occur in some dicotyledons with paedomorphosis, i.e., those with storied cambia and paedomorphosis (e.g., *Macropiper*; Fig. 9.2, part 1; *Begonia*), but in those species, very few transverse divisions occur: the age-on-length curve in *Macropiper* attains stability very quickly and the drop is very slight.

As mentioned earlier, storied cambia have been reported in dicotyledons only, and in very few species that are primitive; storying is unknown in conifers, even those with short tracheids. This correlation appeals as significant. There is a presumptive constraint on shortening of conifer tracheids in that conductive area of the overlap area is reduced when length is reduced. Storying would minimize end wall area of tracheids. If one compares a list of woods with tracheids (Sect. 4.4.8) and woods with storying (Sect. 9.2.4), one finds very little overlap – Lardizabalaceae are the only exception, and in that family, storying is moderate (Carlquist 1985f), and the wide vessels provide such conductive capacity that the conductive significance of tracheid end walls is minimized.

Additional aspects of storied cambia are beginning to be described and are not fully understood. For example, in *Tilia*, the two ends of fusiform cambial initials elongate differentially, and a story with active cell ends tends to

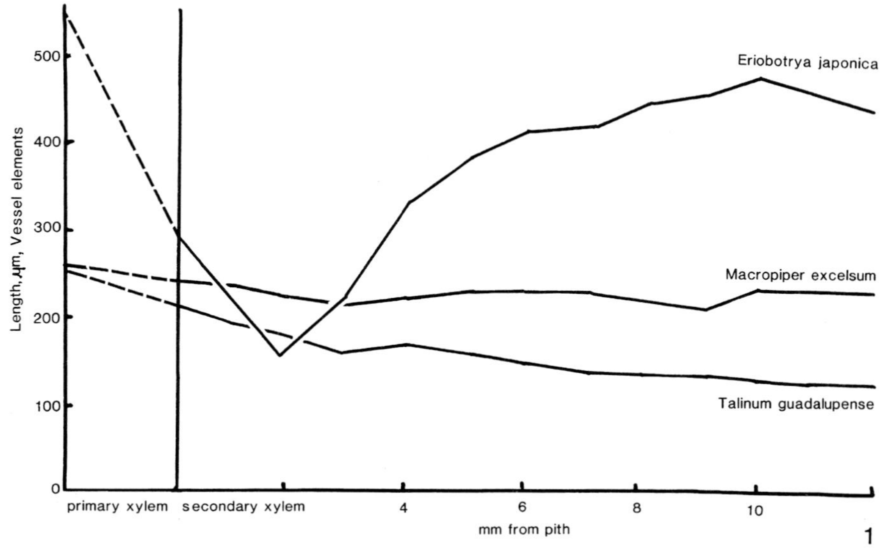

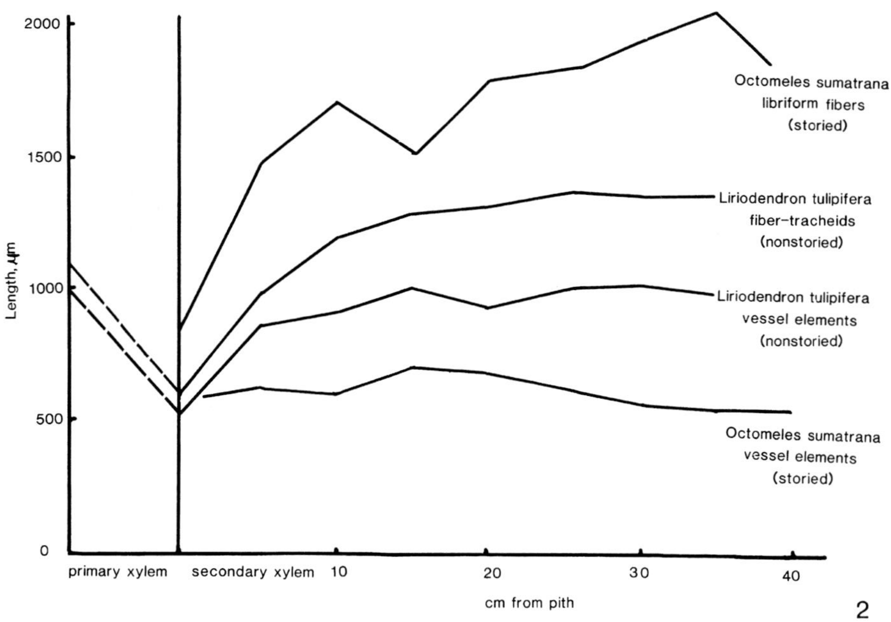

Fig. 9.2. Age-on-length curves for vessel elements (*1*), vessel elements plus imperforate tracheary elements (*2*) of dicotyledon species. *1* Curve that compares a woody dicotyledon (*Eriobotrya japonica*), a succulent with a paedomorphic nonstoried cambium (*Talinum guadalupense*), and a somewhat succulent shrub with a storied paedomorphic cambium (*Macropiper excelsum*); *2* comparison between a tree with a storied cambium (*Octomeles sumatrana*) and a tree with a nonstoried cambium (*Liriodendron tulipifera*). (*1* from Carlquist 1962a; *2* from Bailey and Tupper 1918 and Davidson 1976)

alternate with a story with inactive cell tips; over time, a given cambial cell changes in growth activity (Zagorska-Marek 1975). Loss of fusiform cambial initials from a storied cambium has been demonstrated by Zagorska-Marek (1984). Zagorska-Marek recognizes two orientations of fusiform cambial initial tips, *S* (pointing to the left) and *Z* (to the right), which are not at random and can change, Thus, a storied cambium is not static, although its general appearance may lead us to think that it is.

9.2.4 Systematic Distribution of Storying

As noted above, one must take into account various degrees of storying. Fortunately, most authors who have dealt with storying have recorded pertinent data carefully. In the list below, I have recorded vessel elements and axial parenchyma as storied more commonly than do most authors dealing with this topic. Because vessel elements and axial parenchyma elongate little compared with the fusiform cambial initials from which they are derived, they will conform to the storied pattern very closely. If authors have not noted storying of vessel elements and axial parenchyma, it is perhaps because these are not so conspicuous. More attention has been paid to imperforate tracheary elements and rays when they are storied. However, I think the interests of accuracy are served by taking into account vessels and axial parenchyma, and the listing below has been developed with that convention.

The listing of Record (1919) is the contribution on which all authors who have listed families with storying have built. I have cited literature for only a small portion of the families and genera listed here. The largest number of genera listed are in the family Asteraceae (for listing of papers on wood of Asteraceae, see Carlquist 1966a), and in the family Fabaceae (see Cozzo and Christiani 1950).

Abbreviations are: I = imperforate tracheary elements – mostly libriform fibers in the families below; P = axial parenchyma; R = rays; V = vessel elements. Unless otherwise indicated, these letter codes apply to all genera in a series within a family.

Adoxaceae: *Sambucus* VP (Carlquist and Hoekman 1985b)
Aizoaceae: *Russchia*, *Stayneria* VP (new report)
Amaranthaceae: *Charpentiera*, *Iresine* VPIR (Boureau 1957)
Annonaceae: *Annona*, *Asimina*, *Cathocalyx*, *Goniothalamus*, *Heteropetalum*, *Hexalobus*, *Miliusa*, *Monodora*, *Popowia*, *Rollinia* VP
Asclepiadaceae: *Leptadenia*, *Periploca* VPIR
Asteraceae: *Bigelowia*, *Borrichia*, *Gutierrezia*, *Lepidospartum*, *Stephanomeria*, *Tetradymia* VP; *Argyroxiphium*, *Artemisia*, *Baccharis*, *Bidens*, *Brachyglottis*, *Brickellia*, *Cassinia* (Fig. 9.1, part 3), *Centaurodendron*, *Clibadium*, *Cyclolepis*, *Darwiniothamnus* (Carlquist and Eckhart 1982),

Dendrocacalia (Carlquist 1983d), *Dendroseris*, *Dubautia*, *Eastwoodia*, *Encelia*, *Eriocephalus*, *Espeletia*, *Eupatorium*, *Fitchia*, *Flourensia*, *Gochnatia*, *Gynoxys*, *Haplopappus*, *Hymenoclea*, *Lasthenia*, *Lecocarpus* (Carlquist and Eckhart 1982), *Liabum*, *Loricaria*, *Macraea* (Carlquist and Eckhart 1982), *Mikania*, *Neurolaena*, *Olearia*, *Oparanthus*, *Piqueria*, *Proustia*, *Rhetinodendron*, *Robinsonia*, *Santolina*, *Scalesia* (Carlquist 1982e), *Senecio Sonchus*, *Stephanomeria*, *Symphyochaeta*, *Tetramoloipium*, *Verbesina*, *Vernonia*, *Wilkesia* VPI; *Brachylaena* (part), *Cassinia* (part), *Gochnatia* (part), *Hecastocleis*, *Senecio* (part), *Tarchonanthus* (part) VPIR

Basellaceae: *Anredera* VPI (Carlquist 1999e)

Berberidaceae: *Berberis*, *Mahonia* VPI (Carlquist 1995b)

Bignoniaceae: *Adenocalymma*, *Arrabidaea* VP; *Campsis*, *Enallagma*, *Martinella*, *Paratecoma* VPI; *Cotema*, *Couralia*, *Godmania*, *Tabebuia*, *Tecoma* VPIR

Bixaceae: *Bixa* VPIR (uniseriate rays storied)

Bataceae: *Batis* VPI (Carlquist 1978b)

Begoniaceae: *Begonia* (Carlquist 1985b; Fig. 6.9, part 3)

Bombacaceae: Durioneae except *Camptostemon* VPI; others VP (smaller rays storied in most Bombacaceae except Durioneae)

Boraginaceae: *Cordia* VPI (some species); *Tournefortia* VPIR

Brassicaceae: *Crambe*, *Descurainia*, *Sinapidendron* VPI (Carlquist 1971; see illustrations in that paper)

Capparaceae: *Atamisquea*, *Capparis*, *Crataeva*, *Isomeris*, *Maerua* VPI

Chenopodiaceae: all species with secondary growth VP

Cneoraceae: *Cneorum* VP (Carlquist 1987g)

Cochlospermaceae: *Cochlospermum* VPI (some species; Keating 1968)

Coriariaceae: *Coriaria* VP, VPI (Carlquist 1985e)

Corynocarpaceae: *Corynocarpus* VP

Datiscaceae: *Octomeles*, *Tetrameles* VPI; *Datisca* nonstoried (Davidson 1976)

Ebenaceae: *Diospyros* VPIR (some species)

Elaeagnaceae: *Elaeagnus*, *Shepherdia* VPI; *Hippophae* VPIR

Elaeocarpaceae: *Muntingia* VPI

Fabaceae: *Amburana*, *Anarthrophyllum*, *Ateleia*, *Calpurnia*, *Caragana*, *Coronilla*, *Cytisus*, *Dalea*, *Desmodium*, *Dioclea*, *Erythrina*, *Glycyrhiza*, *Herminiera*, *Laburnum*, *Lotus*, *Medicago*, *Mezoneurum*, *Olneya*, *Poissonia*, *Psoralera*, *Ramorinoa*, *Robinia*, *Sarothamnus*, *Sesbania*, *Sophora*, *Styphnolobium*, *Sutherlandia*, *Wisteria*, *Zuccagnia* VPI; *Afrormosia*, *Albizzia*, *Aeschynomene*, *Andira*, *Arthrocarpum*, *Baphia*, *Bauhinia*, *Belairia*, *Bergeronia*, *Brya*, *Caesalpinia*, *Cassia*, *Castanospermum*, *Centrolobium*, *Cercis*, *Coronilla*, *Coumarouna*, *Dalbergia*, *Dialium*, *Diphysa*, *Enterolobium*, *Ferreirea*, *Genista*, *Gourliea*, *Holocalyx*, *Ichthyomethia*, *Indigofera*, *Inocarpus*, *Koompassia*, *Machaerium*, *Milletia*, *Myrocarpus*, *Myrospermum*, *Neorautenia*, *Ormosia*, *Ostryoderris*, *Ougeinia*, *Parkia*, *Platymiscium*, *Platypodium*, *Platysepalum*, *Poecilanthe*, *Poissonia*,

Pongamia, *Psoralea*, *Pterocarpus*, *Pterogyne*, *Ramorinoa*, *Tamarindus*, *Tipuana*, *Wallaceodendron* VPIR
Frankeniaceae: *Frankenia* VPI (new report)
Geraniaceae: *Viviania* VPI (Carlquist 1985f)
Gesneriaceae: *Cyrtandra* (some species) VPI (Carlquist and Hoekman 1986b)
Gyrostemonaceae: *Codonocarpus*, *Gyrostemon*, *Tersonia* VPI (Carlquist 1978b)
Hippocastanaceae: *Aesculus* VP, VPR
Lactoridaceae: *Lactoris* VPI, plus cells at margin of rays (Carlquist 1990b)
Lardizabalaceae: *Bouquila* VP; *Akebia*, *Holboellia*, *Lardizabala*, *Sinofranchetia* VPI (but storying not pronounced in any genus; Carlquist 1984f)
Lauraceae: *Cryptocarya* VPI (some species), *Silvia* VPI
Loasaceae: *Fuertesia*, *Plakothira* VI (Carlquist 1987d)
Loranthaceae: *Amyema*, *Psittacanthus* VPI
Malvaceae: VP in virtually all; VPR in many (e.g., *Hibiscus* p.p., *Thespesia*)
Meliaceae: *Carapa*, *Cedrela*, *Chickrassia*, *Entendophragma*, *Khaya*, *Pteroxylon*, *Swietenia*, *Xylocarpus* VPIR
Melianthaceae: *Melianthus* VPIR
Misodendraceae: *Misodendron* VPI, VPIR (Carlquist 1983c)
Moraceae: *Ficus* VPI, VPIR; *Paratrophis* VPI (Patel and Bowles 1978b)
Moringaceae: *Moringa* VPIR
Myoporaceae: *Bontia* VP; *Eremophila* (some), *Myoporum* (some) VPI; *Eremophila* (some), *Myoporum* (some) VPIR (Carlquist and Hoekman 1986b)
Nothofagaceae: *Nothofagus antarctica* VR (new report)
Nyctaginaceae: *Bougainvillea*, *Pisonia* VP
Olacaceae: *Schoepfia* VP
Pedaliaceae: *Sesamothamnus rivae* VPI (Carlquist 1987b); *Uncarina leandrii* VPI (slightly; den Outer and van Veenendaal 1983)
Piperaceae: *Macropiper* VPI (Patel and Bowles 1980); *Piper* (some species) VPI
Plantaginaceae: *Plantago* VI (woodier species; Carlquist 1970b)
Plumbaginaceae: *Ceratostigma* VI; *Limonium* VPI (Carlquist and Boggs 1996)
Polemoniaceae: *Loeselia mexicana* VPI (Carlquist et al. 1984)
Polygonaceae: *Calligonum* VP
Ranunculaceae: *Clematis* VP (Sieber and Kucera 1980)
Rutaceae: *Chloroxylon* VPIR
Salvadoraceae: *Azima* VPI, VPIR (den Outer and van Veenendaal 1981)
Sapindaceae: *Sapindus* VPI; *Diplokeleba* VPIR
Scrophulariaceae: *Penstemon* (some species) VPI (Michener 1983)
Simaroubaceae: *Castela*, *Picrasma* VPI; *Kirkia*, *Picraena*, *Simarouba* VPIR
Sphaerosepalaceae: *Rhopalocarpus* VP (Fig. 9.1, part 1; Huard 1965); *Diegodendron* VPIR (Dickison 1988)

Sterculiaceae: *Cola* (part), *Guazuma*, *Pterocymbium*, *Pterospermum* (part), *Sterculia*, *Tarrietia*, *Triplochiton* VPI; *Cola* (part), *Kleinhovia*, *Melochia*, *Pterospermum* (part) VPIR
Stylidiaceae: *Stylidium* VI (Carlquist 1981f)
Stylobasiaceae: *Stylobasium* VPI (Carlquist 1978b)
Surianaceae: *Suriana* VPIR
Tamaricaceae: *Myricaria*, *Tamarix* VPI
Tiliaceae: *Columbia*, *Grewia*, *Heliocarpus*, *Schoutenia*, *Tilia* VPI; *Christiania*, *Nesogordonia*, *Pentace* VPIR
Ulmaceae: *Phyllostylon*, *Zelkova* (same) VP; *Ampelocera*, *Holoptelea* VPIR (small rays only in *Ampelocera*), "more common in temperate genera" (Zhong et al. 1992)
Urticaceae: *Dendrocnide*, *Touchardia*, *Urera* VPI (Bonsen and ter Welle 1984)
Verbenaceae: *Congea* VP; *Citharexylum*, *Gmelina* VPIR
Violaceae: *Viola* (Carlquist 1974; Fig. 6.11, part 3)
Vitaceae: *Parthenocissus* VP (small rays storied); *Tetrastigma* VPI (Metcalfe and Chalk 1950)
Zygophyllaceae: most genera VPIR (Cozzo 1948); VP in *Balanites*

While obviously storying has originated in a number of phylads of dicotyledons independently, presence of storying does tend to characterize certain groups of families commonly thought to be closely related. The families of Malvales (Bombacaceae, Malvaceae, Sterculiaceae, Tiliaceae) can be cited in this regard. Several families of Berberidales (Berberidaceae, Lardizabalaceae, Ranunculaceae) share a storied condition, and this character adds to evidence for placement of Datiscaceae close to Begoniaceae.

9.3 Ray Ontogeny

Ray ontogeny was discussed in relation to ray types (Sect. 6.3), because ray types cannot be understood or classified without reference to ontogeny. Ray ontogeny is equally pertinent in terms of discussions of paedomorphosis in dicotyledon wood, and relevant material is included in the following section.

9.4 Paedomorphosis

9.4.1 Introduction

The dicotyledonous woods sampled by most wood anatomists have been, until recently, woods of tree species. Systematic wood anatomists have begun in recent decades to investigate woods of families that are shrubby to herbaceous or lianoid. As these families are studied, a new picture begins to emerge. The less woody growth forms tend to have wood that can be called juvenilistic. By "juvenilistic" I mean that secondary xylem contains features found more often in primary xylem (particularly metaxylem) than in secondary xylem (as based on secondary xylem of woody species). These juvenile characteristics appear in such seemingly unrelated features as pitting on vessels, age-on-length curve for vessel elements and imperforate tracheary elements (Fig. 9.2, part 1), ray cell histology, and raylessness. The co-occurrence of these features was integrated into a theory of paedomorphosis (Carlquist 1962a). The term "paedomorphosis" is, in this instance, more accurate than the term "neoteny", although the latter has been applied to most situations of juvenilism in plants regardless of their mode of origin.

Barghoorn's (1941a) work on ray ontogeny gives hints about this phenomenon in the instances of *Bocconia frutescens* and *Sesamum radiatum*. These two species have multiseriate rays only; although these species have a few procumbent cells in the rays, most of their ray cells are upright. Thus, as Barghoorn notes, they do not fall into any of Kribs' (1935) ray categories. Kribs (1935) found that elimination of uniseriate rays in certain phylads was concomitant with or preceded by elimination of heterogeneity (i.e., production only of procumbent cells). Rays such as those in *Bocconia* and *Sesamum* thus do not fit into the Kribs scheme. Yet such rays are not scarce if one investigates the woods of the world dicotyledon flora at large instead of arboreal species.

9.4.2 Paedomorphosis and Age-on-Length Curves

The key to the phenomenon (or one should say phenomena) of paedomorphosis in dicotyledon woods is the characteristic age-on-length curve exhibited by these plants. By itself, it can be used as a criterion for recognition of occurrence of paedomorphosis in dicotyledon woods (Sect. 9.4.1). In ontogenetic terms, what the curve for *Talinum guadalupense* represents is protraction into secondary xylem of the transverse divisions that shorten only metaxylem procambial cells in typically woody dicotyledons. In a typically woody dicotyledon such as *Eriobotrya japonica* (Fig. 9.2, part 1), there is a marked drop in length during the primary xylem, but a sharp uptake in the curve thereafter. In *Eriobotrya japonica*, transverse divisions cease in the transition from primary to secondary xylem. If there were no transverse divisions in primary xylem, metaxylem cells would be the same in length as protoxylem cells, and no descent in the curve would occur. The transverse divisions continue for an indefinite time in the secondary xylem of *Talinum guadalupense* and for a shorter time in the secondary xylem of *Macropiper excelsum*. Presumably at the same time, there is relatively little elongation of fusiform cambial initials in these paedomorphic species, or possibly none at all. In *Eriobotrya japonica*, the uptake in the curve represents not merely cessation of transverse divisions shortening the cells derived from the cambium, but also an intrusive growth in the fusiform cambial initials must be occurring.

Species with paedomorphosis tend to belong to families with specialized characteristics, such as Asteraceae or Campanulaceae. Therefore, they tend to have relatively short vessel elements compared with those of primitive woody families (e.g., Illiciaceae). However, within families such as Asteraceae, the species with paedomorphosis have relatively long vessel elements. The gradual shortening of vessel elements in paedomorphic woods may not have a functional explanation. It may be, rather, merely a byproduct of relaxed selection for mechanical strength. Imperforate tracheary elements that are longer tend to be stronger (Wardrop 1951; Wellwood 1962). In this case, paedomorphosis represents minimal elongation of fusiform cambial initials and therefore minimal selection for mechanical strength. Libriform fibers in woods with paedomorphosis may continue to decrease in length for the life of the plant (Anderson 1972). A number of paedomorphic species are relatively succulent forms, such as rosette trees and rosette shrubs, in which thickness of stem, turgor of parenchyma, and lower stature represent decreased selective value of fibrous tissue as a way to achieve sufficient mechanical strength. Some confirmation of this idea is furnished by the fact that some paedomorphic woods (but by no means all) have thin-walled imperforate tracheary elements or highly parenchymatized wood (e.g., *Carica* or *Brighamia*; Carlquist 1962a). There are situations that do not involve paedomorphosis where selection for mechanical strength (for the purpose of self-support) is related, and where parenchyma abundance increases: lianas, for example (Carlquist 1985d).

9.4.3 Paedomorphosis and Lateral Walls of Vessels

Morphology of vessel elements is one aspect that shows paedomorphic tendencies in those woods that have paedomorphic age-on-length curves. These woods often have a higher proportion of lateral wall pitting that is like what one finds in metaxylem: scalariform pitting, for example. In some of these, such as *Macropiper*, one can trace continuity of this pitting pattern from metaxylem into secondary xylem. Scalariform lateral-wall vessel pitting in a family with such specialized features as Begoniaceae (Carlquist 1985b) would be highly unlikely if one did not take into account the concept of paedomorphosis. In some phylads with paedomorphosis, lateral-wall pitting may be pseudoscalariform rather than truly scalariform. Pseudoscalariform pitting, introduced in Section 3.8.2 and illustrated in Fig. 3.14 (parts 1, 2), is also common in less woody dicotyledons.

Lateral wall pitting of vessels with wider-than-normal pit apertures is also common in woods with paedomorphosis, such as cacti. One notes such large pit apertures in vessels of highly parenchymatized woods; where parenchymatous bands occur in woods, vessels with such pit apertures have been reported in those parenchyma bands (Carlquist and Eckhart 1984). One should probably treat the vessels and vasicentric tracheids with wide-band helical secondary wall patterns as a special case (Sect. 6.8.2). These bands are commonly found in globular cacti (Gibson 1973), and have a morphology adapted to pronounced expansion and contraction of the plant body. Note should be taken that some walls of vessels are so large that they appear annular at first glance – the interconnections between bands at the corners of cells may be inconspicuous or, in places, nonexistent (Fig. 3.15, parts 1–4). This phenomenon has been reviewed by Carlquist and Eckhart (1984).

Both the shape of lateral-wall vessel pits and the wide apertures of pits on lateral walls of vessels in woods exhibiting paedomorphosis are indicative of minimal wall-strength configuration. Circular pits with narrow pit apertures have maximal wall strength configuration. Thus the vessel wall pitting and the lowered quantity of fibrous cells and the nature of age-on-length curves in woods exemplifying paedomorphosis seem indicative of lowered selective value for mechanical strength.

9.4.4 Perforation Plates

Paedomorphic woods characteristically occur, with few exceptions, in relatively specialized groups of dicotyledons in which the wood as well as flowers tend to have numerous advanced features. All of the genera cited as having paedomorphic woods have libriform fibers or fiber-tracheids (and mostly libriform fibers), for example; none have tracheids. By far the majority of dicotyledons that show paedomorphosis have simple perforation plates. However, one can find scalariform perforation plates in a few paedomorphic woods. In these particular woods, scalariform perforation plates tend to occur mixed with simple perforation plates, and in the primary xylem, perforation plates are characteristically scalariform. This is a clear illustration of Bailey's (1944b) refugium concept. Bierhorst and Zamora (1965) have documented this concept clearly by listing those families in which perforation plates are simple in secondary xylem but scalariform in primary xylem: the primary xylem is thus a refugium for a primitive feature, the scalariform perforation plate. In a *few* genera that characteristically have scalariform perforation plates in primary xylem and simple perforation plates in secondary xylem, a scattering of scalariform perforation plates occurs in secondary xylem: apparently an extension of the primitive feature into secondary xylem by means of paedomorphosis (this demonstrates the difference between the refugium concept and the paedomorphosis concept). As examples of this paedomorphic presence of some scalariform perforation plates in secondary xylem one can cite *Pentaphragma* of the Pentaphragmataceae (Carlquist 1975a, 1997a); *Canarina canariensis*, *Musschia wollastonii*, and *Platycodon grandiflorum* of the campanuloid Campanulaceae (Shulkina and Zhikov 1980), *Dendroseris* spp.

(Carlquist 1960b) and *Crepidiastrum* (Carlquist 1983d) of the Asteraceae, *Patrinia* of the Valerianaceae (Fig. 3.11, parts 1–5; Carlquist 1983e), and *Mimulus* of the Scrophulariaceae (Michener 1983). Note should be taken that even in these groups, the majority of vessel elements in secondary xylem have simple perforation plates. These facts and other phylogenetic data counter the idea of Shulkina and Zhikov (1980) that the scalariform perforation plates in secondary xylem are evidence for a primitive or relictual phylogenetic position for Campanulaceae. The vast bulk of Campanulaceae have simple perforation plates (Carlquist 1969a), and occurrence of scalariform perforation plates in this group is an indication of paedomorphosis in a rather specialized group (that does retain some moderately primitive features in primary xylem – but that combination is not at all rare, as the data of Bierhorst and Zamora show).

9.4.5 Rays in Paedomorphic Woods

Ray cells are just as sensitive an indicator of paedomorphosis as are the age-on-length curves of paedomorphic woods. If transverse divisions shortening fusiform cambial initials occur at a low rate and are protracted into the secondary xylem in paedomorphic woods, transverse division of ray initials might be expected to follow similar patterns. It does. This was signaled in my paper on paedomorphosis (Carlquist 1962a), and can be demonstrated by numerous groups in which ray cells are upright to square exclusively or with only a few procumbent cells: Asteraceae (Fig. 6.8, part 2; Carlquist 1966a), Begoniaceae (Fig. 6.9, part 3; Carlquist 1985b), Campanulaceae (Figs. 6.8, part 1, 6.9, parts 3 and 4), Goodeniaceae (Carlquist 1969b), Lactoridaceae (Carlquist 1990b), Loasaceae (*Plakothira*; Carlquist 1987d), Lythraceae (Baas and Zweypfenning 1979), Papaveraceae (Barhgoorn 1941a; Cumbie 1983), Pedaliaceae (Barghoorn 1941a; Carlquist 1987b), Piperaceae (Patel and Bowles 1980), Scrophulariaceae (Michener 1983), and Urticaceae (Bonsen and ter Welle 1984).

9.4.6 Raylessness in Paedomorphic Woods

In formal terms, raylessness is a delay (Cumbie 1969) in occurrence of transverse divisions in ray initials, or even an entire absence of such divisions. Thus it can be considered a heightened form of the delay and scarcity of ray initial subdivision that results in the upright ray cell predominance described in the preceding section. The degree of delay of the transverse divisions and their number varies. For example, in *Cyrtandra* (Carlquist and Hoekman 1986a) woods begin rayless but quickly develop rays, whereas other woods remain rayless indefinitely, as in *Plantago* (Carlquist 1970b), *Viola* (Carlquist 1974),

Stylidium (Carlquist 1981f), *Besleria* (Carlquist and Hoekman 1986a), and certain Crassulaceae, such as *Aeonium* and some species of *Kalanchoe* (Fig. 3.2, part 4). The relationship between paedomorphosis and raylessness was noted earlier (Carlquist 1962a). In rayless species, cells in the interfascicular area of the primary body (primary ray areas) are quite vertically elongate, however, more so than in taxa that have predominantly upright ray cells but are not rayless. Thus, at the onset of secondary growth in a rayless species, fusiform cambial initials are of roughly the same length as ray initials. The families that tend to have predominantly upright ray cells tend mostly to be herbaceous, as the listing in the preceding paragraph illustrates. Rayless taxa seem often to exemplify secondary woodiness, as is clearly the case in insular species of *Plantago* (Carlquist 1970b). However, short-lived herbs may also be rayless, as in *Phacelia pedicellata*, an annual (Carlquist and Eckhart 1984). Rayless species appear to be undergoing selection for increased mechanical strength of stems, based on an herbaceous ancestry: substitution of fibers for rays (axial parenchyma is also rare in rayless woods) appears a way in which a plant with short duration of stems can develop mechanically strong tissue; development of parenchyma in short-lived stems is of minimal selective value.

Thus raylessness is not merely an instance of paedomorphosis, but a selection for mechanical strength (involving paedomorphosis) in herbaceous groups.

Barghoorn (1941b) was probably not well advised to cite *Staavia glutinosa*, *Corema conradii*, and *Tetratheca ciliata* as examples of near-raylessness. These are species in which rays are exclusively uniseriate near the pith, but in which juvenilism prevails (ray cells are upright or mostly so in these species), and in which rays remain uniseriate.

9.4.7 Commentary on Paedomorphosis in Dicotyledons

Ideas on paedomorphosis in dicotyledonous woods have been widely accepted (Cumbie 1963, 1967a,b, 1983; Walsh 1975; Bailey 1966; Datta and Deb 1968; Takhtajan 1969; Anderson 1972; Gibson 1973, 1977a,b, 1978a; Koek-Noorman 1976; Baas 1977, 1979a; DeBuhr 1977; Bridgewater and Baas 1978; Baas and Zweypfenning 1979; Michener 1983; Bonsen and ter Welle 1984). Unfortunately, in a few references (e.g., Bierhorst and Zamora 1965; Metcalfe and Chalk 1983, p. 113), the idea of paedomorphosis is incorrectly equated with Bailey's (1944b) refugium hypothesis: the ideas are quite different. Bailey's refugium hypothesis holds that in dicotyledons, primitive features are retained in primary xylem, whereas more specialized features occur in secondary xylem (and presumably specialization began phylogenetically in secondary xylem). The paedomorphosis hypothesis states that in particular woods, features of the primary xylem are ontogenetically protracted into the secondary xylem. Paedomorphosis is not explained by a link between pith diameter and fusiform cambial initial length as proposed by Philipson et al. (1976). Baas (1976) and Koek-Noorman (1976) invalidate the Philipson and Butterfield theory both on the basis of studies of woods and on theoretical grounds.

Patel and Bowles (1980), while not discrediting the theory of paedomorphosis, doubt its applicability to *Macropiper*. Patel and Bowles are incorrect in finding the rays of *Macropiper* as ontogenetically much altered. Compared to the alteration seen in rays of woody dicotyledons, the rays of wood of *Macropiper* show remarkably minimal alteration of the primary ray pattern in that genus: the rays of *Macropiper*, as well as other features in that genus, are clearly paedomorphic, as the figures of Patel and Bowles (1980) show.

Koek-Noorman (1976) claims that *Rubia fruticosa* and *Crucianella maritima* grow under "xeric conditions" and questions whether this is in conflict with my idea that paedomorphic woods tend to occur in relatively uniform climates that do not induce annual cessation of growth in herbaceous dicotyledons. *Rubia fruticosa* grows in dry areas, but the climate is mild (this species occurs at low elevations in the Canaries, where the growing season is uninterrupted by frost). The littoral habitat of *Crucianella maritima* likewise is free from frost. Both of these examples probably represent shrubs evolved from shorter lived ancestors under conditions of mild temperature. This is possible in a dry climate by means of succulence, a feature that *Rubia fruticosa* does have in certain respects.

The idea of Takhtajan (1969) that paedomorphosis is basic to origin of vessels in dicotyledon woods does not appear to be justified. That interpretation was an attempt to integrate the scalariform lateral wall pitting of vessels of woods that show paedomorphosis with the scalariform lateral wall pitting hypothesized as primitive by Frost (1931) in primitive woody dicotyledons and with the scalariform pitting seen in earlywood tracheids of *Tetracentron* and *Trochodendron*. Woods that consisted wholly of scalariformly pitted (both end walls and lateral walls) tracheids would have insufficient mechanical strength in arborescent plants unless the stems also had special fiber zones around bundles, as in tree ferns, or a massive sclerenchymatous cortex, as in the Lepidodendrales (Carlquist 1975a). Woods composed of vessels with scalariform pitting on the lateral walls must also have, together with these vessels, imperforate tracheary elements that compensate for relatively weak vessel walls, just as the latewood tracheids of *Trochodendron* and *Tetracentron* compensate for the mechanically poor earlywood tracheids. The woods I described as instances of paedomorphosis are not only in families with specialized floral features, they occur in plants with growth forms (succulents, rosette trees, annuals, short-lived perennials) that show release from mechanical strength requirements. Such plants would be poor candidates to be ancestors of the angiosperms (see also Soltis et al. 1999). These growth forms are notably lacking in the families with floral features widely regarded as primitive. The wood of dicotyledon families such as Winteraceae, Theaceae, etc. is mechanically strong and does not fall into categories referable to paedomorphosis at all. Conifers and other taxa in which circular bordered pits are present exclusively have no division of labor (except for Gnetales) into vessels and imperforate tracheary elements. In conifers, this pattern has become so pervasive that circular bordered pits are even found on helical elements of the primary xylem (Bierhorst 1960); this is also true in Gnetales (Carlquist 1996d). *Cordaites* did have scalariformly pitted tracheids in metaxylem; this pattern yields to circular bordered pits in secondary xylem, a fact that shows both the selective value of circular pits for improved mechanical strength in arborescent growth forms and the possibility that scalariform pitting may have been widespread in early seed plants, as it probably was in angiosperm ancestors. Rather than being paedomorphic, the wood of vesselless angiosperms shows three solutions, as follows, to the problem of forming a wood strong enough to support a shrubby or arboreal growth form.

1. Division of labor between scalariformly pitted earlywood tracheids and latewood tracheids with circular pits occurs (*Tetracentron*, *Trochodendron*).
2. In a wood with relatively little growth ring activity because of a moderate climate, tracheids form only circular bordered pits as an accommodation to improve mechanical strength (e.g., *Drimys* of the Winteraceae).
3. In a wood with relatively little growth ring activity because of a moderate climate, tracheids form circular bordered pits as an accommodation to improve mechanical strength, but end walls (overlap walls) of tracheids form scalariform pitting, a capability not expressed as an extension from metaxylem as would be true if paedomorphosis were operative: *Bubbia*, *Zygogynum* (Winteraceae), and *Amborella* (Amborellaceae).

All of the above alternatives show selection for optimal mechanical strength, whereas paedomorphic woods do not. In the condition cited for *Bubbia* and *Zygogynum*, ontogenetic studies show that earlier formed secondary xylem has circular pits on tracheid overlap areas, whereas later formed circular xylem bears scalariform pits on overlap areas, presumably concomitant with attainment of canopy status by the tree, which presumably would then experience a requirement for increased pit area on the overlap areas to meet heightened conductive requirements (Carlquist 1983a). This ontogenetic sequence is the reverse of what would be expected if paedomorphosis were operative in this phylad.

One should note, although the reader may regard it as obvious, that "juvenile wood" as used in dendrological literature to denote the wood formed for the first several years of life of a tree, is not to be equated with the concept of paedomorphosis.

9.5 Other Wood Phenomena Related to Cambial Ontogeny

Rumball (1963) studied wood structure with relation to heteroblastism. The New Zealand species he studied are unusual in having marked differences between juvenile and adult foliage. This difference is not related to juvenilism in wood in the sense of paedomorphosis. Rather, in the examples studied by Rumball, a normal (ascending) age-on-length curve is present in both juvenile and adult stems. What is of interest in Rumball's study is that when the transition from juvenile to adult shoots occurs, there is a marked upswing in the age-on-length curve.

Baas et al. (1984) have investigated the effect of dwarfing on wood of tree species. For many of these, age-on-length curves were constructed by these authors. Dwarfed growth proves to have shorter vessel element length, narrower vessels, and narrower growth rings than wood of non-dwarfed specimens of the same respective species.

Fukazawa (1984) has shown differences in density of wood related to age, and has demonstrated these by means of age-on-length curves, permitting definition of what is juvenile wood in woody species. Harzmann (1984) noted shorter libriform fiber length in juvenile wood of *Cedrela odorata*, but this is what one would expect on the basis of age-on-length curves in a typical woody dicotyledon (Fig. 9.2, part 2).

Fiber wall thickness may increase with age or may not (Fukazawa 1984); proportion of different cell types in a wood may change little with age (Fukazawa 1984). There can be sudden shifts from fibrous to nonfibrous zones with age (Mauseth and Plemons 1995) or shifting environmental conditions (Carlquist 1995c). Despite the earlier lack of criteria for juvenile vs. adult woods (Rendle 1958), an understanding does appear to be developing, but perhaps on a species-by-species basis: what is juvenile wood in one species may not have the same features as juvenile wood in another (and, again, the reader should note the distinction between the concept of juvenile wood in the dendrological sense and the concept of paedomorphosis). Age-on-length curves as well as histological studies appear to offer considerable hope for classification of ontogenetic changes in wood.

Vessel diameter is not a function of cambial events per se. However, one may note that change of vessel diameter over time varies with species. Vessel diameter may increase, as in *Bergia* (Carlquist 1984b) or *Begonia* (Carlquist 1985b); it may stay the same, as in *Bocconia* (Cumbie 1983); or it may decrease, as in *Quercus* or *Styrax* (Baas et al. 1984) and many annuals (Hayden and Hayden 1994). One should look for correlations with environmental factors and the effects of size (annuals are confined to quick-drying surface layers of soil, whereas large trees tap deep water tables that are stable).

10 Systematic Application of Wood Data

10.1 Introduction

This book is not concerned directly with wood identification, although ability to recognize the wood features discussed in the preceding chapters is essential for work in wood identification. The present chapter is concerned with how wood characters have been and can be used in taxonomic and phylogenetic studies – and also which wood characters are likely or not likely to be useful, in general. In wood identification, a match of an unknown wood with known woods is usually attempted. In systematic application of wood data, similarities and differences among species, genera, and families are assessed (usually, today, in terms of a cladistic analysis).

One should be especially cautious in interpreting relation between taxa the woods of which have attained similar evolutionary levels. This possibility is a very real one, because so many wood features have evolved similarly in phylads that are not closely related to each other. For example, storied wood structure has been attained in a number of phylads independently. By itself storied wood structure cannot be used as an indicator of relationship between two families, but a cladogram (using a range of phyletically reliable features) might demonstrate that two families acquired storied structure before divergence (a synapomorphy) and thereby a strong case for relationship of the two families is made. Examples of woods that are not closely related but which have similar structural "plans" do exist, and have been cited by Boureau (1957, p. 623). For example, *Maesopsis eminii* (Rhamnaceae) has wood virtually identical in many diagnostic features to that of *Albizia adiantifolia* or *A. gummifera* (Fabaceae). Other pairs of unrelated taxa with similar woods include *Hernandia cordigera* (Hernandiaceae), which resembles *Semecarpus atra* (Anacardiaceae); and *Majidea fosteri* (Sapindaceae), which simulates *Loesenera kalantha* (Fabaceae). In each of these pairs, at least one of the families represented has numerous woody species (Anacardiaceae, Sapindaceae, Fabaceae); the largeness and more particularly the polymorphism of those families explains how such "convergences" could occur. Such convergences are more a concern for wood identification than for systematics, where one already knows the name of the wood studied, but these convergences do indicate the instances of parallel evolution likely to be encountered in woods.

In the first portion of this chapter, examples of achievements (or nonachievements) of comparative wood anatomy in attempting to demonstrate relationships will be presented. In the latter part of the chapter, methods for assessing relationship will be discussed. Finally, methods for presentation of data that express degrees and kinds of relationship will be analyzed.

If wood anatomy has been useful to systematics, the reason is not the self-sufficiency of wood data, but the large number of wood features available. Wood anatomists seem generally well informed about which of these features are the least stable phyletically (e.g., grouped vs. solitary vessels, vessel diameter) and which are strongly indicative of relationship. Note should be taken that most of the studies cited do consider types of data other than wood anatomy, but the examples cited all show a significant if not decisive role played by the wood data.

10.2 Types of Systematic Applications

10.2.1 Placement of Isolated Genera

Particular genera have been cited as discordant within the groups in which they have been included. For example, in reading about the genus *Fitchia* (Asteraceae), I discovered that the genus had originally been assigned to Heliantheae by its author because of presence of two awns per fruit, but Hoffmann (1890) claimed that the corollas are ligulate and therefore the genus should be placed in Lactuceae. Three authors claimed that *Fitchia* should be placed in Mutisieae (see Carlquist 1957b). The corolla of *Fitchia* is only superficially ligulate (lobes lightly adherent to each other), so the key character for Lactuceae is, in fact, not actually present. Comparison of wood as well as other anatomical features of *Fitchia* to wood and other features of other genera representing the three tribes mentioned showed that the preponderance of resemblances was with Heliantheae, and the case of *Fitchia* was resolved very easily (Carlquist 1957b).

Heteropyxis has been considered isolated, and has often been placed in its own family since the genus was described in 1863. Claims have been made for inclusion in, or close relationship to, the families Lythraceae, Myrtaceae, Rhamnaceae, and Rutaceae, respectively. Stern and Brizicky (1958a) demonstrated that *Heteropyxis* belongs in Myrtaceae. Schmid (1980) confirmed the Stern and Brizicky work and also placed another isolated genus, *Psiloxylon*, in Myrtales.

Picrodendron was often considered as constituting a monogeneric family. One should be suspicious about monogeneric families, for while truly isolated genera do exist, some monogeneric families have resulted from stressing one or two characters of a genus that are in conflict with those of the closest family. *Picrodendron* has had a particularly unstable history of systematic placement, and assignments have included Anacardiaceae, Bombacaceae, Euphorbiaceae, Juglandaceae, and Sapindaceae (see Hayden 1977). Wood data were important to Hayden in his resolution: *Picrodendron* belongs to Euphorbiaceae subfamily Oldfieldieae.

Lactoris fernandeziana, sole species of Lactoridaceae, proves to have no fewer than ten features (most of them not widespread in dicotyledons) that occur also in wood of Piperaceae (Carlquist 1990b), showing that *Lactoris* belongs in Piperales rather than Laurales.

Exploration in South America yielded genera that required further work for establishment of systematic position. *Diomma* proved to belong not to Simaroubaceae, but to Rutaceae (Stern and Brizicky 1960). When the genus *Pakaraimaea* was discovered in South America, a group of workers was assembled to help establish the familial placement of this genus; de Zeeuw's (1977) work on wood anatomy helped prove that this genus belongs to Dipterocarpaceae, a family that had not hitherto been found in South America.

A monotypic genus native to South Africa, *Empleuridium*, has been thought to belong to Rutaceae, but data from various disciplines including wood anatomy (Goldblatt et al. 1985) proved the genus to belong to Celastraceae.

Many more examples could be cited. However, each instance investigated has its own profile of evidence, and past achievements in the use of wood in systematics may inspire, but cannot be repeated.

10.2.2 Wood Anatomy Reveals Taxa To Be Isolated

Suriana has been thought to have sapindalean affinities. Gutzwiller (1961) reviews a great deal of material but cannot place the genus more closely than that; consequently, a monogeneric family, Surianaceae, is necessary. Likewise, Dickison and Sweitzer (1970) cannot place *Barbeya* in any existing family. Even if recognized as Barbeyaceae, a close affinity of the genus to such orders as Urticales is not demonstrated by the data at hand. Bridgewater and Baas (1982) find *Xanthophyllum* about equally distant between Polygalaceae and Trigoniaceae. Wood anatomy demonstrates closeness of Coriariaceae with Corynocarpaceae, but Ranunculaceae, Simaroubaceae, and other families have been thought related to Coriariaceae (Carlquist 1985e). Gasson (1999) finds on the basis of wood anatomy that *Taralia* (Fabaceae) must be removed from the tribe Dipterygeae, but its tribal placement is not yet clear.

Flacourtiaceae are a family noted for diversity of its contents. R.B. Miller (1975) has listed some genera as occasionally or frequently assigned to Flacourtiaceae and used data from wood anatomy to decide which genera should be excluded from Flacourtiaceae and where they should be placed. R.B. Miller (1975) decides that *Ancistrothyrsus*, *Barteria*, and *Paropsis* should be transferred to Passifloraceae. *Soyauxia* is excluded but no other family is designated to accept it. *Peridiscus* is segregated as Peridiscaceae (but that family is considered by Miller as close to Flacourtiaceae). *Aphloia* likewise is segregated into a family Miller considers close to Flacourtiaceae and

Neumanniaceae. *Asteropeia* is excluded and recognized in a distinct family, Asteropeiaceae, which is not considered close to Flacourtiaceae. R.B. Miller assigns *Triphyophyllum* to Dioncophyllaceae and returns *Lethodon* to Thymeleaceae. The net effect of Miller's work is to find better positions for many of the genera questionably included in Flacourtiaceae. We note that *Soyauxia* and *Asteropeia* emerge as genera *incertae sedis*, and these offer opportunities for further work. Even though a genus is quite justifiably found to be more isolated than hitherto thought, one can still attempt to locate what its closest relatives are.

Likewise, study of woods of families belonging to Thorne's (1968) concept of Pittosporales revealed that most of the families do have relatively primitive woods, with scalariform perforation plates in vessels and tracheids as the imperforate tracheary element type. However, Bruniaceae (Carlquist 1978a), Geissolomataceae (Carlquist 1975c), and Grubbiaceae (Carlquist 1977a) form a close group that may show affinities to Saxifragales; Byblidaceae (Carlquist 1976b) and Roridulaceae (Carlquist 1976c) do not appear to belong to that grouping. Woods of Balanopaceae (Carlquist 1982f), Buxaceae (Carlquist 1982d), Daphniphyllaceae (Carlquist 1982f), and Myrothamnaceae (Carlquist 1976a) look closer to Hamamelidales or Rosales, although the wood data are not decisive. Rosales and Hamamelidales may prove to be close to each other, in fact, judging from some recent cladograms. Pittosporaceae (Carlquist 1981d) differ from the above families and should be placed near Araliales. Wood data are insufficient to certify the placement of the "pittosporalean" families, and molecular evidence is likely to prove more conclusive – but addition of evidence from wood anatomy to other kinds of data may nevertheless prove helpful to this question.

10.2.3 Constructing Natural Groupings Aided by Wood Anatomy

Some of the examples above suggest that we do have some genuinely isolated phylads the distinctness of which has not been hitherto stressed. However, another possibility is evident: the potential relative may not have been included in the taxa selected for comparison. This latter possibility is particularly troubling, because there is no easy answer. Plant anatomists are often not well versed in the literature on taxonomy and phylogeny. The more one examines this literature, the more manifold the possible relationships of a particular isolated genus may seem. The conclusions of any given paper must be judged in terms of whether a wide range of possibilities is taken into account or not. An example is furnished by the case of Fabaceae. In systems in vogue until recently, Fabaceae was universally considered rosalean, but it proves instead to be sapindalean (Dickison 1981b; Thorne 1983). If one investigated Fabaceae by including only rosalean families, one would probably conclude that Fabaceae was a distinctive offshoot of Rosales, rather than (as it must now be considered) a specialized group of Sapindales. How many "peripheral" or even unlikely groups can be investigated when one is attempting to establish the affinities of an isolated genus or family? When does one decide that the most pertinent potential relatives have been investigated? Until recently, that was a problem not easily resolved by the wood anatomist. However, the situation has been ameliorated by molecular systematics, which has presented a high probability for relating particular families that have not been considered close to each other hitherto, and thus there is a confirming line of evidence previously unavailable. If one finds wood anatomy to be essentially identical between families that have been regarded on macromorphological and molecular evidence (usually plotted in cladistic form), as between Theaceae and Pentaphylacaceae (Carlquist 1984g) or between Malesherbiaceae and Turneraceae or Passifloraceae (Carlquist 1984h), one has an excellent case.

Wood anatomy has been important in definition of an order in the case of Myrtales. Various workers have contributed to this synthesis, which is summarized by van Vliet and Baas (1984). Two anatomical features are of especial value in Myrtales because they are found in only a small number of families of dicotyledons at large: intraxylary phloem and vestured pits. Intraxylary phloem does not occur in secondary xylem, to be sure, but it does seem to form a kind of antecedent to interxylary phloem, which occurs in secondary xylem of a scattering of genera throughout Myrtales, notably Onagraceae (Carlquist 1975b, 1977b, 1983f, 1987i). Vestured pits are of characteristic occurrence throughout Myrtales, although absent in a few genera (e.g., *Mouriri*) or species (Carlquist 1975b; van Vliet 1978; ter Welle and Koek-Noorman 1981). Two isolated genera have been referred to Myrtales: *Psiloxylon* (Schmid 1980) and *Alzatea* (Baas 1979b). The former genus has been segregated as a monogeneric myrtalean family, Psiloxylaceae; the latter has been included in Lythraceae. Myrtales are not exceptionally difficult to define as an order compared to others in angiosperms. The van Vliet and Baas (l984) treatment differs but little from that of Thorne (1976). There are still some families dubiously placed in Myrtales. The occurrence of intraxylary phloem and of vestured pits in Gentianales leads van Vliet and Baas (1984) to hypothesize that Gentianales is the order closest to Myrtales, and these two features also speak for inclusion of Vochysiaceae in Myrtales.

The glucosinolate ("mustard oil")-bearing families were once scattered among a number of orders, but molecular evidence shows that, in fact, this chemical is an excellent indicator of relationship (Rodman et al. 1996), and a radically-revised Capparales has emerged. If one studies the wood anatomy of this new Capparales, one finds hitherto-unappreciated similarities: Akaniaceae and Bretschneideraceae belong to the new Capparales and have wood that is essentially identical (Carlquist 1996b), although these two families were relegated to different orders before the reconstruction of Capparales described in the paper by Rodman et al. (1996).

Comprehensive studies of wood of an entire order are only one approach to phyletic use of wood data. One may, as it were, add particular families to an existing order one family at a time, so that over time a natural grouping is achieved. Thus, Dickison's study of Eucryphiaceae (1978) places that family near Cunoniaceae in Rosales. Strasburgeriaceae is confirmed as a family of Theales near Ochnaceae (Dickison 1981a). Likewise, the monogeneric Oncothecaceae (Carpenter and Dickison 1976) and Paracryphiaceae (Dickison and Baas 1977) are judged to belong to Theales. A recently described family of Central America that appeared to be isolated, Ticodendraceae, has wood and bark that are clearly close to those of coryloid Betulaceae (Carlquist 1992a). Wood anatomy proves critical in delimiting the (Berberidales) Ranunculiflorae (Carlquist 1999a).

10.2.4 Wood Anatomy and Family, Genus, and Species Characters

Wood anatomy can tip the balance as to whether a single family should be recognized or a series of families should be segregated. *Viviania* (Carlquist 1985f) differs from Geraniaceae (sensu stricto) by having bordered pits on imperforate tracheary elements (which are tracheidlike fiber-tracheids), as well as in other wood characters. Some authors have segregated *Viviania* as a monogeneric family, and at first glance wood anatomy might seem to lend support. However, *Dirachma* (sometimes segregated from Geraniaceae as Dirachmaceae), *Biebersteinia* (sometimes segregated from Geraniaceae as Biebersteiniaceae), and *Balbisia* (sometimes segregated from Geraniaceae in Ledocarpaceae, which also includes *Wendtia* and *Rhynchotheca*) must be taken into account. The genera just named form a series of taxa intermediate *to various degrees* between *Viviania* and Geraniaceae sensu stricto (Carlquist 1981f). Because of these degrees of difference, segregation of any of these units as families seems inadvisable. Recognition of these units as infrafamilial categories is still conceivable, however. In Ebenaceae, *Maba* and *Diospyros* have identical wood anatomy, so Record (1928) suggests recognition of only one genus. The genus *Sarcocarpaea* differs from Sapotaceae in its seed type and in having tracheids rather than fiber-tracheids (Marco 1933). However, these differences are apparently the only ones by which *Sarcocarpaea* differs from Sapotaceae, and therefore segregation of *Sarcocarpaea* at a subfamilial rather than a familial level appears the best treatment. Den Outer and van Veenendal (1996) find that woods of *Monopetalanthus* and *Tetraberlinia* are so similar that the genera should be united.

Perhaps the most comprehensive keys to genera using wood features are those of Moll and Janssonius (1909–1936) and Janssonius (1940). Wood features may be relatively uniform throughout a family or may be quite different from species to species. At this latter end of the spectrum, one can cite Janssonius (1950). For example, he said, "In Euphorbiaceae, the woods of *Glochidion* proved to be interspecifically so widely different that it became impossible to subdivide this large genus into smaller groups." Yet in contrast,

Janssonius (1950) finds that "in *Engelhardtia* (Juglandaceae), of which I studied two species and three varieties, I observed a similar variability as that occurring in many large genera." Thus students of wood anatomy should not exclude the possibility of finding an appreciable number of features the variation patterns of which follow species lines. Sometimes the pattern of wood feature distribution within a genus will reveal a few highly distinctive species within a large genus, as for example *Eremophila duttonii* compared with other species of *Eremophila* (Carlquist and Hoekman 1986b).

In a few instances, not merely a few but all or nearly all the species of a genus will show distinctions with respect to wood anatomy. Baas and Zweypfenning (1979) give an excellent example of this in *Lagerstroemia*. Another interesting example is found in the genus *Misodendron*, in which subgenera as well as species can be delineated in a key that uses wood and stem features (Carlquist 1985c):

1 Cambium relatively inactive in ray areas, ray cells basically thin walled and radially stretched; a few tangential divisions occur randomly in ray cells; fibers in fascicular areas in strands, patches, or absent, not forming a homogeneous background; axial parenchyma with thin nonlignified walls, relatively abundant; vessels with laterally much-extended pits, the pitting thereby superficially resembling helical bands; pith cells thin-walled, strands of vessels (not accompanied by phloem) in pith. Subgenus *Angelopogon*

  2 Only a single circle of fascicular areas ("bundles") present as irregularly-shaped patches (as seen in transection) between earlywood and latewood; rhomboidal crystals present in rays. Pith narrow. Section *Angelopogon* . . . . *M. linearifolium*

  2 Fascicular areas develop as a second circle of bundles in pith after the outer circle of fascicular areas is well advanced (pith circle not present uniformly throughout stems); fibers present in fascicular areas as discrete strands, usually in pairs in latewood, or absent; ray cells contain druses or no crystals. Pith wide.

    3 Fibers wholly absent in fascicular areas and pith; no protophloem fibers present, no brachysclereid nests in cortex. Section *Telophyllum* . . . . *M. quadriflorum*

    3 Fibers present in fascicular areas as strands of fibers, usually a pair of strands present in latewood in each fascicular area; fibers present at tips of fascicular areas adjacent to pith; fibers present in protophloem position; brachysclereid nests present in outer cortex. Section *Archiphyllum*.

      4 Fiber strands in both protophloem position and adjacent to pith on fascicular areas slender; druses and sclereid nests absent in ray areas. . . . . *M. brachystachyum*

      4 Fiber strands in protophloem position and adjacent to pith on fascicular areas massive; druses and occasional nests of brachysclereids present in ray areas. . . . . . *M. oblongifolium*

1 Cambium as active in interfascicular areas as in fascicular areas; rays consist wholly of vertically elongate fibers identical to fibers of the fascicular areas; fibers form the background or fascicular areas and are pervasive, merging imperceptibly with fibers of potential ray areas, or absent (*M. gayanum*); axial parenchyma not abundant; vessels with scalariformlike or alternate pits, the pits not much widened laterally so as to cover more than one vessel face; pith consists wholly of sclereids; no strands of vessels are present, and no second cycle of fascicular areas ("bundles") occurs in the pith. Subgenus *Misodendron*

  5 Fibers absent in fascicular areas, which consist wholly of vessels mixed with axial parenchyma; axial parenchyma cells with thin nonlignified walls; pits on lateral walls of vessels all circular, alternate, and with wide apertures. . . . . *M. gayanum*

  5 Fibers present in fascicular areas, merging imperceptibly into the identical fibers of potential ray areas, a rayless condition therefore present; axial parenchyma cells sparse, with thin but lignified walls; pits on vessel walls oval to elliptical, scalariformlike.

6 Protophloem fiber strands present (strands may not be in apparent contact with phloem); pits on lateral walls of vessels frequently with wide gaping apertures.
7 Earlywood vessels forming almost uninterrupted circles around stem. . . . . . . *M. angulatum*
7 Earlywood vessels in patches, interrupted by what appear to be ray areas consisting of fibers exactly like the fibers abundant as the background cell type in latewood areas. . . . . *M. recurvum*
6 Protophloem strands absent, no sclerenchyma of any kind in cortex; lateral wall pits of vessels with narrow slitlike apertures. . . . . *M. imbricatum*, *M. punctulatum*

The preceding key illustrates that one can usefully incorporate features from stem tissues other than wood. Wood keys to species often omit non-wood features, as in the key to *Albizia* offered by Chauhan and Dayal (1985).

Attention should be called to the note of Baas and Zweypfenning (1979), who find that distinctions among species based on wood anatomy may break down if more numerous samples are studied. Variation in wood anatomy within species of eucalypts has been studied by Wilkes (1988). Comparative anatomical studies typically utilize few samples because few samples per species are usually collected and because preparation and observation time are limited. Keys based on wood anatomy cannot be expected to be as exception-free as keys based on gross morphology, which are usually based on observations of numerous specimens.

Also, one should note that provenance can have an effect on wood anatomy in any given species (Stern and Greene 1958; Bissing 1976, 1982). The distinctions to be found within species are related to phenotypic modification and to genetic races. Differences among populations chiefly in quantitative features of cell size are to be expected (Stern and Greene 1958; Bissing 1976).

10.3 Systematic Value of Particular Wood Features

10.3.1 Basis for Systematic Value of Characters

In attempting to find wood features useful for systematic purposes, one wishes for character states that follow ordinal, familial, or generic lines. One also wishes for features that are not markedly subject to phenotypic modification. The fewer the times a feature has evolved in dicotyledons the better, because one is faced less with parallelism and the interpretative problems that presents. Wood features that are more sensitive indicators of ecology are less useful for systematic purposes – a fact that cannot be stressed too strongly. Different species may differ by means of wood features that are based on ecological differences rapidly and easily evolved polyphyletically in relation to ecology, but one must realize in this case that one is essentially contrasting the habitats of the two species. Hopefully, the ecologically sensitive wood features, if incorporated in a cladogram, will turn out to be homoplasies, just as flower color likely might. Features that commonly show parallel evolution and thus would be homoplasies are best omitted from the data matrix of a cladogram. Features that are most useful systematically may be of significance, but they tend to be features that are not unique in what they do. For example, crystal type and distribution are of systematic value in dicotyledons, but their function, that of deterring predators, is performed by a series of chemicals, not just calcium oxalate crystals. Thereby a phylad may retain a particular crystal type even though other chemical defenses exist in that phylad. Wood features of systematic value for families and orders tend to be those that are "conservative," that originated early in a phylad (and may have had a marked selective value then), but which do not currently change in response to selection. One should not think that features not currently of unique selective value (but working in combination with others, like herbivore defenses) or not of obvious selective value may never have had a selective value – they likely once did, and may still have some (although less than at their time of origin perhaps).

At the other end of the spectrum, features that are essentially relictual (e.g., scalariform perforation plates) are not of systematic value if they have been retained independently in groups that are not closely related. However, they may have been retained in families that are close to each other (symplesiomorphy), and the contrast between two families that have retained a primitive feature and the majority that do not have such a feature provides their essential interest with respect to the taxonomic system.

Wood features do evolve at rates different from each other, even if they tend to be, in a statistical sense, synchronized (e.g., the statistical correlations of Frost 1930a,b). Individual species may deviate widely from those correlations: for example, a wood with long scalariform perforation plates may have libriform fibers rather than (as one might expect) tracheids as its imperforate tracheary element type. However, the majority of cases should show more synchrony among features. Those features that are markedly out of synchrony, therefore, are of interest with respect to systematics, and finding the basis for the lack of synchrony may be an interesting pursuit.

Even though the features most sensitive with respect to ecology are the least useful systematically, one should not think that wood features that do not seem to vary with respect to ecology have no ecological interpretation. A feature that evolved as a mechanism for coping with xeric conditions may be retained in relatively mesic circumstances because it is of neutral but not negative selective value in the wetter place. When we view wood, we are essentially viewing an assemblage of features old and new, of minor and of major importance, that have been of selective value to the functioning of wood in a phylad. We should not think of some wood features as having systematic value, with others to be interpreted ecologically. While that compartmentalization is appealing, it is misleading and only delays the evolutionary synthesis of wood features and the ways in which they have evolved in response to the pressures of the environment. If, for example, we say that some species of *Escallonia* have growth rings (Stern 1974), that vessel element length varies among species of *Ficus* (Koek-Noorman et al. 1984b), or that species of *Artemisia* differ with respect to mean vessel diameter (Carlquist 1966b), this can lead to an understanding of the ecological differences among species in the genus even though our knowledge of the systematics of these genera is not improved.

10.3.2 Vessel Grouping

Some families or genera have grouped vessels, some solitary ones. If a wood has tracheids, vessels are solitary or nearly so; if a wood has fiber-tracheids or libriform fibers, the degree of vessel grouping varies according to ecology (Carlquist 1984a). A family broken between genera with tracheids and genera with fiber-tracheids or libriform fibers (e.g., Hydrophyllaceae) might indicate a systematic difference between major portions of the family. Within those families in which fiber-tracheids or libriform fibers are the imperforate tracheary element type, the degree of vessel grouping follows ecology very closely and systematic value is therefore less.

Types as well as degrees of vessel grouping can characterize genera (Cozzo 1950). A listing of families that have diagonal (and tangential) aggregations of vessels is given in Section 3.4.3.3.

10.3.3 Perforation Plates

In some families, some genera have only simple perforation plates, whereas others have scalariform perforation plates or a mixture of the two (e.g., Myrtaceae; Schmid and Baas 1984). Because a relatively small number of genera have retained scalariform perforation plates, a genus that does have such plates may be said to have a systematic distinction from others in the same family that have simple plates. In Buxaceae, *Styloceras* has many bars per plate, the other genera relatively few bars (Carlquist 1982d). Species of *Ilex* differ with respect to number of bars per perforation plate. Situations like this provide systematic criteria. The nature of the perforation plate is the sort of character that is valued in wood identification because of its conspicuousness. A listing of families with scalariform perforation plates is provided (Sect. 3.7.2).

A limited number of families have nonbordered perforation plates. Many of these appear to belong to the order Caryophyllales (Carlquist 2000b), an order that has been redefined so as to include Ancistrocladaceae, Droseraceae, Polygonaceae, Plumbaginaceae, and Santalaceae.

10.3.4 Lateral Wall Pitting of Vessels

Scalariform lateral wall pitting has been retained in a few primitive genera and families such as Hydrangeaceae. Genera within a family may be distinguished by lateral wall pitting type. For example, *Magnolia* has scalariform lateral wall pitting whereas *Liriodendron* has opposite pitting. *Vitis* has scalariform lateral wall pitting of vessels, whereas some other genera of Vitaceae have alternate pits. A few species of *Nothofagus* have transitional pitting, whereas others have opposite or alternate pitting (Patel 1986; Carlquist 1987e). Species of *Opuntia* differ from each other with respect to lateral wall pitting (Gibson 1977a), but this may be indicative of paedomorphosis (Carlquist 1962a). Vessel pit diameter and density differ among species of *Tectona* (Gottwald and Parameswaran 1980).

A listing of families with scalariform lateral wall pitting has been provided (Sect. 3.8.3).

10.3.5 Vesturing

Vestured pits and vesturing are characteristic of the families of Myrtales (van Vliet and Baas 1984). Because vesturing has originated only a few times in woody dicotyledons, the occurrence of this feature is of very considerable systematic importance, and variations in vesturing are also of great significance. For example, the peculiar verrucae on vessel walls of *Cercidium* (Carlquist 1989b) and their presence or absence in particular species offer excellent systematic criteria. In Fabaceae, some genera have vestured pits whereas others have nonvestured pits, and thereby excellent systematic distinctions are available (Quirk and Miller 1983, 1985). A listing of families with vesturing is offered (Sect. 3.10.2).

10.3.6 Helical Sculpture

Stress has been laid on the fact that there is not a single type of helical sculpture on vessel walls, there are several types. Therefore more than one list was offered (Sect. 3.12.2); helical sculpture may occur as grooves interconnecting pit apertures, as pairs of thickening accompanying these grooves, and as bands (either coarse or fine) on vessel walls. Scanning electron microscopy is ideal for observation of helical thickenings, but light microscopy must also be used. SEM used alone is insufficient. One sometimes cannot see the pits underlying helical sculpture, for example. As studies employing both means appear, we will have more examples of systematic distinction of helical sculpture. In the family

Asteraceae, kind and degree of helical sculpture vary with taxon, but we must remember that helical sculpture is relatively sensitive as an indicator of ecology, and thus to a certain extent we are describing ecological differences when differences in helical sculpture are cited. James and Ingle (1956) find presence of helical thickenings in vessels to be characteristic of certain genera of Flacourtiaceae.

10.3.7 Tyloses

The significance of tyloses is essentially a physiological one, although they do characteristically develop in some woods and not in others. For this reason, some authors (e.g., Metcalfe and Chalk 1950) have developed lists of species with tyloses. Presence of tyloses can be of assistance in wood identification, but perhaps less commonly are they of systematic importance. Presence or absence of tyloses and whether or not they are sclerosed are cited by Dickison (1972) as generic and specific criteria in Connaraceae.

10.3.8 Imperforate Tracheary Elements

The type of imperforate tracheary elements – whether tracheids, fiber-tracheids, or libriform fibers – varies according to genus in such families as Hydrophyllaceae (Carlquist et al. 1983; Carlquist and Eckhart 1984) and Polemoniaceae (Carlquist et al. 1984). Listings of families characterized by these three types of imperforate tracheary elements are given in Sections 4.5.8, 4.6.5, and 4.7.6, respectively.

Vasicentric tracheids are a distinctive phenomenon that characterize some orders (e.g., Violales) but are scarce in others (Annonales). They may even characterize particular genera or species. A listing of the families in which they have been found is given in Section 4.12.4. In some of these families, only a single species is known to have vasicentric tracheids. Although the physiological significance of vasicentric tracheids is great, the fact that they occur in particular orders suggests that either vasicentric tracheids or the mechanism to form them characterize particular groups. Alternatively, the ability of vasicentric tracheids to help plants survive drought may have permitted the orders that have them to enter drier situations more commonly. For example, Annonales rarely occur in dry (or exceptionally cold) sites, and this may be correlated with vasicentric tracheid scarcity in the order.

10.3.9 Axial Parenchyma

Some families have a single axial parenchyma type, even despite size and diversity. Such conservatism is shown by Asteraceae, all of which have vasicentric scanty parenchyma (although this may be supplemented by bands of parenchyma related to fiber dimorphism). Families supposedly related to Asteraceae, such as Calyceraceae and Valerianaceae (Carlquist 1983e), also have vasicentric scanty parenchyma. In other families, a wide range of parenchyma types is present. In Boraginaceae, axial parenchyma may be narrow-banded apotracheal, wide-banded apotracheal, vasicentric scanty, aliform, or confluent. Other families with a wide range of axial parenchyma types include Ebenaceae, Loganiaceae, and Rubiaceae. In these families, one might look to axial parenchyma type as a generic character. It is cited as a generic character in *Buchenavia* of the Combretaceae (Alfonso and Richter 1991), as well as in Hernandiaceae and Gyrocarpaceae (Shutts 1960), and the relative abundance of parenchyma in apotracheal bands characterizes species of *Pereskia* and *Quiabentia* of the Cactaceae (Bailey 1962).

Diffuse crystalliferous axial parenchyma cells are found in families of the Sapindales (Rutales): Aceraceae, Anacardiaceae, Connaracerae, Fabaceae, Juglandaceae, Meliaceae, Melianthaceae, Rutaceae, Sapindaceae, and Simaroubaceae. Hippocastanaceae have crystalliferous axial parenchyma, but in a banded distribution. Of the Rutales as defined by Thorne (1976), crystalliferous axial parenchyma is absent only in Burseraceae. This seems a remarkable validation of Thorne's concept by an unusual wood characteristic. Diffuse crystalliferous axial parenchyma does occur in some other families, and there it is indicative not of relationship to Rutales, but of an independent origin. Santalaceae and Olacaceae have diffuse crystalliferous axial parenchyma, and its presence can be used as a feature allying this pair of families with each other.

10.3.10 Rays

As with axial parenchyma types, the ray types designated by Kribs (1935) may be found singly in some families, whereas two or more can be found in others. A notable example of the latter tendency is Fabaceae, in which all six of Kribs' types occur. There are examples in which occurrence of a few types links families. Betulaceae and Fagaceae agree in having Homogeneous type I and Homogeneous type II rays; they also agree in having aggregate rays (Metcalfe and Chalk 1950). Casuarinaceae also have aggregate rays, which could be indicative of relationship to Fagales, but they have Heterogeneous type IIB rays.

One must remember that rays can change markedly in histology during ontogeny, as in *Bursera* (Fig. 6.1). Rays in *Euptelea* begin as Heterogeneous type I but become Heterogeneous type IIB during development.

The presence of tile cells (Sect. 6.12.1) is a feature that links four families of Malvales: Bombacaceae, Malvaceae, Sterculiaceae, and Tiliaceae.

Ray width has been found to characterize genera of Annonaceae (Vander Wyk and Canright 1956). Ray height is a taxonomic character in Vochysiaceae (Quirk 1980).

10.3.11 Cell Contents, Secretory Canals

The example of diffuse crystalliferous axial parenchyma as a feature unifying most families of Sapindales (Rutales) has been cited above. One can view crystals or other cell inclusions either in terms of the tissue in which they occur or in terms of the nature of contents (e.g., calcium oxalate crystals).

Presence of the various major crystal types tends to characterize families and subfamilies quite frequently. This is an example of a "conservative" feature. Crystal presence is not the sole herbivore defense mechanism but one of numerous devices plants have, so that it has tended to evolve in a conservative fashion (e.g., one may imagine that a family with raphides will not evolve some crystal type superior to raphides that would supplant the raphides, but rather add a chemical such as a flavonoid or alkaloid). Because of the systematic value of crystal types, listings have been given here for families (and genera) in which the following types are located: rhomboidal (Sect. 7.1.2.1), elongate (Sect. 7.1.2.2), acicular (Sect. 7.1.2.3), raphides (Sect. 7.1.2.4), crystal sand (Sect. 7.1.2.5), and druses (Sect. 7.1.2.6). Raphides are present in Onagraceae but absent in other Myrtales except for one genus of Melastomataceae. Raphides in rays may signal relationship among Aizoaceae, Nyctaginaceae, and Phytolaccaceae.

Crystal presence or absence defines sections of the genus *Cordia* of the Boraginaceae (Gottwald 1983) and species in Myoporaceae (Carlquist and Hoekman 1986b).

Cystolith presence is related to generic lines in the few (Sect. 7.2) families in which they are present, such as Opiliaceae (Koek-Noorman and van Rijkevorsel 1983).

Silica bodies characterize one entire family (Chrysobalanaceae), but are mostly a generic character in dicotyledon woods, as indicated by the listings of Section 7.4.

Laticifers in wood are not nearly so abundant as laticifers elsewhere in the vegetative body of dicotyledons, so that their presence in wood is not so much a familial character but a generic one. The listing of Section 7.5 illustrates this.

Idioblasts containing oil or mucilage are so characteristic of Magnoliales (Annonales), Illiciales, and Piperales that the families not in these orders yet listed as having these cells by Metcalfe and Chalk (1983, p. 222) can be suspected of having some type of cell with a different chemical content

(Burseraceae, Cactaceae, Myoporaceae, Proteaceae, Rutaceae). The revised listing is found in Section 7.7.

Intercellular canals (secretory canals) in wood may be radial, in rays (Sect. 7.10.1), or vertical, in axial parenchyma (Sect. 7.10.2). The roster of families with radial canals is much greater than the roster of families with vertical ones – doubtless a function of the smaller number of families with abundant axial parenchyma bands suitable for containing vertical canals. If one views these listings as a whole, they seem to contain most of the families of Sapindales (Rutales) – Anacardiaceae, Burseraceae, Connaraceae, Fabaceae, Julianaceae, Meliaceae, Sapindaceae, and Simarubaceae have one or both types of canals. The family Dipterocarpaceae is rich in vertical canals, but three genera lack them – *Marquesia*, *Monotes*, and the recently-described *Pakaraimaea* (de Zeeuw 1977). Thus vertical canal presence in Dipterocarpaceae is a generic feature. It is also a generic feature in the fabaceous genus *Copaifera* (Moens 1955).

10.3.12 Successive Cambia

Successive cambia are very characteristic of families of Caryophyllales (Centrospermae), as noted in Section 8.2. Not all of the families have them, so they may be an ancestral (symplesiomorphic) feature of the order, and perhaps genes suppressing formation of more than one cambium have been evolved in several families or suborders as apomorphies. Presence of successive cambia has been used as a reason for excluding Chloanthaceae and Avicenniaceae from Verbenaceae, in which a single cambium is then characteristic. The genus *Chrysanthemoides* has successive cambia and thereby can be excluded from the genus *Osteospermum* (Adamson 1937).

10.3.13 Interxylary Phloem

Interxylary phloem formed by a single cambium (families listed in Sect. 8.3) is found in a scattering of myrtalean families (Lythraceae, Melastomataceae, Onagraceae). Although one could say this occurrence is indicative of relationship, that would be misleading. All myrtalean families have intraxylary phloem (phloem adjacent to pith), and that is probably the evolutionary antecedent for presence of interxylary phloem in the three families mentioned. This exemplifies the tendency for structures abundant on other portions of a plant to be less commonly present in wood (e.g., laticifers, secretory canals, druses). In viewing such characters as indications of relationship, one should keep in mind the wider occurrence in the plant body.

10.3.14 Storied Structure

The distribution of storied cambia and therefore storied wood structure in wood of dicotyledons is certainly not random, but confined to certain families (Sect. 9.2.4). For example, a number of sapindalean families appear on that list of families (Fabaceae, Hippocastanaceae, Meliaceae, Melianthaceae, Moringaceae, Simaroubaceae). There is also a notable tendency for storying in the malvalean families (Bombacaceae, Malvaceae, Sterculiaceae, Tiliaceae). Storying is present in *Octomeles* and *Tetrameles* but absent in *Datisca* (Davidson 1976), thereby forming a generic criterion in this family. However, the stems of *Datisca* are relatively short-lived because of the herbaceousness of that genus, and one should take that into account. One may note in this connection that storying also occurs in woods of *Begonia*, but only in older stems of rather woody species (Carlquist 1985b); when one considers ontogenetic factors, presence of storying could link Begoniaceae and Datiscaceae, which have been thought related.

10.4 Formats for Reporting Taxonomic Features of Woods

10.4.1 Tables

Tabular representation of quantitative and qualitative data has become commonplace in papers on wood anatomy. One can be genuinely grateful for this, because it makes available large quantities of data in a more easily accessible fashion, suitable for rapid comparisons, than does running text. For descriptions of a small number of species, however, a running text may be a more economical type of presentation (in terms of printing costs).

With respect to quantitative data, comparison of various types of presentations suggests that means, rather than ranges, offer figures more indicative of what is shown by the wood features measured. Extremely narrow and wide vessel elements can occur in a given wood sample almost by chance, but a few extreme cells are not representative of the sample. If one takes a functional viewpoint, the wood physiology is dictated by cells near average dimensions – cells large in number – rather than a few cells of extreme dimensions.

Some workers have added figures for standard deviation of quantitative wood features in addition to the means. If space for presentation of quantitative data is abundant, this nicety is welcome. However, if space is at a premium, the presentation of standard deviation can be questioned. Standard deviations only apply to the cells measured, and thus they are of no value for projecting what one would find if more numerous samples or samples from different parts of a plant were studied. Because these variables can be quite extreme (e.g.,

Stern and Greene 1958), this lack of predictability by standard deviation figures renders them of questionable statistical value.

There is little point in singling out one style of tabular presentation as preferable to others, because I would like to encourage originality, fullness of data, and pertinence of items entered in a table. Comparisons among families with respect to qualitative data offer problems because descriptive phrases consume space. The table by Moseley and Beeks (1955) on Garryaceae shows how this can be done within a single page. Incorporation of features other than those involving wood in comparison tables has the potential merit of presenting the most valuable comparisons, and also showing how features in organs other than wood agree with those in wood.

10.4.2 Keys

Keys have been mentioned above. In presentation of wood data, as in data from macromorphology, keys summarize conspicuous differences but must omit some diagnostic features. Decker's (1966) key to genera of subfamily Luxembergieae (Ochnaceae) was based on gross morphology as well as one based on wood anatomy. From such a comparison, one can see to what extent wood features follow the distribution of gross morphology features.

10.4.3 Cladograms

With the incorporation during the past decade of cladistics into plant taxonomy, its appearance in other systematics-related disciplines is to be expected. Notable in application of cladistic methodology to wood anatomy are the papers of Baas and Zweypfenning (1979), ter Welle and Koek-Noorman (1981), Baas et al. (1988), and Noshiro and Baas (1998). Cladograms have tended to be increasingly comprehensive in their inclusion of data sets: integration of data from DNA analysis (even representing several gene sites) with data from macromorphology, anatomy, and ultrastructure (e.g., Angiosperm Phylogeny Group 1998) is likely to become increasingly common, demanding though such syntheses are.

10.4.4 Graphs and Other Diagrams

Graphs in which values for pairs of features are shown for two or more species are useful in demonstrating degree and kind of divergence among taxa. For example, Bonsen and Welle (1983) give graphs that compared vessel diameter and vessel element length with fiber length for Cecropiaceae, Urticaceae, and *Poikilospermum*. A trio of genera of Annonaceae is similarly compared by ter Welle and van Rooden (1982). Although the space consumed by these graphs is considerable for the number of features presented, there are compensatory values. The oversimplifications that sometimes result from averaging values are avoided. One can demonstrate how two taxa do or do not overlap in quantitative features.

Graphs or diagrammatic devices that show range and frequency distribution for particular wood features of Apiaceae and Araliaceae were reproduced in Carlquist (1961a, p. 53). Graphic representations may well bring to light wood phenomena that otherwise could not be appreciated. For example, age-on-length curves (Fig. 9.2) have revealed phenomena (with far-ranging implications) that tabular data or running text could not have demonstrated as adequately or directly.

11 Evolution in Wood: An Ecological/Functional Synthesis

11.1 Introduction

In the presentation of wood features of Chapters 2–9, comments are given on evolutionary status of wood features where appropriate. What cannot be evident in those chapters are: (1) How do wood features evolve with respect to each other? (2) What kinds of evidence exist for evolutionary status of wood features? (3) How polyphyletic are particular wood features? (4) How conservative are particular wood features? And (5) to what extent is evolution of particular features irreversible or reversible? Features of wood anatomy ought to obey the same neo-Darwinist principles as those operating in the evolution of other features in organisms.

11.2 I.W. Bailey and the "Major Trends of Xylem Evolution"

11.2.1 Nonwoody Plants

Irving Bailey's broad acquaintanceship with xylem of vascular plants, including fossil groups, led him to suspect that shortening of tracheary elements has occurred during phylesis of vascular plants over geological time. Data collected by Bailey and Tupper (1918) validate this idea. However, note should be taken that there are groups that do not follow these trends.

Living ferns are noncambial (except perhaps for Ophioglossaceae), and length of tracheids (or vessels elements) in living ferns is governed not by evolutionary status but by degree of elongation in the organ in which they are borne (Carlquist 1975a). Monocotyledons are like living ferns in lacking cambia, and have much the same range of growth forms as ferns: rosette herbs, "rosette trees," and creeping rhizomatous plants. The various authors who have dealt with evolution of xylem in monocotyledons have stressed vessel presence and perforation plate morphology in various organs but have been notably silent on length of vessel elements and tracheids. On the basis of scattered observations in monocotyledons, I believe that monocotyledons are like ferns in having tracheids and vessel elements that correspond in their length to degree of organ elongation.

11.2.2 Conifers and Vesselless Angiosperms

Conifers and *Ginkgo* have tracheids that correspond in length to tree (or shrub) size (Carlquist 1975a). Tracheids are short in seedlings, but as the stem of a seedling is converted into the trunk of a tree, tracheids become progressively longer (Bailey and Tupper 1918). A branch of that tree will have tracheid length proportional to its stem diameter. This is implicit in a graph of Bailey and Tupper (1918, p. 190) for *Pseudotsuga menziesii* (= *P. taxifolia*). Tracheid lengths in conifers are approximately related to stem diameter and height either on a species-by-species basis or by organ within an individual (Carlquist 1975a). There is no relationship evident between phyletic status and tracheid length in conifers, although the only possible exception to these patterns is in main stems of Araucariaceae, in which tracheid length is somewhat greater than that in other conifer groups.

Tracheid length in vesselless dicotyledons is also proportional to stem size rather than to phyletic status (Carlquist 1975a). However, as soon as the threshold into vessel presence is crossed, the length of vessel elements (which can be taken as fusiform cambial initial length, since vessel elements elongate relatively little during their maturation) decreases with phylesis.

If tracheids have shortened over geological time in vesselless plants, one could say that plant size has become smaller. That is demonstrable in particular phylads (e.g., *Equisetum* is smaller than *Calamites*). Although no explanation for why tracheids should shorten over geological time was offered by Bailey and Tupper (1918), one can hypothesize progressive successful exploitation of drier habitats (in which plant size is accordingly smaller). Perpetually mesic habitats occupy a relatively small portion of the earth, so that successful groups are ones that can range into habitats of a seasonal nature, where smaller plant size is a corollary.

Dry conditions subject water columns to greater tensions. Shorter tracheids are also narrower (Bannan 1965) and narrower tracheids permit water columns to withstand greater tensions without embolizing than do long, wide tracheids. Experimental evidence of that correlation was offered by Rundel and Stecker (1977) and by Lewis and Tyree (1985). Shorter tracheids have less overlap (end wall) area, and therefore have poorer conductive ability compared to longer tracheids. Poorer conductive characteristics can be countered by such features as microphylly (possible in sunny habitats), drought-deciduous habit, or mycorrhizal association. Conifers are microphyllous except for species that are trees in humid rain forest (*Agathis*, *Podocarpus*). This suggests why vesselless angiosperms, none of which are microphyllous, are all confined to situations of abundant moisture availability.

Thick-walled, narrow latewood tracheids are mechanically superior to earlywood tracheids (Boatwright and Garrett 1983), but conductively poorer. Maintenance of higher conductive capacity can be achieved, therefore, by alternate production of earlywood and latewood in seasonal environments.

Uniformly moderately thick-walled tracheids in relatively nonseasonal environments (e.g., *Agathis*, *Araucaria*) are adaptive also, however. Longer tracheids are stronger than shorter ones (Wellwood 1962) as well as greater in conductive capacity. In smaller conifers, mechanical strength is probably not lowered by decreasing cell size, although conductive capacity probably is lowered (while conductive safety is increased). In other words, the feature that explains the selective value of longer tracheids is the increased conductive capacity of longer end-wall overlap areas. This explains why there is a sharp drop in length of all tracheary elements (both imperforate tracheary elements and vessel elements) with the origin of vessels (Carlquist 1975a, p. 141), a conspicuous and curious phenomenon. With origin of the perforation plate, there is no longer any selective value for long tracheids so as to achieve greater conductive area per cell.

11.2.3 Vessel-Bearing Dicotyledons

The Bailey and Tupper (1918) data show that vessel-bearing dicotyledons have a wide range of vessel-element lengths, and that shorter vessel elements occur in more specialized groups (groups with more numerous specialized floral features). Therefore, there has been a continued drop in length of fusiform cambial initials during phylesis of woody dicotyledons. However, that tendency is not readily explainable on the same bases as those that have produced the drop in tracheid length over time. Yet data do show that vessel element length is shorter in vessel-bearing dicotyledons of drier habitats compared with their relatives in wetter habitats. This is most conveniently shown in a single large family, such as Asteraceae (Carlquist 1966a). Shortening of vessel elements in drier habitats can also be shown on a floristic basis (Novruzova 1968; Carlquist 1977d; Baas et al. 1983; Barajas-Morales 1985; Carlquist and Hoekman 1985b).

What has caused the reduction in length of fusiform cambial initials once vessels originated? If related to xeromorphy, several possibilities are available. Shorter vessel elements might be stronger because of the constriction formed by each end wall (even if it contains a simple perforation plate). Presence of these vessel constrictions could resist deformations related to tensions in water columns (Carlquist 1975a). However, another more appealing possibility is that short vessel elements are valuable because they localize air embolisms to a greater extent than long ones because of the constrictions formed by perforation plates, even simple perforation plates (Carlquist 1982c). Slatyer (1967) stated that air bubbles should be so confined by vessel elements, even if perforation plates are simple. Root pressure may expel air bubbles from vessels in lianas and some other woody plants (O'Leary 1965; Putz 1983; Ewers 1985), but in many woody plants that does not happen. Evidence for localization of air bubbles in individual vessel elements has been provided by Sperry (1985), tending to confirm Slatyer's statement. My own observations with a woody plant that has simple perforation plates (Fig. 3.11, part 12) have shown that individual bubbles tend to be confined to individual vessel elements. Where a long bubble extending beyond a single vessel element occurred, I observed that the bubble shrank to the nearest perforation plate, and the tip of the bubble appeared confined by that perforation plate, even though no bars traversed the perforation plate. There is no question that when water stress occurs, air embolisms occur frequently in vessels and are problems for the conductive system; this has been demonstrated effectively and repeatedly (Zimmermann and Milburn 1982; Tyree and Dixon 1986).

Alternative explanations for shortening of vessel elements can be imagined, although there is at present few data to support them. For example, one can suppose that narrow vessels are of selective value because they embolize less readily – a fact that has been demonstrated (Ellmore and Ewers 1985). Production of narrow vessel elements could be hypothesized to lead to production of shorter vessel elements if there is a correlation between diameter and length morphogenetically. But as one can observe in a growth ring of any ring-porous species, narrow vessel elements can be achieved without shortening of vessel elements; in fact, latewood vessel elements are slightly longer than earlywood vessels (Swamy et al. 1960).

Shorter fusiform cambial initials could be hypothesized to be advantageous because divisions in a short fusiform cell (e.g., the vertical radial divisions in fusiform initials of storied cambia) could take place more readily. However, the existence of long fusiform cambial initials in conifers counters that idea quite easily. Storied cambia have notably short fusiform cambial initials, and there is not a strong correlation between storied cambia and dry habitats, whereas there is a strong correlation between short vessel elements and dry habitats.

One could hypothesize that vessel element dimensions are related to the size of plants in which they occur. However, in the southern Californian flora, shrubs have shorter vessel elements than trees, but herbs have longer vessel elements than shrubs (Carlquist and Hoekman 1985b).

One could hypothesize that shortening of fusiform cambial initials in dicotyledons is related to mechanical considerations and that length of imperforate tracheary elements, rather than length of vessel elements, is basic to the trend. Because of intrusiveness of cambial derivatives destined to become imperforate tracheary elements, very long cells can be achieved from products of relatively short fusiform cambial initials. The mean length of imperforate tracheary elements in dicotyledons (1,317 μm) is about twice the mean length of vessel elements in dicotyledons (649 μm) according to the data of Metcalfe and Chalk (1950, pp. 1360–1361), so the average dicotyledon is not achieving long mechanical cells by having long fusiform cambial initials, but rather by intrusive growth of derivatives of the fusiform cambial initials. Length of mechanical cells in wood is probably not as significant a factor as wall thickness, wall chemistry, wall ultrastructure, or amount of mechanical tissue in providing mechanical strength (Boatwright and Garrett 1983).

Zimmermann (1978b) declared vessel element length functionally meaningless. However, the observed pattern of vessel element length distributions with relation to ecological factors is not random, either on an experimental basis (see beginning of this section) or on a comparative basis (Zhang et al. 1992 report that vessel element length in Rosaceae is related to moisture availability).

11.2.4 Features Associated with Vessel Element Length

I. W. Bailey realized that the phenomenon of phyletic shortening of fusiform cambial initials and therefore of the cells derived from them is not an isolated one. Rather than pursuing these correlations himself, Bailey supervised graduate students who worked on them. Frost (1930a,b, 1931) studied evolution of the vessel elements (see Chap. 2). Kribs studied rays (1935; Chapter 6) and axial parenchyma (1937; Chap. 5). As we will see, not all of the features correlated with change in fusiform cambial initial length were studied by Frost and Kribs.

Several features are significant about the studies by Bailey, Frost, and Kribs:

1. The features studied are related to specialization in woods.
2. The features studied show not just extreme expressions but a series of gradations that can be defined (e.g., scalariform, transitional, opposite, and alternate types of lateral wall pitting in vessels).
3. The features these workers studied do not bear a direct relationship to the length of fusiform cambial initials. For example, there is no reason to believe that histology of rays bears any morphogenetic relationship to length of fusiform cambial initials. Even with scalariform perforation plate morphology, the link between length of end wall and the number of bars per perforation plate is very loose (Carlquist 1975a, p. 143), and would not pass statistical significance tests.
4. The features studied by these workers can evolve independently of each other, but that is a matter of independence of the features, not of shifting in direction. In the overall picture, a number of wood features are evolving in similar directions together, even though synchrony among the features is not tight. For example, one can find a species with simple perforation plates that has a primitive ray structure.
5. The features must evolve polyphyletically and must do so in the same way in not a few, but in many dicotyledon groups. For example, the correlations between degrees of perforation plate simplification and fusiform cambial initial length would not hold if only a few phylads had experienced it, or if the simplification proceeded in different directions in different phylads.
6. The features can be hypothesized to relate to gradual improvement of conductive efficiency, and therefore successive levels would represent selective advantage. Higher flow rates in vessels and higher flow rates of photosyn-

thates in axial and ray parenchyma ought to occur in more specialized woods. Imperforate tracheary elements should simultaneously take on maximal strength configurations.

7. Because conductive efficiency is related to success of phylads not so much in permanently mesic habitats as in seasonally dry ones, a higher proportion of a flora of a seasonally dry area should consist of specialized woods than would be true in a flora of a uniformly mesic area.

The reader should note that items 1 through 4 above were tacitly or implicitly realized to be operative by Bailey and his students. Items 5 through 7 were not considered by those workers and represent my analysis of their findings.

The features reviewed below have been described in preceding chapters. The purpose of the present review is to analyze several features conjunctively, to attempt construction of links between morphological evolution and ecological adaptation where feasible, and to lay a groundwork for the discussion of irreversibility (Sect. 11.3) and thereby to understand the features to which irreversibility may apply and those to which it may not.

11.2.5 Perforation Plate Morphology

Correlations between vessel element length and end wall morphology were offered by Frost (1930a,b; see Sect. 3.6). One can consider various aspects of the end wall of a vessel: end wall angle (used by Frost but not currently employed because of its variability and difficulty in measurement); whether scalariform or simple (used by Frost; see Sect. 3.7.3); or bar number (not used by Frost, but usable were one to study opening stages in wood specialization rather than the full gamut of it).

One way of looking at the end wall appealed to Frost (1930a) because of his assumption that the primitive vessel element is essentially tracheidlike: similarity between end wall and lateral wall. Frost devised two categories: (1) imperceptible transition between end wall and lateral wall (i.e., scalariform perforations on end wall merging into scalariform lateral wall pitting); (2) dissimilar end wall and lateral wall (e.g., scalariform perforations on end wall, alternate pitting on lateral walls). This selection of features is interesting in that two vessel features are being used in tandem (Table 11.1).

Table 11.1. Vessel type compared to vessel element length

Vessel type (see above)	No. of spp.	Vessel element length (μm)
1	29	1,270
2	22	870

The interesting thing that emerges from viewing Frost's three papers is that he used all conceivable features in which primitive vessels resemble tracheids, even though some of these did not yield as dramatic results as others. They all do show statistical correlation of a similar sort, however. Among the other features Frost studied were tendency for vessels to shift from angularity in transection to roundness in transection and tendency for borders on bars of perforation plates to be lost during phylesis.

The analysis of angularity of vessels is interesting because more angular vessels are narrower ones, and the angularity is basically a byproduct of the limited number of cell contacts. One can find instances of specialized woods in which narrow vessels are angular. Vessels round in transection are usually surrounded by more numerous cells. Thus, comparing angular with round is, to a certain extent, comparing diameter, and indeed, as we can see by looking at markedly ring-porous species, notably wide vessels are found in more specialized woods. Frost (1930a,b) may not have realized he was involved in a size comparison, but in fact, he was. The shift to a round shape and a wider (viewed statistically – individual cases may vary) diameter is a shift to capability of handling greater volume of water per unit time per vessel (see Sect. 3.13).

Frost's use of bordered bars as compared to nonbordered bars on perforation plates is of interest because one can find both conditions within a single wood, as in *Schisandra chinensis*. Frost fully appreciated this. One may first comment that even though bordered and nonbordered bars may coexist within a single wood, one may, with few exceptions, characterize a particular species as having one type or the other, and Frost was able to make sense out of a complex pattern by just such categorization.

One may also comment that in the *Schisandra chinensis* example, the series from bordered to nonbordered bars Frost (1930a) drew was, in fact, a series from metaxylem vessels (bordered) to secondary xylem (nonbordered). This series brings into play Bailey's (1944b) refugium hypothesis, which states that more primitive expressions tend to persist in primary xylem (in Schisandraceae, metaxylem and early secondary xylem perforation plates often have numerous bars, whereas later formed vessel elements have simple perforation plates; Carlquist 1999c). In fact, Frost (1930a) both analyzed vessel phylesis and proved the validity of Bailey's refugium idea in a single table (Table 11.2).

The above table does confirm the idea, a corollary of Bailey's refugium hypothesis, that vessels originated in the secondary xylem in dicotyledons and progressed into the primary xylem (Bailey 1944b). We may, going further, wonder why this should be true. One notices that vessel diameter in many species increases in diameter over time in the secondary xylem (e.g., Trimeniaceae; Carlquist 1984e; Begoniaceae; Carlquist 1985b). Primary xylem has narrower tracheary elements than secondary xylem, typically. If these trends are applicable, one could say that secondary xylem vessels have greater conductive capacity than primary xylem vessels. However, safety is the inverse

Table 11.2. Primary and secondary xylem compared to tracheary element length

Xylem category	Tracheary element length		
	Protoxylem (μm)	Metaxylem (μm)	Secondary xylem (μm)
a) Tracheids only in primary xylem; vessels with scalariform perforation plates in both primary and secondary xylem	1,948	1,647	709
b) Vessel elements with scalariform perforation plates in primary xylem; plates simple in secondary xylem	1,331	1,203	426
c) Vessel elements with simple perforation plates in both primary and secondary xylem	974	751	338

of conductive capacity, and retention of safety in primary xylem is also a possibility worthy of consideration.

Simplification of the perforation plate in vessels by formation of fewer bars per plate and of wider perforations between bars is a pervasive tendency in evolution of the vessel element (Fig. 11.1). This has been claimed phyletically to represent adaptation to increased conductive efficiency (Haberlandt 1914). The ecological correlates of this seem clear, when one looks at the ecological sites in which scalariform perforation plates are preserved (Sect. 3.7.2). Disappearance of bars on perforation plates in relation to seasonal environments (Carlquist 1975a) is an idea supported by the work of Versteegh (1968), Dickison (1979), and Dickison et al. (1978). Bar disappearance relates to peak conductive demands, not mean conductive rates.

Curiously, Zimmermann (1983) has rejected this hypothesis, claiming that other modifications of vessels suffice to improve conductive characteristics. One can counter Zimmermann's thoughts by saying that certainly evolution is unlikely to leave one aspect of a vessel completely unaltered and only affect others, despite the fact that simplification of a perforation plate would reduce friction. Zimmermann (1983) is doubtless taking this viewpoint in relation to his idea that the scalariform perforation plate is retained as a mechanism for sieving out air bubbles that form when frozen water thaws. That concept is not supported by distribution and kinds of scalariform perforation plates, most of which occur in areas where frost never occurs (see Sect. 3.7.2). Separation of air bubbles into vessel elements separated by scalariform perforation plates has been demonstrated by Sperry (1985), however.

Disappearance of bars on the perforation plate probably does not decrease safety greatly: the great loss of safety has occurred when the threshhold from tracheid to vessel element has been crossed. The functional explanation of scalariform perforation plates with few but wide bars is a curious problem dis-

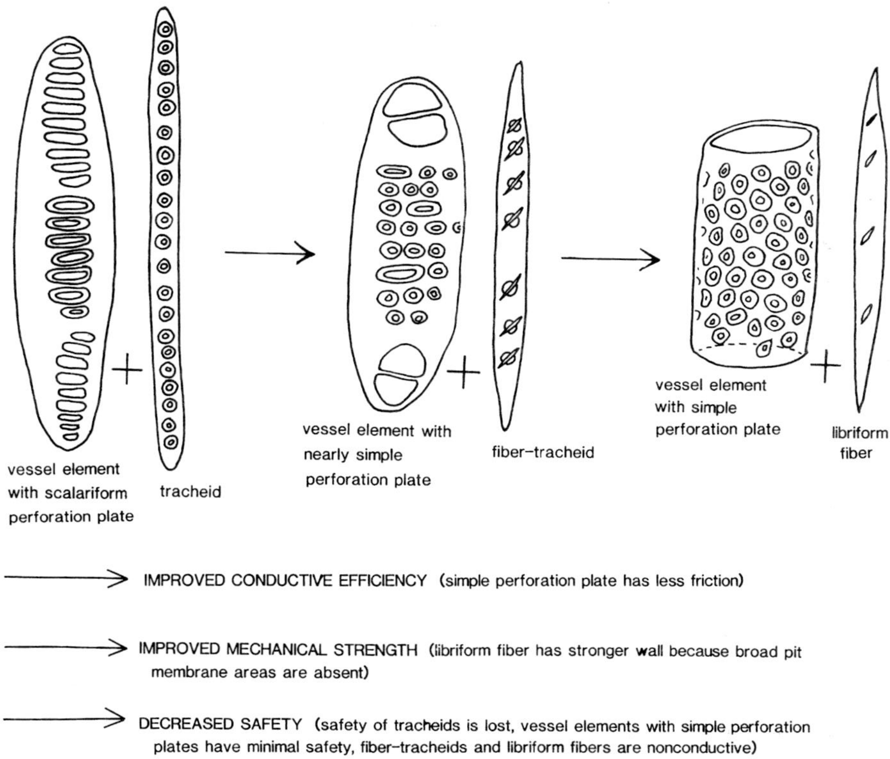

Fig. 11.1. Trends in the evolution of vessel elements and of imperforate tracheary elements. This scheme is based upon the assumption that in a primitive wood, a vessel element with a scalariform perforation plate will tend to be associated with a tracheid, whereas in a specialized wood, a vessel element with a simple perforation plate will tend to be associated with libriform fibers. Although deviations from these conditions certainly occur, the three stages shown do tend to be represented in particular species. The series shows a progressive increase in division of labor between a conductively efficient cell (vessel element) and a mechanically optimal cell (imperforate tracheary element). The evolutionary strategies the series represents are indicated by *arrows* at the *bottom* of the figure. Attention is called to the third of these (decreased safety), which is a consequence of the evolutionary series unless some compensatory phenomenon or an additional condition (see Figs. 11.2, 11.3) occurs

cussed in Section 3.7.2. Such bars (and their associated wide perforations) could not serve for sieving out air bubbles.

The distribution of tracheids in wood of dicotyledons and the xylem of monocotyledons leaves one in little doubt that retention of tracheids evidences a selective value for the safety of tracheids (Fig. 11.2, left column). Tracheids are retained in many species in which only simple perforation plates are found. In the woody flora of southern California, scalariform perforation plates occur in only a small proportion of species (3.9%) – all of them riparian species or species of moist canyons or moist alpine regions. True tracheids, in contrast,

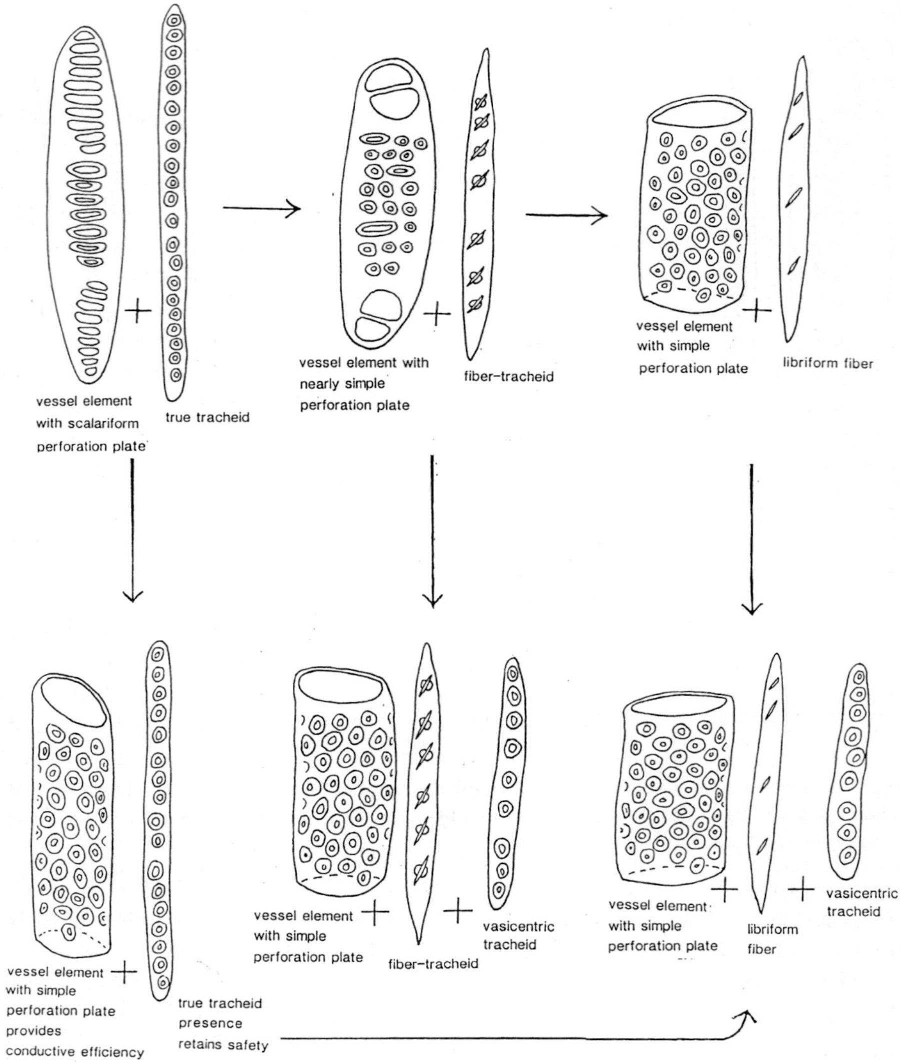

Fig. 11.2. A scheme showing ways in which the three stages shown in Fig. 11.1 can yield products in which conductive efficiency and conductive safety can both be present within a wood. Vessel elements with simple perforation plates produce conductive efficiency in this view, and either true tracheids or vasicentric tracheids produce conductive safety. The *horizontal arrow*, below *bottom row* of drawings, indicates origin of vasicentric tracheids by tracheid dimorphism, from a wood that has true tracheids. Additional conditions that tend to improve conductive safety are shown in Fig. 11.3

occur in 15.3% (vasicentric and vascular tracheids are additional to that figure). Comparison of these two percentages alone would lead one to the conclusion that in a predominantly dryland flora, tracheids are preferentially retained as a safety device, but that interpretation is less likely for the presence of the scalariform perforation plates. All available information leads one to the conclusion that while some modifications of a scalariform perforation plate (e.g., those with wide bars) might have a special significance, persistence of the scalariform perforation plate in woody dicotyledons chiefly represents a retention of primitive conditions in which it is functionally of little negative selective value because of slow, steady transpiration rates, as in tropical upland cloud forest, wet temperate forest, tropical rain forest understory, etc. A confirmation of this can be found in those species in which a very primitive version of a scalariform perforation plate is present, relictual presence of pit membrane fragments or strands (Figs. 3.6, 3.7). These are all from tropical cloud forest situations where transpiration is likely to be very slow and steady. These very primitive perforation plates, coincidentally, counter Zimmermann's idea that retention of scalariform perforation plates serves for sieving out bubbles after thaw of frozen water in vessels.

Simplification of the perforation plate can lead to formation of some perforations that are large or oval. By the very uncommon aberrant or transitional perforation plates of *Comptonia*, *Juglans*, and *Salix*, Muhammad and Sattler (1982) have attempted to show relationship between angiosperms and *Gnetum*, for which they figure similarly transitional or aberrant plates. This selection of unusual perforation plates does not constitute a reason to reject the concept of Bailey (1944b), based on massive evidence, that vessels in angiosperms and Gnetales originated independently (Carlquist 1996d). This theory is mentioned at this juncture because a few selected stages in simplification of the perforation plate (in both *Gnetum* and the selected dicotyledons) are used by Muhammad and Sattler. Recent cladograms based on expanded molecular evidence (Soltis et al. 1999) show Gnetales nested within conifers.

The theory of Young (1981) that angiosperms may be secondarily vesselless has been questioned, and on the basis of the DNA evidence from several gene sites, a vesselless genus, *Amborella*, has been claimed to be basal to all other dicotyledons (Soltis et al. 1999). Young's claim of vessellessness in Cactaceae and Crassulaceae are in error. Dicotyledons would have had to have entered more extreme xeric regions than they have at present to have become vesselless, then they would have had to shift into the most mesic situations possible (leaving no representatives in other habitats) to account for the present highly mesic preferences of vesselless dicotyledons. This theory is cited at this point to show that the vessel, although it has less potential safety as a conductive cell in woods than a tracheid has, is not by itself of sufficient negative selective value to result in its elimination. There may be other possible evolutionary reasons why vessels might be eliminated from woods, but I know of no hypotheses that give such reasons. The fact that narrow vessels embolize less readily than wide ones (Ellmore and Ewers 1985; Hargrave et al. 1994) suggests that the narrower vessels become, the more safety they have, so that in water-stress situations, very narrow vessels are not likely to be eliminated even in the most extreme situations.

11.2.6 Lateral Walls of Vessels

As discussed in Section 3.8.3, Frost (1930b) demonstrated a progression from scalariform to transitional to opposite to alternate pitting (Fig. 11.1). This series can be expressed in terms of change from more tracheidlike to less tracheidlike. As noted in Chapter 3, we need not hypothesize a wood composed wholly of scalariformly pitted tracheids – that is quite unlikely (Sect. 3.6.1). Increase in mechanical strength was suggested as the reason why vessel walls have undergone the shift from scalariform to alternate pitting (Carlquist 1975a). The shift in pitting type cannot be explained on the basis of conductive efficiency or conductive safety – thus, more than one kind of selective factor must be hypothesized to account for the major trends of xylem evolution in dicotyledons.

The reason for increased mechanical strength of vessel walls is not so evident, however. Stronger vessel walls would play a role in increasing mechanical strength of a wood, and possibly stronger vessel walls would prevent vessel deformation when strong tensions occur in vessels. Comparative data are difficult to interpret on this point, because woods that have vessels with minimal mechanical strength, such as *Carica* or *Crassula*, tend to occur in succulent plants where tensions during conduction probably never develop to high values. I see no problem in hypothesizing conductive efficiency as the factor responsible for evolution of the simple perforation plate while hypothesizing mechanical strength as the factor leading to change in lateral wall pitting: conduction through lateral wall pitting is probably so much less than that which traverses the end wall that selection for conductive efficiency in lateral wall pitting would be difficult to imagine. Both mechanical strength and conductive efficiency can operate as selective factors on a single cell type in the case of tracheids. The tracheids of *Bubbia balansae*, for example, show an end wall pattern that represents selection for conductive efficiency, yet the lateral walls are not similarly affected, thereby showing that mechanical strength considerations prevail there (Carlquist 1983b).

11.2.7 Axial Parenchyma

Kribs (1937) used commonly accepted types of axial parenchyma construction as categories when he attempted to correlate axial parenchyma evolution with vessel evolution in wood of dicotyledons (Section 5.2.2). The results are clear in that diffuse parenchyma is clearly the most primitive, whereas other types rank as more specialized to various quantitative degrees. The problematic aspect of Kribs' treatment lies in an attempt to sequence the parenchyma types into an evolutionary series. This attempt is only partly successful, for entirely understandable reasons. Types of parenchyma other than diffuse and absent may represent different pathways that achieve much the same results: more massive sheets of vertical tissue, forming fewer but larger points of contact with the ray system. Both bands of apotracheal parenchyma and bands (or cylindrical sheaths) of paratracheal parenchyma do this. From the Kribs data, one can read that more massive axial parenchyma aggregations, either apotracheal or paratracheal, are more specialized than smaller aggregations.

The significance of paratracheal parenchyma occurrence seems to have been revealed by a series of workers (Braun 1961, 1970, 1983, 1984; Sauter 1966a,b; Sauter et al. 1973; Czaninski 1977; Gregory 1978; Sperry et al. 1988). These workers find that release of sugar into vessels (from starch in the axial or ray parenchyma, particularly axial parenchyma) is a mechanism for renewing and maintaining conductive processes in the xylem. Although these workers have dealt only with trees, there is every reason to think that their results apply to woody herbs (Carlquist and Eckhart 1984), vines (Carlquist 1985d), and other growth forms.

The occurrence of "contact cells" in parenchyma as denoted by Braun (1967a,b, 1968a,b) represents the anatomical mechanism for achieving transfer of sugars into vessels. Braun's concept of "isolation cells," however, has an unfortunate negative connotation. The "isolation cells" are involved in conduction of photosynthates, and the elongate nature of these cells relates to their conductive abilities.

One can entertain the possibility that grouping of axial parenchyma cells is influenced by evolution of imperforate tracheary elements. Theoretically, aggregation of diffuse cells would avoid isolation of parenchyma cells among nonconductive cells such as libriform fibers. In fact, axial parenchyma cells even when diffuse are probably never isolated from each other and from ray cells; axial parenchyma cells probably form an interconnecting network with other parenchyma cells even if they are apparently dispersed in the wood. We tend to have a section-oriented view of cell dispersion in woods.

When axial parenchyma is absent, it often is supplanted by living fibers (which may be designated as nucleated fibers or septate fibers). In a sense, this is the most massive form axial parenchyma can take: conversion of the entire fibrous background of a wood to a quasiparenchyma system. Living fibers are certainly a potential photosynthate storage system, but although they offer massiveness of storage volume, they have the disadvantage of small and sparse pit interconnections with each other and with other cell types, and of lack of contact of many of the fibers with ray cells. A system of living fibers forms a valuable photosynthate storage mechanism where events related to that storage are not sudden (Carlquist 1975b); where sudden flushes of growth and flowering occur, types of photosynthate storage and conduction other than living fibers have a high selective value.

In addition to transfer of sugar into vessels, the axial parenchyma, along with ray parenchyma, represents a site for storage of photosynthates in the plant. These photosynthates do more than regulate conduction; they serve as the basis for growth and flowering events. Axial parenchyma is often rich in storage of starch. Unfortunately, textbooks and reviews place much more stress on tracheary elements than on wood parenchyma, so understanding of wood parenchyma functions has not been well advanced.

11.2.8 Rays

The analysis by Kribs (1935) of ray evolution showed ingenuity in that he had to define types of ray histology that represented probable evolutionary levels in order to compare ray evolution to a standard such as vessel element length. Kribs devised the types on the basis of three kinds of changes: (1) elimination of either multiseriate or uniseriate rays, leaving one or the other exclusively; (2) increase in homogeneity of ray cell shape, which in Kribs' interpretation means an increase in the proportion of procumbent cells in rays; and (3) elimination of uniseriate wings on multiseriate rays. Ray heterogeneity, a manifestation of phyletic primitiveness, consists of presence of both procumbent and upright (erect) ray cells; the procumbent cells are generally in the multiseriate portions of multiseriate rays in primitive woods.

The ray types of Kribs have been widely accepted in their original definitions; a later modification (Kribs 1950) which creates potential confusion has not been followed.

A functional interpretation of why ray evolution in terms of the Kribs' types should take place has been offered earlier (Carlquist 1975a). In this interpretation, both axial and radial parenchyma systems become more efficient at vertical and radial photosynthate conduction, respectively, and the points interconnecting the two systems have become fewer and more massive, enhancing the efficiency of parenchyma conduction (massive rays that maximize radial conduction of photosynthates are not likely to be connected to a system of axial parenchyma that is not similarly efficient). Upright cells in multiseriate rays are present in Heterogeneous type I rays chiefly as uniseriate rays composed of upright cells and upright cells in the uniseriate wings on multiseriate rays. These upright cells are not maximally efficient for radial transportation of photosynthates; they are like a histological transition between the axial and radial parenchyma, a transition made desirable by the numerous and small points of contact between the two systems. As both systems become more specialized, such transitional cells become unnecessary, and upright cells disappear from rays. Massive rays consisting of procumbent cells exclusively are most efficient when connected to similar massive vertical conduits: apotracheal banded and paratracheal banded or vasicentric parenchyma.

11.2.9 Imperforate Tracheary Elements

The shift in imperforate tracheary elements from tracheids to fiber-tracheids to libriform fibers has been linked to the major trends in xylem evolution (Fig. 11.1). Statistical documentation of the Frost-Kribs sort has been offered to support this by Metcalfe and Chalk (1950, p. xlv). Instead of the phrase designation "fibres with distinctly bordered pits" used by Metcalfe and Chalk, the term "tracheid" as suggested by the IAWA Committee on Nomenclature (1964) is used here. The evolutionary trends in imperforate tracheary elements do not involve merely degree of border presence on pits of tracheids. Not only are borders narrowed, but pit diameter tends to be lessened and pit density diminished during the shift to fiber-tracheids and libriform fibers. As stressed in the wording of Fig. 11.1, these trends enhance mechanical strength, because the weak points of the wall represented by the pit membranes are diminished in size and abundance on the cell. As the pit membrane area per cell of imperforate tracheary elements is diminished, the conductive function of the cell is very rapidly cancelled. Circumstantial evidence (Carlquist 1984a) suggests that fiber-tracheids as well as libriform fibers are nonconductive cells: the conductive capacity of fiber-tracheids is insufficient to deter vessel grouping as a mechanism for insuring safety in wood.

Although imperforate tracheary element specialization in a given wood is not always synchronous with evolutionary status of vessel elements (a synchrony is illustrated in Fig. 11.1), one can still describe the shift of vessel elements to more efficient conductive conduits while imperforate tracheary elements become ever more adaptive as mechanical elements in a division of labor. In conifers, such division of labor is minimal; the relatively moderate division of labor that exists in conifer tracheids occurs by virtue of formation of earlywood and latewood, which tends to favor conduction and mechanical strength, respectively. In dicotyledons, growth rings feature conductive efficiency in earlywood, but safety is enhanced in latewood.

11.2.10 Storied Wood Structure

To the best of my knowledge, no tabular data relating storied wood structure to vessel element length have been compiled, although that task could easily be accomplished. However, Bailey (1923) clearly enunciated a relationship between short fusiform cambial initials and storying, and thus storied cambia become part of the major trends of xylem evolution. One cannot cite degrees of storying as intermediate degrees of specialization the way one can pick lateral vessel wall pitting type or ray types that are intermediate between primitive or specialized (degrees of storying in the cambium are ontogenetic rather than phylogenetic). A wood when mature tends to have either a storied cambium or a nonstoried cambium. The manifestations of the storied cambium in wood do vary not because of intermediacy in evolutionary status but because in some species, imperforate tracheary elements (chiefly libriform fibers) have a greater degree of intrusiveness and do not show storying when mature, even though they were derived from storied fusiform cambial initials.

Even though we lack degrees of intermediacy (at least easily perceived ones) in dealing with storying, there can be little doubt of its evolutionary status or the fact that it is not a change that is achieved as readily, in phylogenetic terms, as widening of vessels. If one looks at a list of genera that have storying (Sect. 9.2.4), one finds only genera with simple perforation plates, a clear indication not merely of phyletic specialization of storied wood structure, but of its correlation with the major trends of xylem evolution.

11.3 Irreversibility

The use by Bailey (1944b) of the word "irreversible" with respect to the above features, which I (Carlquist 1961a) termed the "major trends of xylem evolution," has undoubtedly appeared to some workers as an exaggerated claim, because in such a diverse group as angiosperms, some kinds of exceptions to unidirectional trends would not be unexpected. For example, intercalation of paedomorphosis in a wood results in increasing length of fusiform cambial initials despite the overall phyletic tendency in dicotyledons toward length reduction. Statistically, the progressions discussed above show one direction, although exceptions in a particular genus are always conceivable. Discussion of Bailey's viewpoint has been offered earlier (Carlquist 1980a; Baas et al. 1983), and a literal interpretation, allowing no exceptions, has been challenged by ter Welle and Koek-Noorman (1981) and Baas and Wheeler (1996). Baas and Wheeler compare vessellessness to a distribution of vesselless families on a particular cladogram, but in fact, the vesselless dicotyledons are arranged quite differently in a more recent cladogram (Soltis et al. 1999), showing the problems of finding a standard against which the concept of irreversibility – defined narrowly or loosely – can be compared. I do not believe that ter Welle and Koek-Noorman (1981) interpret my (Carlquist 1975a) discussion correctly when they claim I found adaptation to ecological features a way in which reversion can occur in the major trends of xylem evolution. In fact, adaptation to ecology is precisely the force that drives the major trends of xylem evolution. This was realized by Bailey (1966) when he stated, "It should be emphasized here that in future investigations of plants of divergent habits of growth more attention should be devoted to ecological and physiological influences in the habitats in which plants normally grow."

One can cite a kind of "noise level" (Carlquist 1980a) within which variations of the various "irreversible" features can occur. By "noise level," I mean the kinds of variations demonstrated by standard deviation, by prolonged juvenilism (Fig. 9.2), by phenotypic modifiability, or by dwarfing (e.g., Baas et al. 1984). In some cases, vessel elements with many bars per perforation plates occur within the same wood samples as vessel elements with few bars or even simple perforation plates; this is an example of what I call "noise level." The occurrence of both tracheids and fiber-tracheids in *Austrobaileya* (Bailey and Swamy 1949) is not an example of reversion, but of progressive evolution (Carlquist 1988c).

Paedomorphosis can have the effect of masking the major trends, but this is a case of overlay of juvenilistic features, not of reversion. Paedomorphosis, for example, does not convert specialized rays to primitive ones, it converts them to juvenilistic ones. Paedomorphosis may result in production of occasional scalariform perforation plates in a secondary xylem with perforation plates otherwise simple, in species in which primary xylem has scalariform perforation plates (e.g., *Vitex lucens*; Butterfield and Meylan 1975), but surely this situation can be seen for what it is rather than used to symbolize evolutionary chaos.

Fossil woods have been providing evidence that supports the Baileyan ideas (Wheeler and Baas 1991). Scalariform perforation plates are progressively more common as one goes backward in time (Page 1981; Scott and Wheeler 1982). Fusiform cambial initials (judged by the wood cells derived from them) have shortened over geological time (Manchester 1979). More primitive axial and ray parenchyma types were more common in fossil wood assemblages than they are in contemporary floras (Bande and Prakash 1984). We should not be surprised if, in a fossil assemblage, we find some woods that are much more specialized than others (Page 1981), for phylads with specialized wood features have been evolving constantly in dicotyledons, and not all phylads are synchronous in development of advanced features.

Why should some features in wood show progression in a particular direction strongly enough to evoke, with whatever reservations and exceptions one may choose to evoke, the word "irreversible"?

1. If driven by adaptation to ecological features, evolution is constantly providing kinds of woods better adapted to dealing with frost and drought. There is no question that highly primitive dicotyledon woods are, to an extraordinarily large extent, restricted to localities that are mesic and frost-free, or at least mesic (with deciduous habit counteracting the cold season, for example). Climatic events that are more seasonal will tend to encourage evolution toward greater conductive efficiency, such as loss of bars on perforation plates, as contended by Wolfe (1981) on the basis of fossil floras. As noted earlier (Carlquist 1980a), highly seasonal climates are more perilous

than mesic, frost-free climates, and extinction in the former is very likely much higher because of greater frequency of catastrophic climatic events in more seasonal climates. Thus, evolution will continually favor radiation of phylads in saturated mesic habitats into unsaturated seasonal habitats. Specialized woods, however, are not at a disadvantage in the mesic, frost-free habitats, and will be able to enter them under some phytogeographical circumstances. Thus, arborescent Asteraceae with specialized woods are entirely competent in wet Hawaiian forest (Carlquist 1974), for example.

2. Conceding that specialized woods have no disadvantage in mesic situations and succeed better than primitive ones in seasonal habitats, the only feature that prevents entry into more seasonal habitats is the lack of favorable mutations leading to more specialized conditions (assuming wood alone as significant for a plant in this instance). Adding bars is unlikely to have any beneficial effect, and would probably merely increase impedence. Even if scalariform perforation plates were to provide the kind of bubble-sieving action that Zimmermann (1978b, 1983) envisions, other mechanisms must be just as effective, because floras with deep frost have many species with simple plates.
3. There is a functional interrelationship between many of the features described above, so that the direction of evolution of one feature is indissolubly linked to the direction of all features. As an example, the following intricate pattern is suggested. Imperforate tracheary elements in a phylad lose borders on their pits (together with diminution in pit diameter and density). This accompanies greater division of labor, in that not only do the imperforate tracheary elements go from conductive to nonconductive, the vessel elements proceed to improved conductive efficiency. Axial and ray parenchyma become aggregated into larger groupings with more massive and fewer contacts between horizontal and vertical systems. Other details could be added to this scenario, but the syndrome nature of the pattern is evident. The reversion of any one feature would require a simultaneous reversion of the other features, and this is both a functional and a statistical unlikelihood.
4. The changes involved in the major trends of xylem evolution may be hypothesized to have a relatively complicated genetic basis compared to other wood changes. For example, widening of vessels is relatively easily changed, in all likelihood, compared to achievement of a storied condition. If the genetic basis of the features involved in the major trends of xylem evolution is more complicated, reversion becomes progressively more difficult when specialized character states are achieved.
5. The vessel, imperforate tracheary element, axial parenchyma, and ray characters cited above do not all relate to the same function. Thus, perforation plate simplification improves conduction of water in vessels but change of lateral wall pitting (presumably) improves mechanical strength of vessel walls. Mere escape from a tracheidlike conformation cannot explain the evolutionary trends. One can say that the major trends of wood evolution

represent not one ecological or physical (mechanical strength) factor, but a series of different factors to which specialized wood represents numerous adaptations. Even if selective pressure for one of these factors – say, mechanical strength – were reversed, reversion in photosynthate flow efficiency in parenchyma would not be similarly affected. Different selective pressures do occur for the features independently, but they operate in similar directions. Strong selection for improved mechanical strength might accelerate loss of borders on pits of imperforate tracheary elements in a plant that remained in a highly mesic setting and thus did not experience pressure for loss of bars from perforation plates, so synchrony between evolution of the two features would not be present, but note that selective pressure is not running backwards. The genus *Meliosma* is an example of a genus in which the preceding scenario may apply (Carlquist et al. 1993).

11.4 Features Related to the Major Trends of Xylem Evolution, but with Modifications

11.4.1 Ratio Between Imperforate Tracheary Element Length and Vessel Element Length (F/V Ratio)

Although Bailey and his students did not comment on this feature, the Bailey and Tupper (1918) data clearly show that dicotyledonous woods are diverse with respect to this feature. If a plant with vesselless wood can, by definition, be said to have a particular mean tracheid length, a vessel-bearing derivative of that plant would be expected to have a divergence in length between its vessel elements and its imperforate tracheary elements. The divergence between vessel element length and length of imperforate tracheary elements ought to increase with phylesis, given intrusive capabilities of imperforate tracheary elements as they mature. This hypothesis (Carlquist 1975a) proves valid. A sample of dicotyledons with scalariform perforation plates, tracheids, and diffuse parenchyma showed a ratio of 1.40. Particular families of dicotyledons noted for primitive features have a lower ratio, such as Bruniaceae (Carlquist 1978a), but Illiciaceae with a ratio of about 1.40 are more typical of primitive families (Carlquist 1982c). The average dicotyledon has a ratio (termed "F/V ratio" by authors such as van Vliet et al. 1981) around 2.0 (based on mean vessel element length and mean "fibre" length for dicotyledons; Metcalfe and Chalk 1950, pp. 1360–1361). Values mostly fall between 1.3 and 1.8 in Melastomataceae (van Vliet et al. 1981). A sample of dicotyledons with specialized woods has a value of 2.60 (Carlquist 1975a); Asteraceae have a similar average value (Carlquist 1966a).

There are a few dicotyledon families in which a much higher F/V ratio is achieved. One can find a ratio as high as 9.50 (*Zelkova americana*, data from

Bailey and Tupper 1918). However, one cannot impute an extraordinary degree of specialization to *Zelkova americana* because of that value. Progression from 1.00 to about 2.60 represents phylesis toward an optimal mechanical system (and perhaps an optimum vessel element length in proportion) and is a division of labor related to the major trends of xylem evolution. Degrees of intrusiveness appreciably beyond 2.60 in families such as Urticaceae and Ulmaceae may represent special cases in which there is selection for exceptionally long fibers.

11.4.2 Vessel Grouping

After reviewing preparations and description of dicotyledons at large, I found that some genera showed phyletic increase of vessel grouping proportionate to dryness of habitat (e.g., Asteraceae; Carlquist 1966a). In other families, in contrast, no vessel grouping occurs even though the habitat may be very dry (e.g., Krameriaceae). There proves to be a correlation between solitary vessels and tracheids in dicotyledons as a whole (Carlquist 1984a). In brief, families with tracheids always show vessels solitary (or nearly so) in transection. Families in which imperforate tracheary elements are fiber-tracheids or libriform fibers may have solitary vessels or grouped vessels, depending on the degree to which dryness characterizes the habitat of a particular species (Carlquist 1984a). Solitary vessels are not indicative of primitiveness, contrary to what the data in a table of Metcalfe and Chalk (1950, p. xlv) might suggest. The apparent correlation between solitary vessels and primitiveness in woody dicotyledons is an indirect one: tracheids are statistically correlated with primitiveness in dicotyledons (Metcalfe and Chalk 1950, p. xlv – "fibres with bordered pits"); solitary vessels are an indication of tracheid presence, not primitiveness, as further information cited below demonstrates.

The functional significance of vessel grouping is related to conductive capability of imperforate tracheary elements. Tracheids can be hypothesized to be conductive in nature, as suggested by their density of large fully-bordered pits. In woods with tracheids, disabling of vessels by air embolism in vessels does not alter the conductive capability of the wood except in terms of peak volume: the three-dimensional conductive pathways are preserved by the fact that the entire ground tissue of such a wood is composed of conductively functional tracheids. If vessels are disabled in a wood with fiber-tracheids or libriform fibers adjacent to vessels, the three-dimensional conductive pathways cannot be maintained by imperforate tracheary elements adjacent to vessels. The conductive capability of tracheids (as the term is used in this book) is validated by the work of Braun (1970), and fiber-tracheids and libriform fibers as construed here are nonconductive.

If several vessels occur in a group, one or more vessels may not develop air embolisms (particularly if some of the vessels are narrower), so that the intact vessels do maintain the three-dimensional conductive pathways (Fig. 11.3) in woods with nonconductive imperforate tracheary elements. Radial grouping would have the effect of supplying a sequence of intact vessels as earlier ones are disabled.

Some grouping of vessels can occur in woods with tracheids in which growth rings are very narrow (Fig. 2.6, part 3), simply because a high degree of crowding leads to occasional random contacts of vessels. Solitary vessels can occur in woods of dry habitats provided that vasicentric tracheids are abundant, as in *Quercus* (Fig. 4.7, part 1), providing evidence that there is no link between primitiveness and solitary vessels. Only a few instances of vasicentric tracheid presence, however, show sufficient abundance so that the vasicentric tracheids have the effect of deterring vessel grouping. *Quercus* probably exemplifies tracheid dimorphism – the process indicated by the horizontal arrow, lower portion of Fig. 11.2. The data available (Carlquist 1984a) show that presence of tracheids is a more effective mechanism for maintaining conductive pathways than grouping of vessels, since the latter occurs only in woods with the nonconductive imperforate tracheary elements, fiber-tracheids, and libriform fibers.

11.4.3 Vessel Widening

The table of Frost (1930b; Sect. 3.13) correlates angular vessels with narrow vessels; vessels rounded in outline are wider, according to Frost's figures. This correlation has received little comment in literature on wood evolution. The best interpretation of the data in this table seems to be that widening in order to achieve greater conductive capacity is one of the changes by which vessel elements become less tracheidlike and better adapted to conductive efficiency.

However, vessel widening, while apparently operative to a degree at the outset of vessel evolution (vessels are wider than tracheids), should not be considered as a universal measuring-stick of wood evolution. For example, very wide vessels can be achieved in various phylads, and presence of extremely wide vessels in *Quercus* should not lead us to assume that *Quercus* wood is therefore extremely specialized. Once simple perforation plates have been achieved, further widening of vessels is probably controlled by ecological factors independent of the major trends of xylem evolution.

11.5 Features Not Related to the Major Trends of Xylem Evolution: Conductive Efficiency and Safety

The features discussed below are not related statistically to length of fusiform cambial initials. These features are therefore considered readily reversible.

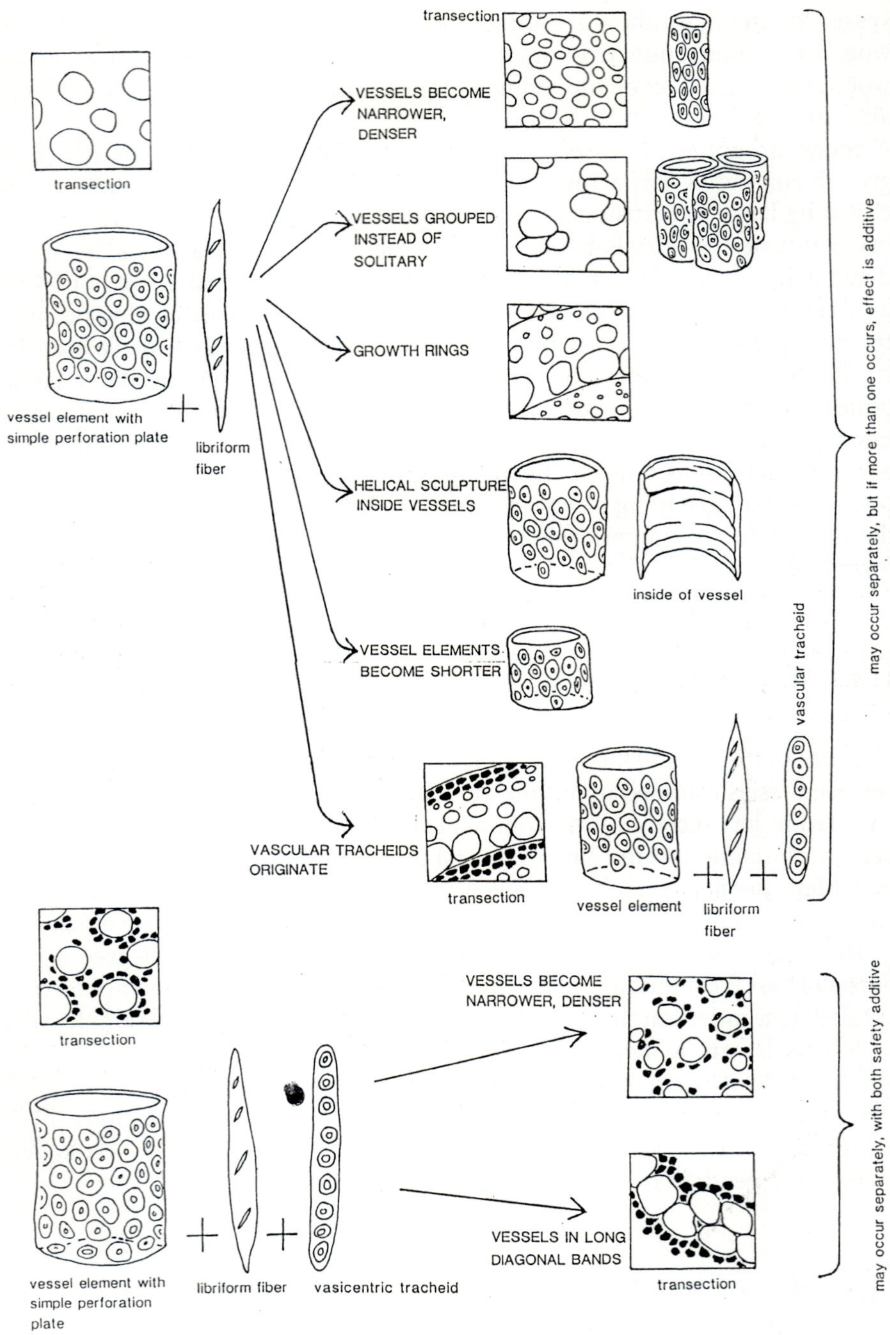

Fig. 11.3. A scheme showing ways – in addition to those shown in Fig. 11.2 – in which conductive safety can be achieved. All of the products on the *right half* of the figure tend to be considerably more abundant in areas where more marked seasonal dryness occurs. See also text

11.5.1 Vessel Diameter

Wide vessel diameter contributes to greater conductive efficiency (Baas 1976; Carlquist 1976a; Zimmermann 1978b). Vines and lianas exemplify this to a marked degree, as noted by various authors beginning with Schenck (1893). Wide vessels offer minimal friction.

Narrow vessels, in contrast, must not be regarded as a form of conductive inefficiency. Narrow vessels tend to develop embolisms much less readily than wide ones (Zimmermann and Brown 1971; Ellmore and Ewers 1985; Ewers 1985; Hargrave et al. 1994; Jarbeau et al. 1995). Wider vessels are less capable of recovery from embolisms than narrow ones (Ewers 1985).

11.5.2 Vessel Length

Zimmermann (1978a) demonstrated that various lengths of *vessels* (note: not to be confused with length of vessel *elements*) may be observed, depending on the species and depending on the portion of the plant. Long vessels characterize the wide earlywood vessels; vessels in latewood are shorter. Zimmermann (1983) concluded that longer vessels confer greater conductive efficiency, shorter vessels confer greater safety (spread of air embolisms is stopped by the terminus of a vessel). Unfortunately, we have information on vessel length in only a few plants, and such data are not easy to obtain.

11.5.3 Vessel Density

Number of vessels per mm^2 might easily be regarded merely as an inverse of vessel diameter, but that would be a mistake. Certainly there are packing constraints. However, the relationship is by no means perfectly inverse. This can be demonstrated by Fig. 3.1 (part 3); in this species (*Nepenthes lowii*), vessel density is half or less what it is in other species with comparable vessel diameter. Greater vessel density confers the potential advantage of greater redundancy (Zimmermann 1978b; Carlquist 1980a).

11.5.4 Growth Rings

Growth rings show the reversibility of vessel diameter and vessel density in a single season. The earlywood characteristics offer conductive efficiency; latewood offers conductive safety (Fig. 11.3). Ring porosity has been considered an adaptation to seasonal climates (Gilbert 1940; Sweitzer 1971; Carlquist 1980a). Evidence from the fossil record shows that in post-Eocene time, ring porosity developed as seasonality in climates increased (Wheeler and Baas 1991; Scott and Wheeler 1982). We have few data on the degree to which growth rings may be obligate or non-obligate in particular species, but growth rings are rather modifiable in a sampling of Californian species (Bissing 1976).

Most conduction occurs in earlywood, and for the most part, wood of the most recent growth ring accounts for most conduction; most vessels in ring-porous woods become nonfunctional after a year, and earlywood vessels become nonfunctional before the end of a single season (Huber 1935; Ellmore and Ewers 1985). Latewood vessels may function for more than a single season (Huber 1935; Zimmermann and Brown 1971; Ellmore and Ewers 1985).

Although ring porosity has often been treated as an adaptation, if only by default, it is in fact an assemblage of phenomena. This was demonstrated earlier (Carlquist 1980a) and has been used in the construction of the types of growth rings recognized in Chapter 2. The potential physiological significance of the various growth ring types is detailed there.

11.5.5 Vessel Dimorphism, Fibriform Vessels

The habit of having wide vessels plus narrow vessels, with relatively few vessels intermediate in size, has been termed "vessel dimorphism" (Carlquist 1981b). Narrow vessels in plants that show this condition are called "fibriform vessels" (Woodworth 1935). This condition is common in lianas and vines (Ayensu and Stern 1964; Carlquist 1984f, 1985d). However, fibriform vessels can be found in nonscandent groups such as shrubby Hydrophyllaceae (Carlquist et al. 1983; Carlquist and Eckhart 1984).

The occurrence of narrow vessels in vines seems related to widening of a few vessels, which thereby forestall the enlargement of others. That explanation, however, may not be applicable to the shrubby Hydrophyllaceae. In functional terms, vessel dimorphism results in abundance of narrow vessels which, as noted in Section 11.5.1, have much greater conductive safety than wide vessels: narrow vessels would continue to conduct even though the wider vessels become embolized.

11.5.6 Conductive Area

Comparative data on conductive areas (mean transectional vessel area times number of vessels per mm^2) have been computed for a scattering of species (Carlquist 1975a; Styer and Stern 1979b). Such data show that vines, lianas, and annuals have a high proportion of their stems devoted to vessels. Relatively low conductive areas characterize succulents, desert shrubs, and rosette shrubs, all of which probably experience low conductive rates. Thus conductive area seems related to conductive efficiency. This calculation does not express conductive safety.

11.5.7 Helical Sculpture

Types of helical sculpture and their systematic and ecological distribution are covered in Section 3.12. The ecological distribution shows higher frequencies of occurrence in dry areas (Webber 1936; Carlquist and Hoekman 1985b), and it is one of a number of adaptations seen in woods of such areas (Fig. 11.3). Within a plant, helical sculpture is more common in latewood (Carlquist and Hoekman 1985b). Distribution of helical sculpture in woody dicotyledons leads one to believe that helical sculpture improves safety by virtue of increasing the surface area of vessels. This might increase the bonding of water to vessel walls (Carlquist 1983f). Jeje and Zimmermann (1979) found that resistance to flow in vessels is lowered and therefore conductive rates might be enhanced by presence of helical sculpture in vessels. I doubt that rapid conduction is the selective reason for development of helical sculpture. Rather, the observations of Jeje and Zimmermann most likely indicate that refilling of embolized vessels is aided by the helical sculpture. However, comparative data do not support the conclusion that increased conductive rates are the factor accounting for evolution of helical sculpture. The greater prominence of helical sculpture in latewood is understandable because latewood vessels represent by their diameter a last line of resistance to cavitation, so presence in them of the maximal wall configuration in addition to their narrowness mechanism has an additive effect, making the latewood maximally secure – as indeed it should be.

11.5.8 Vesturing

Formerly separated into the phenomena of vestured pits and of warted wall surfaces of vessel walls and tracheids, the term vesturing now applies to both phenomena. In fact, both types of occurrence may sometimes be found together and are intercontinuous, as in *Leptospermum ericoides*, *Metrosideros robusta*, and *Persoonia toru* (Meylan and Butterfield 1974). A hypothesis that would explain vestured pits ought to explain vestured walls also. On this account, Zweypfenning's (1978) hypothesis must be discounted, for it deals only with vestured pits. Zweypfenning claimed that the vestures on pits prevent excessive deflection and rupture of pit membranes when there is a greater tension on one side.

Vesturing represents an interesting case in interpretation (Carlquist 1983f) because comparative anatomy holds key information. Vesturing is present in woods from dry places, such as *Eucalyptus* and other Myrtaceae, but not all Myrtaceae with vesturing occur in dry habitats. This caused hesitation by Zweypfenning (1978) in offering his hypothesis, but systematic literature indicates that vesturing is relatively conservative, characterizing whole families, even an entire order, with few exceptions (Myrtales; van Vliet and Baas 1984). Thus, one can imagine that vesturing has originated only a few times and disappears slowly when shift from dry to wet habitats occurs (vesturing probably is of little or no negative selective value in wet environments).

Parham and Baird (1974) found some interesting distributions of vesturing with respect to growth rings. In *Fagus grandifolia*, *F. orientalis*, and *Sassafras albidum*, vesturing is absent in earlywood vessels but present in latewood vessels. In Winteraceae, vesturing is present in tracheids of cold-temperate species (Patel 1974; Carlquist 1988a, 1989c), but absent in subtropical species (Carlquist 1983a, 1988a, 1989c, 2000c). These distributions remind one of distributions of helical sculpture. For these reasons, similar functional hypotheses of embolism prevention or vessel refilling (Carlquist 1983f) may be entertained. In either case, increased safety would result. Vesturing is reversible in phylads, and has been lost in some Melastomataceae (ter Welle and Koek-Noorman 1981) and Onagraceae (Carlquist 1975b). Van Vliet and Baas (1984) suggest origin only a few times (perhaps a single time in Myrtales), followed by development of variations within phylads.

11.5.9 Vasicentric Tracheids

As a review of this phenomenon (Carlquist 1985a) shows, vasicentric tracheids are considerably more common than hitherto realized (see listing; Sect. 4.11.4). Species with vasicentric tracheids tend to occur in Mediterranean-type areas (Carlquist 1985a; Carlquist and Hoekman 1985b) and in dry tropical areas. Vasicentric and true tracheids are more common in vines and lianas than in other growth forms (Carlquist 1985d). Vasicentric tracheids have probably evolved as a safety mechanism, one that permits phylads to enter seasonally quite dry localities (Fig. 11.3). Confirmatory of this hypothesis is the fact that all but two of the families listed as having vasicentric tracheids have simple perforation plates, indicative of seasonally dry climates. Vasicentric tracheids are thus a mechanism for safety much like vessel grouping. Vasicentric tracheids provide a subsidiary conductive system arranged so that when vessels embolize, the vasicentric tracheids maintain the same three-dimensional pattern of conduction that the vessels represented. In addition, vessels can be linked together by means of vasicentric tracheids into diagonal aggregations (Fig. 11.3, near bottom). These aggregations incorporate many to most vessels into intercontinuous groups. These groups unite most vessels in the wood into a single network, so that deactivation of the conductive system in any part of the network does not disable the system as a whole.

Vasicentric tracheids can occur in woods with fiber-tracheids (even in some woods with fiber-tracheids transitional to tracheids), or in woods with libriform fibers (Figs. 11.2, 11.3). This suggests polyphylesis of vasicentric tracheids from woods of various phyletic states, and several pathways have been suggested (Fig. 4.2). In some of the phylads with vasicentric tracheids, only those species in the most stressful environments have vasicentric tracheids. These various circumstances suggest the reversibility of vasicentric tracheids.

11.5.10 Vascular Tracheids

By virtue of a redefinition (Carlquist 1985a), which in fact resembles the usages of many workers closely, vascular tracheids are defined as tracheids at the end of a growth ring that are, in a sense, vessels too narrow to bear perforation plates (hence the name "vascular"). Vascular tracheids occur in woods that also have libriform fibers or fiber-tracheids. As noted in Section 4.10, vascular tracheids potentially provide maximally embolism-resistant cells at the end of a growing season; these can serve as a water-bearing sheath that protects the integrity of the cambium throughout a dry season. Thus, vascular tracheids as defined above give safety to the conductive system (Fig. 11.3), but their distribution bears a relationship to what is protected: cambium, rather than a three-dimensional network that would protect foliage. For this reason, vascular tracheids tend to occur in drought-deciduous species.

11.6 Features Not Related to the Major Trends of Xylem Evolution: Features Related to Photosynthates

11.6.1 Interxylary Phloem

Interxylary phloem (strands of phloem formed internally from a single cambium) is not related to vessels. The distribution of interxylary phloem speaks for a function in storage and translocation of photosynthates (Carlquist 1975b). Systematic distribution of interxylary phloem in Onagraceae is suggestive of relationship to growth and flowering events, and appears to correlate with sudden flushes of growth or flowering (Sect. 8.3). More investigation is needed on functioning of interxylary phloem.

11.6.2 Parenchyma Absence, Raylessness

In rayless stems of limited duration, as in woody species of *Plantago* and *Stylidium*, little or no axial parenchyma or ray tissue is produced. This appears to be a short-term gain in mechanical tissue in a situation where conduction of photosynthates is not of selective importance (Sect. 6.6). Raylessness is infrequent in dicotyledons at large, appears related to secondary woodiness or other short-term selection for increased mechanical strength, and occurs in some plants for finite periods of time (rays are formed in older stems).

In many other cases, loss of axial parenchyma is part of a shift that involves production of a parenchyma substitute, nucleated or septate fibers (Wolkinger 1969, 1970a,b). This has the effect of converting the ground tissue into a parenchymalike tissue (Sect. 4.7). In some instances, living fibers clearly serve for starch storage (Carlquist 1975b). The distribution of living fibers – as a ground tissue in which other cell types are dispersed – suggests that living fibers may not affect conduction in vessels very much (although in some species, fibers are septate near vessels, nonseptate elsewhere). Septate fibers are not ideal (note small pit size) for conduction of photosynthates except at a slow rate.

11.6.3 Parenchyma Bands Produced by Fiber Dimorphism

Axial parenchyma with this mode of origin is not related to the major trends of xylem evolution (Carlquist 1958a, 1961a), but is an innovation that has occurred in a scattering of woody dicotyledon families (Sect. 4.8). The circumstances in which fiber dimorphism occurs are worth investigation, and as more instances are reported, a pattern may emerge. Physiological studies are needed.

11.6.4 Bordered Pits in Ray Cells

Bordered pits tend to occur commonly on tangentially-oriented walls of ray cells. Although not frequently reported by wood anatomists, borders can be seen in radial sections of woods in a surprisingly large number of woods (Sect. 6.12.11). The pattern of occurrence of bordered pits in ray cells suggests that they help optimize flow of photosynthates in a radial direction.

11.7 Features Not Related to the Major Trends of Xylem Evolution: Mechanical Strength of Wood

The anatomical aspects of water conduction in woods have been explored by plant physiologists, especially in recent years. However, mechanical strength is of equal importance. The number of studies that relate to strength of woods as it is evidenced in anatomical details is still relatively few, and papers on this topic tend to appear in journals not familiar to wood anatomists (e.g., Easterling et al. 1982; Boatwright and Garrett 1983). In part, the dimensions and shapes of cells relate to strength characteristics; in part, the chemistry and ultrastructure of the cell wall material is involved.

11.7.1 Wall Thickness of Imperforate Tracheary Elements

The limits within which wall thickness of imperforate tracheary elements can fluctuate are very broad, and we know relatively little about the mechanical significance of these. Certainly some trees do have thin-walled libriform fibers (e.g., *Ochroma*), although most do not. The architecture of branching, the bulkiness of stems, the characteristic weight and stress experienced by stems all affect mechanical characteristics. Although strength characteristics of particular woods are known, few studies correlate strength with anatomy. Certainly succulents lacking in libriform fibers suggest that thickness of secondary walls in imperforate tracheary elements may be one factor related to mechanical strength. In certain woody shrubs with notably slender, wiry stems, the imperforate tracheary elements are thick-walled, as in Stilbaceae and Retziaceae (Carlquist 1986c).

11.7.2 Ray Cells

Ray cells in dicotyledon woods commonly have lignified secondary walls of moderate (2–6 μm) thickness. Woods in which ray cells have thin nonlignified walls are by no means always those of succulent or herbaceous species, but many of them are. Because ray cells have walls comparable in thickness and chemistry to those of imperforate tracheary elements, we can assume that they contribute to wood strength. We do not know at present to what extent ray cell wall characteristics contribute to the strength of a wood, or whether shape and size of rays is involved in wood strength. Some woods with tall, wide rays (e.g., *Piper*) probably do have lower wood strength on account of such rays, although in the case of *Piper*, presence of thick-walled lignified libriform fibers in wood may compensate.

11.7.3 Vessel Wall Thickness

Walls of vessel elements in dicotyledons vary in thickness considerably, and the significance of this has not been clarified. Although one can find species with thick-walled imperforate tracheary elements and relatively thin-walled vessel elements, vessel wall thickness is often greater in species with thick-walled imperforate tracheary elements. Thick-walled vessel elements may contribute to the strength of wood. The possibility that wall thickness may counter deformation of vessels under conditions of strong tensions in water columns seems somewhat less likely. However, thicker walled vessels may be found in plants of dry areas (Carlquist 1980a; Baas et al. 1983).

11.7.4 Paedomorphosis as Lessened Mechanical Strength

The hypothesis that paedomorphosis represents relaxation of selection for mechanical strength (Carlquist 1975a) is based on the idea that descending age-on-length curves (Fig. 9.2, part 1) reflect decreased mechanical strength. Woods with this curve also show other indications of lower mechanical strength: thin-walled ray cells, lessened lignification of rays, taller rays, and parenchymatization of axial xylem.

Another feature often seen in paedomorphic species that equates to lessened mechanical strength is production of a thinner cylinder of secondary xylem; this, in turn, makes preservation of juvenile features more likely.

11.7.5 Raylessness

Raylessness has some aspects of paedomorphosis, but it also represents a short-term gain in mechanical strength in stems of short duration (see Sect. 11.6.2).

11.8 Features Not Related to the Major Trends of Xylem Evolution: Water Storage

11.8.1 Parenchymatization of Axial Xylem

Replacement of mechanically strong imperforate tracheary elements (chiefly libriform fibers) by thin-walled parenchyma is a familiar feature of certain stem succulents, such as *Carica* and *Adenium*, many cacti, and many

Crassulaceae. Additional examples can be found in roots of the Andean rosette plant *Nastanthus* (Carlquist and DeVore 1998). This last example signals the potential role parenchymatization of axial xylem can play in water storage. If one could select, a priori, the xylem plan change that would result in production of a high proportion of water storage tissue, parenchymatization of axial xylem would be selected because this would result in a large volume gain for water-storage parenchyma; rays comprise a smaller portion of the stem, although they obviously represent the second largest potential volume of water-storage parenchyma in wood of a stem. In stems that have parenchymatization ("pervasive axial parenchyma") such as the globular cereoid cacti, the ray tissue is, in cell size and wall characteristics, very much like axial parenchyma (Mauseth 1993). Although secondary xylem is an important locus of water storage, wood is probably subordinate to pith and cortex in this respect.

Axial parenchyma bands are prominent in some species with succulent stems, such as *Pereskia* (Bailey 1962), *Chorisia*, and *Erythrina*. In these and similar genera, wide bands of axial parenchyma undoubtedly play a role in water storage related to dry tropical habitats. Although large zones of parenchyma such as these could be used for storage of photosynthates, the volume devoted to thin-walled parenchyma appears to exceed the volume likely to be involved in a tissue primarily devoted to photosynthate storage.

11.8.2 Rays

Water storage in rays is not served by parenchymatization, because rays are entirely or essentially parenchymatous. Rather, water storage is achieved by minimal change of primary rays as they are extended in the secondary xylem. Obviously wide secondary rays can be attained by this process only if primary rays are wide also. Paedomorphosis is a process that tends to inhibit alteration of large rays, and thereby maximize water storage in secondary xylem.

The role of rays in adapting to expansion and contraction of stems in relation to water loss and gain is worthy of note. Wide, tall rays are better able to adapt to seasonal change in water storage volume than narrow, short ones.

11.8.3 Shrinkable Tracheary Elements

Globular cacti are notable for having vessel elements and vasicentric tracheids that can contrast and expand concordantly with the water storage tissues of the stem (Mauseth 1993). In order to be shrinkable, large pit aperture areas must be present on vessels (Fig. 3.15) or (as in the globular cacti) wide-helix bands (Fig. 4.9, parts 1 and 2). Shrinkable vessel elements are associated with adjacent parenchyma that can shrink concurrently. The wide-helix bands, vessel elements, and tracheids are functionally incompatible with non-shrinkable fibers, so the occurrence of the wide-helix elements in globular cacti, which are devoid of fibers, is understandable; however, some genera of Portulacaceae (Carlquist 1998b) and Calyceraceae (Carlquist and DeVore 1998) also have helical-banded secondary xylem vessels in a background of thin-walled parenchyma.

11.8.4 Minute Lateral Vessel Wall Pits

The diameter of pits on lateral walls of vessels in dicotyledons falls mostly between 5 and 8 μm. Exceptionally wide pits are found in vessels of succulents. Smaller pits are characteristic of certain families and genera (Record 1943d). Lindorf (1994) found vessel pits to be smaller in species of a dry tropical forest. Further studies are needed to see whether ecological correlations with minute pit size are widespread.

11.9 Features Not Related to the Major Trends of Xylem Evolution: Predation Deterrence

Although various features probably functioning in predation deterrence have been discussed in this book, particularly in Chapter 7, stress needs to be laid on the multiplicity of devices and chemicals serving in this capacity, and the multiplicity of organisms that are being deterred.

Toxic heartwood compounds may help woods resist fungi, boring insects, and possibly bacteria as well (Hillis 1987; Blanchette and Biggs 1992). These come into play only incidentally in anatomical studies, when such terms as "resinlike," "gummy," "tanninlike," and "dark-staining" deposits are cited in various cell types. Steward (1966) stresses the toxicity of these compounds, but we have few experimental data on that point.

Readily visible with light microscopy are the calcium oxalate crystals, silica bodies, and cystoliths that occur in wood. These may be considered defenses against chewing and boring insects primarily.

The localities for herbivore defense mechanisms lie primarily in vacuoles of ray cells, secondarily in axial parenchyma cells, and least in imperforate tracheary elements. An exception to this ranking may occur in heartwood in which imperforate tracheary elements may contain a greater bulk of compounds formed as liquids (e.g., the "resinlike compounds"), although these compounds may be synthesized in parenchyma and then transferred to the imperforate tracheary elements. Deposition of defensive compounds and crystals is greater in the periphery of a plant (e.g., cortex, bark) than in wood. The heartwood defenses may deter fungi that cause rot, whereas defenses nearer the surface of the plant may deter chewing or boring insects and foraging vertebrates.

Two categories of structures that occur in a limited number of dicotyledon woods but may have deterrence significance in them are secretory canals and laticifers. The nature of occurrence of these two wood features is characterized by affecting only restricted areas (e.g., a single secretory canal in the center of a ray), and they seem to originate surrounded by parenchyma, whereas widespread deposition of compounds in vacuoles has the advantage of distributing a defense mechanism in massive sheetlike form throughout the wood, a pattern more difficult for predators to avoid .

As noted in the preceding chapter, any given species probably has a multiplicity of defense mechanisms against any given potential predator. Those evolved earlier in time may be continued by a phylad rather than supplanted, so that the effect is additive as new compounds and structures originate. This has the effect of distancing the defense mechanism and the predator, so that a particular category, such as raphides, may characterize a large taxonomic unit (e.g., family) rather than species. This phenomenon makes the defense mechanism deposits, structures, and compounds very useful for systematic purposes.

The phenomenon of tile cells (Sect. 6.12.1) has traditionally been regarded merely as a systematic peculiarity of four families of Malvales. However, tile cells are here regarded as cases of division of labor in rays in which some cells (e.g., the upright cells) contain one category of compounds, whereas other cells contain other compounds. This phenomenon is not unique to Malvales, although tile cells per se may be. The net effect is to alternate cells with different defense mechanisms within a ray.

11.10 Wood Features for Which Evolutionary Interpretation Is Uncertain

11.10.1 Sclerotic Tyloses

Although there is a functional explanation for the formation of tyloses, there is no apparent reason why a small number of taxa have sclerosed tyloses. Sclerosed tyloses might be merely the tendency for tyloses to have longevity in woods in which secondary wall materials continue to be secreted. The functional nature of sclereids elsewhere in the plant body may perhaps be to create textures that resist chewing insects. While one cannot deny the possibility that sclerosed tyloses (or, for that matter, sclerosed ray cells) could have this function in wood, the number of occurrences of both of these sclerosed cell types is so small that an interpretation is difficult.

11.10.2 Rays Uniseriate at Initiation of Secondary Xylem

Bailey and Nast (1948) refer to the tendency of *Schisandra* to have exclusively uniseriate rays early in ontogeny of wood as a specialized tendency, because they regard presence of Heterogeneous type I rays, both at the beginning as well as later in secondary xylem, as primitive. The condition shown by *Schisandra* is also realized in a related family, Illiciaceae (Carlquist 1982c). However, a wide range of other families, both specialized and primitive (if one judges by floral features or wood features), have rays exclusively uniseriate at the beginning of secondary xylem but both multiseriate and uniseriate later: Burseraceae (Barghoorn 1941b; Fig. 6.1), Buxaceae (Carlquist 1982d), and Ericaceae (Wallace 1986), for example. This feature has not attracted attention among wood anatomists, perhaps because one usually does not section both early-formed as well as later-formed xylem unless one is dealing with a group in which material of some wood samples consists only of twigs. The occurrence of this pattern of ray ontogeny is probably much wider than present reports would suggest, and until we know more about its systematic distribution, we cannot readily determine its phylogenetic status or evolutionary significance.

11.11 Sequences of Evolutionary Changes with Respect to Ecology

11.11.1 Sequences Toward Conductive Safety: The Southern California Dicotyledon Wood Flora

11.11.1.1 Effect of Imperforate Tracheary Element Type

From a detailed analysis of the southern Californian flora (Carlquist and Hoekman 1985b), one can develop sequences. This flora can be regarded as predominantly a dryland flora with small mesic pockets. Scalariform perforation plates are present essentially only in three habitats: riparian, moist shady valleys, and alpine scrub. These represent less than 7% of the genera studied, and even a smaller percentage at the species level. Consequently one can see that woods with scalariform perforation plates are relictual. They represent genera that have few species and are apparently unable to radiate out into habitats with strong seasonal fluctuation in water availability.

The phylads that account for virtually all of the radiation in the southern Californian flora are those with simple perforation plates. These can be said to fall into two groups: those with tracheids as the ground mass of the secondary xylem, and those with nonconductive imperforate tracheary elements – fiber-tracheids or libriform fibers (Fig. 11.2). A little more than 20% of the flora falls into the first category. The true-tracheid-bearing woods represent a relatively high percentage and one can easily hypothesize that the safety of tracheid-bearing woods is so considerable that this primitive wood type is advantaged in this region. The tracheids confer safety because of the ability of tracheids to localize embolism. The simple perforation plates developed on tracheid-bearing woods are adaptive in a seasonal climate with strong fluctuations in peak flow. Some conspicuous elements of the southern Californian flora, notably such Rosaceae as *Adenostoma*, *Cercocarpus*, and *Heteromeles*, have woods in this category. Phylesis of woods according to the patterns traditionally accepted (Fig. 11.1) results in production of specialized conducting and mechanical cells, and therefore good mechanical cells but conductive cells that are conductively efficient but poor in safety. By simplification of perforation plates in woods with true tracheids, the safety can be retained (Fig. 11.2).

11.11.1.2 Changes in Woods with True Tracheids

If we view woods with true tracheids, we can find within this group certain additional alterations in wood structure that are helpful in dryland climates. If the theory of vessel grouping (Carlquist 1984a) is correct, the species with true tracheids ought to have fewer vessels per group. This is clearly true: woods with true tracheids average 1.33 vessels per group, whereas woods with fiber-tracheids or libriform fibers average 4.65 vessels per group (Carlquist and Hoekman 1985b). If vessel grouping is superfluous as a means of safety in the tracheid-bearing woods, what other changes are possible? The options include acquisition of growth rings, acquisition of helical sculpture, narrowing of vessels, and shortening of vessels. Interestingly, woods with true tracheids have to a great degree developed growth rings (70.4%), helical sculpture (40.8%), and narrow vessels (26.5 μm, compared with mean vessel diameter of 34.4 μm for the flora as a whole). However, shortening of vessel elements has not occurred in the tracheid-bearing woods so much: they have vessels that average 281 μm in length, compared with 218 μm in woods in which libriform fibers or fiber-tracheids are present. Failure of vessel elements to shorten in the tracheid-bearing woods is not, in my opinion, because these woods are more primitive. Rather, I believe a functional explanation is present. If woods retain tracheids, the functional capabilities of the tracheids must be maintained. The essential part of a tracheid where conduction is concerned is the overlap area (end wall), and the longer the tracheid, the longer the overlap area and the better its conductive capacity. If fusiform cambial initials shorten in a tracheid-bearing phylad, the length of the tracheids would be reduced accordingly, and the conductive capabilities of the tracheids diminished somewhat. Therefore, in tracheid-bearing woods selection for relatively (but not exceptionally) long tracheids has resulted in selection for moderately long fusiform cambial initials, which in turn produce relatively long vessel elements.

If one assumes that genetic information for some evolutionary changes is easier to acquire than for others, one must conclude that growth rings are relatively easily achieved – perhaps the simplest adaptation to a seasonal environment. Helical sculpture (assuming that it does function in a dry environment, as seems obvious from its distribution) is more difficult to develop, and some groups have been unable to develop it, evidently.

11.11.1.3 Woods with Fiber-Tracheids or Libriform Fibers

The woods without true tracheids are held here to lack thereby an important ingredient of conductive safety (Fig. 11.1). So important in a dryland flora such as that of southern California is the selective value of tracheids that they are well represented. Of the woods with libriform fibers or fiber-tracheids, 43.7% of them have vasicentric tracheids, and an additional 17.1% of them have vascular tracheids (these two categories are defined as mutually exclusive, and also excluded are the 20% of the flora that have true tracheids). Thus 60.8% of the woody species that do not have true tracheids have another kind of tracheid – a remarkably high percentage. Although the percentage of the world's flora that has vasicentric or vascular tracheids has not been computed, one might guess that these two tracheid types can be found in only about 10% of the world dicotyledon flora at large.

The great significance of vasicentric tracheids is their accompaniment of vessels, and thereby their maintenance, should the vessels be embolized, of the three-dimensional pattern of conduction in woods. This accounts for the fact that woods with vasicentric tracheids have evergreen foliage – the sclerophylls of the Californian scrublands such as *Arctostaphylos*, *Ceanothus*, and *Dendromecon*. Vascular tracheids are formed at the ends of growth rings, and safeguard the cambium during the dry season, but presumably not the leaves, and so species with vascular tracheids often have drought-deciduous leaves.

Other adaptations for dryland existence developed in woods that have libriform fibers or fiber-tracheids include narrowing of vessels (with concomitant increase in number of vessels per mm^2), development of growth rings, development of helical sculpture, grouping of vessels, and shortening of vessels (Fig. 11.3). The degree to which these features have been developed in the southern Californian woods with libriform fibers or fiber-tracheids can be appreciated if one compares them to southern Californian species from riparian areas, which have minimal adaptation to drought (percentages of riparian species with cited feature are in parentheses in the following sentences). The percentages for woods with libriform fibers or fiber-tracheids are: growth rings, 65.8% (56.3%); helical sculpture present, 39.2% (12.5%); vessels per group, 4.65 (1.77); vessels per mm^2, 216 (163); vessel diameter, 37.9 μm (50.7 μm); vessel element length, 218 μm (418 μm).

11.11.1.4 Comparison to the World Flora

Even though there are mesic pockets (which conveniently serve for contrast with dry habitats) in the southern Californian flora, the dryland species are so predominant that we can use this flora as a whole as an indicator of adaptation to drought. Figures for the world flora from Metcalfe and Chalk (1950, p. 1360–1361) may represent a bias in favor of wet forest trees, but the figures can be considered a representative sample of sorts. World flora figures from this source are placed in parentheses in the following sentence. The southern Californian flora percentages and the comparative figures are: vessel diameter, 35.2 µm (93.9 µm); mean number of vessels per mm^2, 251 (31.25); mean vessel element length 233 µm (649 µm). These figures show that the southern Californian flora represents a marked development of mechanisms that improve conductive safety, oriented toward survival through the long dry season.

One should keep in mind that the Mediterranean-climate areas (including deserts adjacent to them) are not the only areas that show wood xeromorphy. Alpine areas are, if anything, more extreme than lowland dry areas (see figures in Carlquist and Hoekman 1985b), and in alpine areas the effect of cold (which produces physiological drought) may be added to that of seasonal drought if the alpine areas dry during the warmer portion of the year. Other areas that experience considerable cold show wood xeromorphy. Dry tropical lowlands contain species that exhibit a surprising amount of wood xeromorphy.

In assessing wood adaptations of dryland species, one must remember that wood adaptations are by no means the only way of adapting to dryness. They are stressed here precisely because until recently, adaptation in woods was regarded as minimal. However, we must keep in mind the other important adaptations, some of which are more important than wood adaptations: deciduousness, ability to die back and sacrifice large numbers of branches or branchlets, mycorrhizal associations, midday stomatal closure, etc.

11.11.2 Evolution Toward Conductive Efficiency: The Wet Forest

The picture of conductive safety presented above has implied a contrast: features that produce conductive safety are in most cases antipodal to those that produce conductive efficiency (but we should regard dryland woods not as inefficient, but as safety-adapted). Wet forest trees are by no means uniform in their adaptations, and certain elements within the wet forest show moderate selection for safety while still evidencing conductive efficiency. We may consider adaptation by woods in a wet forest essentially consists of conduction of large volumes of water per unit time per transectional area of wood, but the rates of transpiration higher than those of cloud forests are to be expected in hot sunny wet lowland forests.

Interestingly, some lines of increasing conductive efficiency are demonstrated by the major trends of xylem evolution – for example, simplification of the scalariform perforation plate. Species with scalariform perforation plates are restricted to wet areas (or areas with wet microclimates) in which transpiration rates are low (Sect. 3.7.2).

Evolution of ray and axial parenchyma types has occurred in wet forest areas but indirectly in response to conductive efficiency. If the water-conducting system develops characteristics for handling large volumes rapidly, the photosynthate-conducting system (rays + axial parenchyma) may be expected to exhibit similar changes.

The most important single alteration of woods of wet forest areas other than origin of simple perforation plates is widening of vessels. This change reduces friction, compared to the equivalent transectional area in the form of numerous smaller vessels. Among trees, notably wide vessels are found in lowland tropical trees (Baas 1976); there is a concomitant decrease in number of vessels per mm^2. If we were to view earlywood vessels of wet-temperate trees, however, we would see comparable vessel diameter. This means that wet temperate trees have latewood designed for conductive safety. The safety in the latewood of temperate ring-porous trees and shrubs is reminiscent of that in dryland shrubs, whereas the earlywood can be closely compared with conductive characteristics of tropical rain forest trees.

Lianas, a distinctive tropical form, certainly exhibit wide vessels in stems (Schenck 1893; Obaton 1960) that confer maximal conductive efficiency (although vessels are narrower in roots; Ewers et al. 1997). An appreciable number of lianas, however, have true tracheids in their wood also (e.g., Dilleniaceae, Icacinaceae, Menispermaceae); lianas with this wood type may exist because of the safety provided by the true tracheids (Carlquist 1985d). Lianas characteristically have vessels dimorphic in diameter (Carlquist 1985d; Gasson and Dobbins 1991). An appreciable proportion of lianas (and vines) have vasicentric tracheids also (Carlquist 1985d), very likely for the same reason. The occurrence of starch-rich parenchyma in lianas is abundant (Carlquist 1985d) and could represent a mechanism for conductive safety according to the Braun (1970, 1984) considerations. Thus, lianas apparently show strong selection for conductive efficiency and conductive safety simultaneously. Vessel dimorphism, highly characteristic of lianas, provides wide, efficient vessels, together with narrow vessels that would be relatively embolism-resistant. The tradeoff in lianas of having a large conductive area devoted to vessels is that mechanical tissue is proportionately less (Ewers and Fisher 1991).

The major trends of xylem evolution include only one feature that could be said to contribute to safety, shortening of vessel elements, and some workers have discounted the importance of that feature in promoting safety. Retention of tracheids would provide conductive safety in trees of wet forests, but would run counter to maximal division of labor between conductively efficient vessels and mechanically strong imperforate tracheary elements. Modification of tracheids to provide mechanically strong elements is to a certain extent possible, and to the extent it is, tracheids have been retained. The listing of taxa with scalariform perforation plates (Sect. 3.7.2) is much smaller than the listing of taxa with tracheids (Sect. 4.4.8), suggesting that this is what may have happened.

On the whole, however, conductive safety is low in trees of wet forests, although dry margins of wet forests and cold wet forests have heightened percentages of features that confer conductive safety.

Tropical groups that could develop adaptations to cold (e.g., deciduousness, growth rings) without adaptation to drought have spread into moist temperate forests: Aquifoliaceae, Cornaceae, Hamamelidaceae, Lauraceae, and Magnoliaceae, for example. Invasion of dry areas has required evolution of more numerous features involving conductive safety. We can envision that phylads such as Fagaceae (which have vasicentric tracheids) have been able to move readily from wet tropical forests into temperate areas that have cold or drought or both, and this movement has been accompanied by heightening of features that confer conductive safety. The point here is that Fagaceae, by having originated safety features to a greater degree than other families, have succeeded to an exceptional degree in temperate areas. The safety features of Fagaceae are not of negative selective value in wet areas, however.

11.12 Methods for Relating Ecology to Wood Anatomy

11.12.1 Source of Material

Ideally one wishes a wood sample of relatively mature nature from a known plant portion, documented by an herbarium specimen, from a locality for which reasonably good climatic data can be obtained. Such wood samples can be obtained if one collects one's own wood samples, but that is not always convenient. Xylarium samples are lacking in documentation to various degrees, and the finished blocks present in wood collections can obscure size and age of sample. Understandably, some workers would rather use a poorly-documented sample of a particular species than omit that species from a study. Where ecological data are concerned, one should have good data relating to water availability – and often those data are insufficient because microclimatic features are usually not cited. These can be very important: an understory tree exists in a climatic regime quite different from one emergent in the same locality.

Data on roots, branches, and lignotubers, where appropriate, should be added to stem data. Unfortunately, xylaria rarely contain anything but stem wood (branches rather than main stems are undoubtedly often represented in xylaria, but not labeled as such).

Cultivated material is probably reliable for most qualitative wood features, but some quantitative features may be altered to an appreciable (and unknown) extent (Bissing 1976, 1982). Immature (twig) material, often derived from herbarium specimens, is used in comparative studies in order to present a more complete representation of species in a particular study, but the immature nature of twig wood is obvious and should be noted in a paper.

11.12.2 Selection of Wood Data

11.12.2.1 Quantitative Data

With respect to ecology, wood analysis places great emphasis on quantitative features of vessel elements. The most important features are number of bars per perforation plate, vessel diameter, number of vessels per mm^2 of transection, vessel element length, and number of vessels per group.

Use of the mean figure in data presentation appeals to me as advisable for all of these measurements because the mean figure is most likely to present a picture of how most cells are functioning, and physiological function of a wood is dictated by the way in which the vast majority of cells are operating, not by the performance of a few extreme cells. One could with plausibility argue that the occurrence of growth rings mitigates against the use of means, because averaging wide vessels with narrow vessels does not present an accurate image of either earlywood or latewood. Where vessel diameter and vessel density are concerned, this reservation is certainly understandable, and analysis of these features for latewood and earlywood separately could be very informative. Because degree of vessel grouping can alter markedly between earlywood and latewood in certain growth rings (but not in others), one might consider adding that feature to the others where earlywood and latewood are analyzed separately. Obviously a difficulty does arise, however, in delimiting portions of a growth ring, and the author should give criteria for this.

Perforation plate morphology could be considered a qualitative feature, but if scalariform perforation plates are present, mean bar number should be determined, and if any differences occur between earlywood and latewood, that should be recorded.

With respect to vessel diameter, lumen diameter gives a better idea of physiological function than outside diameter of a vessel, and although the latter has been traditionally used, many workers are now using the former method.

Number of vessels per mm^2 should be calculated by counting each vessel (Wheeler 1986), despite a contrary earlier instruction (apparently rarely followed) by Rendle and Clark (1934).

Vessel grouping is best calculated as mean number of vessels per group (solitary = 1, a pair of vessels in contact = 2, etc.).

Vessel element length is typically measured by including the extreme tips of the vessel. I endorse this method, which has been used by most workers.

Vessel element length can only be determined accurately from macerations. These same comments apply even more strongly to determining length of imperforate tracheary elements.

If one views standard deviation figures for the data on the above features, one discovers that standard deviation levels off beyond 20–25 measurements in most instances. Consequently, means based on 25 or more measurements are advisable.

Vessel wall thickness, pit diameter on lateral walls of vessels, and quantitative features of the imperforate tracheary elements (except for length) are probably of significance to comparative studies other than those primarily concerned with ecology. In any case, measurements of these features are best done by obtaining a median or representative condition rather than averaging numerous measurements because of the great quantitative variation in these features within a section.

11.12.2.2 Qualitative Features

Growth ring presence or absence is significant, as well as sharpness of growth ring demarcation. However, the tacit assumption that growth rings are either absent (diffuse-porous), present in marked condition (ring-porous) or in intermediate condition (semi-ring-porous) needs modification. There are quite divergent types of growth rings that represent quite distinctive types of adaptation to ecology, and consequently the growth ring type classification offered in Chapter 2 is recommended.

Presence of helical sculpturing of any kind should be reported. However, workers should go beyond recording mere presence or absence, and if possible record whether the sculpture consists of bands, of grooves interconnecting pit apertures, or both; degree of coarseness of the helical sculpture; and distribution of helical sculpture within a growth ring (there often is much less helical sculpture in earlywood than in latewood). Helical sculpture may be confined to vessels, or present in both vessels and in tracheids.

Vestured pits should be reported, and vesturing of walls also; SEM methodology is now essential. The distribution of degrees of vesturing within a growth ring should be recorded, and the cell types in which it is present.

11.12.3 Groupings To Be Compared

11.12.3.1 Taxonomic Groupings

The inherent interest in comparing species within genera with respect to ecology obviously depends on the range of ecology those species span. One should not expect a wide range of wood expressions in genera such as *Symplocos* (van den Oever et al. 1981), because this genus characteristically occurs in wet forests or in microclimates that are relatively cool and moist. When studying such a genus, one should not claim that wood anatomy does not reveal ecological distinctions; in fact, in such a genus, wood anatomy probably is an accurate indicator that differences between species with respect to ecology are minor.

When analyzing wood data with respect to ecology, one occasionally is surprised that the data do not fit one's ideas about the ecological site represented. In this case, one may wish to consider if one's ecological assessments are, in fact, accurate (water availability can be much modified by microclimatic conditions), or whether important mitigating factors (deciduousness, succulence) alter to an appreciable extent the degree to which wood features function in controlling the water economy of a plant.

11.12.3.2 Floristic Groupings

Some wood has involved comparison of small areas, or florulas (e.g., H.J. Miller 1975; Carlquist 1977d; Barajas-Morales 1985). These areas can have the advantage of relatively uniform ecology – depending on one's accuracy in judging habitat uniformity.

If one considers larger areas, comparable to those covered by floras, one inevitably includes a range of habitats. In this case, one must subdivide the flora into ecological regions in order to make meaningful comparisons (e.g., Baas et al. 1983; Carlquist and Hoekman 1985b). The number and kind of divisions one makes will very much affect one's comparisons.

Note should be taken that habitats alone are not sufficient in analyzing data in floras (or even florulas), because certain growth forms will deal with a given habitat in quite different ways. Obviously an annual or a succulent plant will have a wood formula quite different from that of a woody shrub in a particular habitat. For an example of how both plant associations and plant habitats can be taken into account in analyses relating to ecology, the reader is referred to the study of Carlquist and Hoekman (1985b).

11.13 Ratios, Indices, Equations

Because vessel diameter, vessel density, and vessel element length are all closely related to ecology, I developed a ratio termed "Mesomorphy" (Carlquist 1977b; Carlquist and DeBuhr 1977). Mesomorphy is the vessel diameter times vessel element length divided by number of vessels per mm^2 (means are used for these figures). The use of this ratio has been criticized (van den Oever et al. 1981), largely on the grounds that the Hagen-Poisseuille equation for conductivity is better justified. The use of the Hagen-Poisseuille equation is endorsed by Zimmermann (1983, p. 13). The version of this equation he gives is:

$$L_p(\text{hydraulic conductivity}) = \frac{r^4\pi}{8\cap} \quad (1)$$

where r = vessel radius, and Ç = viscosity of liquid.

On the other hand, a quite different version of the equation has been offered:

$$\text{Conductivity} = \frac{\text{vessel radius}^4}{\text{vessels per mm}^2}$$

A third version of the equation is offered by Gibson et al. (1985).

At first glance a formula related to behavior of ideal capillaries seems logical to apply to vessels. However, in fact, vessels or tracheids of plant species fall variously below the predicted values according to a Hagen-Poiseuille equation (Zimmermann 1983, p. 15; Gibson et al. 1984). Vines come closest, because of the relatively gigantic diameter of their vessels, which are relatively friction-free (Zimmermann 1983), but even here, one suspects that the numerous narrow vessels also present in vine stems (Carlquist 1985d) have been neglected, and that some computations have been based only on the large vessels.

The fact that actual values fall variously below predicted values makes one question the value of the Hagen-Poiseuille equation for anatomical studies. This equation is designed for ideal capillaries, but dicotyledon woods do not have ideal capillaries. The Hagen-Poiseuille equation is much better suited to blood veins and capillaries, which are endless. Vessels have finite length, and terminate blindly, and such a termination is not taken into account in the Hagen-Poiseuille equation. Long vessels, as in earlywood of ring-porous species, come closest to the equation, but many plants have relatively short vessels. Vessels form a network, and network degree and contact degree (Bosshard and Kucera 1973b), important characteristics of the conductive system, are not taken into account in the Hagen-Poiseuille equation.

When computed for a given group of plants, the Hagen-Poiseuille equation has an enormous range of values (Carlquist and Hoekman 1986a,b) that does not correspond to the gamut of ecological factors such as rainfall or other measurements of water availability. Conductivity formulas thus do not have good predictability.

However, in addition to the above limitations of Hagen-Poiseuille equations, the most serious objection one must register is that they deal only with conductive efficiency, whereas wood plans represent compromises between conductive efficiency and conductive safety. If a wood had zero safety, a Hagen-Poiseuille equation could be applied (admittedly with reservations as noted above), but the vast majority of woods have anywhere from a few to many devices for safety of conduction, and none of these devices (e.g., vasicentric tracheid presence) are taken into account at all by the Hagen-Poiseuille equation. The Hagen-Poiseuille equation gives the misleading impression that woods with conductive safety are merely conductively inefficient. Wood is a compromise between conductive efficiency and conductive safety, a concept that is gaining general acceptance (Tyree et al. 1994). The Mesomorphy ratio can easily be criticized as arbitrary, but this ratio was designed to reflect the compromise between conductive efficiency and conductive safety well, and therefore more accurately indicates ecology than do Hagen-Poiseuille equations. Vessel element length, although omitted by the Hagen-Poiseuille equation and dismissed by some workers as irrelevant to ecology (van den Oever et al. 1981) shows a very close relationship to ecology (Baas and Werker 1981), and may prove to be operative in confining air embolisms in vessels or otherwise affecting conductive safety. The apparent lack of correlation between vessel element length and ecology in *Symplocos* (van den Oever et al. 1981) is probably due to the relative uniformity of ecological tolerance in that genus: factors other than wood may represent prime adaptations to ecology in *Symplocos*. One must take into account mitigating features such as drought deciduousness, leaf size, diffusive resistance, succulence, etc. However, if one does, one finds that the Mesomorphy ratio reflects ecology closely in case after case (Carlquist 1981d; Carlquist and Hoekman 1985b; Patterson and Tanowitz 1989). Except for the rather small number of vesselless dicotyledons, the three quantitative features in this ratio are represented in wood of all dicotyledons, and are thus almost universally applicable.

However, many other wood features doubtless contribute to conductive safety and should be taken into account: vasicentric tracheid presence, vessel grouping, vesturing, etc. Careful examination of all applicable factors and their degrees of representation and distribution in woods must be undertaken. Both ecology and plant habit offer complexities in interpreting the way in which wood features are adapted to the environment. These must be judged by the investigator and reported in relation to data, indices, or equations.

11.14 Measuring Ecology

Obviously ecologists have much more sophistication in measuring climatic factors and physiological functions than most plant anatomists do, but as hinted at the end of the preceding paragraph, plant anatomists must look for the most relevant ones. Because different species exploit a single habitat in quite different ways and with different wood formulas, wood anatomists must look for the most relevant ones. Latitude and altitude have been used (van der Graaf and Baas 1974). Latitudes and altitudes are available on or can be calculated on the basis of herbarium labels, but are not direct indicators of ecology. The most significant factors are moisture availability and freezing. One cannot use rainfall without reference to the way in which other factors modify water availability of the locality (e.g., north or south slope exposure) or water economy within the plant (drought deciduousness, microphylly, seasonal dieback of stems). Observations on the functioning of a plant in its natural habitat may give one a better concept of how ecological factors influence a plant in its occupancy of a given habitat. Convenient rough categories can be used, but each taxonomic group and each geographical area provides its own problems. Ideally, one would wish for studies involving collaboration between a comparative plant anatomist (who studies aspects other than wood as well as wood) and a comparative physiological plant ecologist. Such syntheses will ultimately depict the evolutionary significance of wood anatomy more effectively than studies of limited methodology and scope, useful though they have been.

References

Adamson R S 1934 Anomalous secondary thickening in Compositae. Ann Bot 48:505–514

Adamson R S 1936 Note on the stem structure of Boscia rehmanniana Post. Trans R Soc S Afr 23:297–301

Adamson R S 1937 Anomalous secondary thickening in Osteospermum. Trans R Soc S Afr 24:303–312

Akachuku A E, Burley J 1979 Variation of wood anatomy of Gmelina arborea Roxb. in Nigerian plantations. IAWA Bull 4:94–99

Alfonso V A, Richter H G 1991 Wood and bark of Buchenavia Eichl. (Combretaceae). IAWA J 12:123–141

Aloni R, Zimmermann M H 1983 The control of vessel size and density along a plant axis. A new hypothesis. Differentiation 24:203–208

Amobi G L 1973 Periodicity of wood formation in some trees of lowland forest in Nigeria. Ann Bot NS 37:211–218

Anderson L C 1972 Studies on Bigelowia (Asteraceae). II. Xylary comparisons, woodiness, and paedomorphosis. J Arnold Arb 53:499–514

Angiosperm Phylogeny Group 1998 An ordinal classification for the families of flowering plants. Ann Mo Bot Gard 85:531–553

Armstrong J E, Wilson T K 1980 Wood anatomy of Horsfieldia (Myristicaceae). IAWA Bull NS 1:121–129

Artschwager E F 1920 On the anatomy of Chenopodium album L. Am J Bot 7:252–260

Ashworth V E T M, Dos Santos G 1997 Wood anatomy of four Californian mistletoe species (Phoradendron, Viscaceae). IAWA J 18:229–245

Ayensu E S, Stern W C 1964 Systematic anatomy and ontogeny of the stem in Passifloraceae. Contrib US Nat Herb 34:45–73

Baas P 1973 The wood anatomical range in Ilex (Aquifoliaceae) and its ecological and phylogenetic significance. Blumea 21:193–258

Baas P 1976 Some functional and adaptive aspects of vessel member morphology. Leiden Bot Ser 3:157–181

Baas P 1977 The peculiar wood structure of Leptospermum crassipes Lehm. (Myrtaceae). IAWA Bull 1977/2:25–30

Baas P 1979a The peculiar wood structure of Vaccinium lucidum (Bl) Miq (Ericaceae). IAWA Bull 1979/1:11–16

Baas P 1979b The anatomy of Alzatea Ruiz & Pavon (Myrtales). Acta Bot Neerl 28:156–158

Baas P 1980 Reliability and citation of wood specimens. IAWA Bull NS 1:72

Baas P 1986a Terminology of imperforate tracheary elements – in defence of libriform fibres with minutely bordered pits. IAWA Bull NS 7:82–86

Baas P 1986b Wood anatomy of Lythraceae – additional genera (Capuronia, Galpinia, Haitia, "Orias", and Pleurophora). Ann Mo Bot Gard 73:810–819

Baas P, Gregory M 1985 A survey of oil cells in the dicotyledons with comments on their replacement by and joint occurrence with mucilage cells. Isr J Bot 34:167–196

Baas P, Vetter R E (eds) 1989 Growth rings in tropical trees. IAWA J 19:95–174

Baas P, Werker E 1981 A new record of vestured pits in Cistaceae. IAWA Bull NS 2:41–42

Baas P, Wheeler E A 1996 Parallelism and reversibility in xylem evolution: a review. IAWA J 17:351–364

Baas P, Zhang X 1986 Wood anatomy of trees and shrubs of China. I. Oleaceae. IAWA Bull NS 7:195–220
Baas P, Zweypfenning R C V J 1979 Wood anatomy of the Lythraceae. Acta Bot Neerl 28:117–155
Baas P, Werker E, Fahn A 1983 Some ecological trends in vessel characters. IAWA Bull NS 4:141–159
Baas P, Lee C, Zhang X, Cui K, Deng Y 1984 Some effects of dwarf growth on wood structure IAWA Bull NS 5:45–63
Baas P, Esser P M, van der Westen M E T, Zandee M 1988 Wood anatomy of the Oleaceae. IAWA Bull NS 9:103–182
Back E L 1969 Intercellular spaces along the ray-parenchyma gas canal system of living wood? Wood Sci 1:31–34
Bailey D C 1980 Anomalous growth and vegetative anatomy of Simmondsia chinensis. Am J Bot 67:147–161
Bailey I W 1911 The relation of the leaf-trace to the formation of compound rays in the lower dicotyledons. Ann Bot 25:225–241
Bailey I W 1912 The evolutionary history of the foliar ray in the wood of the dicotyledons and its phylogenetic singificance. Ann Bot 26:647–661
Bailey I W 1919 Structure, development, and distribution of so-called rims or bars of Sanio. Bot Gaz 47:449–468
Bailey I W 1923 The cambium and its derivative tissues. IV. The increase in girth of the cambium. Am J Bot 10:499–509
Bailey I W 1925 Some salient lines of specialization in tracheary pitting. I. Gymnospermae. Ann Bot 39:587–598
Bailey I W 1933 The cambium and its derivative tissues. VIII. Structure, distribution and diagnostic significance of vestured pits in dicotyledons. J Arnold Arb 14:259–273
Bailey I W 1936 The problem of differentiation and classification of tracheids, fiber-tracheids, and libriform fibers. Trop Woods 45:18–23
Bailey I W 1944a The comparative morphology of the Winteraceae. III. Wood. J Arnold Arbor 25:97–103
Bailey I W 1944b The development of vessels in angiosperms in morphological research. Am J Bot 31:421–428
Bailey I W 1957 Additional notes on the vesselless dicotyledon, Amborella trichopoda Baill. J Arnold Arbor 38:374–378
Bailey I W 1962 Comparative anatomy of the leaf-bearing Cactaceae. VI. The xylem of Pereskia saccharosa and Pereskia aculeata. J Arnold Arbor 43:376–383
Bailey I W 1964 Comparative anatomy of the leaf-bearing Cactaceae. XI. The xylem of Pereskiopsis and Quiabentia. J Arnold Arbor 45:140–157
Bailey I W 1966 The significance of reduction of vessels in the Cactaceae. J Arnold Arbor 46:288–292
Bailey I W, Howard R A 1941a The comparative morphology of the Icacinaceae. I. Anatomy of the node and internode. J Arnold Arbor 22:125–132
Bailey I W, Howard R A 1941b The comparative morphology of the Icacinaceae. III. Imperforate tracheary elements and xylem parenchyma. J Arnold Arbor 22:432–442
Bailey I W, Nast C G 1945 Morphology and relationships of Trochodendron and Tetracentron. I. Stem, root, and leaf. J Arnold Arbor 26:143–154
Bailey I W, Nast C G 1948 Morphology and relationships of Illicium, Schisandra and Kadsura. J Arnold Arbor 29:77–89
Bailey IW, Sinnott EW 1914 Investigation on the phylogeny of the angiosperms. II. Anatomical evidence of reduction in certain of the Amentiferae. Bot Gaz 58:36–60
Bailey I W, Swamy B G L 1948 Amborella trichopoda Baill., a new morphological type of dicotyledon. J Arnold Arbor 29:245–254
Bailey I W, Swamy B G L 1949 The morphology and relationships of Austrobaileya. J Arnold Arbor 30:211–226

Bailey I W, Swamy B G L 1950 Sarcandra, a vesselless genus of the Chloranthaceae. J Arnold Arbor 31:117–129

Bailey I W, Thompson W P 1918 Additional notes upon the angiosperms Tetracentron, Trochodendron, and Drimys, in which vessels are absent from the wood. Ann Bot 32:503–512

Bailey I W, Tupper W W 1918 Size variation in tracheary cells. I. A comparison between the secondary xylems of vascular cryptogams, gymnosperms, and angiosperms. Proc Am Acad Arts Sci 54:149–204

Baird W V, Blackwell W H 1980 Secondary growth in the axis of Halogeton glomeratus (Bieb.) Meyer (Chenopodiaceae). Bot Gaz 141:269–276

Bande M B, Prakash U 1984 Evolutionary trends in the secondary xylem of woody dicotyledons from the Tertiary of India. Paleobotanist 32:44–75

Bannan M W 1965 The length, tangential diameter, and length/width ratio of conifer tracheids. Can J Bot 43:967–984

Barajas-Morales J 1985 Wood structural differences between trees of two tropical forests in Mexico. IAWA Bull NS 6:355–364

Barghoorn E S 1940 The ontogenetic development and phylogenetic specialization of rays in the xylem of dicotyledons. I. Primitive ray structure. Am J Bot 27:918–928

Barghoorn E S 1941a The ontogenetic development and phylogenetic specialization of rays in the xylem of dicotyledons. II. Modification of the multiseriate and uniseriate rays. Am J Bot 28:273–282

Barghoorn E S 1941b The ontogenetic development and phylogenetic specialization of rays in the xylem of dicotyledons. III. The elimination of rays. Bull Torrey Bot Club 68:317–325

Basson P W, Bierhorst D W 1967 An analysis of differential lateral growth in the stem of Bauhinia surinamensis. Bull Torrey Bot Club 94:404–411

Behm M 1895 Beiträge zur anatomischen Characteristik der Santalaceen. Bot Cbl 62:65–74, 97–107, 129–139, 161–170, 193–210

Bhambie S 1972 Correlation between form, structure, and habit in some lianas. Proc Indian Acad Sci B 75(5):246–256

Bierhorst D W 1960 Observations on tracheary elements. Phytomorphology 10:249–305

Bierhorst D W, Zamora P M 1965 Primary xylem elements and element associations of angiosperms. Am J Bot 52:657–710

Bissing D R 1976 The effect of cultivation on the expression of anatomical features of the wood of selected dicotyledons. PhD Thesis, Claremont Graduate School, Claremont, California, 117 pp

Bissing D R 1982 Variation in qualitative anatomical features of the xylem of selected dicotyledonous woods in relation to water availability. Bull Torrey Bot Club 109:371–384

Blanchette R A, Biggs A R (eds) 1992 Defense mechanisms of woody plants against fungi. Springer, Berlin Heidelberg New York, 458 pp

Blank F 1939 Beitrag zur Morphologie von Caryocar nuciferum L. Ber Schweiz Bot Ges 49:437–494

Bliss M C 1921 The vessel in seed plants. Bot Gaz 71:314–326

Boatwright S W, Garrett G G 1983 The effect of microstructure and stress state on the fracture behavior of wood. J Mater Sci 18:2181–2199

Bonsen K J 1991 Gefässverschluss-Mechanismen in Laubbäumen. Vierteljahrschr Naturforsch Ges Zürich 136(1):12–50

Bonsen K J, ter Welle B J H 1983 Comparative wood and leaf anatomy of the Cecropieae (Urticales). Bull Mus Nat Hist Nat Paris B Adansonia 5:151–177

Bonsen K J, ter Welle B J H 1984 Systematic wood anatomy and affinities of the Urticaceae. Bot Jahrb Syst 105:49–71

Bosshard H H, Kucera L J 1973a Über Anfangs- und Endzonen der Jahrringe. Holz Roh Werkst 31:484–486

Bosshard H H, Kucera L J 1973b Dreidimensionale Strukturanalyse des Holzes – erste Mitteilung: die Vernetzung des Gefässsystems in Fagus sylvatica L. Holz 31:437–445

Bosshard H H, Kucera L J, Stocker U 1982 Das Gefäss-System in präjuvenilem Holz von Fraxinus excelsior L. Vierteljahrschr Naturforsch Ges Zürich 127/1:29–48

Botosso DC, Vidal Gomes A 1982 Radial vessels and series of perforated ray cells in Annonaceae. IAWA Bull NS 3:39–44
Boureau E 1957 Anatomie végétale, vol 3. Presses Universitaires de France, Paris, pp 525–752
Braun H J 1955 Beiträge zur Entwicklungsgeschichte der Markstrahlen. Bot Studien 4:73–131
Braun H J 1961 The organization of the hydrosystem in the stemwood of trees and shrubs. IAWA Bull 1:2–9
Braun H J 1967a Entwicklung und Bau der Holzstrahlen unter dem Aspekt der Kontakt-Isolations-Differenzierung gegenüber dem Hydrosystem. I. Das Prinzip der Kontakt-Isolations-Differenzierung. Holzforschung 23(3):3–37
Braun H J 1967b Entwicklung und Bau der Holzstrahlen unter dem Aspekt der Kontakt-Isolations-Differenzierung gegenüber dem Hydrosystem. II. Die Typen der Kontakt-Holzstrahlen. Holzforschung 21:145–153
Braun H J 1968a Entwicklung und Bau der Holzstrahlen unter dem Aspekt der Kontakt-Isolations-Differenzierung gegenüber dem Hydrosystem. III. Die Typen der Kontakt-Isolations-Holzstrahlen und der Isolations-Holzstrahlen. Holzforschung 22:53–60
Braun H J 1968b Entwicklung und Bau der Holzstrahlen unter dem Aspekt der Kontakt-Isolations-Differenzierung gegenüber dem Hydrosystem. IV. Die Organisation der Holzstrahlen. Holzforschung 22:153–157
Braun H J 1970 Funktionelle Histologie der sekundären Sprossachse. I. Das Holz. Handbuch der Pflanzenanatomie, vol 9(1). Gebrüder Borntraeger, Berlin, pp 1–190
Braun H J 1983 Zur Dynamik der Wassertransportes in Bäumen. Ber Dtsch Bot Ges 96:29–47
Braun H J 1984 The significance of the accessory tissues of the hydrosystem for osmotic water shifting as the second principle of water ascent, with some thoughts concerning the evolution of trees. IAWA Bull NS 5:275–294
Braun H J, Wolkinger F 1970 Zur funktionellen Anatomie des axialen Holzparenchyms und Vorschläge zur Reform seiner Terminologie. Holzforschung 24:19–26
Bridgewater S D, Baas P 1978 Wood anatomy of the Punicaceae. IAWA Bull 1978/1:3–6
Bridgewater S D, Baas P 1982 Wood anatomy of Xanthophyllum Roxb. IAWA Bull NS 3:115–125
Brown H P, Panshin A J, Forsaith C C 1949 Textbook of wood technology. I. Structure, identification, defects and uses of the commercial woods of the United States. McGraw Hill, New York
Butterfield B G 1972 Developmental changes in the vascular cambium of Aeschynomene hispida Willd. N Z J Bot 10:373–386
Butterfield B G 1973 Variation in the size of fusiform cambial initials and vessel members in Hoheria angustifolia Raoul. N Z J Bot 11:391–410
Butterfield B G, Meylan B A 1972a Scalariform perforation plate development in Laurelia novae-zelandiae A Cunn.: a scanning electron microscope study. Aust J Bot 20:253–259
Butterfield B G, Meylan B A 1972b Trabeculae in a hardwood. IAWA Bull 1972/1:3–6
Butterfield B G, Meylan B A 1975 Simple to scalariform combination perforation plates in Vitex lucens Kirk (Verbenaceae) and Brachyglottis repanda J R & G Forst (Compositae). IAWA Bull 1975/3:39–42
Butterfield B G, Meylan B A 1976 The occurrence of septate fibres in some New Zealand woods. N Z J Bot 14:123–130
Butterfield B G, Meylan B A 1979 Observations of trabeculae in New Zealand hardwoods. Wood Sci Technol 13:59–65
Butterfield B G, Meylan B A 1980 Three-dimensional structure of wood. An ultrastructural approach, 2nd edn. Chapman and Hall, London, 103 pp
Butterfield B G, Meylan B A 1982 Cell wall hydrolysis in the tracheary elements of the secondary xylem. In: Baas P (ed) New perspectives in wood anatomy. Martinus Nijhoff/Junk, The Hague, pp 71–84
Butterfield B G, Philipson W R, Meylan B A, Ohtani J 1984 Comparative morphology of the vessel elements in the wood of Pseudopanax C. Koch (Araliaceae). N Z J Bot 22:509–514
Canright J E 1955 The comparative morphology and relationships of the Magnoliaceae. IV. Wood. J Arnold Arbor 36:119–140
Carlquist S 1957a Wood anatomy of Mutisieae (Compositae). Trop Woods 106:29–45

Carlquist S 1957b The genus Fitchia (Compositae). Univ Calif Publ Bot 29:1–144
Carlquist S 1958a Wood anatomy of Heliantheae (Compositae). Trop Woods 108:1–30
Carlquist S 1958b The woods and flora of the Florida Keys. Compositae. Trop Woods 109:1–37
Carlquist S 1958c Anatomy and systematic position of Centaurodendron and Yunquea (Compositae). Brittonia 10:78–93
Carlquist S 1959 Wood anatomy of Helenieae (Compositae). Trop Woods 111:19–39
Carlquist 1960a Wood anatomy of Astereae (Compositae). Trop Woods 113:54–84
Carlquist 1960b Wood anatomy of Cichorieae (Compositae). Trop Woods 112:65–91
Carlquist S 1961a Comparative plant anatomy. Holt Rinehart Winston, New York, 146 pp
Carlquist S 1961b Wood anatomy of Inuleae (Compositae). Aliso 5:21–37
Carlquist S 1962a A theory of paedomorphosis in dicotyledonous woods. Phytomorphology 12:30–45
Carlquist S 1962b Wood anatomy of Senecioneae (Compositae). Aliso 5:123–146
Carlquist S 1964a Wood anatomy of Vernonieae (Compositae). Aliso 5:451–467
Carlquist S 1964b Morphology and relationships of Lactoridaceae. Aliso 5:421–435
Carlquist S 1965a Wood anatomy of Eupatorieae (Compositae). Aliso 6(1):89–103
Carlquist S 1965b Wood anatomy of Cynareae (Compositae). Aliso 6(1):13–24
Carlquist S 1966a Wood anatomy of Compositae: a summary, with comments on factors controlling wood evolution. Aliso 6(2):25–44
Carlquist S 1966b Wood anatomy of Anthemideae, Ambrosieae, Calenduleae, and Arctotideae (Compositae). Aliso 6(2):1–23
Carlquist S 1969a Wood anatomy of Lobelioideae (Campanulaceae). Biotropica 1:47–72
Carlquist 1969b Wood anatomy of Goodeniaceae and the problem of insular woodiness. Ann Mo Bot Gard 56:358–390
Carlquist S 1970a Wood anatomy of Hawaiian, Macaronesian, and other species of Euphorbia. Bot J Linnean Soc 63 [Suppl 1]:181–193
Carlquist S 1970b Wood anatomy of insular species of Plantago and the problem of raylessness. Bull Torrey Bot Club 97:353–361
Carlquist S 1970c Wood anatomy of Echium (Boraginaceae). Aliso 7:183–199
Carlquist S 1970d Hawaii: a natural history. Nat Hist Press, Garden City, 463 pp
Carlquist S 1971 Wood anatomy of Macaronesian and other Brassicaceae. Aliso 7:365–384
Carlquist S 1974 Island biology. Columbia Univ Press, New York, 660 pp
Carlquist S 1975a Ecological strategies of xylem evolution. Univ California Press, Berkeley, 259 pp
Carlquist S 1975b Wood anatomy of Onagraceae, with notes on alternative modes of photosynthate movement in dicotyledon woods. Ann Mo Bot Gard 62:386–424
Carlquist S 1975c Wood anatomy and relationships of the Geissolomataceae. Bull Torrey Bot Club 102:128–134
Carlquist S 1976a Wood anatomy of Myrothamnnus flabellifolia (Myrothamnaceae) and the problem of multiperforate perforation plates. J Arnold Arbor 57:119–126
Carlquist S 1976b Wood anatomy of Byblidaceae. Bot Gaz 137:35–38
Carlquist S 1976c Wood anatomy of Roridulaceae: ecological phylogenetic implications. Am J Bot 63:1003–1008
Carlquist S 1977a Wood anatomy of Grubbiaceae. J S Afr Bot 43:129–144
Carlquist S 1977b Wood anatomy of Onagraceae: additional species and concepts. Ann Mo Bot Gard 64:627–637
Carlquist S 1977c Wood anatomy of Tremandraceae: phylogenetic and ecological implications. Am J Bot 64:704–713
Carlquist S 1977d Ecological factors in wood evolution: a floristic approach. Am J Bot 64:887–896
Carlquist S 1978a Wood anatomy of Bruniaceae: correlations with ecology, phylogeny, and organography. Aliso 9:323–364
Carlquist S 1978b Wood anatomy and relationships of Bataceae, Gyrostemonaceae, and Stylobasiaceae. Allertonia 5:297–330
Carlquist S 1980a Further concepts in ecological wood anatomy, with comments on recent work in wood anatomy and evolution. Aliso 9:499–553

Carlquist S 1980b Anatomy and systematics of Balanopaceae. Allertonia 2:191–246
Carlquist S 1981a Wood anatomy of Zygogynum (Winteraceae); field observations. Bull Mus Nat Hist Nat Ser 4 Sect B Adansonia 3:281–292
Carlquist S 1981b Wood anatomy of Nepenthaceae. Bull Torrey Bot Club 108:324–330
Carlquist S 1981c Wood anatomy of Chloanthaceae (Dicrastylidacea). Aliso 10:19–34
Carlquist S 1981d Wood anatomy of Pittosporaceae. Allertonia 7:355–391
Carlquist S 1981e Wood anatomy of Cephalotaceae. IAWA Bull NS 2:175–178
Carlquist S 1981f Types of cambial activity and wood anatomy of Stylidium (Stylidiaceae). Am J Bot 68:778–785
Carlquist S 1982a Exospermum stipitatum (Winteraceae): observations on wood, leaves, flowers, pollen, and fruit. Aliso 10:277–289
Carlquist S 1982b Wood anatomy of Dipsacaceae. Taxon 31:443–450
Carlquist S 1982c Wood anatomy of Illicium (Illiciaceae): phylogenetic, ecological, and functional interpretations. Am J Bot 69:1587–1598
Carlquist S 1982d Wood anatomy of Buxaceae: correlations with ecology and phylogeny. Flora 172:463–491
Carlquist S 1982e Wood and bark anatomy of Scalesia (Asteraceae). Aliso 10:301–312
Carlquist S 1982f Wood anatomy of Daphniphyllaceae: ecological and phylogenetic considerations, review of pittosporalean families. Brittonia 34:252–266
Carlquist S 1982g The use of ethylenediamine in softening hard plant structures for paraffin sectioning. Stain Technol 57:311–317
Carlquist S 1983a Wood anatomy of Belliolum (Winteraceae); a note on floral opening. J Arnold Arbor 64:161–169
Carlquist S 1983b Wood anatomy of Bubbia (Winteraceae), with comments on origin of vessels in dicotyledons. Am J Bot 70:578–590
Carlquist S 1983c Wood anatomy of Calycanthaceae: ecological and systematic implications. Aliso 10:427–441
Carlquist S 1983d Observations on the vegetative anatomy of Crepidiastrum and Dendrocacalia (Asteraceae). Aliso 10:383–395
Carlquist S 1983e Wood anatomy of Calyceraceae and Valerianaceae, with comments on aberrant perforation plates in predominantly herbaceous groups of dicotyledons. Aliso 10:413–425
Carlquist S 1983f Wood anatomy of Onagraceae: further species; root anatomy; significance of vestured pits and allied structures in dicotyledons. Ann Mo Bot Gard 69:755–769
Carlquist S 1984a Vessel grouping in dicotyledon woods: significance and relationship to imperforate tracheary elements. Aliso 10:505–525
Carlquist S 1984b Wood and stem anatomy of Bergia suffruticosa: relationship of Elatinaceae and broader significance of vascular tracheids, vasicentric tracheids, and fibriform vessel elements. Ann Mo Bot Gard 71:232–242
Carlquist S 1984c Wood anatomy of some Gentianaceae: systematic and ecological conclusions. Aliso 10:573–582
Carlquist S 1984d Wood anatomy of Loasaceae with relation to systematics, habit, and ecology. Aliso 10:583–602
Carlquist S 1984e Wood anatomy of Trimeniaceae. Plant Syst Evol 144:103–118
Carlquist S 1984f Wood and stem anatomy of Lardizabalaceae, with comments on the vining habit, ecology, and systematics. Bot J Linnean Soc 88:257–277
Carlquist S 1984g Wood anatomy and relationships of Pentaphylacaceae: significance of vessel features. Phytomorphology 34:84–90
Carlquist S 1984h Wood anatomy of Malesherbiaceae. Phytomorphology 34:180–190
Carlquist S 1985a Vasicentric tracheids as a drought survival mechanism in the woody flora of southern California and similar regions; review of vasicentric tracheids. Aliso 11:37–68
Carlquist S 1985b Wood anatomy of Begoniaceae, with comments on raylessness, paedomorphosis, relationships, vessel diameter, and ecology. Bull Torrey Bot Club 112:59–69
Carlquist S 1985c Wood and stem anatomy of Misodendraceae: systematic and ecological conclusions. Brittonia 37:58–75

Carlquist S 1985d Observations on functional wood histology of vines and lianas: vessel dimorphism, tracheids, vasicentric tracheids, narrow vessels, and parenchyma. Aliso 11: 139–157
Carlquist S 1985e Wood anatomy of Coriariaceae: phylogenetic and ecological implications. Syst Bot 10:174–183
Carlquist S 1985f Wood anatomy and familial status of Viviania. Aliso 11:159–165
Carlquist S 1986a Terminology of imperforate tracheary elements. IAWA Bull NS 7:75–81
Carlquist S l986b Terminology of imperforate tracheary elements. A reply. IAWA Bull NS 7:168–170
Carlquist S 1986c Wood anatomy of Stilbaceae and Retziaceae: ecological and systematic implications. Aliso 11:299–316
Carlquist S 1987a Wood anatomy of Nolanaceae; relationship to Solanaceae. Aliso 11:463–471
Carlquist S 1987b Wood anatomy of Martyniaceae and Pedaliaceae. Aliso 11:473–483
Carlquist S 1987c Diagonal and tangential vessel aggregations in wood: function and relationship to vasicentric tracheids. Aliso 11:451–462
Carlquist S 1987d Wood anatomy of Plakothira (Loasaceae). Aliso 11:563–569
Carlquist S 1987e Pliocene Nothofagus wood from the Transantarctic mountains. Aliso 11:571–583
Carlquist S 1987f Wood anatomy and relationships of Stackhousiaceae. Bot Jahrb Syst 108:473–480
Carlquist S 1987g Wood anatomy of Cneoraceae: ecology, relationships, and generic definition. Aliso 12:7–16
Carlquist S 1987h Wood anatomy of Scytopetalaceae. Aliso 12:65–75
Carlquist S 1987i Wood anatomy of noteworthy species of Ludwigia (Onagraceae) with relation to ecology and systematics. Ann Mo Bot Gard 75:889–896
Carlquist S 1987j Adaptive wood anatomy of chaparral shrubs. In: Keeley S (ed) The California chaparral: paradigms reexamined. Contrib Los Angeles Co Mus Nat Hist Ser 34:25–35
Carlquist S 1987k Presence of vessels in Sarcandra (Chloranthaceae); comments on vessel origins in angiosperms. Am J Bot 74:1765–1771
Carlquist S 1988a Wood anatomy of Drimys (Winteraceae). Aliso 12:81–95
Carlquist S 1988b Wood anatomy and relationships of Duckeodendraceae and Goetzeaceae. IAWA Bull NS 9:3–12
Carlquist S 1988c Tracheid dimorphism: a new pathway in evolution of imperforate tracheary elements. Aliso 12:103–118
Carlquist S 1989a Wood and bark anatomy of Empetraceae: comments on paedomorphosis in woods of certain small shrubs. Aliso 12:497–515
Carlquist S 1989b Wood anatomy of Cercidium (Fabaceae), with emphasis on vessel wall sculpture. Aliso 12:235–255
Carlquist S 1989c Wood anatomy of Tasmannia; summary of wood anatomy of Winteraceae. Aliso 12:257–275
Carlquist S 1989d Wood anatomy and relationships of Montinia. Aliso 12:369–378
Carlquist S 1989e Wood and bark anatomy of Degeneria. Aliso 12:485–495
Carlquist S 1990a Wood anatomy of *Ascarina* (Chloranthaceae) and the tracheid-vessel element transition. Aliso 12:667–684
Carlquist S 1990b Wood anatomy and relationships of Lactoridaceae. Am J Bot 11:1498–1505
Carlquist S 1991 Wood and bark anatomy of Ticodendron: comments on relationships. Ann Mo Bot Gard 78:96–104
Carlquist S 1992a Wood anatomy of Hedyosmum (Chloranthaceae) and the tracheid-vessel element transition. Aliso 13:447–462
Carlquist S 1992b Wood anatomy and stem of Chloranthus; summary of wood anatomy of Chloranthaceae, with comments on relationships, vessellessness, and the origin of monocotyledons. IAWA Bull NS 13:3–16
Carlquist S 1992c Pit membrane remnants in perforation plates of primitive dicotyledons and their significance. Am J Bot 79:660–672

Carlquist S 1992d Vegetative anatomy and relationships of Eupomatiaceae. Bull Torrey Bot Club 119:167–180
Carlquist S 1992e Wood anatomy of Lamiaceae. A survey, with comments on vascular and vasicentric tracheids. Aliso 13:309–338
Carlquist S 1992f Wood anatomy of sympetalous dicotyledon families: a summary, with comments on systematic relationships and the origin of the woody habit. Ann Mo Bot Gard 79:303–332
Carlquist S 1992g Wood anatomy of selected Cucurbitaceae and its relationship to habit and systematics. Nord J Bot 12:347–355
Carlquist S 1992h Wood anatomy of Solanaceae. A survey. Allertonia 6:279–326
Carlquist S 1993 Wood and bark anatomy of Aristolochiaceae; systematic and habital correlations. IAWA J 14:341–357
Carlquist S 1995a Wood and bark anatomy of Ranunculaceae (including Hydrastis) and Glaucidiaceae. Aliso 14:65–84
Carlquist S 1995b Wood anatomy of Berberidaceae: ecological and phylogenetic considerations. Aliso 14:85–103
Carlquist S 1995c Wood anatomy of Caryophyllaceae: ecological, habital, systematic, and phylogenetic implications. Aliso 14:1–17
Carlquist S 1995d Wood anatomy of Ranunculiflorae: a summary. Plant Syst Evol [Suppl] 9:11–24
Carlquist S 1996a Materials for anatomical and ultrastructural studies: pollen, wood, and pickled materials. In: Stuessy T F, Sohmer S H (eds) Sampling the green world. Columbia Univ Press, New York, pp 143–156
Carlquist S 1996b Wood anatomy of Akaniaceae and Bretschneideraceae: a case of near-identity and its systematic implications. Syst Bot 21:607–616
Carlquist S 1996c Wood anatomy of primitive angiosperms: new perspectives and syntheses. In: Taylor D W, Hickey L J (eds) Flowering plant origin, evolution, and phylogeny Chapman and Hall, New York, pp 68–90
Carlquist S 1996d Wood, bark, and stem anatomy of Gnetales: a summary. Int J Plant Sci 157 [Suppl 6]:S58–S76
Carlquist S 1996e Wood and stem anatomy of Menispermaceae. Aliso 14:155–170
Carlquist S 1996f Wood anatomy of Limnanthaceae and Tropaeolaceae in relation to habit and phylogeny. Sida 17:333–342
Carlquist S 1997a Pentaphragma: a unique wood and its significance. IAWA J 18:3–12
Carlquist S 1997b Wood anatomy of Buddlejaceae. Aliso 15:41–56
Carlquist S 1997c Wood anatomy of Argyroxiphium (Asteraceae): adaptive radiation and ecological correlations. J Torrey Bot Soc 124:1–10
Carlquist S 1998a Wood and stem anatomy of Petiveria and Rivina (Caryophyllales): systematic implications. IAWA J 19:383–391
Carlquist S 1998b Wood anatomy of Portulacaceae and Hectorellaceae: ecological, habital, and systematic implications. Aliso 16:137–153
Carlquist S 1998c Wood anatomy of Resedaceae. Aliso 16:127–135
Carlquist S 1998d Wood and bark anatomy of Caricaceae: correlations with systematics and habit. IAWA J 19:191–206
Carlquist S 1998e Wood anatomy of Dubautia (Asteraceae: Madiinae) in relation to adaptive radiation. Pac Sci 52:356–368
Carlquist S 1998f Wood anatomy of Wilkesia (Asteraceae) with relation to systematics, organography, and habit. J Torrey Bot Soc 125:261–267
Carlquist S 1999a Wood anatomy, stem anatomy, and cambial activity of Barbeuia (Caryophyllales). IAWA J 20:431–440
Carlquist S 1999b Wood and stem anatomy of Stegnosperma (Caryophyllales): phylogenetic relationships; nature of lateral meristems and successive cambial activity. IAWA J 20:149–163
Carlquist S 1999c Wood and bark anatomy of Schisandraceae: implications for phylogeny, habit, and vessel evolution. Aliso 18:45–55

Carlquist S 1999d Wood anatomy of Agdestis (Caryophyllales): systematic position and nature of the successive cambia. Aliso 18:35–43
Carlquist S 1999e Wood, stem, and root anatomy of Basellaceae with relation to habit, systematics, and cambial variants. Flora 194:1–12
Carlquist S 2000a Wood and bark anatomy of Achatocarpaceae. Sida 19:71–78
Carlquist S 2000b Wood and stem anatomy of phytolaccoid and rivinoid Phytolaccaceae (Caryophyllales): ecology, systematics, nature of systematic cambia. Aliso 19:13–29
Carlquist S 2000c Wood and bark anatomy of Takhtajania (Winteraceae); phylogenetic and ecological implications. Ann Mo Bot Gard 87:317–322
Carlquist S, Boggs C J 1996 Wood anatomy of Plumbaginaceae. Bull Torrey Bot Club 123:135–147
Carlquist S, DeBuhr L 1977 Wood anatomy of Penaeaceae (Myrtales): comparative, phylogenetic, and ecological implications. Bot J Linnean Soc 75:211–227
Carlquist S, DeVore M 1998 Wood anatomy of Calyceraceae with reference to ecology, habit, and systematic relationships. Aliso 17:63–76
Carlquist S, Eckhart V M 1982 Wood anatomy of Darwiniothamnus, Lecocarpus, and Macraea (Asteraceae). Aliso 10:291–300
Carlquist S, Eckhart V M 1984 Wood anatomy of Hydrophyllaceae. II. Genera other than Eriodictyon, with comments on parenchyma bands containing vessels with large pits. Aliso 10:397–412
Carlquist S, Grant M L 1963 Studies in Fitchia (Compositae): novelties from the Society Islands; anatomical studies. Pac Sci 17:282–298
Carlquist S, Hanson M A 1991 Wood and stem anatomy of Convolvulaceae: a survey. Aliso 13:51–94
Carlquist S, Hoekman D A 1985a Wood anatomy of Staphyleaceae: ecology, statistical correlations, and systematics. Flora 117:195–216
Carlquist S, Hoekman D A 1985b Ecological wood anatomy of the woody southern California flora. IAWA Bull NS 6:319–347
Carlquist S, Hoekman D A 1986a Wood anatomy of Gesneriaceae. Aliso 11:279–297
Carlquist S, Hoekman D A 1986b Wood anatomy of Myoporaceae: ecological and systematic considerations. Aliso 11:317–335
Carlquist S, Miller R B 1999 Vegetative anatomy and relationships of Setchellanthus caeruleus (Setchellanthaceae). Taxon 48:289–302
Carlquist S, Wilson E J 1995 Wood anatomy of Drosophyllum (Droseraceae): ecological and phylogenetic considerations. Bull Torrey Bot Club 122:185–189
Carlquist S, Zona S 1988a Wood anatomy of Acanthaceae: a survey. Aliso 12:201–227
Carlquist S, Zona S 1988b Wood anatomy of Papaveraceae, with comments on vessel restriction patterns. IAWA Bull NS 9:253–267
Carlquist S, Eckhart V M, Michener D C 1983 Wood anatomy of Hydrophyllaceae. I. Eriodictyon. Aliso 10:397–412
Carlquist S, Eckhart V M, Michener D C 1984 Wood anatomy of Polemoniaceae. Aliso 10:547–572
Carlquist S, Morrell P L, Manchester S R 1993 Wood anatomy of Sabiaceae (s.l.); ecological and systematic implications. Aliso 13:521–549
Carlquist S, Schneider E L, Miller R B 1994 Wood and bark anatomy of Argemone (Papaveraceae). IAWA J 15:247–255
Carlquist S, Dauer K, Nishimura S Y 1995 Wood and stem anatomy of Saururaceae, with reference to ecology, phylogeny, and origin of the monocotyledons. IAWA J 16:133–150
Carpenter C S, Dickison W C 1976 The morphology and relationships of Oncotheca balansae. Bot Gaz 137:141–153
Cassens D L 1980 Vestured pits in the New World Pithecellobium (sensu lato). IAWA Bull NS 1:59–64
Castro M A 1985 Structure of the vessel-parenchyma pit membrane in some species of Lauraceae. IAWA Bull NS 6:35–38
Castro M A 1991 Ultrastructure of vestures on the vessel wall in some species of Prosopis (Leguminosae-Mimosoideae). IAWA Bull NS 12:425–430

Chalk L 1933 Multiperforate plates in vessels, with special reference to the Bignoniaceae. Forestry 7:16–25
Chalk L 1955 Ray volumes in hardwoods. Trop Woods 101:1–10
Chalk L 1970 Short fibres with clearly defined intrusive growth, with special reference to Fraxinus. Bot J Linnean Soc 63 [Suppl 1]:163–168
Chalk L, Chattaway M M 1933 Perforated ray cells. Proc R Soc Lond B 113:82–92
Chalk L, Chattaway M M 1934 Measuring the length of vessel members. Trop Woods 40:19–26
Chalk L, Chattaway M M 1937 Identification of woods with included phloem. Trop Woods 50:1–31
Chattaway M M 1933a Ray development in the Sterculiaceae. Forestry 7:93–108
Chattaway M M 1933b Tile-cells in the rays of the Malvales. New Phytol 32:261–173
Chattaway M M 1936 The relation between fibre and cambial initial length in dicotyledonous woods. Trop Woods 46:16–20
Chattaway M M 1939 Crystalliferous strands. Rodriguesia 4:55–58
Chattaway M M 1948 Note on the vascular tissue in the rays of Banksia. J Comm Sci Ind Res Org 21:275–278
Chattaway M M 1951 The development of horizontal canals in rays. Aust J Sci Res 4:1–11
Chattaway M M 1955 Crystals in woody tissues. 1. Trop Woods 102:55–74
Chattaway M M 1956 Crystals in woody tissues. 2. Trop Woods 104:100–124
Chauhan L, Dayal R 1985 Wood anatomy of Indian Albizias. IAWA Bull NS 6:213–218
Cheadle V I, Esau K 1964 Secondary phloem of Liriodendron tulipifera. Univ Calif Publ Bot 36:143–252
Chowdhury K A 1939 The formation of growth rings in Indian trees, part I. Indian For Rec Ns Util 1:1–34
Chowdhury K A 1940a The formation of growth rings in Indian trees, part II. Indian For Rec Ns Util 2:41–57
Chowdhury K A 1940b The formation of growth rings in Indian trees, part III. A study of the effect of locality. Indian For Rec Ns Util 2:59–75
Chowdhury K A 1964 Growth rings in tropical trees and taxonomy. J Indian Bot Soc 43:334–342
Chowdhury K A, Purkayastha S K, Ghosh S S, Sond J S 1956 Septate fibres in angiosperms and their length. Nature 177:1243–1244
Cockrell R A 1941 A comparative study of the wood of several South American species of Strychnos. Am J Bot 28:32–41
Côté W A (ed) 1965 Cellular ultrastructure of woody plants. Syracuse Univ Press, Syracuse, 603 pp
Côté W A, Day A C 1962 Vestured pits – fine structure and apparent relationship with warts. Techn Assoc Pulp Paper Ind 45:906–910
Cozzo D 1948 Anatomía del leño de las especies argentinas de la tribu “Zygophylleae” (Zigofilaceas). Rev Inst Nac Sci Cienc Bot 1:57–85
Cozzo D 1950 Anatomía del leño secundario de las leguminosas papilionoideas Argentinas silvestres y cultivadas. Rev Inst Nac Invest Cienc Nat Ciencias Bot 1:223–361
Cozzo D 1951 Investigaciones anatomicas de maderas de Sapotaceas Argentinas. Rev Inst Nac Sci Ciencias Bot 2:263–290
Cozzo D 1953 The structure and diagnostic significance of crateriform borders in the vessels of Cercidium. J Arnold Arb 24:187–190
Cozzo D, Christiani L Q 1950 Los generos de fanerogamas Argentinas con estructura leñosa estratificada. Rev Inst Nac Invest Cienc Nat 1:363–405
Cozzo D, Rodriguez E M 1959 Anatomia comparada de la madera de 14 especies de Eucalyptus cultivadas en la Argentina. Rev Fac Agron Vet 14(3):1–27
Cristiani L Q 1948 Anatomía del leño secundario de las especies Argentinas del genero “Monttea.” Comm Inst Nac Invest Cienc Nat Cienc Bot 1:1–6
Cumbie B G 1960 Anatomical studies in the Leguminosae. Trop Woods 113:1–47
Cumbie B G 1963 The vascular cambium and xylem development in Hibiscus lasiocarpus. Am J Bot 50:944–951
Cumbie B G 1967a Developmental changes in the vascular cambium of Leitneria floridana. Am J Bot 54:414–424

Cumbie B G 1967b Development and structure of the xylem in Canavalia (Leguminosae). Bull Torrey Bot Club 94:162–175
Cumbie B G 1969 Developmental changes in the vascular cambium of Polygonum lapathifolium. Am J Bot 56:139–146
Cumbie B G 1983 Developmental changes in the wood of Bocconia vulcanica Donn. Smith. IAWA Bull NS 4:131–140
Cumbie B G, Mertz D 1962 Xylem anatomy of Sophora (Leguminosae) in relation to habit. Am J Bot 49:33–40
Cutler D F, Gregory M 1998 Anatomy of the dicotyledons, vol IV, 2nd edn. Saxifragales. Oxford Science Publications, Oxford, 324 pp
Czaninski Y 1977 Vessel-associated cells. IAWA Bull 3:51–55
Dadswell H E, Eckersley A M 1940 The wood anatomy of some Australian Lauraceae with methods of their identification. CSIRO Bull 132. CSIRO, Melbourne
Dadswell H E, Ingle H D 1948 The anatomy of timbers of the south-west Pacific area. I. Anacardiaceae. Aust J Sci Res Ser B 1:391–415
Dadswell H E, Wardrop A B 1949 What is reaction wood? Aust For 13:22–33
Datta P C, Deb A 1968 Secondary xylem, floral traces, and phylogeny of common Indian species of Rumex. Beitr Biol Pflanzen 45:69–90
Davidson C 1976 Anatomy of xylem and phloem of the Datiscaceae. Nat Hist Mus Los Angeles County Contrib Sci 280:1–28
Davis S D, Sperry J S, Hacke E G 1999 The relationship between xylem conduit diameter and cavitation caused by freezing. Am J Bot 86:1367–1372
Dayal R, Rao R V, Sharma B 1984 Perforated ray cells in woods of Indian Myrsinaceae and Loganiaceae. IAWA Bull NS 5:2225–2228
DeBuhr L E 1977 Wood anatomy of the Sarraceniaceae: ecological and evolutionary implications. Plant Syst Evol 128:159–169
Decker J M 1966 Wood anatomy and phylogeny of Luxembergieae (Ochnaceae). Phytomorphology 16:39–54
Denne M P 1973 Tracheid dimension in relation to shoot vigour in Picea. Forestry 46:117–124
den Outer R W 1986 Storied structure of the secondary phloem. IAWA Bull NS 7:47–51
den Outer R W, Schutz P R 1981 Wood anatomy of Apeiba (Tiliaceae). IAWA Bull NS 2:187–192
den Outer R W, van Veenendaal W L H 1980 Wood and bark anatomy of Alluaudia (Didieriaceae) from Madagascar. IAWA Bull NS 1:133–139
den Outer R W, van Veenendaal W L H 1981 Wood and bark anatomy of Azima tetracantha Lam. (Salvadoraceae) with description of its included phloem. Acta Bot Neerl 30:199–207
den Outer R D, van Veenendaal W L H 1983 Wood anatomy of Uncarina leandrii H. Humb. (Pedaliaceae) and its relation to Bignoniaceae. IAWA Bull NS 4:53–59
den Outer R D, van Veenendaal W L H 1996 Wood anatomy of the Aphanocalyx-Monopetalanthus complex (Caesalpinioideae). IAWA J 17:205–223
de Zeeuw C 1977 Pakaraimea, Dipterocarpacea of the Western Hemisphere. III. Stem anatomy. Taxon 26:368–380
Dickison W C 1972 Anatomical studies in the Connaraceae. II. Wood anatomy. J Elisha Mitchell Sci Soc 88:120–136
Dickison W C 1978 Comparative anatomy of the Eucryphiaceae. Am J Bot 65:722–735
Dickison W C 1979 A note on the wood anatomy of Dillenia (Dilleniaceae). IAWA Bull 1979/2-3: 57–60
Dickison W C 1981a Contributions to the morphology and anatomy of Strasburgeria and a discussion of the taxonomic position of the Strasburgeriaceae. Brittonia 33:564–580
Dickison W C 1981b The evolutionary relationships of the Leguminosae. In: Polhill R M, Raven P H (eds) Advances in legume systematics: proceedings of the international legume conference, Kew, 24–29 July 1978. Royal Botanic Gardens, Kew, pp 35–54
Dickison W C 1984 On the occurrence of silica grains in woods of Hibbertia (Dilleniaceae). IAWA Bull NS 5:341–343
Dickison W C 1986 Wood anatomy and affinities of the Alseuosmiaceae. Syst Bot 11:214–221

Dickison W C 1988 Xylem anatomy of Diegodendron humbertii. IAWA Bull NS 9:332–336
Dickison W C, Baas P 1977 The morphology and relationships of Paracryphia (Paracryphiaceae). Blumea 23:417–438
Dickison W C, Phend K D 1985 Wood anatomy of the Styracaceae: evolutionary and ecological considerations. IAWA Bull NS 6:3–22
Dickison W C, Sweitzer E M 1970 The morphology and relationships of Barbeya oleoides. Am J Bot 57:468–476
Dickison W C, Rury P M, Stebbins G L 1978 Xylem anatomy of Hibbertia in relation to ecology and systematics. J Arnold Arbor 59:32–49
Diettert R A 1938 The morphology of Artemisia tridentata Nutt. Lloydia 1:3–74
Dobbins D R 1969 Studies on anomalous cambial activity in the Bignoniaceae. I. Development of the vascular pattern. Can J Bot 47:2101–2106
Dobbins D R 1970 The influence of leaves on cambial activity in the cross vine. Am J Bot 57:735
Dobbins D R 1971 Studies on anomalous cambial activity in the Bignoniaceae. II. A case of differential production of vascular tissues. Am J Bot 58:697–705
Dobbins D R 1981 Anomalous secondary growth in lianas of the Bignoniaceae is correlated with the vascular pattern. Am J Bot 68:142–144
Dobbins D R, Fisher J B 1986 Wound responses in girdled stems of lianas. Bot Gaz 147:278–289
Dong Z, Baas P 1993 Wood anatomy of trees and shrubs from China. V. Anacardiaceae. IAWA J 14:87–102
Duke N C, Birch W R, Williams W T 1981 Growth rings and rainfall correlation in a mangrove tree of the genus Diospyros (Ebenaceae). Aust J Bot 29:135–142
Dute R R, Rushing A E 1988 Notes on torus development in the wood of Osmanthus americanus (L.) Benth. and Hook. ex Gray (Oleaceae). IAWA Bull NS 9:237–244
Dute R R, Rushing A E 1990 Torus structure and development in the woods of Ulmus alta Michx., Celtis laevigata Willd., and Celtis occidentalis L. IAWA Bull NS 11:71–83
Dute R R, Freeman J D, Henning F, Barnard L D 1996 Intervascular pit membrane structure in Daphne and Wikstroemia – systematic implications. IAWA J 17:161–181
Eames A J 1910 On the origin of the broad ray in Quercus. Bot Gaz 49:161–167
Easterling K E, Harryson R, Gibson L J, Ashby M F 1982 On the mechanics of balsa and other woods. Proc R Soc Lond A 383:31–41
Eckstein D, Baas P (eds) 1999 Dendrochronology in monsoon Asia. IAWA J 20:223–350
Eckstein D, Sass U, Baas P (eds) 1995 Growth periodicity in tropical trees. IAWA J 16:323–442
Eeckhout L-E 1951 On the definition of compound rays in wood anatomy. Med Lab Hout Technol Gent 6:1–16
Ellmore G S, Ewers F W 1985 Hydraulic conductivity in trunk xylem of elm, Ulmus americana. IAWA Bull NS 6:303–307
Erak S 1971 Diameter, wall width, and length of vessel elements and diameter and wall width of fibres in beech wood grown on the same parent rock at different altitudes in Bosnia (in Serbo-Croatian). Pregled Naucnoteh Rad Inform Zavod Tehn Drveta 8:19–26
Esau K 1965 Plant anatomy, 2nd edn. Wiley, New York, 767 pp
Esau K 1977 Anatomy of seed plants, 2nd edn. Wiley, New York
Esau K, Cheadle VI 1969 Secondary growth in Bougainvillea. Ann Bot NS 33:807–819
Essiamah S, Eschrich W 1985 Changes of starch content in the starch tissues of deciduous trees during winter and spring. IAWA Bull NS 6:97–106
Evert R F 1961 Some aspects of cambial development in Pyrus communis. Am J Bot 48:479–488
Ewers F W 1985 Xylem structure and water conduction in conifer trees, dicot trees, and lianas. IAWA Bull NS 6:309–317
Ewers F W, Fisher J B 1991 Why vines have narrow stems: histological trends in Bauhinia (Fabaceae). Oecologia 88:233–237
Ewers F W, Carlton M R, Fisher J B, Kolb K J, Tyree M T 1997 Vessel diameter in roots versus stems of tropical lianas and other growth forms. IAWA J 18:261–279
Fagerlind F, Dunbar A 1973 Some electron microscopical methods for solving wood anatomical problems. Bot Notis 126:519–533
Fahn A 1933 Annual wood ring development in maquis trees of Israel. Palest J Bot Jerus Ser 6:1–26

Fahn A 1955 The development of the growth ring in wood of Quercus infectoria and Pistacia lentiscus in the hill region of Israel. Trop Woods 101:52–59
Fahn A 1958 Xylem structure and annual rhythm of development in trees and shrubs of the desert. I. Tamarix aphylla, T. jordanis var. negevensis, T. gallica var. maris-mortui. Trop Woods 109:81–94
Fahn A 1959a Xylem structure and annual rhythm of development in trees and shrubs of the desert. II. Acacia tortilis and A. raddiana. Bull Res Counc Isr 7D:23–28
Fahn A 1959b Xylem structure and annual rhythm of development in trees and shrubs of the desert. III. Eucalyptus camaldulensis and Acacia cyanophylla. Bull Res Counc Isr 7D:122–131
Fahn A 1959c Annual rhythm of xylem development in trees and shrubs of Israel. Proc IX Bot Congr, Montreal, p 110
Fahn A 1962 Xylem structure and the annual rhythm of cambial activity in woody species of the east Mediterranean region. IAWA Bull 1962/1:2–6
Fahn A 1979 Secretory tissues in plants. Academic Press, London
Fahn A 1985 The development of the secondary body in plants with interxylary phloem. In: Kucera L J (ed) Xylorama. Birkhäuser, Basel, pp 58–67
Fahn A, Leshem B 1963 Wood fibres with living protoplasts. New Phytol 62:91–98
Fahn A, Schchori Y 1968 The organization of the secondary conductive tissues in some species of the Chenopodiaceae. Phytomorphology 17:147–154
Fahn A, Zimmermann M H 1982 Development of the successive cambia in Atriplex halimus (Chenopodiaceae). Bot Gaz 143:353–357
Fahn A, Werker E, Baas P 1986 Wood anatomy and identification of trees and shrubs from Israel and adjacent regions. Isr Acad Sci Human, Jerusalem, 212 pp
Ferreirinha M P 1965 Appraisal of the variation in the micrographic fibre characteristics within and between trees of tropical hardwood species. Garcia de Orta 13:383–390
Fineran B A 1985 Graniferous tracheary elements in haustoria of root parasitic angiosperms. Bot Rev 51:389–441
Fisher J B 1980 The vegetative and reproductive structure of papaya (Carica papaya). Lyonia 1:191–208
Fisher J B 1981 Wound healing by exposed secondary xylem in Adansonia (Bombacaceae). IAWA Bull NS 2:193–199
Fisher J B 1982 A survey of buttresses and aerial roots of tropical trees for presence of reaction wood. Biotropica 14:56–61
Fisher J B, Stevenson J W 1981 Occurrence of reaction wood in branches of dicotyledons and its role in tree architecture. Bot Gaz 142:82–95
Fisher J B, Ewers F W 1989 Wound-healing in stems of lianas after twisting and girdling injuries. Bot Gaz 150:251–265
Foster R C 1967 Fine structure of tyloses in three species of the Myrtaceae. Aust J Bot 15:25–34
Francis W D 1924 The development of buttresses in Queensland trees. Proc R Soc Queensl 36:21–37
Francis W D 1926 The development of corrugated stems of some Eastern Australian trees. Proc R Soc Queensl 38:62–67
Frey A 1929 Calcium-monohydrat und Trihydrat. Handb Pflanzenanat III/1a(8):1–130
Frey-Wyssling A 1981 Crystallography of the two hydrates of crystalline calcium oxalate in plants. Am J Bot 68:130–141
Frison E 1948 De la présence d'amidon dans le lumen des fibres du bois. Bull Agric Congo Belge 39:869–874
Fritts H C 1976 Tree rings and climate. Academic Press, New York, 567 pp
Frost F H 1930a Specialization in secondary xylem in dicotyledons. I. Origin of vessel. Bot Gaz 89:67–94
Frost F H 1930b Specialization in secondary xylem in dicotyledons. II. Evolution of end wall of vessel segment. Bot Gaz 90:198–212
Frost F H 1931 Specialization in secondary xylem in dicotyledons. III. Specialization of lateral wall of vessel segment. Bot Gaz 91:88–96

Fujii T 1993 Application of a resin casting method to wood anatomy of some Japanese Fagaceae species. IAWA J 14:273–288
Fukazawa K 1984 Juvenile wood of hardwoods judged by density variation. IAWA Bull NS 5:65–73
Gale R 1982 Some pitfalls in wood identification, with reference to Nothofagus. IAWA Bull NS 3:179–184
Garratt G A 1933 Systematic anatomy of the woods of the Myristicaceae. Trop Woods 35:6–48
Gasson P 1999 Wood anatomy of the tribe Dipterygeae with comments on related papilionoid and caesalpinoid Leguminosae. IAWA J 20:441–455
Gasson P, Dobbins D R 1991 Wood anatomy of the Bignoniaceae, with a comparison of trees and lianas. IAWA Bull NS 12:389–417
Ghosh S S, Purkayastha S K 1960 Characteristic arrangement of septate fibers in wood of Averrhoa sp. Sci Cult 25:690–691
Ghouse A K M, Yunus M 1973 Some aspects of cambial development in the shoots of Dalbergia sissoo. Flora 162:549–558
Ghouse A K M, Yunus M 1974a Cambial structure in Dalbergia. Phytomorphology 24:152–158
Ghouse A K M, Yunus M 1974b The ratio of ray and fusiform initials in some woody species of the ranalian complex. Bull Torrey Bot Club 101:363–366
Gibson A C 1973 Comparative anatomy of secondary xylem in Cactoideae (Cactaceae). Biotropica 5:29–65
Gibson A C 1975 Another look at the cactus research of Irving Widmer Bailey. Cactus Succ J Suppl 1975:76–85
Gibson A C 1977a Wood anatomy of Opuntias with cylindrical to globular stems. Bot Gaz 138: 334–351
Gibson A C 1977b Vegetative anatomy of Maihuenia (Cactaceae) with some theoretical discussion of ontogenetic changes in xylem cell types. Bull Torrey Bot Club 104:35–48
Gibson A C 1978a Dimorphism of secondary xylem in two species of cacti. Flora 167:403–408
Gibson A C 1978b Wood anatomy of platyopuntias. Aliso 9:279–307
Gibson A C 1978c Structure of Pterocactus tuberosus, a cactus geophyte. Cactus Succ J 50:41–43
Gibson A C 1978d Rayless secondary xylem of Halophytum. Bull Torrey Bot Club 105:39–44
Gibson A C 1981 Vegetative anatomy of Pachycormus (Anacardiaceae). Bot J Linnean Soc 83: 273–284
Gibson A C, Calkin H W, Nobel P S 1984 Xylem anatomy, water flow, and hydraulic conductance in the fern Cyrtomium falcatum. Am J Bot 71:564–574
Gibson A C, Calkin H W, Nobel P S 1985 Hydraulic conductance and xylem structure in tracheid-bearing plants. IAWA Bull NS 6:293–302
Giebel K P, Dickison W C 1976 Wood anatomy of Clethraceae. J Elisha Mitchell Sci Soc 92:17–26
Gilbert S G 1940 Evolutionary significance of ring porosity in woody angiosperms. Bot Gaz 102:105–120
Gin A 1909 Réchérches sur les Lythracées. Trav Lab Mat Med Paris 6:1–166
Ginzburg C 1963 Some anatomic features of splitting of desert shrubs. Phytomorphology 13: 92–97
Giraud B 1980 Correlations between wood anatomical characters in Entandophragma utile (Meliaceae). IAWA Bull NS 1:73–75
Giraud B 1981a On the special nature of vessel perforations in Hieronyma andina P. and H. (Euphorbiaceae). IAWA Bull NS 2:43–44
Giraud B 1981b Structure des perforations des éléments de vaisseaux dans le genre Hieronyma Allem. (Euphorbiaceae). Bull Mus Nat Hist Nat Paris 4 Ser Sec B Adansonia 3:95–104
Giraud B 1983 Les cellules perforées de rayons ligneux chez les Euphorbiacées. Bull Mus Nat Hist Nat Paris 4–5 Ser Sec B Adansonia 2:213–221
Goldblatt P, Tobe H, Carlquist S, Patel V C 1985 Familial position of the Cape genus Empleuridium. Ann Mo Bot Gard 72:167–183
Gomes A V, Teixeira L L, Schaitza E G, Hofmeister R M 1989 Perforation plates in vessels of Citharexylum myrianthum Cham. (Verbenaceae). IAWA Bull NS 10:27–34
Gottwald H 1972 Tyloses in fibre tracheids. Wood Sci Technol 6:121–127

Gottwald H 1980 "Louro preto" – found to be the first silica-bearing Cordia (Cordia glabrata, Boraginaceae). IAWA Bull NS 1:55–58
Gottwald H 1983 Wood anatomical studies of Boraginaceae (s.l.) I. Cordioideae. IAWA Bull NS 4:161–178
Gottwald H, Parameswaran N 1964 Vielfache Gefäßdurchbrechungen in der Familie Dipterocarpaceae. Z Bot 52:321–334
Gottwald H, Parameswaran N 1966 Das sekundäre Xylem der Familie Dipterocarpaceae, anatomische Untersuchungen zur Taxonomie und Phylogenie. Bot Jahrb 85:410–508
Gottwald H, Parameswaran N 1967 Beiträge zur Anatomie und Systematik der Quiinaceae. Bot Jahrb 87:361–381
Gottwald H, Parameswaran N 1968 Das sekundäre Xylem und die systematische Stellung der Ancistrocladaceae und Dioncophyllaceae. Bot Jahrb 88:49–69
Gottwald H, Parameswaran N 1980 Anatomy of wood and bark of Tectona (Verbenaceae) in relation to taxonomy. Bot Jahrb 101:363–384
Gould E M Jr 1984 Annual report of activities at the Harvard Forest. Harvard University Press, Cambridge, 14 pp
Govindarajalu E, Swamy B G L 1955 Size variations of septate wood fibers in Pithecellobium dulce Benth. Nature 176:315
Gregory M 1994 Bibliography of systematic wood anatomy of dicotyledons. IAWA J Suppl 1:1–265
Gregory M, Baas P 1989 A survey of mucilage cells in vegetative organs of the dicotyledons. Isr J Bot 38:125–174
Gregory R A 1978 Living elements of the conducting secondary xylem of sugar maple (Acer saccharum Marsh). IAWA Bull 4:65–69
Gregory R A 1981 A rapid method of estimating the relative amount of vascular ray tissue. IAWA Bull NS 2:77–78
Greguss P 1959 Holzanatomie der Europäischen Laubhölzer und Sträucher. Akademiai Kiado, Budapest, 330 pp
Grew N 1682 The anatomy of plants. Rawlings, London (reprinted 1965 by Johnson Reprint, New York), 304 pp
Gunning B E S, Pate J S 1974 Transfer cells. In: Robards A W (ed) Dynamic aspects of plant ultrastructure. McGraw Hill, Maidenhead, pp 441–480
Gunning B E S, Pate J S, Briarty L G 1968 Specialized "transfer cells" in main veins of leaves and their possible significance in phloem translocation. J Cell Biol 37:C7–C12
Gutzwiller M-A 1961 Die phylogenetische Stellung von Suriana maritima L. Bot Jahrb 81:1–49
Haberlandt G 1914 Physiological plant anatomy (trans by M. Drummond), 4th edn. Macmillan, London, 777 pp
Hale J D, Clermont L D 1963 Influence of parenchyma cell-wall morphology on basic physical and chemical characteristics of wood. J Polymer Sci C 2:253–261
Handa T 1937 Anomalous secondary growth in Bauhinia japonica Maxim. Jpn J Bot 9:37–53
Handa T 1938 Anomalous secondary growth in the axis of Bauhinia championi Benth. Jpn J Bot 9:303–311
Hargrave K R, Kolb K J, Ewers F W, Davis S D 1994 Conduit diameter and drought-induced embolism in Salvia mellifera Greene (Labiatae). New Phytol 126:695–705
Harrar E S 1946 Notes on starch grains in septate fiber-tracheids. Trop Woods 85:1–9
Harzmann L J 1984 Zur Entwicklung der Faserzellänge während des Jugendwachstums bei Cedrela odorata L. Flora 175:183–193
Haskins M L, Hayden W J 1987 Anatomy and affinities of Penthorum. Am J Bot 74:164–177
Hayden W J 1977 Comparative anatomy and systematics of Picrodendron, genus incertae sedis. J Arnold Arbor 58:257–279
Hayden S M, Hayden W J 1994 Stem development, medullary bundles and wood anatomy of Croton glandulosus var. septentrionalis (Euphorbiaceae). IAWA J 15:51–63
Heimsch C 1942 Comparative anatomy of the secondary xylem in xylem in the "Gruinales" of Wettstein, with reference to taxonomic groupings. Lilloa 8:83–198

Hejnowicz Z, Zagorska-Marek B 1974 Mechanism of changes in grain inclination in wood produced by storeyed cambium. Acta Soc Bot Pol 18:381–397
Hess R W 1936 Occurrence of raphides in woods. Trop Woods 46:22–31
Hess R W 1946a Keys to American woods. XVII. Woods with vasicentric tracheids. Trop Woods 85:13–14
Hess R W 1946b Keys to American woods. XVIII. Fibers with spiral thickenings. Trop Woods 85:14–15
Hess R W 1946c Keys to American woods. XIX. Special fibers in parenchyma-like arrangement. Trop Woods 85:15–16
Hess R W 1946d Keys to American woods. Woods with unilaterally paratracheal parenchyma. Trop Woods 85:16–19
Hess R W 1950 Classification of the wood parenchyma in dicotyledons. Trop Woods 96:1–20
Hillis W E 1987 Heartwood and tree exudates. Springer, Berlin Heidelberg New York, 240 pp
Hoffmann O 1890 Compositae. In: Engler A, Prantl K (eds) Die natürlichen Pflanzenfamilien, vol 4/5 Verlag Wilhelm Engelmann, Leipzig, pp 87–391
Horak K E 1981 Anomalous secondary thickening in Stegnosperma (Phytolaccaceae). Bull Torrey Bot Club 108:189–197
Horner H T, Tiffany L H, Cody A M, Knaphu G 1985 Development of fungal calcium oxalate crystals associated with the basidiocarps of Geastrum minus (Lycoperdales). Scanning Electron Microsc 1985 II:789–801
Höster H-R, Liese W 1966 über das Vorkommen von Reaktions-Gewebe in Wurzeln und Asten der Dikotyledonen. Holzforschung 20:80–90
Huard J 1965 Anatomie des Rhopalocarpacees. Adansonia Ser 2, 5:103–123
Huber B 1935 Die physiologische Bedeutung der Ring-und Zerstreutporigkeit. Ber Dtsch Bot Ges 53:711–719
IAWA Committee 1981 Standard list of characters suitable for computerized hardwood identification. IAWA Bull NS 2:99–145
IAWA Committee 1989 List of microscopic features for hardwood identification. IAWA Bull NS 10:219–332
IAWA Committee on Nomenclature 1964 Multilingual glossary of terms used in wood anatomy. Verlagsanstalt Buchdruckerei Konkordia, Winterthur, 185 pp
Ingle H D, Dadswell H E 1953a The anatomy of the timbers of the south-west Pacific area. II. Apocynaceae and Annonaceae. Aust J Bot 1:1–26
Ingle H D, Dadswell H E 1953b The anatomy of the timbers of the south-west Pacific area. III. Myrtaceae. Aust J Bot 1:353–401
Ingle H D, Dadwell H E 1956 The anatomy of the timbers of the south-west Pacific area. IV. Cunoniaceae, Davidsoniaceae, Eucryphiaceae. Aust J Bot 4:125–151
Iqbal M, Ghouse A K M 1983 An analytical study on cell size variation in some arid zone trees of India: Acacia nilotica and Prosopis spicigera. IAWA Bull NS 4:46–52
James C F, Ingle H D 1956 The anatomy of the timbers of the south-west Pacific Area. V. Flacourtiaceae. Aust J Bot 4:200–215
Jane F W 1934 Terminal and initial parenchyma in wood. Nature 133:534
Jane F W 1956 The structure of wood. Macmillan, New York, 427 pp
Jansen S, Smets E 1998 Vestured pits in some woody Gentianaceae. IAWA J 19:35–42
Jansen S, Smets E, Baas P 1998 Vestures in woody plants: a review. IAWA J 19:347–382
Janssonius H H 1940 Bestimmungstabelle fur die javanischen Hölzer. Brill, Leiden
Janssonius H H 1950 Wood anatomy and relationship. Taxonomic notes in connection with the key to the Javanese woods. Blumea 6:407–461
Jarbeau J A, Ewers F W, Davis S D 1995 The mechanism of water-stress-induced embolism in two species of chaparral shrubs. Plant Cell Environ 18:189–196
Jayme G, Azzola F I 1964 Zur Morphologie der Tracheiden im Rotbuchenholz (Fagus sylvatica L.). Holzforschung 18:9–14
Jeffrey E C 1917 The anatomy of woody plants. Univ Chicago Press, Chicago, 478 pp
Jeje A A, Zimmermann M H 1979 Resistance to water flow in vessels. J Exp Bot 30:817–827

Johnson M A, Truscott F H 1956 On the anatomy of Serjania. I. Path of the bundles. Am J Bot 43:509–518

Jones C S 1984 The effect of axis splitting on xylem pressure potential and water movement in the desert shrub Ambrosia dumosa (Gray) Payne (Asteraceae). Bot Gaz 145:125–131

Jost L 1891 Ueber Dickenwachstum und Jahresringbildung. Z Bot 49:485–500, 501–510, 525–531, 541–547, 557–563, 573–579, 589–596, 605–611, 625–630

Keating R C 1968 Comparative morphology of Cochlospermaceae. I. Synopsis of the family and wood anatomy. Phytomorphology 18:379–392

Keefe J M, Moseley M F 1978 Wood anatomy and phylogeny of Paeonia section Moutan. J Arnold Arb 59:274–297

Keith C T 1971 Observations on the anatomy and fine structure of the trabeculae of Sanio. IAWA Bull 3:3–10

Kerr A J 1976 Ash, silica, and lignin in New Zealand beech. N Z J For Sci 6:108–113

Khan S 1980 Studies on the seasonal activity of vascular cambium and secondary phloem in some Myrtaceae. PhD Thesis, Aligarh Muslim University

Kingsley M A 1911 On the anomalous splitting of the rhizome and root of Delphinium scaposum. Bull Torrey Bot Club 38:307–317

Klaassen R K W M 1999 Wood anatomy of the Sapindaceae. IAWA J Suppl 2:1–214

Klein G 1923 Zur Aetiologie der Thyllen. Z Bot 15:418–439

Koek-Noorman J 1969 A contribution to the wood anatomy of South American (chiefly Suriname) Rubiaceae I. Acta Bot Neerl 18:108–123

Koek-Noorman J 1970 A contribution to the wood anatomy of the Cinchoneae, Coptosapeteae, and Naucleae (Rubiaceae). Acta Bot Neerl 19:154–164

Koek-Noorman J 1972 The wood anatomy of Gardenieae, Ixoreae and Mussaendeae (Rubiaceae). Acta Bot Neerl 21:301–320

Koek-Noorman J 1976 Juvenile characters in the wood of certain Rubiaceae with special reference to Rubia fruticosa Ait. IAWA Bull 1976/3:38–42

Koek-Noorman J, Hogeweg P 1974 The wood anatomy of Vanguerieae, Cionchoneae, Condamineae, and Rondeletieae (Rubiaceae). Acta Bot Neerl 23:627–653

Koek-Noorman J, Puff C 1983 The wood anatomy of Rubiaceae tribes Anthospermeae and Paederieae. Plant Syst Evol 143:17–45

Koek-Noorman J, Hogeweg P, van Maanen W H M, ter Welle B J H 1979 Wood anatomy of the Blakeeae (Melastomataceae). Acta Bot Neerl 28:21–43

Koek-Noorman J, van Rijkevorsel P 1983 Wood and leaf anatomy of Opiliaceae. Willdenowia 13:147–174

Koek-Noorman J, Topper S M C, ter Welle B J H 1984a The systematic wood anatomy of the Moraceae (Urticales). I. Tribe Castilleae. IAWA Bull NS 5:183–195

Koek-Noorman J, Topper S M C, ter Welle B J H 1984b The systematic wood anatomy of the Moraceae (Urticales). II. Tribe Dorsteinieae. IAWA Bull NS 5:317–329

Koek-Noorman J, Topper S M C, ter Welle B J H 1984c The systematic wood anatomy of the Moraceae (Urticales). III. Tribe Ficeae. IAWA Bull NS 5:330–334

Koeppen R C 1980 Silica bodies in wood of arborescent Leguminosae. IAWA Bull NS 1:180–184

Krawczyszyn J 1971 Unidirectional splitting and uniting of rays in the cambium of Platanus accompanying the formation of interlocked grain in wood. Acta Soc Bot Pol 40:57–79

Krawczyszyn J 1972 Movement of the cambial domain pattern and mechanism of formation of interlocked grain in Platanus. Acta Soc Bot Pol 41:443–461

Kribs D A 1935 Salient lines of structural specialization in the wood rays of dicotyledons. Bot Gaz 96:547–557

Kribs D A 1937 Salient lines of structural specialization in the wood parenchyma of dicotyledons. Bull Torrey Bot Club 64:177–186

Kribs D A 1950 Commercial foreign woods on the American market. Edwards, Ann Arbor

Kucera L J 1981 Cutting wood specimens for observation under the scanning electron microscope. J Microsc 124:319–325

Kucera L J 1985a Zur Morphologie der Interzellularen in den Markstrahlen, part 1. Stand der Kenntnisse über die Interzellularen. Vierteljahrschr Naturforsch Ges Zurich 130:35–74
Kucera L J 1985b Zur Morphologie der Interzellularen in den Markstrahlen, part 2. Rasterelektronmikroskopishe Untersuchungen. Vierteljahrschr Naturforsch Ges Zurich 130:157–198
Kucera L J 1985c Zur Morphologie der Interzellularen in den Markstrahlen, part 3. Gewebeanalytische Untersuchungen. Vierteljahrschr Naturforsch Ges Zurich 130:251–310
Kucera L J 1985d Zur Morphologie der Interzellularen in den Markstrahlen, part 4. Deltamikroskopische Untersuchungen und Gesamtschau. Vierteljahrschr Naturforsch Ges Zurich 130:374–397
Kucera L J, Philipson W R 1977 Growth eccentricity and reaction anatomy in branchwood of Pseudowintera colorata. Am J Bot 65:601–607
Kucera L J, Meylan B A, Butterfield B G 1977 Vestured simple perforation plates. IAWA Bull NS 1:3–6
Kukachka B F 1977 Sectioning refractory woods for anatomical studies. USA For Serv Res Note FPL 0236:1–9 (also Microsc Acta 80:301–307, 1978)
Kukachka B F 1978 Wood anatomy of the neotropical Sapotaceae. I. Bumelia. For Prod Lab Res Pap FPL 325:1–9
Kukachka B F, Miller R B 1980 A chemical spot-test for aluminum and its value in wood identification. IAWA Bull NS 1:104–109
Ladefoged K 1952 The periodicity of wood formation. Dansk Biol Skrifter 7(3):1–98
Larson P R 1960 A physiological consideration of the springwood-summerwood transition in red pine. For Sci 6:110–122
Larson P R 1964 Some indirect effects of environment on wood formation. In: Zimmermann M H (ed) The formation of wood in forest trees. Academic Press, New York, pp 345–465
Larson P R 1969 The indirect effect of photoperiod on tracheid diameter in Pinus resinosa. Am J Bot 49:132–137
Läuchli A 1972 Translocation of organic solutes. Annu Rev Plant Physiol 23:197–218
Leandri J 1931 Observations sur les rhizomes à structure anormales des quelques espèces africaines de Gnidia. Bull Soc Bot Fr 78:307–312
Lee C L 1961 Crystallinity of wood cellulose fibers studied by X-ray methods. For Prod J 11:8–12
Lemesle R 1946 Les divers types de fibres à ponctuations areolées chez les dicotyledones apocarpiques archaïques et leur role dans la phylogénie. Ann Sci Nat Bot Ser 11(7):19–52
Lemesle R 1956a Revues d'histologie vegetale. Les éléments du xylème dans les angiospermes a caractères primitifs. Bull Soc Bot Fr 103:629–677
Lemesle R 1956b Mise en évidence de pseudo-trachéides dans le bois secondaire de quelques Bruniacées. C R Acad Sci Paris 242:165–167
Lemesle R, Duchaigne A 1955a Contribution à l'étude histologique et phylogénétique de Degeneria vitiensis. I. Bailey and Smith. Rev Gen Bot 62:708–719
Lemesle R, Duchaigne A 1955b Particuliarités histologiques des constituants du bois chez les Degeneriacées. C R Acad Sci Paris 240:1122–1123
Lersten N R 1982 Tracheid bars and vestured pits in legume seeds (Leguminosae:Papillionoideae). Am J Bot 69:98–107
Lewis A M, Tyree M T 1985 The relative susceptibility to embolism of larger vs. smaller tracheids in Thuja occidentalis. IAWA Bull NS 6:93
Lindorf H 1994 Eco-anatomical wood features of species from a very dry tropical forest. IAWA J 15:361–376
Linnemann G 1953 Untersuchungen über den Markstrahlanteil am Holz der Buche. Ber Dtsch Bot Ges 66:37–63
Lohr E 1969 Jahresverlust bei Laubbäumen mit ringporigen Holz. Allg Forst Jagdztg 140:18–22
Lörcher H, Weberling F, Hildebrand M 1987 Transferzellen im Xylem der Valerianaceae. Plant Syst Evol 156:69–89
Lowell C, Lucansky T W 1990 Vegetative anatomy and morphology of Ipomoea quamoclit (Convolvulaceae). Bull Torrey Bot Club 117:232–246
MacDougal D T 1939 Features of growth control in trees. Proc Am Philos Soc 81:421–445
MacDuffie R C 1921 Vessels of the gnetalean type in angiosperms. Bot Gaz 71:438–445

Maheshwari P 1930 Contributions to the morphology of Boerhaavia diffusa. J Indian Bot Soc 19:42–61
Mainieri C 1959 Identificão das principais madeiras de comerciais na Brasil. Inst Pesquesas Tech Sao Paulo Bot 46:1–189
Manchester S R 1979 Triplochitoxylon (Sterculiaceae): a new genus of wood from the Eocene of Oregon and its bearing on xylem evolution in the extant genus Triplochiton. Am J Bot 66: 699–708
Manchester S R 1980 Chattawaya (Sterculiaceae): a new genus of woods from the Eocene of Oregon and its implications for xylem evolution of the extant genus Pterospermum. Am J Bot 67:59–67
Manchester S R, Miller R B 1978 Tile cells and their occurrence in Malvalean fossil woods. IAWA Bull 2–3:23–28
Marco H F 1933 The wood of Sarcosperma paniculatum. Trop Woods 33:1–4
Mark R 1965 Tensile strength analysis of the cell walls of coniferous tracheids. In: Côté W A (ed) Cellular ultrastructure of woody plants. Syracuse Univ Press, Syracuse, pp 493–533
Matthew L, Shah G L 1983 Vestured pits and warts in Verbenaceae. IAWA Bull NS 4:39–40
Mauseth J D 1993 Water-storing and cavitation-preventing adaptations in wood of cacti. Ann Bot 72:81–89
Mauseth J D, Plemons B J 1995 Developmentally variable, polymorphic woods in cacti. Am J Bot 82:1199–1205
Mauseth J D, Plemons-Rodriguez B J 1997 Presence of paratracheal water storage tissue does not alter vessel characteristics in cactus wood. Am J Bot 84:815–822
Mauseth J D, Uozumi Y, Plemons B J, Landrum J V 1995 Structural and systematic study of an unusual tracheid type in cacti. J Plant Res 108:517–526
McDonald J A 1992 Evolutionary implications of typical and anomalous secondary growth in arborescent Ipomoea (Convolvulaceae). Bull Torrey Bot Club 119:262–267
McLean J D, Richardson P E 1973 Vascular ray cells in woody stems. Phytomorphology 23:59–64
Mennega A M W 1969 The wood structure of Dicranostyles (Convolvulaceae). Acta Bot Neerl 18:173–179
Mennega A M W 1972a A survey of the wood anatomy of the New World Hippocrateaceae. In: Ghouse A K M, Yunus M (eds) Research trends in plant anatomy – KA Chowdhury Commemorative Volume. Tata McGraw Hill, New Delhi
Mennega A M W 1972b Wood structure of the genus Talisia (Sapindaceae). Acta Bot Neerl 21:578–586
Mennega A M W 1973 An unusual type of parenchyma strand occurring in the wood of Cedrilinga catenaeformis Ducke. IAWA Bull 1:3–5
Mennega A M W 1980 Anatomy of the secondary xylem. In: Leweuwenberg A J M (ed) Angiospermae: Ordnung Gentiales fam. Loganiaceae. Die natürlichen Pflanzenfamilien, 2nd edn, 28bI. Duncker and Humblot, Berlin, pp 112–161
Mennega A M W 1997 Wood anatomy of the Hippocrateoideae (Celastraceae). IAWA J 18:331–368
Metcalfe C R, Chalk L 1950 Anatomy of the dicotyledons, 2 vols. Clarendon Press, Oxford, 1500 pp
Metcalfe C R, Chalk L 1983 Anatomy of the dicotyledons, vol 2, 2nd edn. Wood structure and conclusion of the general introduction. Clarendon Press, Oxford, 297 pp
Meylan B A 1981 Reaction wood in Pseudowintera colorata – a vessel-less dicotyledon. Wood Sci Technol 15:81–92
Meylan B A, Butterfield B G 1973 A trabecula with a vestured pit. IAWA Bull 3:12–14
Meylan B A, Butterfield B G 1974 Occurrence of vestured pits in the vessels and fibres of New Zealand woods. N Z J Bot 12:3018
Meylan B A, Butterfield B G 1975 Occurrence of simple, multiple, and combination plates in the vessels of New Zealand woods. N Z J Bot 13:1–18
Meylan B A, Butterfield B G 1978a The structure of New Zealand woods. DSIR Bull 122. N Z DSIR, Wellington, 250 pp
Meylan B A, Butterfield B G 1978b Occurrence of helical thickenings in the vessels of New Zealand woods. New Phytol 81:139–146

Meylan B A, Butterfield B G 1982 Pit membrane structure in the vessel-less woods of Pseudowintera Dandy. IAWA Bull NS 3:167–174
Michener D C 1981 Wood and leaf anatomy of Keckiella (Scrophulariaceae): ecological considerations. Aliso 10:39–57
Michener D C 1983 Systematic and ecological wood anatomy of Californian Scrophulariaceae. I. Antirrhinum, Castilleja, Galvezia, and Mimulus sect. Diplacus. Aliso 10:471–487
Middleton T M 1987 Aggregate rays in New Zealand Nothofagus Blume (Fagaceae) stem wood and their influence on vessel distribution. IAWA Bull NS 8:53–57
Mikesell J E 1979 Anomalous secondary thickening in Phytolacca americana L. (Phytolaccaceae). Am J Bot 66:997–1005
Mikesell J E, Popham R H 1976 Ontogeny and correlative relationship of the primary thickening meristem in four-o'clock plants (Nyctaginaceae) maintained under long and short photoperiods. Am J Bot 63:427–437
Milanez F R 1938 On crystalliferous strands. Arch Inst Biol Veg Rio de Janeiro 4(1):79–85
Milanez F R 1939 On crystalliferous strands. II. Rodriguesia 4:59–65
Milanez F R 1944 Nota sobre a classificao do parenquima do lenho. Rodriguesia 8(17):1–3
Miller H J 1975 Anatomical characteristics of some woody plants of the Angmassalik district of southeast Greenland. Medd Om Gronland 198(6):1–30
Miller R B 1975 Systematic anatomy of the xylem and comments on the relationships of Flacourtiaceae. J Arnold Arbor 56:20–102
Miller R B 1976a Reticulate thickenings in some species of Juglans Am J Bot 63:898–901
Miller R B 1976b Wood anatomy and identification of species of Juglans. Bot Gaz 137:368–377
Miller R B 1977 Vestured pits in Boraginaceae. IAWA Bull 3:43–48
Miller R B 1978 Potassium calcium sulfate crystals in the secondary xylem of Capparis. IAWA Bull 2/3:50
Moens P 1955 Les formations sécrétrices des copaliers Congolais. La Cellule 57:35–64
Moll J W, Janssonius H H 1909–1936 Mikrographie des Holzes der auf Java vorkommenden Baumarten, 6 vols. Brill, Leiden
Moseley M F 1948 Comparative anatomy and phylogeny of the Casuarinaceae. Bot Gaz 119: 231–280
Moseley M F, Beeks R M 1955 Studies of the Garryaceae. I. The comparative morphology and phylogeny. Phytomorphology 5:314–346
Moss E H 1934 Rings of cork in the wood of herbaceous perennials. Nature 133:598
Moss E H 1936 The ecology of Epilobium angustifolium with reference to rings of periderm in the wood. Am J Bot 23:14–20
Moss E H 1940 Intraxylary cork in Artemisia with reference to its significance. Am J Bot 27:762–768
Moss E H, Gorham A L 1953 Interxylary cork and fission of stems and roots. Phytomorphology 3:285–294
Muhammad A F 1984 Perforation plate structure in Comptonia peregrina (Myricaceae). IAWA Bull NS 5:217–224
Muhammad A F, Micko M M 1984 Accumulation of calcium crystals in the decayed wood of aspen attacked by Fomes igniarius. IAWA Bull NS 5:237–241
Muhammad A F, Sattler R 1982 Vessel structure of Gnetum and the origin of angiosperms. Am J Bot 69:1004–1021
Mullenders W 1947 L'origine du phloème interxylarien chez Stylidium et Thunbergia. Étude anatomique. La Cellule 51:7–48
Myer J E 1922 Ray volumes of the commerical woods of the United States and their significance. J For 20:337–351
Nagai S, Ohtani J, Fukazawa K, Wu J 1994 SEM observations on perforated ray cells. IAWA J 15:293–300
Nair M N B, Mohan Ram H Y 1989 Vestured pits and vestured vessel member walls in some Indian dicotyledonous woods. Bot J Linnean Soc 100:323–336

Nair M N B, Mohan Ram H Y 1992 Structure of sola wood: the traditional Indian art material. Curr Sci 62:746–751
Nair M N B, Shah J J, Subramanyam S V 1983 Ultrastructure and histochemistry of traumatic gum ducts in the wood of Azadirachta indica A. Juss. IAWA Bull NS 4:103–112
Nazma B, Rao S, Rao R V 1981 Occurrence of perforated ray cells in the wood of Drypetes roxburghi (Wall.) Hurusawa. IAWA Bull NS 2:201–203
Netolitzky F 1929a Die Kieselkörper. Handbuch der Pflanzenanatomie III/1a:1–19. Gebrüder Borntraeger, Berlin
Netolitzky F 1929b Die Kalksalze als Zellinhaltskörper. Handbuch der Pflanzenanatomie III/1a: 20–80. Gebrüder Borntraeger, Berlin
Norberg P H, Meier H 1966 Physical and chemical properties of the gelatinous layer in tension wood fibre of aspen (Populus tremula). Holzforschung 20:174–178
Noshiro S, Baas P 1998 Systematic wood anatomy of Cornaceae and allies. IAWA J 19:43–97
Novruzova Z A 1968 The water conducting system of trees and shrubs in relation to ecology. Baku (USSR) Izdatel'stvo Akademii Nauk Azerbaijan, 229 pp
Novruzova Z A 1972 The effect of ecological conditions on relation of ring width to solid matter in trees and shrubs. For Abstr 34:4200
Obaton M 1960 Les lianes ligneuses à structure anomales des forêts denses d'Afrique occidentale. Ann Sci Nat Bot 12 Ser 1:1–220
Ogata K 1988 Wood anatomy of the Caprifoliaceae of Japan. IAWA Bull NS 9:299–316
Ohtani J 1987 Vestures in septate wood fibres. IAWA Bull NS 8:59–67
Ohtani J, Ishida S 1978a Pit membrane with torus in dicotyledonous woods. J Jpn Wood Res Soc 24:673–675
Ohtani J, Ishida S 1978b An observation on the perforation plates in Japanese dicotyledonous woods using scanning electron microscopy. Res Bull Coll Exp For Coll Agr Hokkaido Univ 335:65–98
Ohtani J, Meylan B A, Butterfield B G 1983 Occurrence of warts in the vessel elements and fibres of New Zealand woods. N Z J Bot 21:359–372
Ohtani J, Meylan B A, Butterfield B G 1984a Vestures or warts – proposed terminology. IAWA Bull NS 5:3–9
Ohtani J, Meylan B A, Butterfield B G 1984b A note on vestures on helical thickenings. IAWA Bull NS 5:9–11
Ohtani J, Wu J, Fukazawa K, Xiao S 1989 Multiple perforation plates in Gmelina arborea Roxb. (Verbenaceae). IAWA Bull NS 10:35–41
Ohtani J, Saitoh Y, Wu J, Fukazawa K, Xiao S 1992 Perforation plates in Knema furfuracea (Myristicaceae). IAWA Bull 13:301–306
O'Leary J W 1965 Root-pressure exudation in woody plants. Bot Gaz 126:108–115
Page V M 1981 Dicotyledonous wood from the upper Cretaceous of central California. III. Conclusions. J Arnold Arbor 62:437–455
Pajouh D P, Schweingruber F H 1985 Atlas des bois du Nord de l'Iran. Institut Federal de Réchérches Forestières, Birmendorf, Switzerland, 133 pp
Paliwal G S, Prasad N U S R K 1970 Seasonal activity of cambium in some tropical trees. I. Dalbergia sissoo. Phytomorphology 20:333–339
Paliwal G S, Srivastava L M 1969 The cambium of Alseuosmia. Phytomorphology 19:5–8
Panshin A J, de Zeeuw C 1980 Textbook of wood technology. I. Structure, identification, uses, and properties of the commercial woods of the United States and Canada, 4th edn. McGraw Hill, New York, 722 pp
Pant D D, Bhatnagar S 1975 Morphological studies in Argyreia. Bot J Linnean Soc 70:45–69
de Paolis D 1948 Ciclo de accrescimento e differenziazione delle gemme in piante perenni nel teritorio de Bari III: l'evoluzione della cerchia lignosa in Prunus amygdalus Stokes dal Diciembre 1947 al Diciembre 1948. Nuova G Bot Ital NS 55:214–234
de Paolis D 1950 Considerazioni sulla variabilita della cerchia legnosa del Mandorlo. La luce e un fattore importante? Novo G Bot Ital NS 57:210–222

Parameswaran N 1964 Über die Länge der septierten Fasern bei Tectona grandis. Naturwissenschaften 51:317–318

Parameswaran N, Conrad H 1982 Wood and bark anatomy of Balanites aegyptica in relation to ecology and taxonomy. IAWA Bull NS 3:75–88

Parameswaran N, Gomes A V 1981 Fine structural aspects of helical thickenings and pits in vessels of Ligustrum lucidum Ait. (Oleaceae). IAWA Bull NS 2:179–185

Parameswaran N, Liese W 1969 On the formation and fine structure of septate wood fibres of Ribes sanguineum. Wood Sci Technol 14:372–382

Parameswaran N, Liese W 1973 Scanning electron microscopy of multiperforate perforation plates. Holzforschung 27:181–186

Parameswaran N, Richter H-G 1984 The ultrastructure of crystalliferous cells in some Lecythidaceae with a discussion of their terminology. IAWA Bull NS 5:229–236

Parameswaran N, Knigge H, Liese W 1985 Electron microscopic demonstration of a suberised layer in the tylosis wall of beech and oak. IAWA Bull NS 6:269–271

Parham R A, Baird W M 1974 Warts in the evolution of angiosperm wood. Wood Sci Technol 8:1–10

Parham R A, Kaustinen N 1973 On the morphology of spiral thickenings. IAWA Bull 2:8–17

Patel J D, Shah J J 1980 Sapwood and heartwood in Dalbergia latifolia. Phytomorphology 30:140–149

Patel R N 1964 On the occurrence of gelatinous fibres with special reference to root wood. J Inst Wood Sci 12:67–80

Patel R N 1965 A comparison of the anatomy of the secondary xylem in roots and stems. Holzforschung 19:72–79

Patel R N 1973 Wood anatomy of the dicotyledons indigenous to New Zealand. 1. Cornaceae. N Z J Bot 11:3–22

Patel R N 1974 Wood anatomy of the dicotyledons indigenous to New Zealand. 4. Winteraceae. N Z J Bot 12:19–32

Patel R N 1978 Wood anatomy of the dicotyledons indigenous to New Zealand. 11. Oleaceae. N Z J Bot 16:1–6

Patel R N 1986 Wood anatomy of the dicotyledons indigenous to New Zealand. 15. Fagaceae. N Z J Bot 24:189–202

Patel R N, Bowles A 1978a Wood anatomy of the dicotyledons indigenous to New Zealand. 12. Icacinaceae. N Z J Bot 16:7–12

Patel R N, Bowles A 1978b Wood anatomy of the dicotyledons indigenous to New Zealand. 13. Moraceae. N Z J Bot 16:13–19

Patel R N, Bowles A 1980 Wood anatomy of the dicotyledons indigenous to New Zealand. 14. Piperaceae. N Z J Bot 18:507–513

Patterson R, Tanowitz B D 1989 Evolutionary and geographic trends in adaptive wood anatomy in Eriastrum densifolium (Polemoniaceae). Am J Bot 76:706–713

Pearson R G, Wheeler E A 1981 Computer identification of hardwood species. IAWA Bull NS 2:37–40

Petrić B, Šćukanec V 1975 Ray tissue percentages in wood of Yugoslavian hardwoods. IAWA Bull 3:43

Pfeiffer H 1926 Das abnorme Dickenwachstum. Handbuch der Pflanzenanatomie, vol 9(2). Gebrüder Borntraeger, Berlin, pp 1–272

Phelps J E, Isebrands J G, Jowett D 1982 Raw material quality of short-rotation, intensively cultured Populus clones. I. A comparison of stem and branch properties at three spacings. IAWA Bull NS 3:193–206

Philipson W R, Butterfield B G 1967 A theory on the causes of size variation in wood elements. Phytomorphology 17:155–159

Philipson W R, Studholme W D 1966 Unusual forms of secondary thickening in New Zealand trees and shrubs. Tuatara 14:139–144

Philipson W R, Ward J M, Butterfield B G 1971 The vascular cambium. Its development and activity. Chapman and Hall, London, 162 pp

Pizzolato T D 1982 Anatomical modification by 2,4-DB of vascular cambium and secondary xylem in soybean internodes. Can J Bot 60:2142–2146

Pobeguin T 1943 Les oxalates de calcium chez quelques angiospérmes. Ann Sci Nat Bot Ser 11(4):1–95

Preston C E 1901 Structural studies in southwestern Cactaceae. Bot Gaz 32:35–55

Priya P B, Bhat K M 1999 Influence of rainfall, irrigation and age on the growth periodicity and wood structure in teak (Tectona grandis). IAWA J 20:181–193

Purkayastha S K, Nigam P N, Ghosh S S 1959 Characteristic distribution of septate fibers in some Indian woods and their structure. Proc 46th Indian Sci Congr 3:278

Putz F E 1983 Liana biomass and leaf area in a "terra firma" forest in the Rio Negro Basin, Venezuela. Biotropica 15:185–189

Quirk J T 1980 Wood anatomy of the Vochysiaceae. IAWA Bull NS 1:172–179

Quirk J T, Miller R B 1983 Nonvestured pits in Kompassia Maingay (Leguminosae). IAWA Bull NS 4:191–195

Quirk J T, Miller R B 1985 Vestured pits in the tribe Cassieae Bronn (Leguminosae). IAWA Bull NS 6:200–212

Radlkofer L 1875 Sopra i varii tipi delle anomalie dei tronchi nelle Sapindaceae. Atti Congr Bot Internaz tenuto a Firenze nel Mese di Maggio 1874, pp 60–65

Ragonese A M 1976 Consideraciones sobre el problema de la classificaion de los elementos traqueales no perforados de las dicotiledoneas y en especial de algunas Mirtaceas. Darwiniana 20:476–490

Ragonese A M 1977 Caracteres anatomicos de parenquima radial y axial en el leño de las Mirtaceas. Darwiniana 21:27–41

Rao R V, Sharma B, Dayal R 1984 Occurrence of perforated ray cells in Santalaceae. IAWA Bull NS 5:313–316

Rao R V, Sharma B, Dayal R 1989 Anatomy of aerial heartwood of Sonneratia caseolaris (L.) Engler (Sonneratioideae). IAWA Bull NS 10:374–378

Record S J 1919 Storied or tier-like structure of certain dicotyledonous woods. Bull Torrey Bot Club 46:253–273

Record S J 1925a Cystoliths in wood. Trop Woods 3:10–12

Record S J 1925b Occurrence of intercellular canals in dicotyledonous woods. Trop Woods 4:1720

Record S J 1925c Note on sclerosed or stone cell tyloses. Trop Woods 4:11

Record S J 1927 Occurrence of calcium carbonate deposits in wood. Trop Woods 12:22–26

Record S J 1928 Is Maba distinct from Diospyros? Trop Woods 13:6

Record S J 1933 The woods of Rhabdodendron and Duckeodendron. Trop Woods 33:6–10

Record S J 1942a Keys to American woods I. Ring-porous woods. Trop Woods 72:19–29

Record S J 1942b Keys to American woods. II. Pores in ulmiform or wavy tangential arrangement. Trop Woods 72:29–32

Records S J 1942c Keys to American woods. III. Pores in flame-like or dendritic arrangement. Trop Woods 32:35

Record S J 1943a Keys to American woods. IV. Vessels virtually all solitary. Trop Woods 73:24–33

Record S J 1943b Keys to American woods. V. Vessels with spiral thickenings. Trop Woods 73:33–42

Record S J 1943c Keys to American woods. VI. Vessels with scalariform perforation plates. Trop Woods 74:18–31

Record S J 1943d Keys to American woods. VII. Vessels with very fine pitting. Trop Woods 74:31–43

Record S J 1943e Keys to American woods. X. Woods with storied structure. Trop Woods 76:32–47

Record S J 1944a Keys to American woods. XI. Woods with resin or gum ducts. Trop Woods 77:20–26

Record S J 1944b Keys to American woods. XII. Parenchyma reticulate. Trop Woods 77:26–38

Record S J 1944c Keys to American woods. XIII. Woods with septate fibers. Trop Woods 78:33–45

Record S J 1944d Keys to American woods. XIV. Dicotyledonous woods with xylem rays virtually all uniseriate. Trop Woods 79:15–34

Record S J 1944e Keys to American woods. XV. Fibers with conspicuous bordered pits. Trop Woods 80:11–13

Record S J 1944f Keys to American woods. XVI. Woods with oil (or similar) cells. Trop Woods 80:13–15

Record S J, Chattaway M M 1939 List of anatomical features used in classifying dicotyledonous woods. Trop Woods 57:11–16

Reinders E 1935 Fiber-tracheids, libriform fibers, and systematics in wood anatomy. Trop Woods 44:30–36

Rendle B J 1958 A note on juvenile and adult wood. IAWA Bull 2:1–6

Rendle B J, Clarke S H 1934 The diagnostic value of measurements. Trop Woods 40:27–37

Richter H G 1981 Anatomie des sekundären Xylems und der Rinde der Lauraceae. Sonderb Naturwiss Verh Hamburg 5:1–148

Richter H G 1982 The wood structure of Couratari Aubl. and Couroupita Aubl. (Lecythidaceae). IAWA Bull NS 3:45–55

Rodman J E, Karol K G, Price R A, Sytsma K 1996 Molecules, morphology, and Dahlgren's expanded order Capparales. Syst Bot 21:289–307

Rodman J E, Olivere M K, Nakamura R R, McClammer J U, Bledsoe A H 1984 A taxonomic analysis and revised classification of Centrospermae. Syst Bot 9:297–323

Rudall P J 1981 Wood anatomy in the Hyptidinae (Labiatae). Kew Bull 35:735–741

Rudall P J 1982 An unusual type of perforation plate in Canthium barbatum Seem. (Rubiaceae). IAWA Bull NS 3:127–129

Rudall P J 1985 Perforated ray cells in Hyptis hogei – a new record for Labiatae. IAWA Bull NS 6:161–162

Rudall P J 1989 Laticifers in vascular cambium and wood of Croton spp. (Euphorbiaceae). IAWA Bull NS 10:379–383

Rumball W 1963 Wood structure in relation to heteroblastism. Phytomorphology 13:206–214

Rundel P W, Stecker R E 1977 Morphological adaptations of tracheid structure to water stress gradients in the crown of Sequoiadendron giganteum. Oecologia 27:135–139

Saitoh T, Ohtani J, Fukazawa K 1993 The occurrence and morphology of tyloses and gums in the vessels of Japanese hardwoods. IAWA J 14:359–371

Sarkany S, Stieber J, Fillo J 1957a Investigations on the wood of Hungarian Populus species by means of quantitative xylotomy (in Russian). Ann Univ Sci Budapest (Sect Biol) 1:219–229

Sarkany S, Stieber J, Fillo J 1957b A quantitative xylotomical analysis of Populus species in Hungary. Acta Biol Budapest Suppl 1:404–441

Sastrapadja D S, Lamoureux C 1969 Variations in wood anatomy of Hawaiian Metrosideros (Myrtaceae). Ann Bogorienses 5(1):1–83

Sauter J J 1966a Untersuchungen zur Physiologie der Pappelholzstrahlen. I. Jahresperiodischer Verlauf der Stärkespeicherung im Holzstrahlparenchym. Z Pflanzenphysiol 55:246–258

Sauter J J 1966b Untersuchungen zur Physiologie der Pappelholzstrahlen. II. Jahresperiodische Änderungen der Phosphatase-Aktivität im Holzstrahlparenchym und ihre mögliche Bedeutung für den Kohlehydratstoffwechsel und den aktiven Assimilattransport. Z Pflanzenphysiol 55:349–382

Sauter J J, Iten W I, Zimmermann M H 1973 Studies on the release of sugar into the vesels of sugar maple (Acer saccharum). Can J Bot 51:1–8

Schenck H 1893 Beiträge zur Biologie und Anatomie der Lianen. II. Schimpers Bot Mittheil Trop 5:1–271

Schmid R 1980 Comparative anatomy and morphology of Psiloxylon and Heteropyxis, and the subfamilial and tribal classification of Myrtaceae. Taxon 29:559–595

Schmid R, Baas P 1984 The occurrence of scalariform perforation plates and helical wall thickenings in wood of Myrtaceae. IAWA Bull NS 5:197–215

Scholander P F, Hammel H T, Hemmingsen E A, Garey W 1962 Salt balance in mangroves. Plant Physiol 37:722–729

Schweingruber F H 1990 Anatomie europäischer Hölzer. Haupt, Bern, 802 pp

Schweingruber F H 1996 Tree rings and environment dendroecology. Haupt, Bern, 467 pp

Scott D H, Brebner G 1889 On the anatomy and histology of Strychnos. Ann Bot 3:275–302

Scott R A, Wheeler E A 1982 Fossil woods from the Eocene Clarno Formation of Oregon. IAWA Bull NS 3:135–154
Šćukanec V, Petrić B 1977 The relationship between wood ray shape and ray volume percentage in beech. IAWA Bull 3:57–59
Sharma B L, Rao R V, Bisen S S, Dayal R 1985 Modified scalariform and reticulate perforation plates in species of Euodia (Rutaceae). IAWA Bull NS 6:39–41
Shulkina T V, Zhikov S E 1980 The anatomical structure of the stem in the family Campanulaceae s. str. in relation to the evolution of life forms. Bot Zhurn 65:627–639
Shutts C F 1960 Wood anatomy of Hernandiaceae and Gyrocarpaceae. Trop Woods 113:85–123
Siddiqi M, Wilson T K 1974 Wood anatomy of the genus Knema (Myristicaceae). Bull Torrey Bot Club 101:354–362
Sidiyasa K, Baas P 1998 Ecological and systematic wood anatomy of Alstonia (Apocynaceae). IAWA J 19:207–229
Sieber M, Kucera L J 1980 On the stem anatomy of Clematis vitalba L. IAWA Bull NS 1:49–54
Singh B 1943 The origin and distribution of interxylary and intraxylary phloem in Leptadenia. Proc Indian Acad Sect B 18:14–19
Singh B 1944 A contribution to the anatomy of Salvadora persica L. with special reference to the origin of the included phloem. J Indian Bot Soc 23:71–78
Slatyer R O 1967 Plant-water relationships. Academic Press, London, 366 pp
Solereder H 1908 Systematic anatomy of the dicotyledons (trans Boodle and Fritsch), 2 vols. Clarendon Press, Oxford, 1182 pp
Soltis P S, Soltis D E, Chase M W 1999 Angiosperm phylogeny inferred from multiple genes as a tool for comparative biology. Nature 402:402–403
Spackman W, Swamy B G L 1949 The nature and occurrrence of septate fibers in dicotyledons. Am J Bot 36:804 (Abstr)
Sperry J S 1985 Xylem embolism in the palm Rhapis excelsa. IAWA Bull NS 6:283–292
Sperry J S, Donnelly J R, Tyree M T 1988 Seasonal occurrence of xylem embolism in sugar maple (Acer saccharum). Am J Bot 75:1212–1218
Stahel B 1972 Callus-like tissue in Piptadeniastrum africanum (Hook. f.) Brenan. IAWA Bull 1:10–12
Stern W L 1954 Comparative anatomy of xylem and phylogeny of Lauraceae. Trop Woods 100:1–72
Stern W L 1967 Kleinodendron and xylem anatomy of Cluytieae. Am J Bot 54:663–676
Stern W L 1974 Comparative anatomy and systematics of woody Saxifragaceae. Escallonia. Bot J Linnean Soc 68:1–20
Stern W L 1978a Comparative anatomy and systematics of woody Saxifragaceae. Hydrangea. Bot J Linnean Soc 76:83–113
Stern W L 1978b Index Xylariorum. Institutional wood collections of the world, 2. Taxon 27:233–269
Stern W L, Brizicky G K 1958a The comparative anatomy and taxonomy of Heteropyxis. Bull Torrey Bot Club 85:111–123
Stern W L, Brizicky G K 1958b The woods and flora of the Florida Keys. Passifloraceae. Trop Woods 109:45–53
Stern W L, Brizicky G K 1960 The morphology and relationships of Diomma, gen. inc. sed. Memoirs N Y Bot Gard 10:38–57
Stern W L, Chambers K L 1960 The citation of wood specimens herbarium vouchers in anatomical research. Taxon 9:7–13
Stern W L, Greene S 1958 Some aspects of variation in wood. Trop Woods 108:65–71
Stern W L, Brizicky G K, Eyde R H 1969 Comparative anatomy and relationships of the Columelliaceae. J Arnold Arbor 50:36–75
Stern W L, Sweitzer E M, Phipps R E 1970 Comparative anatomy and systematics of woody Saxifragaceae. Bot J Linnean Soc 63 [Suppl 1] (New Res Plant Anat):215–237
Stewart C M 1966 The chemistry of secondary growth in trees. CSIRO Div For Prod Technol Pap 43:107

Studholme W P, Philipson W R 1966 A comparison of the cambium in two woods with included phloem: Heimerliodendron brunonianum and Avicennia resinifera. N Z J Bot 4:355–365

Styer C H, Stern W L 1979a Comparative anatomy and systematics of woody Saxifragaceae. Philadelphus. Bot J Linnean Soc 79:267–289

Styer C H, Stern W L 1979b Comparative anatomy and systematics of woody Saxifragaceae. Deutzia. Bot J Linnean Soc 79:291–319

Suzuki M, Yoda K, Suzuki H 1996 Phenological comparison of the onset of vessel formation between ring-porous and diffuse-porous trees in a Japanese temperate forest. IAWA J 17: 431–444

Swamy B G L, Bailey I W 1950 Sarcandra, a vesselless genus of the Chloranthaceae. J Arnold Arbor 31:117–129

Swamy B G L, Govindarajalu E 1957 Anatomical studies in polyploid strains of Parthenium argentatum. J Asiatic Soc Sci 23:43–54

Swamy B G L, Parameswaran N, Govindarajalu E 1960 Variation in vessel length within one growth ring of certain arborescent dicotyledons. J Indian Bot Soc 39:163–170

Sweitzer E M 1971 Comparative anatomy of Ulmaceae. J Arnold Arbor 52:523–585

Taylor F W 1966 A study of the natural variation of certain properties of wood of yellow popular (Liriodendron L.) within trees and between geographical areas. Diss Abstr 26:6948

Taylor F W 1969 Variation of wood properties in sycamore (Platanus occidentalis). Res Pap Prod Util Lab Mo State Univ 7:1–18

Teixeria L L 1983 Some unusual features in the wood of Sloanea lasiocoma K. Schum. (Elaeocarpaceae) and Casearia obliqua Spreng. (Flacourtiaceae). IAWA Bull NS 4:213–217

ter Welle B J H 1975 Spiral thickenings in the axial parenchyma of Chrysobalanaceae. Acta Bot Neerl 24:397–405

ter Welle B J H 1976a On the occurrence of silica grains in the secondary xylem of Chrysobalanaceae. IAWA Bull 2:19–29

ter Welle B J H 1976b Silica grains in woody plants of the neotropics, especially Surinam. In: Baas P, Bolton A J, Catling D M (eds) Wood structure in biological and technological research. Leiden Bot Ser 3. Leiden Univ Press, Leiden, pp 107–142

ter Welle B J H 1980 Cystoliths in the secondary xylem of Sparattanthelium (Hernandiaceae). IAWA Bull NS 1:43–48

ter Welle B J H, Koek-Noorman J 1981 Wood anatomy of the neotropical Melastomataceae. Blumea 27:335–394

ter Welle B J H, Mennega A M W 1977 On the presence of large styloids in the secondary xylem of the genus Henriettea (Melastomataceae). IAWA Bull 2:31–35

ter Welle B J H, van Rooden J 1982 Systematic wood anatomy of Desmopsis, Sapranthus, and Stenanoa (Annonaceae). IAWA Bull NS 3:15–23

Thompson W P, Bailey I W 1916 Are Tetracentron, Trochodendron, and Drimys specialized or primitive types? Mem N Y Bot Gard 6:27–32

Thorne R F 1968 Synopsis of a putatively phylogenetic classification of the flowering plants. Aliso 6:57–66

Thorne R F 1976 A phylogenetic classification of the Angiospermae. Evolutionary Biol 9:35–106

Thorne R F 1983 Proposed new realignments in the angiosperms. Nordic J Bot 3:85–117

Timell T E 1983 Origin and evolution of compression wood. Holzforschung 37:1–10

Tippo O 1938 The comparative anatomy of the secondary xylem of Moraceae and their presumed allies. Bot Gaz 100:1–99

Tomlinson P B, Craighead F C 1972 Growth-ring studies on the native trees of sub-tropical Florida. In: Ghouse A K M, Yunus M (eds) Research trends in plant anatomy – KA Chowdhury commemoration volume. Tata McGraw Hill, New Delhi, pp 39–51

Topper S M C, Koek-Noorman J 1980 The occurrence of axial latex tubes in the secondary xylem of Artocarpus JR and G Forster (Moraceae). IAWA Bull NS 1:113–119

Tupper W W 1913 An unuusal seasonal growth ring in Eucalyptus. Pap Mich Acad Sci Arts Lett 13:217–219

Tyree M T, Dixon M A 1986 Water stress induced cavitation and embolism in some woody plants. Physiol Plant 66:397–405
Tyree M T, Davis S G, Cochard H 1994 Biophysical perspectives of xylem evolution: is there a tradeoff of hydraulic efficiency for vulnerability to dysfunction? IAWA J 15:335–360
Utsumi Y, Sano Y, Ohtani J, Fujikawa S 1996 Seasonal changes in distribution of water in the outer growth rings of Fraxinus mandshurica var. japonica: a study by cryo-scanning electron microscopy. IAWA J 17:113–124
Vales M A 1983 Verzierte einfache Gefäßdurchbrechungen im Holz einiger endemischer Rubiaceen-Arten aus Kuba. Feddes Repert 94:493–495
Vales M A, Babos K 1977 Wood anatomy of Ceratopyxis Hooker f. ex Hooker (Rubiaceae), a monotypic endemic genus of West Cuba. Acta Bot Acad Sci Hung 23:275–283
van den Oever L, Baas P, Zandee M 1981 Comparative wood anatomy of Symplocos and latitude of provenance. IAWA Bull NS 2:3–24
van der Graaff N A, Baas P 1974 Wood anatomical variations in relation to latitude and altitude. Blumea 22:101–123
van der Walt J J A, van der Schiff H P, Schweickerdt HG 1973 Anomalous growth in the stems of the lianes Mikania cordifolia (Burm. f.) Robins (Compositae) and Paullinia pinnata Linn. (Sapindaceae). Kirkia 9:109–138
Vander Wyk RW, Canright J E 1956 The anatomy and relationships of the Annonaceae. Trop Woods 104:1–24
van Veenendal W L H, den Outer R W 1993 Development of included phloem and organisation of the phloem network in the stem of Strychnos millepunctata (Loganiaceae). IAWA J 14: 253–265
van Vliet G J C M 1976a Wood anatomy of the Rhizophoraceae. Leiden Bot Ser 3:20–75
van Vliet G J C M 1976b Radial vessels in rays. IAWA Bull NS 3:35–37
van Vliet G J C M 1978 The vestured pits of Combretaceae and allied families. Acta Bot Neerl 27:273–285
van Vliet G J C M 1979 Wood anatomy of the Combretaceae. Blumea 25:141–223
van Vliet G J C M 1981 Wood anatomy of the palaeotropical Melastomataceae. Blumea 27:395–462
van Vliet G J C M, Baas P 1984 Wood anatomy and classification of the Myrtales. Ann Mo Bot Gard 71:783–800
van Vliet G J C M, Koek-Noorman J, ter Welle B J H 1981 Wood anatomy, classification and phylogeny of the Melastomataceae. Blumea 27:463–473
Venugopal V, Krishnamurthy K V 1987 Seasonal production of secondary xylem in the twigs of certain tropical trees. IAWA Bull NS 8:31–40
Versteegh C 1968 An anatomical study of some woody plants in the mountain flora in the tropics (Indonesia). Acta Bot Neerl 17:151–159
Vestal P A, Vestal M R 1940 The formation of septa in fiber-tracheids of Hypericum androsaemum L. Leafl Bot Mus Harvard Univ 8:169–188
Wagenführ R 1961 Vergleichende anatomische Untersuchungen an Tropenholzern aus Africa. Holztechnologie 8:281–282
Wagner K A 1946 Notes on the anomalous stem structure of a species of Bauhinia. Am Midl Nat 36:151–256
Waisel Y, Fahn A 1965 The effects of environment on wood formation and cambial activity in Robinia pseudoacacia L. New Phytol 64:436–442
Waisel Y, Lipschitz N, Kuller A 1972 Patterns of water movement in trees and shrubs. Ecology 53:520–523
Wallace G D 1986 Wood anatomy of Cassiope (Ericaceae). Aliso 11:393–415
Walsh M A 1975 Xylem anatomy of Hibiscus (Malvaceae) in relation to habit. Bot Gaz 136:30–40
Wardrop A B 1948 The nature of reaction wood. I. The structure and properties of tension wood fibres. Aust J Sci Res B 1:3–16
Wardrop A B 1951 Cell wall organization and the properties of xylem. I. Cell wall organization and the variation of breaking load in tension of xylem in conifer stems. Aust J Sci Res B 4:391–414

Wardrop A B 1955 The nature of reaction wood. IV. Variations in cell wall organization of tension wood fibres. Aust J Bot 3:177–189
Wardrop A B 1956 The nature of reaction wood. V. The distribution and formation of tension wood in some species of eucalypts. Aust J Bot 4:152–166
Webber I E 1934 Bearing of upright ray cells in the wood of Hibiscus mutabilis L. on usage of the term tile-cell. Trop Woods 37:9–13
Webber I E 1936 The woods of sclerophyllous and desert shrubs and desert plants of California. Am J Bot 23:181–188
Webber I E 1938 Intercellular cavities in the rays of dicotyledonous woods. Lilloa 2:465–469
Webber I E 1945 The wood of Ambelania laxa. Lilloa 11:49
Wellwood R W 1962 Tensile testing of small wood samples. Pulp Pap Mag Can 63(2):T61–T67
Werker E, Baas P 1981 Trabeculae of Sanio in secondary tissues of Inula viscosa (L.) Desf. and Salvia fruticosa Mill. IAWA Bull NS 2:69–76
Whalen M A 1987 Wood anatomy of the American Frankenias (Frankeniaceae): systematic and evolutionary implications. Am J Bot 74:1211–1223
Wheeler E A 1981 Intervascular pitting in Fraxinus americana L. IAWA Bull NS 2:169–174
Wheeler E A 1982 Ultrastructural characteristics of red maple (Acer rubrum L.) wood. Wood Fiber 14:43–53
Wheeler E A 1983 Intervascular pit membranes in Ulmus and Celtis native to the United States. IAWA Bull NS 4:79–88
Wheeler E A 1986 Vessels per square millimetre or vessel groups per square millimetre? IAWA Bull NS 7:73–74
Wheeler E A, Baas P 1991 A survey of the fossil record for dicotyledonous wood and its significance for evolutionary and ecological wood anatomy. IAWA Bull NS 12:275–332
Wheeler E A, Baas P 1998 Wood identification – a review. IAWA J 19:241–264
Wheeler E A, Pearson R G 1985 A critical review of the IAWA standard list of characters formatted for the IDENT programs. IAWA Bull NS 6:151–161
Wheeler E A, Thomas R J 1981 Ultrastructural characteristics of mature wood of southern red oak (Quercus falcata Michx.) and white oak (Quercus alba L.). Wood Fiber 13:169–181
Wheeler E A, Pearson R G, La Pasha C A, Zack T, Hatley W 1986 Computer-aided wood identification. N C Agric Res Serv Bull 474:1–160
Wilkes J 1988 Variations in wood anatomy within species of Eucalyptus. IAWA Bull NS 9:13–23
Wilkinson H P 1994 Leaf and twig anatomy of the Pterostemonaceae (Engl.) Small: ecological and systematic features. Bot J Linnean Soc 115:115–131
Wilson T K 1960 The comparative morphology of the Canellaceae. Trop Woods 112:1–27
Winstead J E 1972 Fiber tracheid length and wood specific gravity of seedlings as ecotypic characters in Liquidambar styraciflua L. Ecology 53:165–172
Wloch W, Zagorska-Marek B 1982 Reconstruction of storeyed cambium in the linden. Acta Soc Bot Pol 51:215–228
Wolfe J A 1981 Paleoclimatic significance of the Oligocene and Neogene floras of the northwestern United States. In: Niklas K J (ed) Paleobotany, paleoecology, and evolution. Praeger, New York, pp 79–101
Wolkinger F 1969 Morphologie und systematische Verbreitung lebender Holzfasern bei Sträuchern und Bäumen. I. Zur Morphologie und Zytologie. Holzforschung 23:138–144
Wolkinger F 1970a Morphologie und systematische Verbreitung lebender Holzfasern bei Sträuchern und Bäumen. II. Zur Histologie. Holzforschung 24:141–151
Wolkinger F 1970b Das Vorkommen lebender Holzfasern in Sträuchern und Bäumen. Phyton (Austria) 14:55–67
Wolkinger F 1971 Morphologie und systematische Verbreitung lebender Holzfasern bei Sträuchern und Bäumen. III. Systematische Verbreitung. Holzforschung 25:29–30
Woodcock D W 1989a Distribution of vessel diameter in ring-porous trees. Aliso 12:287–293
Woodcock D W 1989b Relationships among wood variables in two species of ring-porous trees. Aliso 12:543–554
Woodworth R H 1934 Perforated fiber-tracheids in the passion flower. Science 80:449–450

Woodworth R H 1935 Fibriform vessel members in the Passifloraceae. Trop Woods 41:8–16

Yaghmaie M, Catling D 1984 The occurrence of vascular tracheids in Betula and some other Betulaceae genera. Plant Syst Evol 147:125–131

Yarrow G L, Popham R A 1981 The ontogeny of the primary thickening meristem of Atriplex hortensis L. (Chenopodiaceae). Am J Bot 68:1042–1049

Yatsenko-Khmelevsky A A 1954 Research and methods of anatomical investigation of woody plants (in Russian). Isdatelistvo Akademii Nauk SSR, Moscow, 337 pp

Young D A 1981 Are the angiosperms primitively vesselless? Syst Bot 6:313–330

Zagorska-Marek B 1975 Growth activity of fusiform initials in storeyed cambium. Acta Soc Bot Pol 14:537–551

Zagorska-Marek B 1984 Pseudotransverse divisions and intrusive elongation of fusiform cambial initials in the storeyed cambium of Tilia. Can J Bot 62:20–27

Zamski E 1979 The mode of secondary growth and the three-dimensional structure of the phloem in Avicennia. Bot Gaz 140:67–76

Zamski E, Azenkot A 1981 Sugarbeet vasculature. I. Cambial development and the three-dimensional structure of the vascular system. Bot Gaz 142:334–343

Zhang S-Y, Baas P 1992 Wood anatomy of trees and shrubs from China. III. Rosaceae. IAWA Bull NS 13:21–91

Zhang S-Y, Baas P, Zandee M 1992 Wood structure of the Rosaceae in relation to ecology, habit and phenology. IAWA Bull NS 13:307–349

Zhong Y, Baas P, Wheeler E A 1992 Wood anatomy of trees and shrubs from China. IV. Ulmaceae. IAWA Bull NS 13:419–453

Ziglstra K 1909 Die Gestalten der Markstrahlen im sekundären Holze. Rec Trav Bot Neerl 5:17–50

Zimmerman A 1922 Die Cucurbitaceen, vols 1, 2. Fischer, Jena, 205 pp, 186 pp

Zimmermann M H 1971 Transport in xylem. In: Zimmermann M, Brown C L (eds) Trees: structure and function. Springer, Berlin Heidelberg New York, pp 169–220

Zimmermann M H 1978a Vessel ends and the disruption of water flow in plants. Phytopathology 68:253–255

Zimmermann M H 1978b Structural requirements for optimal water conduction in tree stems. In: Tomlinson P B, Zimmermann M H (eds) Tropical trees as living systems. Cambridge Univ Press, Cambridge, pp 517–532

Zimmermann M H 1983 Xylem structure and the ascent of sap. Springer, Berlin Heidelberg New York, 143 pp

Zimmermann M H, Brown C L 1971 Trees: structure and function. Springer, Berlin Heidelberg New York, 336 pp

Zimmermann M H, Jeje A A 1981 Vessel-length distribution in stems of some American woody plants. Can J Bot 59:1882–1892

Zimmermann M H, Milburn J A 1982 Transport and storage of water. In: Lange O L, Nobel P S, Osmond C B, Zeigler H (eds) Physiological plant ecology. II. Encyclopedia of plant physiology, vol 12B. Springer, Berlin Heidelberg New York, pp 135–151

Zimmermann M H, Potter D 1982 Vessel-length distribution in branches, stems, and roots of Acer rubrum L. IAWA Bull Ns 3:103–109

Zürcher E, Kucera L J, Bosshard H H 1985 Bildung und Morphologie der Thyllen. Eine Literaturübersicht. Vierteljahrschr Naturforsch Ges Zürich 130:311–333

Zweypfenning R C V J 1978 A hypothesis on the function of vestured pits. IAWA Bull 1978/1:13–15

Subject Index

Family and genus names are indexed together for families in which only one genus is covered in this book. For characters widespread in a large family (e.g., rhomboidal crystals and storied structure in Fabaceae), only the family name is indexed – additional genera with these features very likely exist in such families. Genera are indexed where a limited number of genera and families have a particular anatomical feature, or where definitions of terms may differ from those of other authors. Genera illustrated are indexed.)

Printing and Binding: Stürtz AG, Würzburg